VOLUME ONE HUNDRED AND FOURTEEN

CURRENT TOPICS IN DEVELOPMENTAL BIOLOGY

Apoptosis and Development

CURRENT TOPICS IN DEVELOPMENTAL BIOLOGY

"A meeting-ground for critical review and discussion of developmental processes"
A.A. Moscona and Alberto Monroy (Volume 1, 1966)

VOLUME ONE HUNDRED AND FOURTEEN

CURRENT TOPICS IN DEVELOPMENTAL BIOLOGY

Apoptosis and Development

Edited by

HERMANN STELLER
Howard Hughes Medical Institute
The Rockefeller University
New York, NY, USA

ELSEVIER

AMSTERDAM • BOSTON • HEIDELBERG • LONDON
NEW YORK • OXFORD • PARIS • SAN DIEGO
SAN FRANCISCO • SINGAPORE • SYDNEY • TOKYO
Academic Press is an imprint of Elsevier

Academic Press is an imprint of Elsevier
125 London Wall, London, EC2Y 5AS, UK
525 B Street, Suite 1800, San Diego, CA 92101-4495, USA
225 Wyman Street, Waltham, MA 02451, USA
The Boulevard, Langford Lane, Kidlington, Oxford OX5 1GB, UK

First edition 2015

ISBN: 978-0-12-410425-9
ISSN: 0070-2153

For information on all Academic Press publications
visit our website at store.elsevier.com

CONTENTS

CONTRIBUTORS

Allyson L. Anding
Department of Molecular, Cell, and Cancer Biology, University of Massachusetts Medical School, Worcester, Massachusetts, USA

Eric H. Baehrecke
Department of Molecular, Cell, and Cancer Biology, University of Massachusetts Medical School, Worcester, Massachusetts, USA

Andreas Bergmann
Department of Molecular, Cell, and Cancer Biology, University of Massachusetts Medical School, Worcester, Massachusetts, USA

Meike Broemer
German Center for Neurodegenerative Diseases (DZNE), Bonn, Germany

Katrin Domsch
Centre for Organismal Studies (COS) Heidelberg, Heidelberg University, Heidelberg, Germany

Caitlin E. Fogarty
Department of Molecular, Cell, and Cancer Biology, University of Massachusetts Medical School, Worcester, Massachusetts, USA

Estee Kurant
Department of Genetics and Developmental Biology, The Rappaport Family Institute for Research in the Medical Sciences, Faculty of Medicine, Technion—Israel Institute of Technology, Haifa, Israel

Flonia Levy-Adam
Department of Genetics and Developmental Biology, The Rappaport Family Institute for Research in the Medical Sciences, Faculty of Medicine, Technion—Israel Institute of Technology, Haifa, Israel

Ingrid Lohmann
Centre for Organismal Studies (COS) Heidelberg, Heidelberg University, Heidelberg, Germany

Jennifer Zuckerman Malin
Laboratory of Developmental Genetics, The Rockefeller University, New York, USA

Kimberly McCall
Department of Biology, Boston University, Boston, Massachusetts, USA

Pascal Meier
The Breakthrough Toby Robins Breast Cancer Research Centre, Chester Beatty Laboratories, Institute of Cancer Research, London, UK

Masayuki Miura
Department of Genetics, Graduate School of Pharmaceutical Sciences, The University of Tokyo, Bunkyo-ku, and Core Research for Evolutional Science and Technology, Japan Agency for Medical Research and Development, Chiyoda-ku, Tokyo, Japan

Albert A. Mondragon
Department of Biology, Boston University, Boston, Massachusetts, USA

Bruno Monier
Université de Toulouse, UPS, and CNRS, LBCMCP, Toulouse, France

Otto Morris
The Breakthrough Toby Robins Breast Cancer Research Centre, Chester Beatty Laboratories, Institute of Cancer Research, London, UK

Shigekazu Nagata
Laboratory of Biochemistry and Immunology, Immunology Frontier Research Center, Osaka University, Osaka, Japan

Chihiro Nishi
Laboratory of Biochemistry and Immunology, Immunology Frontier Research Center, Osaka University, Osaka, Japan

Fani Papagiannouli
Centre for Organismal Studies (COS) Heidelberg, Heidelberg University, Heidelberg, Germany

Jeanne S. Peterson
Department of Biology, Boston University, Boston, Massachusetts, USA

Hyung Don Ryoo
Department of Cell Biology, New York University School of Medicine, New York, New York, USA

Mahendra Seervi
Department of Molecular, Cellular, and Developmental Biology, University of Colorado, Boulder, Colorado, USA, and DBT-PU-IPLS Program, Department of Biotechnology/Botany, Patna University, Patna, Bihar, India

Katsumori Segawa
Laboratory of Biochemistry and Immunology, Immunology Frontier Research Center, Osaka University, Osaka, Japan

Shai Shaham
Laboratory of Developmental Genetics, The Rockefeller University, New York, USA

Jeny Shklover
Department of Genetics and Developmental Biology, The Rappaport Family Institute for Research in the Medical Sciences, Faculty of Medicine, Technion—Israel Institute of Technology, Haifa, Israel

Magali Suzanne
Université de Toulouse, UPS, and CNRS, LBCMCP, Toulouse, France

Allison K. Timmons
Department of Biology, Boston University, Boston, Massachusetts, USA

Satoshi Toda
Laboratory of Biochemistry and Immunology, Immunology Frontier Research Center, Osaka University, Osaka, Japan

Deepika Vasudevan
Department of Cell Biology, New York University School of Medicine, New York, New York, USA

Ding Xue
Department of Molecular, Cellular, and Developmental Biology, University of Colorado, Boulder, Colorado, USA

Yoshifumi Yamaguchi
Department of Genetics, Graduate School of Pharmaceutical Sciences, The University of Tokyo, Bunkyo-ku, and Precursory Research for Embryonic Science and Technology, Japan Science and Technology Agency, Chiyoda-ku, Tokyo, Japan

Yuichi Yanagihashi
Laboratory of Biochemistry and Immunology, Immunology Frontier Research Center, Osaka University, Osaka, Japan

PREFACE

Programmed cell death (PCD) plays important roles in animal development, and its original discovery dates back to 1842 when Karl Vogt reported that cell death is associated with the removal of the notochord during amphibian metamorphosis. During the past two decades, dramatic progress has been made in understanding the biochemical basis of the most prominent form of naturally occurring PCD, termed apoptosis. At the same time, much remains to be learned about the physiological roles and regulation of apoptosis *in vivo*, and how diverse signaling pathways are integrated to select specific cells fated to die during organismal development. Work primarily from three major model systems, the nematode *Caenorhabditis elegans,* the fruit fly *Drosophila melanogaster*, and the mouse *Musmusculus*, has revealed a remarkable complexity in how apoptotic cells interact with their cellular environment during normal development. These interactions play critical roles in the decision of cells to die, and in the recognition and clearance of apoptotic cells by phagocytes. Surprisingly, apoptotic cells were also found to be the source of diverse signals that profoundly influence the fate of surviving neighbors and can affect tissue growth and morphogenesis. This issue of *Current Topics in Developmental Biology* reviews some of the exciting recent advances in this field. The first two chapters focus on the nematode *C. elegans,* where the conserved core pathway for apoptosis was originally discovered. Zuckerman Malin and Shai Shaham review both the classic work and also discuss new findings indicating that the interaction between dying and engulfing cells is more complex than originally thought, and that PCD, at least in some cells, requires more than one death program. The next chapter by Seervi and Xue details the roles of mitochondrial proteins in the execution of cell death in *C. elegans*, including chromosome fragmentation, phosphatidylserine externalization, and elimination of mitochondria. We then have a series of chapters devoted to work on *Drosophila.* Anding and Baehrecke review the complex roles of macroautophagy in cell survival and death, and what is known about autophagic cell death *in vivo*. Peterson et al. highlight advances in our understanding of cell death in the germline, focusing on the *Drosophila* ovary. This is followed by a review from Domsch et al. on the regulation of apoptosis by HOX genes, covering work from worms, flies, and mice. Miura and Yamaguchi discuss the role of caspases during neural development, both in the context of apoptosis and the nonlethal action of caspases for the pruning of axons and dendrites. This is followed by two chapters on the regulation of apoptosis by the ubiquitin-proteasome system. Vasudevan and Ryoo review regulation of cell death by Inhibitor of Apoptosis Proteins, a conserved family of E3 ubiquitin ligases that includes the major known direct negative regulators of caspases. Meier et al. expand on the roles of the ubiquitin system in the regulation of cell death, inflammation, and immunity. The next chapter by Fogarty and Bergmann summarizes our current knowledge about signaling by apoptotic cells, an unexpected and rapidly expanding field. The following two chapters are devoted to the final step in the execution of apoptosis, the engulfment, and clearance of apoptotic cells. The chapter by Nagata and colleagues provides a comprehensive review on the clearance of apoptotic cells by macrophages, with an emphasis on mammals and human disease. Shklover et al. discuss advances from different genetic model organisms, such as *C. elegans*, *Drosophila*, zebrafish, and mouse. Despite a wealth of information on these pathways, many fundamental questions regarding apoptotic cell clearance during development still remain unclear. The final chapter by Monier and Suzanne discusses the morphogenetic role of apoptosis and the crucial role of live imaging for these studies. One of the unexpected new concepts emerging here is the idea that apoptosis may be a source of mechanical force for morphogenetic movements.

Collectively, this series of reviews illustrates both the remarkable advances that have been made in elucidating pathways and detailed biochemical mechanisms to explain key events during apoptosis, but they also reveal the many unresolved questions and unexpected new opportunities that we still face for understanding the physiological regulation and function of PCD.

Hermann Steller

CHAPTER ONE

Cell Death in *C. elegans* Development

Jennifer Zuckerman Malin, Shai Shaham[1]

Laboratory of Developmental Genetics, The Rockefeller University, New York, USA
[1]Corresponding author: e-mail address: shaham@rockefeller.edu

Contents

Abstract

Cell death is a common and important feature of animal development, and cell death defects underlie many human disease states. The nematode *Caenorhabditis elegans* has proven fertile ground for uncovering molecular and cellular processes controlling programmed cell death. A core pathway consisting of the conserved proteins EGL-1/BH3-only, CED-9/BCL2, CED-4/APAF1, and CED-3/caspase promotes most cell death in the nematode, and a conserved set of proteins ensures the engulfment and degradation of dying cells. Multiple regulatory pathways control cell death onset in *C. elegans*, and many reveal similarities with tumor formation pathways in mammals, supporting the idea that cell death plays key roles in malignant progression. Nonetheless, a number of observations suggest that our understanding of developmental cell death in

Current Topics in Developmental Biology, Volume 114
ISSN 0070-2153
http://dx.doi.org/10.1016/bs.ctdb.2015.07.018

C. elegans is incomplete. The interaction between dying and engulfing cells seems to be more complex than originally appreciated, and it appears that key aspects of cell death initiation are not fully understood. It has also become apparent that the conserved apoptotic pathway is dispensable for the demise of the *C. elegans* linker cell, leading to the discovery of a previously unexplored gene program promoting cell death. Here, we review studies that formed the foundation of cell death research in *C. elegans* and describe new observations that expand, and in some cases remodel, this edifice. We raise the possibility that, in some cells, more than one death program may be needed to ensure cell death fidelity.

1. INTRODUCTION

Cell death is a widespread process that is essential for life. Tissue sculpting, organ morphogenesis, and organ size control are but a few of the developmental events that integrally utilize programmed cell death to generate a functioning adult animal. It is therefore not surprising that many things go wrong when cell death goes awry (Fuchs & Steller, 2011). Indeed, neurodegeneration and tumorigenesis, disease states against which armies of researchers have been amassed, result from too much or too little cell culling, respectively (Hanahan & Weinberg, 2011; Youle & van der Bliek, 2012). While the hypothesis that cell death is a regulated phenomenon in animal development was first experimentally addressed in vertebrates (Hamburger & Levi-Montalcini, 1949) and insects (Lockshin & Williams, 1965), the first systematic studies aimed at deciphering the molecular program promoting cell demise employed the free-living soil nematode *Caenorhabditis elegans* (Horvitz, 2003). Early observations of the cellular complement of adult *C. elegans* revealed little variation in the number and position of cells between individuals of similar ages, leading to the proposal that cell lineage in this animal may be invariant. This prediction was largely borne out by taking advantage of the transparent cuticle of the animal to observe cell divisions *in vivo* (Kimble & Hirsh, 1979; Sulston, Albertson, & Thomson, 1980; Sulston & Horvitz, 1977; Sulston, Schierenberg, White, & Thomson, 1983). This heroic effort culminated in a complete cell lineage tree documenting a generally predictable pattern of divisions that generate adult somatic tissue from the zygote. These studies demonstrated that precisely 1090 and 1178 somatic cells must be generated to produce a *C. elegans* hermaphrodite and male, respectively.

Among the generated cells, a small but substantial set (~12%) are eliminated. These cells become refractile under differential interference contrast (DIC) optics (Fig. 1), acquire a rounded morphology, and eventually disappear. Ultrastructural studies reveal that these dying cells are engulfed by neighboring cells (Sulston et al., 1983), and possess characteristics of apoptotic cell death, such as condensed nuclear chromatin, and reduced cytoplasmic volume (Shaham & Horvitz, 1996b; Sulston et al., 1983) (Fig. 1). Like the lineage itself, these cell death events are essentially invariant between individuals and target the same cells at the same time in development. In the hermaphrodite and male, 131 and 147 somatic cells are eliminated, respectively. Subsequent studies demonstrated that cell death is highly prevalent during germline development and maintenance, with roughly 50% of female meiosis products succumbing to apoptosis (Gumienny, Lambie, Hartwieg, Horvitz, & Hengartner, 1999). Developmental death of germ

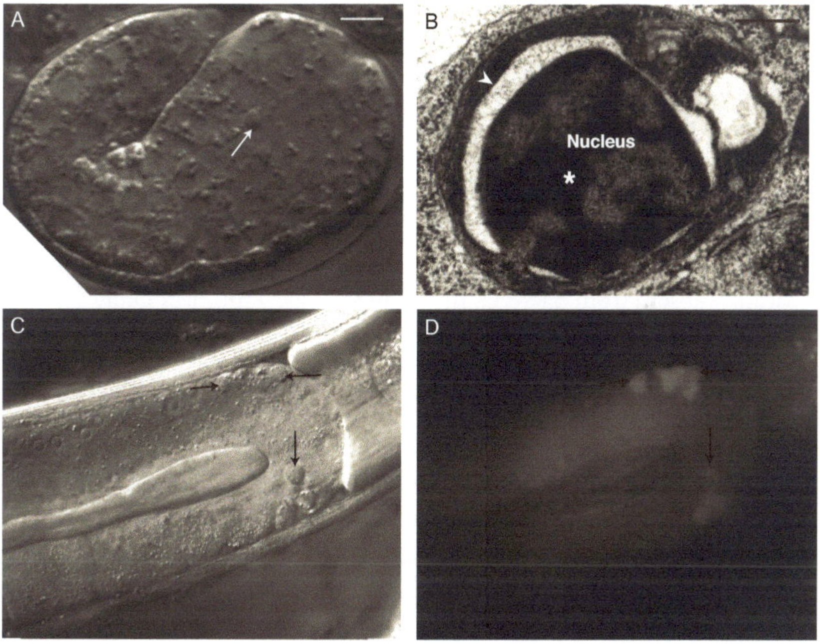

Figure 1 Apoptotic developmental cell death in *C. elegans*. (A) Differential interference contrast (DIC) image of a dying cell (arrow) in a developing embryo. Scale bar, 2.5 μm. (B) Electron micrograph of a dying embryonic cell. Note dark staining, condensed chromatin (asterisk) within the nucleus and shrunken cytoplasm. Arrowhead, nuclear envelope. Scale bar, 400 nm. (C) DIC image of dying cells in the developing gonad (arrows). (D) Fluorescence image of animal in (C) stained with the cell death marker SYTO12. Scale bar, 8 μm.

cells in *C. elegans* differs from somatic cell death in that the identities of dying cells are not ascribed to their lineage (Gumienny et al., 1999; Sulston et al., 1983). *C. elegans* therefore offers two arenas for understanding cell death control: one in which cell death and lineage are tightly coupled, and one in which stochastic processes apparently determine life and death. Studies of the former revealed a core pathway controlling apoptotic cell death from *C. elegans* to mammals.

Here, we describe these key components and their interactions and explore current understanding of the lineage-dependent mechanisms that trigger the activation of these killer genes and proteins. We also discuss a group of genes important for the clearance of dying cells and their relation to cell death execution and delve into a number of mysteries that remain unanswered and which have the potential to expand and modify our understanding of why and how cells die. We end by describing a novel non-apoptotic *C. elegans* cell death program that promotes dismantling of the male-specific linker cell.

2. CORE APOPTOSIS REGULATORS IN *C. ELEGANS*

Most cell death in *C. elegans* is controlled by the proteins CED-3, CED-4, CED-9, and EGL-1, whose functions and interactions have been worked out in some detail (Fig. 2). All four components of this canonical cell death pathway are conserved across disparate animal species, but are apparently absent from bacteria, fungi, and plants. Thus, it is likely that this pathway arose early on in the animal lineage.

2.1 CED-3/Caspase

The most downstream core component of the apoptotic cell death pathway is the protein CED-3, encoded by the *ced-3* gene. The role of *ced-3* in cell death was initially revealed from genetic studies. Animals mutant for the gene *ced-1* (see below) accumulate unengulfed cell corpses during development that are easily detectable using DIC microscopy. A suppressor screen for animals lacking these corpses identified the recessive mutant *ced-3(n717)*, in which most cells fated to die fail to do so. Characterization of the mutant revealed widespread inhibition of cell death, resulting in animals with extra cells (Ellis & Horvitz, 1986). While lineage studies suggest that few, if any, of these extra cells divide (Hoeppner, Hengartner, & Schnabel, 2001), some are able to differentiate (Shaham & Bargmann, 2002), incorporate into neural circuits (White, Southgate, & Thomson, 1991), and even substitute for their

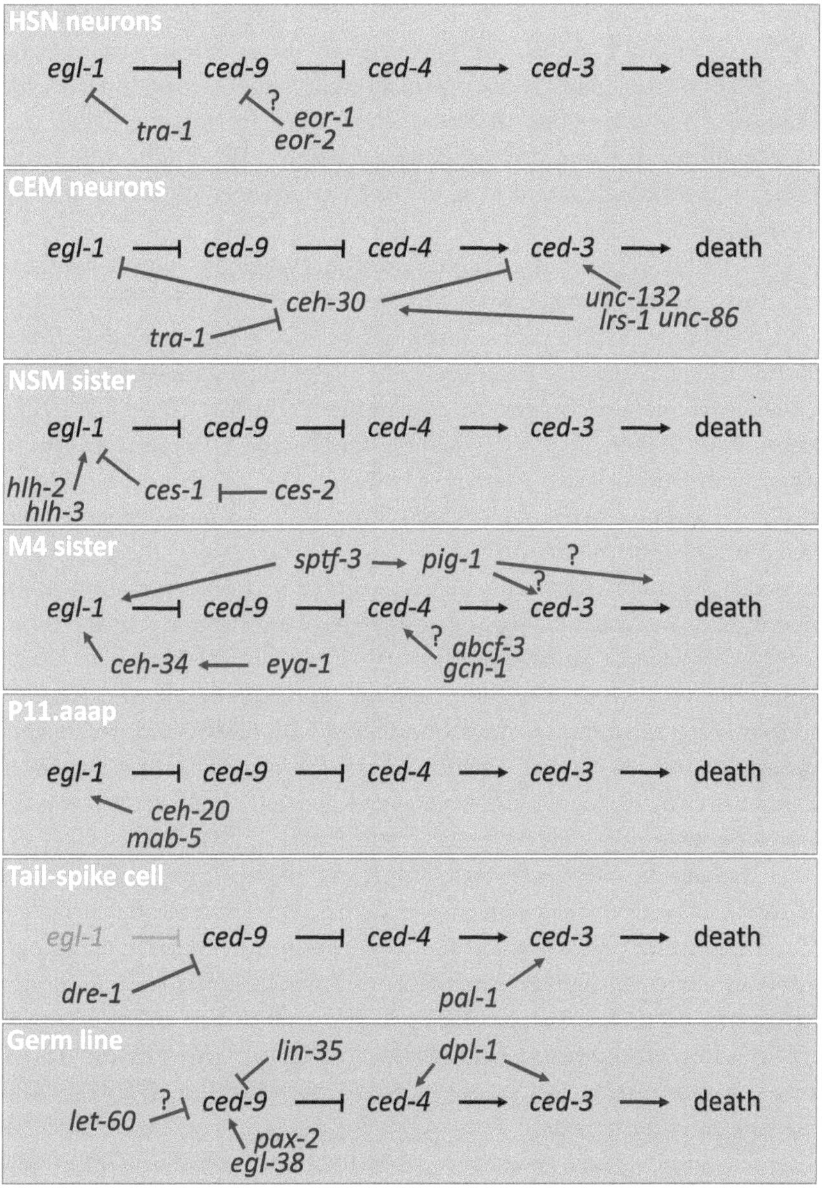

Figure 2 Apoptotic cell death control in *C. elegans*. Core cell death components (black) interact with a variety of regulators (blue (gray in the print version)) in different cells (white). Only cells for which direct control of the core apoptotic pathway is established are shown.

surviving sister cells, if those are experimentally ablated (Avery & Horvitz, 1987). *ced-3* mutant animals are alive, suggesting that at least under laboratory conditions cell death is not essential (Ellis & Horvitz, 1986). However, some *ced-3* animals exhibit defects in chemotaxis to attractive odors, and some exhibit pronounced developmental delay (Ellis, Yuan, & Horvitz, 1991), suggesting that in the wild, cell death likely confers a survival advantage.

The CED-3 protein is a founding member, together with mammalian caspase-1, of the caspase family of proteases (Yuan, Shaham, Ledoux, Ellis, & Horvitz, 1993). Active CED-3 is derived from a precursor that is cleaved to generate three fragments. The N-terminal fragment, which has sequence homology to caspase recruitment domains (CARD), is not required for protease activity, while the middle and C-terminal fragments associate in pairs to form a tetramer, which is the active protease (Huang et al., 2013; Thornberry et al., 1992). Protease function is effected by an active-site cysteine residue that cleaves target proteins after the amino acid aspartate. Aspartate residues also define the junctions between subunits in the precursor form, and biochemical studies have revealed a role for cross-catalysis in protease activation (Qi et al., 2010; Thornberry et al., 1992). Supporting this notion, the ability of the CED-3 precursor to trigger cell death *in vivo* is apparently positively correlated with CED-3 protein concentration (Shaham & Horvitz, 1996b). Expression of middle and C-terminal fragments from separate mRNAs does not generally yield active protease, suggesting that cross-catalysis yields concerted structural changes that form the active moiety. However, for CED-3, and caspase-1, tagging each subunit with an N-terminal leucine zipper domain does result in the formation of active protease (Chelur & Chalfie, 2007), presumably because the leucine zipper facilitates co-translational interactions that promote folding into the active structure. As with caspase-3, a key mammalian effector caspase required for cell elimination in many natural and disease settings, CED-3 shows a preference for cleavage after the sequence DEVD (Xue, Shaham, & Horvitz, 1996).

Insertion of the gene encoding green fluorescent protein (GFP) into a genomic fragment containing the *ced-3* locus reveals widespread expression in early embryos prior to the onset of most cell death events (Maurer, Chiorazzi, & Shaham, 2007). Expression then recedes and disappears in most cells, but turns on in the tail-spike cell (see below), which undergoes apoptosis just prior to hatching. Expression in larvae is generally undetectable, even though *ced-3*-dependent cell death takes place in early larval stages. Just before molting into the adult stage, GFP expression can be detected in the male tail (Maurer et al., 2007). As in the embryo, GFP here is expressed in many cells not fated to die. With few exceptions, therefore, expression of GFP is not detected in cells that die. This observation seemingly contradicts results that suggest a cell-autonomous role for CED-3 in cell death execution. Mosaic analysis demonstrates that *ced-3* expression is required in lineages of dying cells (Yuan & Horvitz, 1990), and ectopic expression studies suggest that only *ced-3*-expressing cells, but not their neighbors, can be induced to die (Shaham & Horvitz, 1996b). While it is possible that the genomic fragment used to drive GFP expression in the studies of Maurer et al. is missing enhancer elements, and therefore generates expression below the threshold of detection, this seems unlikely as the fragment confers rescue in the absence of GFP, and GFP can be detected at high levels in many cells (Maurer et al., 2007). This puzzling observation may, therefore, suggest that turnover of GFP and the CED-3 precursor protein differs markedly, with CED-3 having a much longer half-life. If this is the case, then CED-3 protein expressed in early embryos must be maintained in most or all cells through early larval development.

While caspase enzymology and activity have been studied in *C. elegans* and many other organisms, the mechanism by which these proteases bring upon cellular demise is an enduring mystery. Are there many or few relevant targets? Are there many or few cellular processes that must be blocked or activated to ensure cell death? While a plethora of vertebrate caspase substrates have been described (Poreba, Strozyk, Salvesen, & Drag, 2013), their relevance to cell death is generally not understood. One exception is the ICAD DNase inhibitory subunit, whose inactivation by caspase-mediated cleavage leads to DNA degradation in mammalian cells (Enari et al., 1998; Liu et al., 1998). Proteins related to ICAD or CAD, the active nuclease, are not encoded by the *C. elegans* genome. An RNAi screen for increased TUNEL staining isolated a number of nucleases (designated *crn-1* to *crn-6*) that may be involved in DNA degradation following death. While RNAi against each gene weakly affects DNA end accumulation and cell death kinetics, no surviving cells are seen, suggesting that these nucleases are not essential for cell death (Parrish & Xue, 2003). The DNAse II homolog, NUC-1, however, does play a key role in *C. elegans* DNA degradation during apoptosis (Hevelone & Hartman, 1988; Sulston, 1976; Wu, Stanfield, & Horvitz, 2000). *C. elegans* apoptotic DNA is cleaved into 10 bp fragments (Aruscavage, Hellwig, & Bass, 2010), and *nuc-1* mutants lack these fragments and accumulate larger DNA in dying cells, as well as undegraded bacterial DNA in the gut (Aruscavage et al., 2010; Sulston, 1976). *nuc-1* mutant embryos stained using the TUNEL method (Gavrieli, Sherman, & Ben-Sasson, 1992) display an accumulation of DNA ends (Wu et al., 2000), suggesting that NUC-1 degrades TUNEL-positive intermediates produced by another nuclease. One candidate may be the microRNA processing enzyme Dicer, which can be cleaved by CED-3, and may, surprisingly, itself promote DNA cleavage (Nakagawa, Shi, Kage-Nakadai, Mitani, & Xue, 2010). It is therefore possible that a DNA degradation system analogous to that in mammals is targeted by caspases in *C. elegans*.

Mutations in *C. elegans ced-8*, encoding an Xk-family protein, result in transient accumulation of cell corpses during embryogenesis (Ellis, Jacobson, & Horvitz, 1991; Stanfield & Horvitz, 2000). Two recent studies suggest that CED-8 protein is also a caspase target and that its cleavage allows exposure of phosphatidyl serine (PS) on the surface of dying cells (Chen, Mapes, Lee, Skeen-Gaar, & Xue, 2013; Suzuki, Denning, Imanishi, Horvitz, & Nagata, 2013). PS then acts as one of multiple eat-me signals that promote the engulfment of dying cells by their neighbors (Venegas & Zhou, 2007). A similar set of events takes place in mammalian cells, where the CED-8-related protein Xkr8 is a target of caspase cleavage (Suzuki et al., 2013).

A recent study also suggests that the fidelity of some cell death in *C. elegans* is enhanced by CED-3 cleavage of the GTPase-activating protein CNT-1 (Nakagawa, Sullivan, & Xue, 2014). Cleaved CNT-1 can translocate to the cell membrane where it may control apoptotic signaling. Although transient effects on cell death are observed in the relevant mutants, nearly all cells fated to die eventually succumb. Thus, it remains unclear whether caspase cleavage of CNT-1 is an integral aspect of cell death execution.

C. elegans CED-11 may also function downstream of CED-3 caspase. Defects in this protein, a TRPM-related channel, do not block cell death, but result in a profound change in dying cell appearance under DIC optics (Strange, 2003). Instead of acquiring a refractile appearance, cell corpses appear swollen and vacuolated, a defect that is perhaps not unexpected from a membrane channel mutant. Whether CED-11 is activated by CED-3 cleavage is not known.

In addition to CED-3, the *C. elegans* genome encodes three other caspase-related proteins: CSP-1, -2, and -3 (Shaham, 1998). The *csp-1* locus generates at least two functional transcripts, one encoding a zymogen with a long N-terminal domain (CSP-1A) and one with a short N-terminal region (CSP-1B). *In vitro*, CSP-1B can cleave itself and CED-3, presumably resulting in active enzymes. However, substrate specificities of CED-3 and CSP-1B appear to be different, as CED-3 cannot cleave the CSP-1B precursor. Furthermore, CED-3 can cleave the fluorescent substrate Ac-DEVD-AFC, but not Z-YVAD-AFC, and CSP-1B has the opposite substrate profile (Shaham, 1998). CSP-1 has been suggested to promote cell death during embryonic development (Denning, Hatch, & Horvitz, 2013). However, the effects of *csp-1* mutants are subtle, raising the possibility that CSP-1 may not play a key cell death role.

The *csp-2* locus also generates at least two transcripts, *csp-2*A, encoding a protein with a large N-terminal domain, and *csp-2*B, encoding a short domain. While the active-site cysteine of this protease is conserved, surrounding sequences are not, suggesting that this protease may be inactive. Indeed, *in vitro* activity of the protease could not be demonstrated (Shaham, 1998). While caspase inhibitor proteins are known in mammals and *Drosophila* (e.g., IAP proteins (Goyal, McCall, Agapite, Hartwieg, & Steller, 2000; Hay, Wassarman, & Rubin, 1995; Silke & Vucic, 2014)), *C. elegans* proteins related to these appear not to have any role in cell death (Speliotes, Uren, Vaux, & Horvitz, 2000). CSP-2 has been suggested to inhibit CED-3 activity in the germ line (Geng et al., 2009), perhaps functioning analogous to IAP proteins.

The *csp-3* gene encodes only a C-terminal caspase domain and apparently arose from a partial duplication of the *csp-2* locus (Shaham, 1998). CSP-3 does not contain an active site, suggesting that it could act as a dominant-negative caspase inhibitor. It has been suggested that CSP-3 can inhibit CED-3 (Geng et al., 2008); however, this observation has been contested (Denning et al., 2013).

In contrast to mammals, therefore, *C. elegans* seems to employ only a single major caspase, CED-3, to control cell death, and while other caspase-related genes are expressed in this animal, they likely serve other functions. The *csp-1*, *-2*, and *-3* caspase genes are not well conserved among nematode species, suggesting that whatever roles they have are likely to be tailored specifically to the biology of *C. elegans*. In this context, it is intriguing that the strongest *ced-3* mutations block nearly all cell death in *C. elegans*, and most also have effects as heterozygotes. However, a deletion eliminating the protease domain fails to block nearly 20% of cell death events that occur in the animal's pharynx (Shaham, Reddien, Davies, & Horvitz, 1999). This surprising observation suggests that CED-3 caspase activity may not account for all cell death in the animal. An alternative effector mechanism may be found through analysis of the strongest *ced-3* alleles, which appear to interfere with both CED-3 protease activity and this hypothetical alternative effector pathway (Shaham et al., 1999).

2.2 CED-4

Recessive mutations in the *ced-4* gene, like those in *ced-3*, promote the survival of somatic and germ cells slated to die during *C. elegans* development, suggesting a key role for CED-4 in promoting cell death (Ellis & Horvitz, 1986). CED-4 protein is similar to the mammalian cell death gene Apaf-1 and is composed of four structural domains: an N-terminal CARD domain, a nucleotide-binding domain found in NTPases, a helical domain, and a C-terminal winged-helix domain (Yuan & Horvitz, 1992; Zou, Henzel, Liu, Lutschg, & Wang, 1997). Structural studies suggest that CED-4 monomers can form an octameric complex with fourfold symmetry (Qi et al., 2010) in which CARD domains seem to mediate much of the contact. This is in contrast to the proposed hepatmer formed by Apaf-1 (Acehan et al., 2002). Although explanations based on structural data for the different stoichiometries have been put forward (Qi et al., 2010), the difference remains an intriguing mystery. The nucleotide-binding domain of CED-4, together with its overall structure, bears resemblance to AAA+ ATPases (Qi et al., 2010), although significant deviations in multimer interfaces are observed compared with *bona fide* members of this family. Nonetheless, it is intriguing and perhaps functionally relevant that the *C. elegans* protein MAC-1, an AAA+ family ATPase, can bind CED-4 (as well as Apaf-1), and that overexpression of MAC-1 can prevent some natural cell deaths in the animal (Wu et al., 1999). The CED-4 nucleotide-binding domain binds ATP-Mg^{2+} more strongly than dATP, in contrast to Apaf-1 (Li et al., 1997; Seiffert, Vier, & Hacker, 2002). Whether ATP binding *per se* or exchange of ADP for ATP is required for CED-4 function is still not clear.

Expression studies of CED-4, like those of CED-3, pose a conundrum. Polyclonal antibodies against CED-4 detect the protein in most cells in the early embryo, well before any cell death takes place (Chen et al., 2000). Staining gradually recedes and is absent from larvae, even though *ced-4* is required for the handful of early larval cell death events that occur in *C. elegans* (Sulston & Horvitz, 1977). CED-4 expression is also broadly detected in the germ line (Pourkarimi, Greiss, & Gartner, 2012). As with CED-3, several studies suggest that CED-4 functions within dying cells to promote their demise, and is also required for postembryonic induction of cell death in cells that usually survive (Shaham & Horvitz, 1996b; Yuan & Horvitz, 1990). Why CED-4 is broadly expressed in the embryo, but not detectable in larvae is not at all clear. As with CED-3, technical concerns may resolve this paradox.

Staining of early embryos with a polyclonal CED-4 antibody suggests that CED-4 is associated with mitochondria (Chen et al., 2000). However, a different antibody detects perinuclear staining that does not overlap with mitochondria in later embryos and in the germ line (Pourkarimi et al., 2012). Consistent with the latter localization pattern, RNA interference directed against the inner nuclear membrane protein SUN-1 appears to block CED-4 accumulation at the nuclear periphery. Furthermore, CED-4 and SUN-1 proteins purified from bacteria can physically interact (Tzur, Margalit, Melamed-Book, & Gruenbaum, 2006). Whether CED-4 and SUN-1 bind *in vivo* is unclear, as they are normally segregated to different cellular compartments (cytoplasm vs. inner nuclear membrane). The antibody-staining studies indicate a shift in CED-4 localization during development from mitochondria to perinuclear regions. However, it is also possible that the antibodies used are detecting different subpopulations of CED-4, or that technical issues, such as sample preparation, differ between experiments, providing only incomplete views of CED-4 localization.

The *ced-4* genomic locus encodes at least two alternative transcripts, *ced-4*L and *ced-4*S, that differ by inclusion of an alternate splice site acceptor upstream of exon 4 of the gene, resulting in a 24 amino acid insertion in CED-4L (Shaham & Horvitz, 1996a). Overexpression of CED-4L potently inhibits cell death, suggesting that this protein may normally function as a dominant interfering protein, consistent with the predicted multimeric structure of CED-4S. A number of observations suggest that alternative splicing of *ced-4* may have physiological importance. The alternative splice site and the sequence of the additional exon segment it introduces are highly conserved among nematodes, indicating functional significance. A mutation, *ced-4(n2273)*, which removes the *ced-4*S-specific splice-acceptor site, appears to perturb not only the killer but also a protective function encoded by the locus (Shaham & Horvitz, 1996a). Finally, weak mutations in *ced-4*, which only partially block cell death, are suppressed by mutations in the *spk-1* gene, which encodes an SR protein kinase (Galvin, Denning, & Horvitz, 2011). This family of kinases phosphorylates SR proteins that mediate splicing. Nonetheless, Northern blot and PCR studies demonstrate that the inhibitory *ced-4*L transcript is expressed at one-tenth the level of *ced-4*S in whole animal assays (Shaham & Horvitz, 1996a). It is therefore possible that *ced-4*S is expressed at high levels only in a subset of cells. The isoform-specific expression patterns of CED-4L and CED-4S are not known.

CED-4/Apaf-1-related proteins have been described in many animal species, but not outside the animal kingdom, suggesting the possibility that CED-3 and CED-4 evolved together. Nonetheless, loss of *C. elegans* ICD-1, a protein similar to the beta-subunit of the nascent polypeptide-associated complex, results in widespread cell death that appears, at least in part, to depend on CED-4, but not on CED-3 (Bloss, Witze, & Rothman, 2003). Also, mutations in the gene *pvl-5*, whose identity is still unknown, result in cell death that is inhibited by *ced-3* mutations but not by *ced-4* alleles (Joshi & Eisenmann, 2004). Thus, under at least some circumstances, *ced-3* and *ced-4* may operate independently.

2.3 CED-9

Unlike *ced-3* and *ced-4*, the *ced-9* gene encodes a negative regulator of cell death. Increased CED-9 function, either through overexpression or by a mutation in a conserved domain, promotes survival of cells that normally die during *C. elegans* development (Hengartner, Ellis, & Horvitz, 1992; Hengartner & Horvitz, 1994b). Loss-of-function mutations in *ced-9* are generally lethal. Some alleles result in inappropriate and widespread death of embryonic cells not fated to die. However, strong loss-of-function mutations result in arrest very early in embryonic development with no evidence of ectopic cell death (Hengartner et al., 1992). While the phenotypic discrepancy between these allele classes is not understood, it may suggest that CED-9 has a vital cellular role independent of its role in cell death. Yeast two-hybrid studies identified the centrosome maturation and mitotic spindle assembly factor SPD-5 as a CED-9-binding protein, an interaction also validated by co-affinity purification from transfected human HEK293T cells (Dreze et al., 2009). *spd-5* is required for early embryonic events and *spd-5* loss of function is also embryonic lethal. Thus, CED-9 may contribute to the essential functions of this gene. The *ced-9* mRNA is generated from a longer RNA encoded by a splicing operon that also expresses the gene *mev-1*, encoding the mitochondrial succinate dehydrogenase b560 subunit (Ishii et al., 1998). A functional relationship between CED-9 and MEV-1 has not been established. However, RNAi against *mev-1* produces embryonic lethality (Gonczy et al., 2000) and accumulation of cell corpses in the germ line (Sendoel, Kohler, Fellmann, Lowe, & Hengartner, 2010), raising the untested possibility that strong loss-of-function lesions (such as early stop codon insertions) in *ced-9* may interfere with *mev-1* expression or mRNA processing, possibly through the nonsense-mediated mRNA decay pathway.

CED-9 protein is similar in sequence to mammalian BCL2 and related proteins (Hengartner & Horvitz, 1994b). The protein consists of seven alpha helices. Helix 1 corresponds to sequences that comprise BCL2 homology region 4 (BH4) in other BCL2 family members, but it is not conserved in the primary sequence of CED-9. Helix 3 comprises the BH3 domain, helices 4 and 5 make up the BH1 domain, and helices 6 and 7 overlap the BH2 domain (Woo et al., 2003; Yan et al., 2004). Antisera against CED-9 reveal that the protein is localized to mitochondria in embryos and in the germ line (Chen et al., 2000; Pourkarimi et al., 2012). It has been suggested that the C-terminal region of CED-9, which contains a transmembrane region like other BCL2 family members, contributes to this localization. Indeed, expression in *C. elegans* muscle cells of GFP fused to the transmembrane domain confers mitochondrial localization, and a GFP–CED-9 fusion protein lacking the transmembrane domain fails to localize to mitochondria (Tan, Fire, & Hill, 2007). However, *ced-9* mutants can still be rescued by expression of cytosolic CED-9(ΔTM) or by expression of an endoplasmic reticulum (ER)-targeted CED-9 (Tan et al., 2007). Thus, the transmembrane region appears not essential for function, but may facilitate mitochondrial localization.

BCL2 family members have been suggested to form membrane pores, and *in vitro* experiments demonstrate that they can assemble into channels of variable conductivity (Antonsson et al., 1997; Basanez et al., 2002). CED-9 can also associate with lipid membranes *in vitro*, inducing changes in membrane permeability, and this association appears independent of the C-terminal transmembrane domain (Tan, Zuckerman, Wells, & Hill, 2011). Whether pore formation is required for the *in vivo* function of CED-9, or any BCL-2 family member, is an important outstanding question.

Like CED-3 and CED-4, CED-9 is expressed ubiquitously early in the embryo; however, anti-CED-9 antibodies do not detect the protein in late embryos or in early larvae (Chen et al., 2000; Pourkarimi et al., 2012). Nonetheless, a number of studies suggest that CED-9 acts cell autonomously in larvae to protect cells from cell death (Hengartner et al., 1992; Hengartner & Horvitz, 1994b). Loss of *ced-9* leads to abnormalities in cell death associated with male tail development in L4 larvae (Hengartner et al., 1992) and facilitates ectopic larval cell death induced by CED-3 overexpression (Shaham & Horvitz, 1996b). Thus, even though protein expression has not been confirmed, CED-9 activity is detected postembryonically in both dying and living cells.

Although the death-inhibitory role of CED-9 has been extensively explored, genetic evidence suggests that the *ced-9* gene also encodes a death-promoting activity. Overexpression of CED-9 can suppress inappropriate cell survival in animals carrying the *ced-9(1950)* gain-of-function mutation, and fewer cells survive inappropriately in *ced-9(n1950)* heterozygotes balanced by a wild-type copy of *ced-9* compared with animals balanced with a deletion spanning the *ced-9* locus. Also, double mutants harboring a weak *ced-3*/caspase mutation and a *ced-9* loss-of-function lesion exhibit more ectopic cell survival than weak *ced-3* mutants alone (Hengartner & Horvitz, 1994a). The physical nature of this death-promoting activity is unknown and could be due to a protein arising from alternative splicing or transcription/translation initiation, or a protein modification that competes with the activity of unmodified CED-9.

Finally, while CED-9 and perhaps CED-4L are the best-characterized cell death inhibitors in *C. elegans*, others may exist. DAD-1 is a core component of the multisubunit oligosaccharyltransferase that attaches mannose oligosaccharides to asparagine residues on proteins traveling through the ER. Overexpression of DAD-1 can prevent normally occurring cell death in *C. elegans* (Sugimoto, Hozak, Nakashima, Nishimoto, & Rothman, 1995), and its loss is embryonic lethal (Fraser et al., 2000). Whether DAD-1 is an integral component of the cell death machinery, or mediates modification of core cell death proteins, remains to be determined.

2.4 EGL-1

Genetic screens seeking mutants regulating vulval development in *C. elegans* identified a host of genes required for egg laying, including the gene *egl-1* (Trent, Tsuing, & Horvitz, 1983). Dominant gain-of-function mutations in *egl-1* result in the inappropriate death of the two HSN neurons that regulate hermaphrodite egg laying (Desai, Garriga, McIntire, & Horvitz, 1988). That these neurons normally die during male development suggested initially that *egl-1* might be involved in sex determination; however, the mutant revealed no pleiotropies that might indicate sexual transformation (Desai et al., 1988; Trent et al., 1983). While *egl-1*(gf)/*egl-1*(gf) mutants are 100% egg laying defective, about 30% of animals heterozygous for an *egl-1*(gf) allele and a deletion spanning the locus lay eggs normally (Conradt & Horvitz, 1998). Using this observation, homozygous *egl-1*(gf) animals were mutagenized, and a loss-of-function mutation in the gene was obtained by screening for animals with normal egg laying. Remarkably, animals carrying such *egl-1*(lf) mutations are unable to execute the majority of developmental cell death in the animal, and contain many extra cells (Conradt & Horvitz, 1998).

The *egl-1* gene was cloned using complementation rescue and mapping of the loss-of-function allele of *egl-1* (Conradt & Horvitz, 1998). The predicted EGL-1 protein is only 106 amino acids long and is homologous to the BH3 domain common to many mammalian cell death regulators (Kelekar & Thompson, 1998). By sequence, EGL-1 is most similar to the BH3-only class of cell death regulators that promote cell death in a variety of settings. This similarity is also evident in structural studies of the protein, which has been shown to adopt an amphipathic alpha helix conformation (Yan et al., 2004).

EGL-1 expression differs markedly from that of the other core apoptotic regulators in *C. elegans*. During development, *egl-1* expression is upregulated in cells that die (Conradt & Horvitz, 1999; Thellmann, Hatzold, & Conradt, 2003), although another study suggests that the gene may also be functional elsewhere (Jagasia, Grote, Westermann, & Conradt, 2005). Furthermore, while CED-9, CED-4, and CED-3 all control germ cell death, EGL-1 appears to be dispensable for this stochastic form of cell death (Gumienny et al., 1999), perhaps indicating that EGL-1 can act only to guarantee deterministic activation of the apoptotic program. Consistent with its expression pattern, *egl-1* gene regulatory sequences are extensive and are found both upstream and downstream of the gene. The *egl-1*(gf) lesion that leads to the specific demise of the HSN neurons is a point mutation in a regulatory site 5.6 kb downstream of the *egl-1* transcription unit, which prevents the binding of the inhibitory sex-determination factor TRA-1A (Conradt & Horvitz, 1999). In the absence of TRA-1A binding, *egl-1* is now expressed ectopically only in HSNs, leading to their removal.

Besides EGL-1, two other BH3-only proteins are made by *C. elegans*. DCT-1 is similar to mammalian BNIP3 and can cooperate with CED-3 caspase to induce cell death in cultured mammalian cells (Cizeau, Ray, Chen, Gietz, & Greenberg, 2000). However, a role in *C. elegans* cell death has not been established. CED-13, a protein more similar to EGL-1, is upregulated in the *C. elegans* germ line in response to radiation and may participate in radiation-induced cell death, but does not appear to play a significant role in developmental cell death either in the soma or in the germ line (Schumacher et al., 2005). Thus, unlike mammals, in which a host of BH3-only proteins controls cell death in a myriad of contexts and conditions, *C. elegans* appears to have a single member of this family dedicated to somatic developmental cell death.

3. THE CORE APOPTOTIC PATHWAY OF *C. ELEGANS*

The current model for apoptotic cell death execution in *C. elegans* is shown in Fig. 2. The apoptotic cascade begins when EGL-1 inhibits the activity of CED-9. CED-9 inhibition of CED-4 is then relieved, allowing CED-4 to activate CED-3, promoting cell death. The model is supported by genetic, biochemical, and structural data and predicts similarities and differences with pathway interactions described in mammals.

3.1 CED-4 → CED-3

Genetic studies support the notion that CED-3 functions downstream of CED-4. Overexpression of CED-3 in neurons not fated to die can result in their demise, and this killing is reduced, but not eliminated, in animals harboring a *ced-4*(lf) mutation. Ectopic expression of CED-4 in the same neurons can also promote their death; however, this is nearly entirely abrogated in strong *ced-3* mutants (Shaham & Horvitz, 1996b). A parsimonious explanation for these genetic studies is that CED-4 facilitates the activity of CED-3. While direct binding of the two proteins *in vivo* has not yet been demonstrated, binding studies in 293T cells show that CED-4 and CED-3 can physically interact (Chinnaiyan, Chaudhary, O'Rourke, Koonin, & Dixit, 1997; Chinnaiyan, O'Rourke, Lane, & Dixit, 1997). This interaction seems to occur between the CARD domain in CED-4 and the N-terminal domain of CED-3 (Irmler, Hofmann, Vaux, & Tschopp, 1997), although binding could also be achieved in the absence of the CED-3 N-terminal domain through the L2′ loop of its C-terminal subunit (Huang et al., 2013).

Steady-state binding of CED-4 to CED-3 is correlated with activation of the procaspase. For example, in insect cells, co-expression of CED-4 and CED-3 leads to increased caspase activity dependent on the CED-3 N-terminal domain and the ATP-binding domain of CED-4 (Seshagiri, Chang, & Miller, 1998; Seshagiri & Miller, 1997). Conversely, a CED-4 mutation disrupting a hydrophobic-binding surface between CED-4 and

CED-3 reduces the ability of CED-4 to stimulate CED-3 activity (Huang et al., 2013). Mutations blocking CED-4 oligomerization appear to block processing of CED-3 in cultured cells, suggesting that oligomerization is likely important for activity (Yang, Chang, & Baltimore, 1998). CED-4 has also been suggested to promote the activity of the mature CED-3 caspase *in vitro* (Qi et al., 2010). However, the stoichiometry of CED-4:CED-3 here is 8:2 and not 8:8 as perhaps expected, raising the possibility that this complex may not be the only source of *in vivo* activity.

3.2 CED-9-|CED-4

The lethality observed in *ced-9*(lf) mutants is suppressed by loss-of-function mutations in either *ced-3* or *ced-4* (Hengartner et al., 1992), and cell death induced by CED-4 ectopic expression is reduced by CED-9 overexpression (Shaham & Horvitz, 1996b). These observations, and the discovery that *ced-9*(gf) mutations block cell death, suggest that CED-4 and CED-3 function downstream of CED-9, and that CED-4 may be a direct target of CED-9. This hypothesis is bolstered by the demonstration that CED-9 and CED-4 physically interact (Spector, Desnoyers, Hoeppner, & Hengartner, 1997; Wu, Wallen, Inohara, & Nunez, 1997), and that mutations reducing CED-9 function *in vivo* also reduce CED-9–CED-4 binding (Spector et al., 1997). Expression studies in the early embryo of wild type and mutant *C. elegans* further corroborate this idea. While CED-4 protein, detected using anti-CED-4 sera, appears mitochondria-bound in wild-type animals, it is mainly localized to the nuclear periphery in *ced-9*(lf); *ced-3*(lf) double mutants in which the *ced-3* mutation is used to keep *ced-9*(lf) animals alive (Chen et al., 2000). Thus, CED-9 appears to serve as a mitochondrial anchor for CED-4.

How CED-9 inhibits CED-4 from forming an active octamer is not understood; however, X-ray structures of CED-9 bound to CED-4 reveal a complex in which one CED-9 protein is bound to an asymmetric CED-4 dimer (Yan et al., 2005). Thus, it is possible that one mechanism by which CED-9 inhibits CED-4 activity is by preventing its oligomerization. While the contacts observed in this study are specific, they do conflict with solution studies suggesting that bacterially purified CED-9 and CED-4 form a 2:2 complex (Fairlie et al., 2006; Yan et al., 2005). Resolution of these differences is not yet at hand.

In vitro assays demonstrate that CED-9 protein can serve as a substrate for CED-3 caspase (Xue & Horvitz, 1997), raising the possibility that CED-9 might also block cell death by competitively inhibiting CED-3 caspase activity. While global overexpression of CED-9 yields broad cell survival, mutations of the CED-9 aspartate residues targeting cleavage by CED-3 reduce the ability of CED-9 to inhibit death, consistent with this model. However, such lesions could also affect protein conformation and stability. It therefore remains unclear whether CED-9 cleavage by CED-3 has physiological significance.

3.3 EGL-1-|CED-9

Genetic studies demonstrate that EGL-1 functions upstream of both CED-4 and CED-3. Specifically, ectopic expression of EGL-1 in neurons that normally live can induce their death. This killing activity is entirely blocked in animals lacking either CED-3 or CED-4 function (Conradt & Horvitz, 1998). Likewise, inappropriate death of the HSN neurons in animals carrying *egl-1*(gf) mutations is suppressed by loss-of-function mutations in *ced-3* or *ced-4* (Ellis & Horvitz, 1986).

EGL-1 also appears to function upstream of CED-9. *egl-1*(lf) mutations, which normally block programmed cell death, fail to do so in *ced-9*(lf); *ced-3*(weak lf) mutants that entirely lack CED-9 activity, suggesting that EGL-1 normally inhibits CED-9 activity (Conradt & Horvitz, 1998). CED-9 tagged with glutathione-*S*-transferase (GST) can bind to *in vitro* translated EGL-1 (Conradt & Horvitz, 1998), and CED-9 and EGL-1 can be co-crystallized to form a bound complex (Yan et al., 2004), raising the possibility that EGL-1 inhibition of CED-9 might occur through direct physical contact. The EGL-1–CED-9 crystal structure reveals that binding of EGL-1 results in large conformational changes to CED-9, introducing steric hindrance and misalignment of key residues involved in CED-9–CED-4 binding. Thus, EGL-1 likely acts by preventing the association of CED-9 with CED-4.

Consistent with this idea, overexpression of EGL-1 in early embryos leads to the movement of CED-4 from mitochondria to the nuclear periphery as occurs in animals lacking CED-9 (Chen et al., 2000). Titration studies with purified proteins also demonstrate dissolution of the CED-4–CED-9 complex upon addition of EGL-1 (Yan et al., 2004) but not mammalian BH3-domain proteins (Fairlie et al., 2006), as do co-immunoprecipitation studies of the proteins from cultured mammalian cells (del Peso, Gonzalez, & Nunez, 1998). Further support for this model is provided by the CED-9 gain-of-function G169E mutation that results, *in vivo*, in cell survival

(Hengartner & Horvitz, 1994a). This mutation reduces the binding affinity of CED-9 for EGL-1 (del Peso, Gonzalez, Inohara, Ellis, & Nunez, 2000; Parrish, Metters, Chen, & Xue, 2000) and introduces a bulky amino acid into the binding pocket for EGL-1, leaving the CED-4-binding interface intact. Thus, blocking EGL-1 binding can allow CED-9 to remain bound to CED-4.

The components of the core apoptosis pathway and their signs of interaction appear conserved from *C. elegans* to humans. Nonetheless, even within the core pathway, distinctions are obvious. In mammals, though the BCL2-related protein Diva/Boo can bind Apaf-1 (Inohara et al., 1998; Song, Kuang, Dixit, & Vincenz, 1999), this seems to be the exception. Most BCL2 family proteins appear to control Apaf-1 activity by regulating the release of cytochrome *c*, an Apaf-1-activating factor, from mitochondria. To date, no evidence implicating cytochrome *c* in *C. elegans* apoptosis has been revealed. Indeed, the cytochrome *c*-binding domain in Apaf-1 appears not to be present in CED-4 (Li et al., 1997). Unlike CED-4, which is bound to mitochondria at least at some stages of *C. elegans* development, Apaf-1 is localized to the cytoplasm in a variety of cell types (Hausmann et al., 2000), consistent with a different mode of activation. Why the mammalian and *C. elegans* pathways appear to use similar proteins in different ways is not at all understood.

Whether mitochondria in general play a role in *C. elegans* cell death is also unclear. Some mitochondrial factors have been examined, including endonuclease G (CPS-6) (Parrish et al., 2001), apoptosis-inducing factor (WAH-1) (Wang, Yang, Chai, Shi, & Xue, 2002), and the mitochondrial fission and fusion proteins DRP-1 and FZO-1 (Breckenridge et al., 2008; Jagasia et al., 2005; Lu, Rolland, & Conradt, 2011). While animals carrying lesions in the genes encoding these factors have been reported in some cases to have cell death defects, the effects are rather weak. It is therefore unclear whether any of these proteins specifically interact with the cell death machinery or if the effects are due to more generic changes in cell state. Mutations in core apoptotic components do not obviously affect mitochondrial shape, fission, or fusion in homozygous early embryos (Breckenridge, Kang, & Xue, 2009). Perhaps more relevant, however, in dying cells, mitochondrial fragmentation is observed, is dependent on *egl-1*, and is blocked by a *ced-9*(gf) allele (Jagasia et al., 2005). Importantly, fission is independent of *ced-3* and *ced-4*. Thus, while mitochondrial fragmentation is not required for cell death, it does accompany the process.

4. REGULATING APOPTOSIS

4.1 Control of *egl-1* Expression

The induction of *egl-1* expression in dying cells, and the relatively large and complex regulatory region required for this expression, suggests that the decision to live or die may be mediated by cell-specific transcription factors that bind to the *egl-1* locus to promote and/or inhibit its transcription. As described above, this indeed appears to be the case in HSN neurons, where the sex-determination factor TRA-1A binds in the environs of *egl-1* to block transcription in hermaphrodites (Conradt & Horvitz, 1999) (Fig. 2). The CEM neurons are also sexually dimorphic cells. In this case, the cells survive in males but die in hermaphrodites (Sulston et al., 1983), and sex-specific survival here is also mediated in part by control of *egl-1* expression. The CEH-30 homeodomain transcription factor inhibits CEM death in males by acting with UNC-37/Groucho to block expression of *egl-1* in these cells. In hermaphrodites, the TRA-1 protein promotes CEM neuron death by inhibiting transcription of the *ceh-30* gene, thus allowing *egl-1* expression (Nehme et al., 2010; Peden, Kimberly, Gengyo-Ando, Mitani, & Xue, 2007; Schwartz & Horvitz, 2007) (Fig. 2).

Recessive mutations in the gene *ces-2*, encoding a bZIP transcription factor, and a dominant allele of the gene *ces-1*, encoding a Snail-like bHLH transcription factor, block the deaths of the NSM sister, and NSM sister and I2 sister cells, respectively (Ellis & Horvitz, 1991; Metzstein, Hengartner, Tsung, Ellis, & Horvitz, 1996; Metzstein & Horvitz, 1999). In the NSM sisters, survival correlates with lack of induction of a P*egl-1*::GFP reporter in the cells (Thellmann et al., 2003). CES-1 binds *in vitro* to Snail elements in an *egl-1* genomic fragment that is required *in vivo* to mediate CES-1 activity. These elements overlap with E box sequences that can bind the HLH-2 and HLH-3 transcription factors, which promote NSM death and also promote *egl-1* expression (Thellmann et al., 2003). Thus, CES-1 apparently inhibits *egl-1* expression by blocking access of HLH-2/3 to the *egl-1* promoter (Fig. 2).

Mutations in the *ceh-34* and *eya-1* genes, both encoding homeodomain transcription factors, block the deaths of the M4 and I3 sister cells (Fig. 2). CEH-34 and EYA-1 proteins physically interact, and CEH-34 binds to a regulatory site 5′ of the *egl-1* transcription unit. In both mutants, *egl-1* expression in the M4 sister cell is abrogated (Hirose, Galvin, & Horvitz, 2010). The Sp1 transcription factor SPTF-3 is also required for expression of *egl-1* in and death of the M4 sister cell (Hirose & Horvitz, 2013). In a similar mode of regulation, a complex consisting of the CEH-20/Pbx1 and MAB-5/Hox transcription factors controls the death of the postembryonic cell P11.aaap (Fig. 2). In this case, a site approximately 6 kb downstream of the *egl-1* start codon binds the CEH-20-MAB-5 complex, and mutations in that site or in its cognate-binding factors eliminate *egl-1* expression in the cell (Liu, Strauss, Potts, & Cameron, 2006).

Regulation of *egl-1* expression may also govern the death of the embryonic cells ABprpppapp and ABplpppapp, whose demise appears to be weakly dependent on EGF signaling (Fig. 2). At least one of these cells survives inappropriately in about 30% of animals deficient in either the EGF ligand LIN-3 or its receptor, LET-23/EGFR, whose site of action is not firmly established. Regardless, a ~50% reduction in the levels of a P*egl-1*::GFP reporter is seen in *lin-3*(lf) mutants (Jiang & Wu, 2014).

4.2 *egl-1* Is Not Alone

While *egl-1* is a hub for cell death decision making, more is likely going on. In animals carrying both *ced-9*(strong lf) and *ced-3*(weak lf) mutations, cells destined to live do so, and only a few cells destined to die survive inappropriately (Hengartner & Horvitz, 1994a). In these animals, many cells fated to die appear to do so on cue. Yet, since CED-9 activity is gone, *egl-1* cannot be the relevant death effector in these cells, unless it also has CED-9-independent functions. A role for *egl-1* in starvation-induced autophagy has been proposed, based on the observation that animals homozygous for an *egl-1*(gf) mutation accumulate the autophagy marker LGG-1::DsRED even in the presence of food (Maiuri et al., 2007). However, autophagy genes are not required for cell death in *C. elegans* (Takacs-Vellai et al., 2005), making this an unlikely mechanism for cell death control. However, cell death regulators that function in place of EGL-1 have been uncovered.

In HSN neurons, the EOR-1 protein, similar to the mammalian PLZF transcription factor, is required together with its novel cofactor, EOR-2, for death (Hoeppner et al., 2004; Howard & Sundaram, 2002; Howell, Arur, Schedl, & Sundaram, 2010; Rocheleau et al., 2002). Neither gene promotes expression of *egl-1*. Furthermore, the effects of *eor-1/2* mutations appear to require an intact *ced-9* gene, suggesting that EOR-1/2 may work upstream of or in parallel to CED-9 and in parallel to EGL-1 (Fig. 2). In CEM neurons, three proteins, UNC-86, a POU homeodomain transcription factor,

LRS-1, a leucyl t-RNA synthetase, and the kinase PIM-1 are required for cell death (Nehme et al., 2010; Peden et al., 2007). In hermaphrodites, these genes promote expression of the *ced-3* caspase gene but not *egl-1*. In males, they promote *ceh-30* expression and therefore indirectly inhibit *egl-1* expression (Fig. 2).

In M4 sister cells, two proteins, GCN-1, which may be involved in controlling translation initiation factor levels, and the interacting protein ABCF-3, an AAA+ ATPase, appear to function together to promote cell death, and mutants in these genes have a weak survival defect. Here also, both proteins act independently of CED-9, and hence of EGL-1, to promote cell death (Hirose & Horvitz, 2014). The mechanism of action of this protein complex is not known, but the similarity of ABCF-3 to AAA+ ATPases raises the possibility that they may interact with CED-4 (Fig. 2).

In addition to regulating *egl-1* transcription in M4 sister cells, the SPTF-3/Sp1 protein also promotes expression of the gene *pig-1* (Fig. 2). PIG-1, similar to MELK kinase, has been implicated in the control of a number of cell death events in *C. elegans* and also appears to function in parallel to EGL-1 and CED-9. PIG-1 has been suggested to function in a *ced-3*-independent cell death pathway (Denning, Hatch, & Horvitz, 2012; Hirose & Horvitz, 2013). However, the data are also consistent with a function within the canonical cell death pathway. Indeed, PIG-1 function in asymmetric cell division has been previously described (Chien, Brinkmann, Teuliere, & Garriga, 2013; Cordes, Frank, & Garriga, 2006), raising the possibility that PIG-1 may control the segregation of death effectors downstream of CED-9, such as CED-3 or CED-4. Such segregation could explain how an initially broad expression pattern of these genes leads to the death of specific cells. Supporting this notion, the polarity proteins PAR-4/LKB1 kinase, STRAD-1/STRAD, and MOP-25/MO25 appear to function together with PIG-1 in the death of the ABplpappap cell (Denning et al., 2012).

Parallel control of death is also seen in descendents of the Pn.p blast cells that normally die in the ventral cord of early larva and in pharyngeal cells. Here, components of the conserved retinoblastoma (Rb)/E2F pathway are required for cell death and also function independently of CED-9 and likely EGL-1 (Reddien, Andersen, Huang, & Horvitz, 2007).

4.3 *egl-1* Is Not Always Required

While *egl-1* control is essential for some cell death initiation, in some cases EGL-1 function is not required (Fig. 2). The tail-spike cell is a binucleate cell with a posterior process believed to function as a scaffold for developing tail epithelial cells. Five hours after it is born, the tail-spike cell dies (Sulston et al., 1983). While death requires *ced-3* and *ced-4*, *ced-9*(gf) mutations do not block cell death, and *egl-1*(lf) animals only have a weak defect, with about 30% exhibiting tail-spike cell survival (Maurer et al., 2007), suggesting that EGL-1–CED-9 interactions are not important for the demise of this cell. In the tail-spike cell, transcription of the *ced-3* caspase gene is temporally controlled. Expression of a P*ced-3*::GFP reporter is off until about 25 min before the cell rounds up and dies, suggesting that *ced-3* transcription is an important regulatory node. The transcription factor PAL-1, similar to vertebrate Cdx2, is an important regulator of tail-spike cell death. PAL-1 is required for *ced-3* expression and can bind to sequences upstream of the *ced-3* start codon *in vitro* and probably *in vivo*, and mutations in *pal-1* block tail-spike cell death (Maurer et al., 2007).

Besides PAL-1, the F-Box protein DRE-1 also regulates tail-spike cell death. *dre-1* mutants have inappropriately surviving tail-spike cells, and like *pal-1*, *dre-1* acts in parallel to *egl-1*. However, *dre-1* requires a wild-type copy of *ced-9* to carry out its function. Genetic and protein interaction studies suggest that DRE-1 physically interacts with SKR-1/SKP, which controls protein ubiquitylation, to promote cell death. DRE-1 also interacts with CED-9, which may, therefore, be its *in vivo* target (Chiorazzi et al., 2013).

egl-1(lf) and *ced-9*(gf) mutations block germ cell death in response to ionizing radiation (Gartner, Milstein, Ahmed, Hodgkin, & Hengartner, 2000); however, as in the tail-spike cell, neither mutation affects normal physiological cell death in the germ line. Other regulators appear to take charge here (Fig. 2). Mutants in *pax-2* and *egl-38*, encoding Pax2/5/8 family transcriptional regulators, have increased rates of germ cell death. Overexpression of these genes reduces germline apoptosis (Park, Jia, Rajakumar, & Chamberlin, 2006). Mosaic analysis shows that *egl-38* is required in the germ line for germ cell survival, and RT-PCR, deletion, and chromatin immunoprecipitation studies suggest that both *egl-38* and *pax-2* act by activating *ced-9* transcription (Park et al., 2006). This function may also extend to embryonic cell death, as *egl-38* and *pax-2* single mutants also have increased cell corpses in embryos.

The Rb pathway also plays a role in developmental germ cell death control. LIN-35/Rb appears to inhibit *ced-9* transcription in these cells, while the DP homolog DPL-1 and the E2F-like EFL-2 promote transcription of *ced-3* and *ced-4* (Schertel & Conradt, 2007). Another tumor control pathway, the RAS/MAPK signaling pathway, also appears to be important. Mutations in the *let-60*/Ras, *lin-45*/Raf, and *mek-2* and *mpk-1* MAPK genes block germ cell death (Gumienny et al., 1999) and also lead to other developmental defects in the germ line, suggesting that they likely act early in cell fate determination (Fig. 2).

It is remarkable that so many of the regulators of the apoptotic pathway in *C. elegans* are related in sequence to mammalian oncogenes and tumor suppressors (Table 1). This supports the notion that manipulation of cell death pathways is an important aspect of tumor formation.

5. THE ENGULFMENT GENES AND THEIR ROLES IN CELL DEATH

Apoptosis in *C. elegans* is tightly coupled to engulfment. For many cells destined to die, engulfment by neighboring cells begins even before anaphase of the parent cell division is complete (Sulston et al., 1983). Thus, cells destined to die often do not have an opportunity to function or interact with other cells. Only three *C. elegans* cells live for an extended period of time before death and engulfment ensue: the tail-spike cell (see above), the male-specific linker cell (see below), and the sister of Z1, a gonadal precursor whose function is not known (Sulston et al., 1983). Genetic screens for mutants defective in the engulfment or degradation of cell corpses have identified a host of proteins required for these processes, and these have been extensively reviewed (Lu & Zhou, 2012; Pinto & Hengartner, 2012); we only briefly summarize the findings here.

Two parallel and partially redundant pathways control recognition of dying cells (Ellis, Jacobson, & Horvitz, 1991). One consists of the membrane receptor CED-1, which appears to recognize the lipid PS on the surface of dying cells (Zhou, Hartwieg, & Horvitz, 2001), CED-6, an adapter protein that may function in signal transduction (Liu & Hengartner, 1998, 1999), and CED-7, a membrane ABC transporter that may be involved in PS presentation (Wu & Horvitz, 1998). CED-1 and CED-6 function in the engulfing cell, and CED-7 is required in both dying and engulfing cells. Unlike CED-1 and CED-6, CED-7 is required only in the soma, whereas the phospholipid scramblase PLSC-1 exposes PS in dying germ cells (Venegas & Zhou, 2007). The other pathway consists of the proteins CED-2/CrkII, CED-5/Dock180, CED-10/Rac, and CED-12/Elmo (Gumienny et al., 2001; Reddien & Horvitz, 2000; Wu, Tsai, Cheng, Chou, & Weng, 2001; Zhou, Caron, Hartwieg, Hall, & Horvitz, 2001). CED-2, -5, and -12 appear to function together to regulate CED-10

Table 1 Some *C. elegans* Apoptosis Regulators are Similar to Mammalian Tumor Factors

C. elegans Protein	Function	Regulated Cell (Type)	Mammalian Homolog	Tumor[a]	References
ABCF-3	AAA+ ATPase	M4 sister (neuron)	ABCF3	Liver cancer	Hirose and Horvitz (2014)
CEH-20	Transcription factor	P11.aaap (neuron)	PBX1	Leukemia	Liu et al. (2006)
CEH-34	Transcription factor	M4 sister (neuron)	SIX1	Breast cancer, gastric cancer, others	Hirose et al. (2010)
CES-1	Transcription factor	NSM, I2 (neurons)	SLUG	Acute lymphoblastic leukemia	Metzstein and Horvitz (1999)
CES-2	Transcription factor	NSM (neuron)	HLF	Acute lymphoblastic leukemia	Metzstein et al. (1996)
EFL-2	Transcription factor	Germ cells, pharynx, others	E2F	Colorectal cancer, pancreatic cancer, many others	Reddien et al. (2007) and Schertel and Conradt (2007)
EGL-38	Transcription factor	Germ cells, embryonic cells (?)	Pax2/5/8	Renal cancer, colon cancer, many others	Park et al. (2006)
EOR-1	Transcription factor	HSN (neuron)	PLZF, BCL6	Leukemia, lymphoma, others	Hoeppner et al. (2004) and Howard and Sundaram (2002)
EYA-1	Transcription factor	M4 sister (neuron)	EYA	Breast cancer, gastric cancer, others	Hirose et al. (2010)

Continued

Table 1 Some *C. elegans* Apoptosis Regulators are Similar to Mammalian Tumor Factors—cont'd

C. elegans Protein	Function	Regulated Cell (Type)	Mammalian Homolog	Tumor	References
DRE-1	F-Box, ubiquitylation	Tail-spike cell (epithelial)	FBXO10/11	B-cell lymphoma	Chiorazzi et al. (2013)
LET-23	Receptor tyrosine kinase	Germ cells, ABpl/rpppapp (epithelial)	EGFR	Lung cancer, medulloblastoma, others	Jiang and Wu (2014)
LET-60	GTPase	Germ cells, ABpl/rpppapp (epithelial)	RAS	Many tumors	Jiang and Wu (2014)
LIN-3	Receptor tyrosine kinase ligand	Germ cells, ABpl/rpppapp (epithelial)	EGF	Lung cancer, medulloblastoma, others	Jiang and Wu (2014)
LIN-35	Transcription factor	Germ cells, Pn.p descendents (neurons)	Rb	Retinoblastoma, many tumors	Reddien et al. (2007) and Schertel and Conradt (2007)
MAB-5	Transcription factor	P11.aaap (neuron)	HOX	Breast cancer, many tumors	Liu et al. (2006)
PAL-1	Transcription factor	Tail-spike cell (epithelial)	Cdx2	Intestinal cancer	Maurer et al. (2007)

PAR-4	kinase	ABalapapaa, ABplpappap ABalppaaaa ABalppaapa ABaraaaapp ABplppaaap ABplpppapp ABalpapappa (epithelial, excretory cell) M4 sister (neuron)	LKB1	Liver cancer, leukemia, others	Chien et al. (2013) and Denning et al. (2012)
PAX-2	Transcription factor	Germ cells, embryonic cells (?)	Pax2/5/8	Renal cancer, colon cancer, many others	Park et al. (2006)
PIG-1	kinase	ABalapapaa, ABplpappap ABalppaaaa ABalppaapa ABaraaaapp ABplppaaap ABplpppapp ABalpapappa (epithelial, excretory cell) M4 sister (neuron)	MELK	Glioblastoma, astocytoma, pediatric brain tumor, colon tumor, breast cancer, melanoma, rectal cancer	Chien et al. (2013), Cordes et al. (2006), Denning et al. (2012), and Hirose and Horvitz (2013)

Continued

Table 1 Some *C. elegans* Apoptosis Regulators are Similar to Mammalian Tumor Factors—cont'd

C. elegans Protein	Function	Regulated Cell (Type)	Mammalian Homolog	Tumor	References
UNC-86	Transcription factor	CEM (neurons)	BRN3	Neuroepithelial cancer	Nehme et al. (2010) and Peden et al. (2007)
UNC-132	Kinase	CEM (neuron)	PIM-1	Leukemia, lung cancer, pancreatic cancer, ovarian cancer	Nehme et al. (2010)
TRA-1	Transcription factor	HSN, CEM (neurons)	Gli	Basal cell carcinoma, medulloblastoma,breast cancer, glioma, leukemia, prostate cancer, GI cancers, others	Nehme et al. (2010), Peden et al. (2007), and Schwartz and Horvitz (2007)

[a] Tumors reporting associated changes in function of the indicated mammalian protein.

activity. CED-10 likely acts to control assembly of the actin cytoskeleton during the extension of engulfing cell protrusions around the dying cell.

It has been suggested that CED-1 and CED-6 also function through CED-10 (Kinchen et al., 2005), although this has been debated (Yu, Odera, Chuang, Lu, & Zhou, 2006). Nonetheless, recent studies show that the dynamin protein DYN-1 mediates signaling downstream of CED-1 in a pathway involving clathrin and epsin, which promote actin assembly at the phagocytic cup (Shen et al., 2013). Thus, CED-1 may indirectly impact actin remodeling.

Once an apoptotic cell has been incorporated into a phagosome within the engulfing cell, a series of events alter the phagosomal surface, culminating in lysosomal targeting. Phosphatidylinositol 3-phosphate (PtdIns(3)P) coats the phagosomal membrane, allowing binding of BAR-domain sorting nexins to recruit lysosomes (Lu, Shen, Mahoney, Liu, & Zhou, 2011). In parallel, RAB-5, which binds initially to the phagosomal surface, is replaced by RAB-7, through the activity of the SAND-1 and CCZ-1 complex (Kinchen & Ravichandran, 2010; Nieto et al., 2010), and phagolysosome formation ensues through activities of the conserved HOPs complex (Kinchen et al., 2008) and the proteins UNC-108/RAB2 and RAB-14, which promote acidification of the compartment (Guo, Hu, Zhang, Jiang, & Wang, 2010; Mangahas, Yu, Miller, & Zhou, 2008).

While engulfment has generally been viewed as independent of cell death execution, some observations suggest a more intimate connection. In animals carrying weak *ced-3* caspase mutations, some cells destined to die in the developing ventral cord of *C. elegans* larvae succumb and are removed, while others survive inappropriately. Occasionally, cells that by DIC microscopy appear far gone recover and survive. Thus, in these mutants, cells slated to die teeter on the edge between life and death. Remarkably, in animals also homozygous for mutations in genes controlling engulfment, significantly more cells survive and develop (Reddien, Cameron, & Horvitz, 2001). These results demonstrate that at least under some conditions, engulfment may promote cell death. Similarly, in *C. elegans* males, the cells B.alapaav and B.arapaav have been reported to survive in *ced-1* and *ced-2* mutants, presumably because they fail to be engulfed by the P12.pa cell (Hedgecock, Sulston, & Thomson, 1983).

While the mechanism by which engulfing cells promote cell death of their neighbors is not clear, some molecular insight may be gleaned from studies of DNA degradation accompanying cell death. As described above, mutations in the DNase II gene *nuc-1* result in accumulation of

TUNEL-staining DNA in developing embryos, revealing an activity for an early-acting DNAse. Mutations in the engulfment genes *ced-1* and *ced-7*, but not *ced-2*, *-5*, *-6*, or *-10*, prevent accumulation of TUNEL-reactive fragments (Wu et al., 2000). Thus, CED-1 and CED-7 are required for initiating the degradation of DNA in dying cells. While this could be taken to mean that DNAse activity resides in the engulfing cells, this interpretation is unlikely, since in *ced-2*, *-5*, *-6*, and *-10* mutants, engulfment also fails to take place, but DNase activation still occurs. Rather, barring technical issues with TUNEL staining in these mutants, it seems that a signal unrelated to engulfment *per se*, but dependent on *ced-1* and *ced-7*, promotes DNase activation in dying cells. Although it is likely that *ced-1* and *ced-7* function in the engulfing cell to activate DNA degradation, this has not been directly tested and would be a prerequisite for defining the functions of these genes in DNA degradation.

6. LINKER CELL DEATH

The linker cell is a male-specific leader cell born in the ventral midbody in the second larval stage (L2). The cell guides the elongation of the gonad, and upon reaching the cloacal region in the late L4 stage, it dies. Death may allow fusion of the gonad with the cloaca, resulting in an open germ system competent for sperm transfer during fertilization. Anecdotal reports suggested that linker cell death may require the neighboring engulfing cells, U.l/rp (Hedgecock et al., 1983); however, later studies demonstrated that the cell can die following ablation of these dedicated engulfing cells, or in animals in which the linker cell fails to migrate properly (Abraham, Lu, & Shaham, 2007). Anecdotal reports also suggested initially that linker cell death is partially dependent on *ced-3* and *ced-4* (Ellis & Horvitz, 1986). However, further studies of the genetic requirements for death, coupled with extensive morphological and ultrastructural observations, revealed that linker cell death does not require *ced-3* or *ced-4*. Indeed, the cell death process could proceed in its entirety in animals carrying mutations in any of the core apoptotic pathway genes (Abraham et al., 2007). Thus, linker cell death proceeds by a previously unexplored mode of cell death. Consistent with these novel genetic requirements, the ultrastructural changes accompanying the cell death process are very different from those seen in apoptotic cells. While apoptosis is characterized by nuclear condensation, cytoplasmic shrinkage, and morphologically intact organelles (until the very end) (Fig. 1), the dying linker cell exhibits none of these features.

Instead, the cell displays pronounced nuclear envelope invagination (crenellation), as well as swelling of ER and mitochondria (Abraham et al., 2007) (Fig. 3). These changes are often observed in cells dying during vertebrate development and are prevalent in degenerating cells of patients with polyQ diseases.

Genetic and molecular screens have uncovered genes involved in linker cell death. Initiation of the death process appears to be dependent on both temporal and positional cues. The LIN-29 Zn-finger transcription factor, a downstream component of the *C. elegans* heterochronic pathway controlling developmental timing, promotes linker cell demise (Abraham et al.,

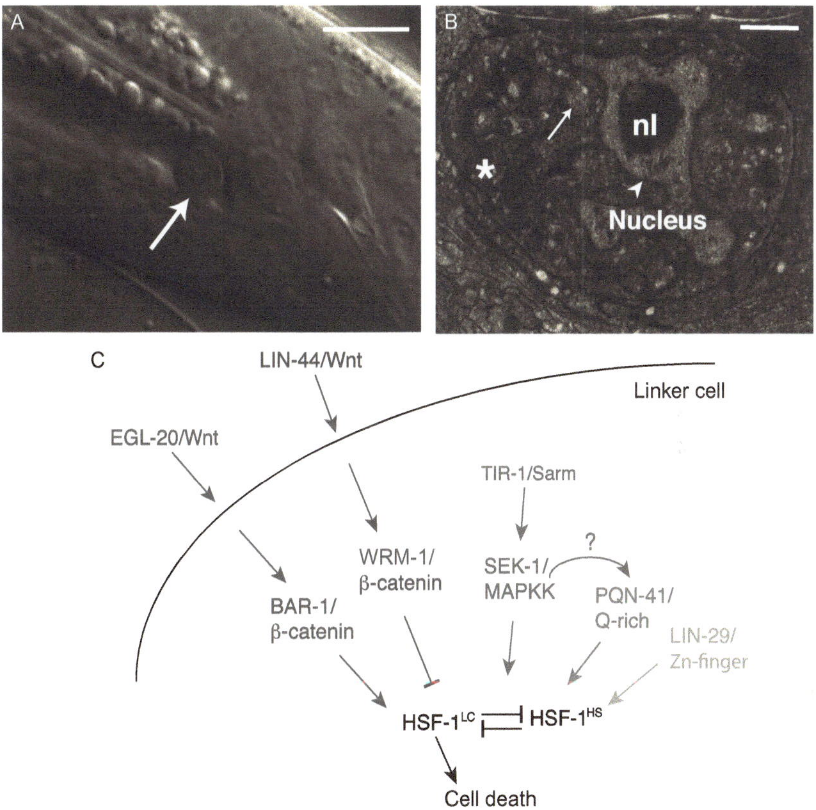

Figure 3 Linker cell death in *C. elegans*. (A) DIC image of a dying linker cell (arrow). Scale bar, 5 μm. (B) Electron micrograph of a dying linker cell. Note crenelated nucleus (arrowhead), intact nucleolus (nl), and swollen organelles (asterisk and arrow). Scale bar, 700 nm. (C) A model for linker cell death. Proteins in the same regulatory pathway are color-matched. How regulators impinge on HSF-1 function is not known. (See the color plate.)

2007). Two opposing Wnt pathways mediate the spatial signal (M. Kinet & S. Shaham, submitted). The Wnt LIN-44 is secreted by male tail cells and acts through the MIG-1 and CFZ-2 Frizzled-like receptors, LIT-1/Nemo-like kinase, and WRM-1/beta-catenin—all of which function in the linker cell—to inhibit linker cell death. This protective pathway is antagonized by the Wnt EGL-20, also expressed in neighboring cells, and its linker cell effectors LIN-17/Frizzled, MOM-5/Frizzled, MIG-5/Dishevelled, and BAR-1/beta-catenin (Fig. 3).

The Wnt-responsive transcription factor POP-1/TCF/LEF does not appear to play a role in linker cell death (M. Kinet & S. Shaham, submitted), suggesting other possible targets. One candidate is the Q-rich protein PQN-41. A C-terminal polypeptide encoded by the locus, and consisting of 35% glutamine residues, is sufficient to rescue linker cell survival exhibited by *pqn-41* mutants (Blum, Abraham, Yoshimura, Lu, & Shaham, 2012). At least some PQN-41 isoforms appear to be localized to nuclei, suggesting that they could function to mediate a transcriptional output of Wnt signaling. Another candidate is the transcription factor HSF-1, which in other contexts transduces heat and stress signals to promote a protective cellular response. In the linker cell, however, *hsf-1*(lf) mutations block linker cell death, suggesting that it, instead, promotes cell dismantling (M. Kinet & S. Shaham, submitted). Supporting this observation, stress targets of HSF-1 are not induced in the linker cell. Importantly, an HSF-1 gain-of-function mutation can rescue inappropriate linker cell survival in Wnt pathway mutants, as well as all other known mutants that block linker cell death (M. Kinet & S. Shaham, submitted). HSF-1, therefore, may be a key downstream regulatory node, and determining its targets is an important next step.

The kinase SEK-1 and its adapter protein TIR-1 also promote linker cell death, and their loss can also be compensated for by the *hsf-1*(gf) allele. Genetic interaction studies suggest that these proteins might function in the same pathway as PQN-41 (Blum et al., 2012). *Drosophila* and murine homologs of TIR-1, dSarm, and Sarm, respectively, have been implicated in axon distal segment degeneration following axotomy (Osterloh et al., 2012), raising the possibility that the dismantling program of the linker cell may share conserved features with nonapoptotic degenerative programs in other systems. Supporting this idea, mutations in ubiquitin-dependent protein degradation system components block linker cell death (J. Malin & S. Shaham, in preparation) and have likewise been implicated in axon degeneration (Korhonen & Lindholm, 2004).

Linker cell engulfment also has novel genetic requirements. Mutations in apoptotic corpse engulfment genes do not affect engulfment of the linker cell (Abraham et al., 2007), suggesting that alternative pathways must exist. Genetic screens have identified a number of mutants in which engulfment and/or corpse degradation appear blocked (L. Kutscher & S. Shaham, unpublished data). Characterization of these mutants will likely reveal aspects of this new machinery.

7. LOOKING AHEAD

In *C. elegans*, mutations in *ced-3* or *ced-4* block most developmental cell death that takes place in the animal. While the mammalian counterparts of these genes clearly participate in apoptotic events, mutations in these genes have surprisingly weak effects on developmental cell death. Caspase-3 and -9 mutants are viable, as are Apaf-1 knockout mice (Honarpour et al., 2000). In some specific instances, the mammalian mutants exhibit a slight delay in cell death timing; however, the process eventually gains traction, and cells give up the ghost. Why are CED-3 and CED-4, therefore, essential for most cell death in nematodes but not in mammals? Why do mammals apparently have parallel programs ensuring cellular demise even if apoptosis fails? While the answers remain unknown, one explanation invokes the inherent developmental differences between *C. elegans* and mammals. The nematode lineage is essentially invariant, and cell identity and survival are nearly always determined cell autonomously by lineal history. Honing of cell differentiation programs therefore leads to reproducible development and cell death that is not very error prone. In mammals, however, cell identity and cell death are very often regulated by cell–cell communication. While cell-autonomous processes are confined by the cell membrane, cell–cell signaling is, by its nature, less spatially contained. Thus, to ensure robust death, alternative death processes may have evolved. In this light, it is intriguing that the only *C. elegans* cell in which cell–cell signaling has been unambiguously demonstrated to control survival, the linker cell, also employs an alternative cell death form. If this hypothesis is correct, two predictions may be made. First, *ced-3* and *ced-4* may have a role in linker cell death, but this may only be elicited in sensitized genetic backgrounds. Second, the nonapoptotic death program governing linker cell death may very well be conserved in mammals. Experiments testing both predictions are currently underway and may eventually reveal an even deeper similarity in cell death processes among animals.

ACKNOWLEDGMENTS

We are grateful to members of our laboratory for comments and for sharing unpublished results. Every attempt to include all relevant citations was made, and we apologize to those authors whose work was not cited either because of space constraints or because of our oversight. This study was supported by NIH Grants HD078703 and NS081490 to S.S.

REFERENCES

Abraham, M. C., Lu, Y., & Shaham, S. (2007). A morphologically conserved non-apoptotic program promotes linker cell death in C. elegans. *Developmental Cell, 12*(1), 73–86.

Acehan, D., Jiang, X., Morgan, D. G., Heuser, J. E., Wang, X., & Akey, C. W. (2002). Three-dimensional structure of the apoptosome: Implications for assembly, procaspase-9 binding, and activation. *Molecular Cell, 9*, 423–432.

Antonsson, B., Conti, F., Ciavatta, A., Montessuit, S., Lewis, S., Martinou, I., et al. (1997). Inhibition of Bax channel-forming activity by Bcl-2. *Science, 277*, 370–372.

Aruscavage, P. J., Hellwig, S., & Bass, B. L. (2010). Small DNA pieces in C. elegans are intermediates of DNA fragmentation during apoptosis. *PLoS One, 5*. e11217.

Avery, L., & Horvitz, H. R. (1987). A cell that dies during wild-type C. elegans development can function as a neuron in a ced-3 mutant. *Cell, 51*, 1071–1078.

Basanez, G., Sharpe, J. C., Galanis, J., Brandt, T. B., Hardwick, J. M., & Zimmerberg, J. (2002). Bax-type apoptotic proteins porate pure lipid bilayers through a mechanism sensitive to intrinsic monolayer curvature. *The Journal of Biological Chemistry, 277*, 49360–49365.

Bloss, T. A., Witze, E. S., & Rothman, J. H. (2003). Suppression of CED-3-independent apoptosis by mitochondrial betaNAC in Caenorhabditis elegans. *Nature, 424*, 1066–1071.

Blum, E. S., Abraham, M. C., Yoshimura, S., Lu, Y., & Shaham, S. (2012). Control of non-apoptotic developmental cell death in Caenorhabditis elegans by a polyglutamine-repeat protein. *Science, 335*, 970–973.

Breckenridge, D. G., Kang, B. H., Kokel, D., Mitani, S., Staehelin, L. A., & Xue, D. (2008). Caenorhabditis elegans drp-1 and fis-2 regulate distinct cell-death execution pathways downstream of ced-3 and independent of ced-9. *Molecular Cell, 31*, 586–597.

Breckenridge, D. G., Kang, B. H., & Xue, D. (2009). Bcl-2 proteins EGL-1 and CED-9 do not regulate mitochondrial fission or fusion in Caenorhabditis elegans. *Current Biology, 19*, 768–773.

Chelur, D. S., & Chalfie, M. (2007). Targeted cell killing by reconstituted caspases. *Proceedings of the National Academy of Sciences of the United States of America, 104*, 2283–2288.

Chen, F., Hersh, B. M., Conradt, B., Zhou, Z., Riemer, D., Gruenbaum, Y., et al. (2000). Translocation of C. elegans CED-4 to nuclear membranes during programmed cell death. *Science, 287*, 1485–1489.

Chen, Y. Z., Mapes, J., Lee, E. S., Skeen-Gaar, R. R., & Xue, D. (2013). Caspase-mediated activation of Caenorhabditis elegans CED-8 promotes apoptosis and phosphatidylserine externalization. *Nature Communications, 4*, 2726.

Chien, S. C., Brinkmann, E. M., Teuliere, J., & Garriga, G. (2013). Caenorhabditis elegans PIG-1/MELK acts in a conserved PAR-4/LKB1 polarity pathway to promote asymmetric neuroblast divisions. *Genetics, 193*, 897–909.

Chinnaiyan, A. M., Chaudhary, D., O'Rourke, K., Koonin, E. V., & Dixit, V. M. (1997). Role of CED-4 in the activation of CED-3. *Nature, 388*, 728–729.

Chinnaiyan, A. M., O'Rourke, K., Lane, B. R., & Dixit, V. M. (1997). Interaction of CED-4 with CED-3 and CED-9: A molecular framework for cell death. *Science, 275*, 1122–1126.

Chiorazzi, M., Rui, L., Yang, Y., Ceribelli, M., Tishbi, N., Maurer, C. W., et al. (2013). Related F-box proteins control cell death in Caenorhabditis elegans and human lymphoma. *Proceedings of the National Academy of Sciences of the United States of America, 110*, 3943–3948.

Cizeau, J., Ray, R., Chen, G., Gietz, R. D., & Greenberg, A. H. (2000). The C. elegans orthologue ceBNIP3 interacts with CED-9 and CED-3 but kills through a BH3- and caspase-independent mechanism. *Oncogene, 19*, 5453–5463.

Conradt, B., & Horvitz, H. R. (1998). The C. elegans protein EGL-1 is required for programmed cell death and interacts with the Bcl-2-like protein CED-9. *Cell, 93*, 519–529.

Conradt, B., & Horvitz, H. R. (1999). The TRA-1A sex determination protein of C. elegans regulates sexually dimorphic cell deaths by repressing the egl-1 cell death activator gene. *Cell, 98*, 317–327.

Cordes, S., Frank, C. A., & Garriga, G. (2006). The C. elegans MELK ortholog PIG-1 regulates cell size asymmetry and daughter cell fate in asymmetric neuroblast divisions. *Development, 133*, 2747–2756.

del Peso, L., Gonzalez, V. M., Inohara, N., Ellis, R. E., & Nunez, G. (2000). Disruption of the CED-9.CED-4 complex by EGL-1 is a critical step for programmed cell death in Caenorhabditis elegans. *The Journal of Biological Chemistry, 275*, 27205–27211.

del Peso, L., Gonzalez, V. M., & Nunez, G. (1998). Caenorhabditis elegans EGL-1 disrupts the interaction of CED-9 with CED-4 and promotes CED-3 activation. *The Journal of Biological Chemistry, 273*, 33495–33500.

Denning, D. P., Hatch, V., & Horvitz, H. R. (2012). Programmed elimination of cells by caspase-independent cell extrusion in C. elegans. *Nature, 488*, 226–230.

Denning, D. P., Hatch, V., & Horvitz, H. R. (2013). Both the caspase CSP-1 and a caspase-independent pathway promote programmed cell death in parallel to the canonical pathway for apoptosis in Caenorhabditis elegans. *PLoS Genetics, 9*. e1003341.

Desai, C., Garriga, G., McIntire, S. L., & Horvitz, H. R. (1988). A genetic pathway for the development of the Caenorhabditis elegans HSN motor neurons. *Nature, 336*, 638–646.

Dreze, M., Charloteaux, B., Milstein, S., Vidalain, P. O., Yildirim, M. A., Zhong, Q., et al. (2009). 'Edgetic' perturbation of a C. elegans BCL2 ortholog. *Nature Methods, 6*, 843–849.

Ellis, H. M., & Horvitz, H. R. (1986). Genetic control of programmed cell death in the nematode C. elegans. *Cell, 44*, 817–829.

Ellis, R. E., & Horvitz, H. R. (1991). Two C. elegans genes control the programmed deaths of specific cells in the pharynx. *Development, 112*, 591–603.

Ellis, R. E., Jacobson, D. M., & Horvitz, H. R. (1991). Genes required for the engulfment of cell corpses during programmed cell death in Caenorhabditis elegans. *Genetics, 129*, 79–94.

Ellis, R. E., Yuan, J. Y., & Horvitz, H. R. (1991). Mechanisms and functions of cell death. *Annual Review of Cell Biology, 7*, 663–698.

Enari, M., Sakahira, H., Yokoyama, H., Okawa, K., Iwamatsu, A., & Nagata, S. (1998). A caspase-activated DNase that degrades DNA during apoptosis, and its inhibitor ICAD. *Nature, 391*, 43–50.

Fairlie, W. D., Perugini, M. A., Kvansakul, M., Chen, L., Huang, D. C., & Colman, P. M. (2006). CED-4 forms a 2:2 heterotetrameric complex with CED-9 until specifically displaced by EGL-1 or CED-13. *Cell Death and Differentiation, 13*, 426–434.

Fraser, A. G., Kamath, R. S., Zipperlen, P., Martinez-Campos, M., Sohrmann, M., & Ahringer, J. (2000). Functional genomic analysis of C. elegans chromosome I by systematic RNA interference. *Nature, 408*, 325–330.

Fuchs, Y., & Steller, H. (2011). Programmed cell death in animal development and disease. *Cell, 147*, 742–758.

Galvin, B. D., Denning, D. P., & Horvitz, H. R. (2011). SPK-1, an SR protein kinase, inhibits programmed cell death in Caenorhabditis elegans. *Proceedings of the National Academy of Sciences of the United States of America*, *108*, 1998–2003.

Gartner, A., Milstein, S., Ahmed, S., Hodgkin, J., & Hengartner, M. O. (2000). A conserved checkpoint pathway mediates DNA damage-induced apoptosis and cell cycle arrest in C. elegans. *Molecular Cell*, *5*, 435–443.

Gavrieli, Y., Sherman, Y., & Ben-Sasson, S. A. (1992). Identification of programmed cell death in situ via specific labeling of nuclear DNA fragmentation. *The Journal of Cell Biology*, *119*, 493–501.

Geng, X., Shi, Y., Nakagawa, A., Yoshina, S., Mitani, S., & Xue, D. (2008). Inhibition of CED-3 zymogen activation and apoptosis in Caenorhabditis elegans by caspase homolog CSP-3. *Nature Structural & Molecular Biology*, *15*, 1094–1101.

Geng, X., Zhou, Q. H., Kage-Nakadai, E., Shi, Y., Yan, N., Mitani, S., et al. (2009). Caenorhabditis elegans caspase homolog CSP-2 inhibits CED-3 autoactivation and apoptosis in germ cells. *Cell Death and Differentiation*, *16*, 1385–1394.

Gonczy, P., Echeverri, C., Oegema, K., Coulson, A., Jones, S. J., Copley, R. R., et al. (2000). Functional genomic analysis of cell division in C. elegans using RNAi of genes on chromosome III. *Nature*, *408*, 331–336.

Goyal, L., McCall, K., Agapite, J., Hartwieg, E., & Steller, H. (2000). Induction of apoptosis by Drosophila reaper, hid and grim through inhibition of IAP function. *The EMBO Journal*, *19*, 589–597.

Gumienny, T. L., Brugnera, E., Tosello-Trampont, A. C., Kinchen, J. M., Haney, L. B., Nishiwaki, K., et al. (2001). CED-12/ELMO, a novel member of the CrkII/Dock180/Rac pathway, is required for phagocytosis and cell migration. *Cell*, *107*, 27–41.

Gumienny, T. L., Lambie, E., Hartwieg, E., Horvitz, H. R., & Hengartner, M. O. (1999). Genetic control of programmed cell death in the Caenorhabditis elegans hermaphrodite germline. *Development*, *126*, 1011–1022.

Guo, P., Hu, T., Zhang, J., Jiang, S., & Wang, X. (2010). Sequential action of Caenorhabditis elegans Rab GTPases regulates phagolysosome formation during apoptotic cell degradation. *Proceedings of the National Academy of Sciences of the United States of America*, *107*, 18016–18021.

Hamburger, V., & Levi-Montalcini, R. (1949). Proliferation, differentiation and degeneration in the spinal ganglia of the chick embryo under normal and experimental conditions. *The Journal of Experimental Zoology*, *111*, 457–501.

Hanahan, D., & Weinberg, R. A. (2011). Hallmarks of cancer: The next generation. *Cell*, *144*, 646–674.

Hausmann, G., O'Reilly, L. A., van Driel, R., Beaumont, J. G., Strasser, A., Adams, J. M., et al. (2000). Pro-apoptotic apoptosis protease-activating factor 1 (Apaf-1) has a cytoplasmic localization distinct from Bcl-2 or Bcl-x(L). *The Journal of Cell Biology*, *149*, 623–634.

Hay, B. A., Wassarman, D. A., & Rubin, G. M. (1995). Drosophila homologs of baculovirus inhibitor of apoptosis proteins function to block cell death. *Cell*, *83*, 1253–1262.

Hedgecock, E. M., Sulston, J. E., & Thomson, J. N. (1983). Mutations affecting programmed cell deaths in the nematode Caenorhabditis elegans. *Science*, *220*, 1277–1279.

Hengartner, M. O., Ellis, R. E., & Horvitz, H. R. (1992). Caenorhabditis elegans gene ced-9 protects cells from programmed cell death. *Nature*, *356*, 494–499.

Hengartner, M. O., & Horvitz, H. R. (1994a). Activation of C. elegans cell death protein CED-9 by an amino-acid substitution in a domain conserved in Bcl-2. *Nature*, *369*, 318–320.

Hengartner, M. O., & Horvitz, H. R. (1994b). C. elegans cell survival gene ced-9 encodes a functional homolog of the mammalian proto-oncogene bcl-2. *Cell*, *76*, 665–676.

Hevelone, J., & Hartman, P. S. (1988). An endonuclease from Caenorhabditis elegans: Partial purification and characterization. *Biochemical Genetics, 26,* 447–461.
Hirose, T., Galvin, B. D., & Horvitz, H. R. (2010). Six and Eya promote apoptosis through direct transcriptional activation of the proapoptotic BH3-only gene egl-1 in Caenorhabditis elegans. *Proceedings of the National Academy of Sciences of the United States of America, 107,* 15479–15484.
Hirose, T., & Horvitz, H. R. (2013). An Sp1 transcription factor coordinates caspase-dependent and -independent apoptotic pathways. *Nature, 500,* 354–358.
Hirose, T., & Horvitz, H. R. (2014). The translational regulators GCN-1 and ABCF-3 act together to promote apoptosis in C. elegans. *PLoS Genetics, 10.* e1004512.
Hoeppner, D. J., Hengartner, M. O., & Schnabel, R. (2001). Engulfment genes cooperate with ced-3 to promote cell death in Caenorhabditis elegans. *Nature, 412,* 202–206.
Hoeppner, D. J., Spector, M. S., Ratliff, T. M., Kinchen, J. M., Granat, S., Lin, S. C., et al. (2004). eor-1 and eor-2 are required for cell-specific apoptotic death in C. elegans. *Developmental Biology, 274,* 125–138.
Honarpour, N., Du, C., Richardson, J. A., Hammer, R. E., Wang, X., & Herz, J. (2000). Adult Apaf-1-deficient mice exhibit male infertility. *Developmental Biology, 218,* 248–258.
Horvitz, H. R. (2003). Nobel lecture. Worms, life and death. *Bioscience Reports, 23,* 239–303.
Howard, R. M., & Sundaram, M. V. (2002). C. elegans EOR-1/PLZF and EOR-2 positively regulate Ras and Wnt signaling and function redundantly with LIN-25 and the SUR-2 Mediator component. *Genes & Development, 16,* 1815–1827.
Howell, K., Arur, S., Schedl, T., & Sundaram, M. V. (2010). EOR-2 is an obligate binding partner of the BTB-zinc finger protein EOR-1 in Caenorhabditis elegans. *Genetics, 184,* 899–913.
Huang, W., Jiang, T., Choi, W., Qi, S., Pang, Y., Hu, Q., et al. (2013). Mechanistic insights into CED-4-mediated activation of CED-3. *Genes & Development, 27,* 2039–2048.
Inohara, N., Gourley, T. S., Carrio, R., Muniz, M., Merino, J., Garcia, I., et al. (1998). Diva, a Bcl-2 homologue that binds directly to Apaf-1 and induces BH3-independent cell death. *The Journal of Biological Chemistry, 273,* 32479–32486.
Irmler, M., Hofmann, K., Vaux, D., & Tschopp, J. (1997). Direct physical interaction between the Caenorhabditis elegans 'death proteins' CED-3 and CED-4. *FEBS Letters, 406,* 189–190.
Ishii, N., Fujii, M., Hartman, P. S., Tsuda, M., Yasuda, K., Senoo-Matsuda, N., et al. (1998). A mutation in succinate dehydrogenase cytochrome b causes oxidative stress and ageing in nematodes. *Nature, 394,* 694–697.
Jagasia, R., Grote, P., Westermann, B., & Conradt, B. (2005). DRP-1-mediated mitochondrial fragmentation during EGL-1-induced cell death in C. elegans. *Nature, 433,* 754–760.
Jiang, H. S., & Wu, Y. C. (2014). LIN-3/EGF promotes the programmed cell death of specific cells in Caenorhabditis elegans by transcriptional activation of the pro-apoptotic gene egl-1. *PLoS Genetics, 10.* e1004513.
Joshi, P., & Eisenmann, D. M. (2004). The Caenorhabditis elegans pvl-5 gene protects hypodermal cells from ced-3-dependent, ced-4-independent cell death. *Genetics, 167,* 673–685.
Kelekar, A., & Thompson, C. B. (1998). Bcl-2-family proteins: The role of the BH3 domain in apoptosis. *Trends in Cell Biology, 8,* 324–330.
Kimble, J., & Hirsh, D. (1979). The postembryonic cell lineages of the hermaphrodite and male gonads in Caenorhabditis elegans. *Developmental Biology, 70,* 396–417.
Kinchen, J. M., Cabello, J., Klingele, D., Wong, K., Feichtinger, R., Schnabel, H., et al. (2005). Two pathways converge at CED-10 to mediate actin rearrangement and corpse removal in C. elegans. *Nature, 434,* 93–99.

Kinchen, J. M., Doukoumetzidis, K., Almendinger, J., Stergiou, L., Tosello-Trampont, A., Sifri, C. D., et al. (2008). A pathway for phagosome maturation during engulfment of apoptotic cells. *Nature Cell Biology*, *10*, 556–566.
Kinchen, J. M., & Ravichandran, K. S. (2010). Identification of two evolutionarily conserved genes regulating processing of engulfed apoptotic cells. *Nature*, *464*, 778–782.
Korhonen, L., & Lindholm, D. (2004). The ubiquitin proteasome system in synaptic and axonal degeneration: A new twist to an old cycle. *The Journal of Cell Biology*, *165*, 27–30.
Li, P., Nijhawan, D., Budihardjo, I., Srinivasula, S. M., Ahmad, M., Alnemri, E. S., et al. (1997). Cytochrome c and dATP-dependent formation of Apaf-1/caspase-9 complex initiates an apoptotic protease cascade. *Cell*, *91*, 479–489.
Liu, Q. A., & Hengartner, M. O. (1998). Candidate adaptor protein CED-6 promotes the engulfment of apoptotic cells in C. elegans. *Cell*, *93*, 961–972.
Liu, Q. A., & Hengartner, M. O. (1999). Human CED-6 encodes a functional homologue of the Caenorhabditis elegans engulfment protein CED-6. *Current Biology*, *9*, 1347–1350.
Liu, X., Li, P., Widlak, P., Zou, H., Luo, X., Garrard, W. T., et al. (1998). The 40-kDa subunit of DNA fragmentation factor induces DNA fragmentation and chromatin condensation during apoptosis. *Proceedings of the National Academy of Sciences of the United States of America*, *95*, 8461–8466.
Liu, H., Strauss, T. J., Potts, M. B., & Cameron, S. (2006). Direct regulation of egl-1 and of programmed cell death by the Hox protein MAB-5 and by CEH-20, a C. elegans homolog of Pbx1. *Development*, *133*, 641–650.
Lockshin, R. A., & Williams, C. M. (1965). Programmed cell death—I. Cytology of degeneration in the intersegmental muscles of the pernyi silkmoth. *Journal of Insect Physiology*, *11*, 123–133.
Lu, Y., Rolland, S. G., & Conradt, B. (2011). A molecular switch that governs mitochondrial fusion and fission mediated by the BCL2-like protein CED-9 of Caenorhabditis elegans. *Proceedings of the National Academy of Sciences of the United States of America*, *108*, E813–E822.
Lu, N., Shen, Q., Mahoney, T. R., Liu, X., & Zhou, Z. (2011). Three sorting nexins drive the degradation of apoptotic cells in response to PtdIns(3)P signaling. *Molecular Biology of the Cell*, *22*, 354–374.
Lu, N., & Zhou, Z. (2012). Membrane trafficking and phagosome maturation during the clearance of apoptotic cells. *International Review of Cell and Molecular Biology*, *293*, 269–309.
Maiuri, M. C., Le Toumelin, G., Criollo, A., Rain, J. C., Gautier, F., Juin, P., et al. (2007). Functional and physical interaction between Bcl-X(L) and a BH3-like domain in Beclin-1. *The EMBO Journal*, *26*, 2527–2539.
Mangahas, P. M., Yu, X., Miller, K. G., & Zhou, Z. (2008). The small GTPase Rab2 functions in the removal of apoptotic cells in Caenorhabditis elegans. *The Journal of Cell Biology*, *180*, 357–373.
Maurer, C. W., Chiorazzi, M., & Shaham, S. (2007). Timing of the onset of a developmental cell death is controlled by transcriptional induction of the C. elegans ced-3 caspase-encoding gene. *Development*, *134*, 1357–1368.
Metzstein, M. M., Hengartner, M. O., Tsung, N., Ellis, R. E., & Horvitz, H. R. (1996). Transcriptional regulator of programmed cell death encoded by Caenorhabditis elegans gene ces-2. *Nature*, *382*, 545–547.
Metzstein, M. M., & Horvitz, H. R. (1999). The C. elegans cell death specification gene ces-1 encodes a snail family zinc finger protein. *Molecular Cell*, *4*, 309–319.
Nakagawa, A., Shi, Y., Kage-Nakadai, E., Mitani, S., & Xue, D. (2010). Caspase-dependent conversion of Dicer ribonuclease into a death-promoting deoxyribonuclease. *Science*, *328*, 327–334.

Nakagawa, A., Sullivan, K. D., & Xue, D. (2014). Caspase-activated phosphoinositide binding by CNT-1 promotes apoptosis by inhibiting the AKT pathway. *Nature Structural & Molecular Biology, 21*, 1082–1090.

Nehme, R., Grote, P., Tomasi, T., Loser, S., Holzkamp, H., Schnabel, R., et al. (2010). Transcriptional upregulation of both egl-1 BH3-only and ced-3 caspase is required for the death of the male-specific CEM neurons. *Cell Death and Differentiation, 17*, 1266–1276.

Nieto, C., Almendinger, J., Gysi, S., Gomez-Orte, E., Kaech, A., Hengartner, M. O., et al. (2010). ccz-1 mediates the digestion of apoptotic corpses in C. elegans. *Journal of Cell Science, 123*, 2001–2007.

Osterloh, J. M., Yang, J., Rooney, T. M., Fox, A. N., Adalbert, R., Powell, E. H., et al. (2012). dSarm/Sarm1 is required for activation of an injury-induced axon death pathway. *Science, 337*, 481–484.

Park, D., Jia, H., Rajakumar, V., & Chamberlin, H. M. (2006). Pax2/5/8 proteins promote cell survival in C. elegans. *Development, 133*, 4193–4202.

Parrish, J., Li, L., Klotz, K., Ledwich, D., Wang, X., & Xue, D. (2001). Mitochondrial endonuclease G is important for apoptosis in C. elegans. *Nature, 412*, 90–94.

Parrish, J., Metters, H., Chen, L., & Xue, D. (2000). Demonstration of the in vivo interaction of key cell death regulators by structure-based design of second-site suppressors. *Proceedings of the National Academy of Sciences of the United States of America, 97*, 11916–11921.

Parrish, J. Z., & Xue, D. (2003). Functional genomic analysis of apoptotic DNA degradation in C. elegans. *Molecular Cell, 11*, 987–996.

Peden, E., Kimberly, E., Gengyo-Ando, K., Mitani, S., & Xue, D. (2007). Control of sex-specific apoptosis in C. elegans by the BarH homeodomain protein CEH-30 and the transcriptional repressor UNC-37/Groucho. *Genes & Development, 21*, 3195–3207.

Pinto, S. M., & Hengartner, M. O. (2012). Cleaning up the mess: Cell corpse clearance in Caenorhabditis elegans. *Current Opinion in Cell Biology, 24*, 881–888.

Poreba, M., Strozyk, A., Salvesen, G. S., & Drag, M. (2013). Caspase substrates and inhibitors. *Cold Spring Harbor Perspectives in Biology, 5*, a008680.

Pourkarimi, E., Greiss, S., & Gartner, A. (2012). Evidence that CED-9/Bcl2 and CED-4/Apaf-1 localization is not consistent with the current model for C. elegans apoptosis induction. *Cell Death and Differentiation, 19*, 406–415.

Qi, S., Pang, Y., Hu, Q., Liu, Q., Li, H., Zhou, Y., et al. (2010). Crystal structure of the Caenorhabditis elegans apoptosome reveals an octameric assembly of CED-4. *Cell, 141*, 446–457.

Reddien, P. W., Andersen, E. C., Huang, M. C., & Horvitz, H. R. (2007). DPL-1 DP, LIN-35 Rb and EFL-1 E2F act with the MCD-1 zinc-finger protein to promote programmed cell death in Caenorhabditis elegans. *Genetics, 175*, 1719–1733.

Reddien, P. W., Cameron, S., & Horvitz, H. R. (2001). Phagocytosis promotes programmed cell death in C. elegans. *Nature, 412*, 198–202.

Reddien, P. W., & Horvitz, H. R. (2000). CED-2/CrkII and CED-10/Rac control phagocytosis and cell migration in Caenorhabditis elegans. *Nature Cell Biology, 2*, 131–136.

Rocheleau, C. E., Howard, R. M., Goldman, A. P., Volk, M. L., Girard, L. J., & Sundaram, M. V. (2002). A lin-45 raf enhancer screen identifies eor-1, eor-2 and unusual alleles of Ras pathway genes in Caenorhabditis elegans. *Genetics, 161*, 121–131.

Schertel, C., & Conradt, B. (2007). C. elegans orthologs of components of the RB tumor suppressor complex have distinct pro-apoptotic functions. *Development, 134*, 3691–3701.

Schumacher, B., Schertel, C., Wittenburg, N., Tuck, S., Mitani, S., Gartner, A., et al. (2005). C. elegans ced-13 can promote apoptosis and is induced in response to DNA damage. *Cell Death and Differentiation, 12*, 153–161.

Schwartz, H. T., & Horvitz, H. R. (2007). The C. elegans protein CEH-30 protects male-specific neurons from apoptosis independently of the Bcl-2 homolog CED-9. *Genes & Development, 21*, 3181–3194.

Seiffert, B. M., Vier, J., & Hacker, G. (2002). Subcellular localization, oligomerization, and ATP-binding of Caenorhabditis elegans CED-4. *Biochemical and Biophysical Research Communications*, *290*, 359–365.

Sendoel, A., Kohler, I., Fellmann, C., Lowe, S. W., & Hengartner, M. O. (2010). HIF-1 antagonizes p53-mediated apoptosis through a secreted neuronal tyrosinase. *Nature*, *465*, 577–583.

Seshagiri, S., Chang, W. T., & Miller, L. K. (1998). Mutational analysis of Caenorhabditis elegans CED-4. *FEBS Letters*, *428*, 71–74.

Seshagiri, S., & Miller, L. K. (1997). Caenorhabditis elegans CED-4 stimulates CED-3 processing and CED-3-induced apoptosis. *Current Biology*, *7*, 455–460.

Shaham, S. (1998). Identification of multiple Caenorhabditis elegans caspases and their potential roles in proteolytic cascades. *The Journal of Biological Chemistry*, *273*, 35109–35117.

Shaham, S., & Bargmann, C. I. (2002). Control of neuronal subtype identity by the C. elegans ARID protein CFI-1. *Genes & Development*, *16*, 972–983.

Shaham, S., & Horvitz, H. R. (1996a). An alternatively spliced C. elegans ced-4 RNA encodes a novel cell death inhibitor. *Cell*, *86*, 201–208.

Shaham, S., & Horvitz, H. R. (1996b). Developing Caenorhabditis elegans neurons may contain both cell-death protective and killer activities. *Genes & Development*, *10*, 578–591.

Shaham, S., Reddien, P. W., Davies, B., & Horvitz, H. R. (1999). Mutational analysis of the Caenorhabditis elegans cell-death gene ced-3. *Genetics*, *153*, 1655–1671.

Shen, Q., He, B., Lu, N., Conradt, B., Grant, B. D., & Zhou, Z. (2013). Phagocytic receptor signaling regulates clathrin and epsin-mediated cytoskeletal remodeling during apoptotic cell engulfment in C. elegans. *Development*, *140*, 3230–3243.

Silke, J., & Vucic, D. (2014). IAP family of cell death and signaling regulators. *Methods in Enzymology*, *545*, 35–65.

Song, Q., Kuang, Y., Dixit, V. M., & Vincenz, C. (1999). Boo, a novel negative regulator of cell death, interacts with Apaf-1. *The EMBO Journal*, *18*, 167–178.

Spector, M. S., Desnoyers, S., Hoeppner, D. J., & Hengartner, M. O. (1997). Interaction between the C. elegans cell-death regulators CED-9 and CED-4. *Nature*, *385*, 653–656.

Speliotes, E. K., Uren, A., Vaux, D., & Horvitz, H. R. (2000). The survivin-like C. elegans BIR-1 protein acts with the Aurora-like kinase AIR-2 to affect chromosomes and the spindle midzone. *Molecular Cell*, *6*, 211–223.

Stanfield, G. M., & Horvitz, H. R. (2000). The ced-8 gene controls the timing of programmed cell deaths in C. elegans. *Molecular Cell*, *5*, 423–433.

Strange, K. (2003). From genes to integrative physiology: Ion channel and transporter biology in Caenorhabditis elegans. *Physiological Reviews*, *83*, 377–415.

Sugimoto, A., Hozak, R. R., Nakashima, T., Nishimoto, T., & Rothman, J. H. (1995). dad-1, an endogenous programmed cell death suppressor in Caenorhabditis elegans and vertebrates. *The EMBO Journal*, *14*, 4434–4441.

Sulston, J. E. (1976). Post-embryonic development in the ventral cord of Caenorhabditis elegans. *Philosophical Transactions of the Royal Society of London. Series B, Biological Sciences*, *275*, 287–297.

Sulston, J. E., Albertson, D. G., & Thomson, J. N. (1980). The Caenorhabditis elegans male: Postembryonic development of nongonadal structures. *Developmental Biology*, *78*, 542–576.

Sulston, J. E., & Horvitz, H. R. (1977). Post-embryonic cell lineages of the nematode, Caenorhabditis elegans. *Developmental Biology*, *56*, 110–156.

Sulston, J. E., Schierenberg, E., White, J. G., & Thomson, J. N. (1983). The embryonic cell lineage of the nematode Caenorhabditis elegans. *Developmental Biology*, *100*, 64–119.

Suzuki, J., Denning, D. P., Imanishi, E., Horvitz, H. R., & Nagata, S. (2013). Xk-related protein 8 and CED-8 promote phosphatidylserine exposure in apoptotic cells. *Science, 341*, 403–406.

Takacs-Vellai, K., Vellai, T., Puoti, A., Passannante, M., Wicky, C., Streit, A., et al. (2005). Inactivation of the autophagy gene bec-1 triggers apoptotic cell death in C. elegans. *Current Biology, 15*, 1513–1517.

Tan, F. J., Fire, A. Z., & Hill, R. B. (2007). Regulation of apoptosis by C. elegans CED-9 in the absence of the C-terminal transmembrane domain. *Cell Death and Differentiation, 14*, 1925–1935.

Tan, F. J., Zuckerman, J. E., Wells, R. C., & Hill, R. B. (2011). The C. elegans B-cell lymphoma 2 (Bcl-2) homolog cell death abnormal 9 (CED-9) associates with and remodels LIPID membranes. *Protein Science, 20*, 62–74.

Thellmann, M., Hatzold, J., & Conradt, B. (2003). The Snail-like CES-1 protein of C. elegans can block the expression of the BH3-only cell-death activator gene egl-1 by antagonizing the function of bHLH proteins. *Development, 130*, 4057–4071.

Thornberry, N. A., Bull, H. G., Calaycay, J. R., Chapman, K. T., Howard, A. D., Kostura, M. J., et al. (1992). A novel heterodimeric cysteine protease is required for interleukin-1 beta processing in monocytes. *Nature, 356*, 768–774.

Trent, C., Tsuing, N., & Horvitz, H. R. (1983). Egg-laying defective mutants of the nematode Caenorhabditis elegans. *Genetics, 104*, 619–647.

Tzur, Y. B., Margalit, A., Melamed-Book, N., & Gruenbaum, Y. (2006). Matefin/SUN-1 is a nuclear envelope receptor for CED-4 during Caenorhabditis elegans apoptosis. *Proceedings of the National Academy of Sciences of the United States of America, 103*, 13397–13402.

Venegas, V., & Zhou, Z. (2007). Two alternative mechanisms that regulate the presentation of apoptotic cell engulfment signal in Caenorhabditis elegans. *Molecular Biology of the Cell, 18*, 3180–3192.

Wang, X., Yang, C., Chai, J., Shi, Y., & Xue, D. (2002). Mechanisms of AIF-mediated apoptotic DNA degradation in Caenorhabditis elegans. *Science, 298*, 1587–1592.

White, J. G., Southgate, E., & Thomson, J. N. (1991). On the nature of undead cells in the nematode Caenorhabditis elegans. *Philosophical Transactions of the Royal Society of London. Series B, Biological Sciences, 331*, 263–271.

Woo, J. S., Jung, J. S., Ha, N. C., Shin, J., Kim, K. H., Lee, W., et al. (2003). Unique structural features of a BCL-2 family protein CED-9 and biophysical characterization of CED-9/EGL-1 interactions. *Cell Death and Differentiation, 10*, 1310–1319.

Wu, D., Chen, P. J., Chen, S., Hu, Y., Nunez, G., & Ellis, R. E. (1999). C. elegans MAC-1, an essential member of the AAA family of ATPases, can bind CED-4 and prevent cell death. *Development, 126*, 2021–2031.

Wu, Y. C., & Horvitz, H. R. (1998). The C. elegans cell corpse engulfment gene ced-7 encodes a protein similar to ABC transporters. *Cell, 93*, 951–960.

Wu, Y. C., Stanfield, G. M., & Horvitz, H. R. (2000). NUC-1, a caenorhabditis elegans DNase II homolog, functions in an intermediate step of DNA degradation during apoptosis. *Genes & Development, 14*, 536–548.

Wu, Y. C., Tsai, M. C., Cheng, L. C., Chou, C. J., & Weng, N. Y. (2001). C. elegans CED-12 acts in the conserved crkII/DOCK180/Rac pathway to control cell migration and cell corpse engulfment. *Developmental Cell, 1*, 491–502.

Wu, D., Wallen, H. D., Inohara, N., & Nunez, G. (1997). Interaction and regulation of the Caenorhabditis elegans death protease CED-3 by CED-4 and CED-9. *The Journal of Biological Chemistry, 272*, 21449–21454.

Xue, D., & Horvitz, H. R. (1997). Caenorhabditis elegans CED-9 protein is a bifunctional cell-death inhibitor. *Nature, 390*, 305–308.

Xue, D., Shaham, S., & Horvitz, H. R. (1996). The Caenorhabditis elegans cell-death protein CED-3 is a cysteine protease with substrate specificities similar to those of the human CPP32 protease. *Genes & Development, 10*, 1073–1083.

Yan, N., Chai, J., Lee, E. S., Gu, L., Liu, Q., He, J., et al. (2005). Structure of the CED-4-CED-9 complex provides insights into programmed cell death in Caenorhabditis elegans. *Nature, 437*, 831–837.

Yan, N., Gu, L., Kokel, D., Chai, J., Li, W., Han, A., et al. (2004). Structural, biochemical, and functional analyses of CED-9 recognition by the proapoptotic proteins EGL-1 and CED-4. *Molecular Cell, 15*, 999–1006.

Yang, X., Chang, H. Y., & Baltimore, D. (1998). Essential role of CED-4 oligomerization in CED-3 activation and apoptosis. *Science, 281*, 1355–1357.

Youle, R. J., & van der Bliek, A. M. (2012). Mitochondrial fission, fusion, and stress. *Science, 337*, 1062–1065.

Yu, X., Odera, S., Chuang, C. H., Lu, N., & Zhou, Z. (2006). C. elegans Dynamin mediates the signaling of phagocytic receptor CED-1 for the engulfment and degradation of apoptotic cells. *Developmental Cell, 10*, 743–757.

Yuan, J. Y., & Horvitz, H. R. (1990). The Caenorhabditis elegans genes ced-3 and ced-4 act cell autonomously to cause programmed cell death. *Developmental Biology, 138*, 33–41.

Yuan, J., & Horvitz, H. R. (1992). The Caenorhabditis elegans cell death gene ced-4 encodes a novel protein and is expressed during the period of extensive programmed cell death. *Development, 116*, 309–320.

Yuan, J., Shaham, S., Ledoux, S., Ellis, H. M., & Horvitz, H. R. (1993). The C. elegans cell death gene ced-3 encodes a protein similar to mammalian interleukin-1 beta-converting enzyme. *Cell, 75*, 641–652.

Zhou, Z., Caron, E., Hartwieg, E., Hall, A., & Horvitz, H. R. (2001). The C. elegans pH domain protein CED-12 regulates cytoskeletal reorganization via a Rho/Rac GTPase signaling pathway. *Developmental Cell, 1*, 477–489.

Zhou, Z., Hartwieg, E., & Horvitz, H. R. (2001). CED-1 is a transmembrane receptor that mediates cell corpse engulfment in C. elegans. *Cell, 104*, 43–56.

Zou, H., Henzel, W. J., Liu, X., Lutschg, A., & Wang, X. (1997). Apaf-1, a human protein homologous to C. elegans CED-4, participates in cytochrome c-dependent activation of caspase-3. *Cell, 90*, 405–413.

CHAPTER TWO

Mitochondrial Cell Death Pathways in *Caenorhabiditis elegans*

Mahendra Seervi*,†,1, Ding Xue*,1

*Department of Molecular, Cellular, and Developmental Biology, University of Colorado, Boulder, Colorado, USA

†DBT-PU-IPLS Program, Department of Biotechnology/Botany, Patna University, Patna, Bihar, India

[1]Corresponding authors: e-mail address: mseervi12@gmail.com; ding.xue@colorado.edu

Contents

Abstract

Programmed cell death is an evolutionarily conserved process essential for animal development and tissue homeostasis. Mitochondria have been demonstrated to play a central role in regulating both the activation and the execution of apoptosis. In particular, mitochondria release multiple proapoptotic factors from its intermembrane space, leading to both caspase-dependent and -independent cell death. Despite the pivotal roles of invertebrate animal models, *Caenorhabiditis elegans* and *Drosophila melanogaster*, in deciphering conserved pathways and mechanisms of programmed cell death, the importance of mitochondria to apoptosis of invertebrates remains elusive and largely unexplored. Recent studies have corroborated significant association between mitochondria and apoptosis in *C. elegans*, making it a thrust area of investigations. In this review, we detail the roles of mitochondrial proteins in mediating execution of cell death in *C. elegans*, including chromosome fragmentation, phosphatidylserine externalization, and elimination of mitochondria, and discuss the potential roles of mitochondria in the activation of *C. elegans* cell death. The combination of traditional

Current Topics in Developmental Biology, Volume 114

ISSN 0070-2153

http://dx.doi.org/10.1016/bs.ctdb.2015.07.019

powerful genetic tools and the emergence of the multiple new reverse genetic techniques, including the highly efficient CRISPR/Cas9 gene-editing method, should make *C. elegans* an ideal animal model for analyzing mitochondrial cell death pathways and associated regulatory mechanisms.

1. INTRODUCTION

Programmed cell death is a highly conserved process in multicellular organisms and is crucial for proper animal development, defense mechanisms, and tissue homoeostasis (Danial & Korsmeyer, 2004). It is a tightly regulated physiological process that rapidly removes damaged, mutated, or virus-infected cells within an organism. Dysregulation of cell death can lead to several human diseases, including cancer, neurodegenerative disease, and autoimmune disorders (Adams, 2003).

Apoptosis is generally viewed as the major and classical type of programmed cell death, although several other forms of cell death such as autophagy and necrosis also play important physiological roles. Apoptosis is so far the best studied one among various types of cell death and shares common hallmarks at the cellular and molecular levels in diverse organisms (Danial & Korsmeyer, 2004; Pennell & Lamb, 1997). Invertebrates such as *Caenorhabiditis elegans* and *Drosophila melanogaster* have been extensively studied in the past decades to investigate the complex mechanisms of apoptosis, partly because of the powerful genetic tools available in these organisms. Genetic studies in nematodes and fruit flies have identified a large number of genes and important signaling networks that are conserved between invertebrates and mammals (Metzstein, Stanfield, & Horvitz, 1998; Steller, 1995).

The major features of apoptosis at the cellular levels are nuclear and cytoplasmic condensation, cell shrinkage, phosphatidylserine (PS) externalization, and chromosome fragmentation (Adams, 2003; Danial & Korsmeyer, 2004; Fadok, Xue, & Henson, 2001; Steller, 1995). At the molecular levels, apoptosis is executed by a conserved family of proteases, named caspases, and is tightly regulated by the Bcl-2 (B-cell lymphoma-2) family of proteins (Adams & Cory, 2001; Boyce, Degterev, & Yuan, 2004). In mammals, signals that induce apoptosis can either originate from inside of the cell (intrinsic pathway) or from outside of the cell, through activation of death receptors on the plasma membrane, which is termed the extrinsic pathway (Krammer, 2000; Strasser, O'Connor, & Dixit, 2000;

Wang, 2001). Both signaling cascades ultimately lead to caspase activation, which in many cases define the point of no return for cell death.

Importantly, these death-signaling events appear to funnel through mitochondria before the execution of death by caspases in mammalian cells. Several proapoptotic molecules, including cytochrome *c*, are released from mitochondria during apoptosis (Wang, 2001). Released cytochrome *c* initiates the formation of a protein complex, apoptosome, which is also comprised of Apaf-1, procaspase-9, and ATP (Kim, Du, Fang, & Wang, 2005; Li et al., 1997; Zou, Henzel, Liu, Lutschg, & Wang, 1997). The apoptosome forms a platform that is conducive to the activation of procaspase-9, which proteolytically activates procaspase-3, a pivotal executioner caspase (Li et al., 1997). Caspase-3 activation subsequently leads to cleavage of many protein substrates, which either are activated to promote different aspects of cell killing or are inactivated to facilitate cell killing because these substrates are essential for the maintenance and survival of the cell (Boyce et al., 2004; Crawford & Wells, 2011). In mammalian cells, this cytochrome *c*-initiated mitochondrial cell death pathway is indispensable for the activation of the intrinsic cell death pathway executed by caspase-3 (Green & Kroemer, 1998; Wang, 2001).

This, however, does not appear to be the case in *C. elegans*, as the activation of the CED-3 caspase, which shares substrate specificity with mammalian caspase-3 (Xue, Shaham, & Horvitz, 1996), is independent of cytochrome *c*. The activation of the CED-3 caspase is tightly controlled by the BH3-only cell death initiator EGL-1, the Bcl-2-like protein CED-9, and the Apaf-1 homologue CED-4 (Conradt & Xue, 2005; Horvitz, 2003). Numerous studies provide evidence to support a model, in which CED-9 binds to and sequesters a CED-4 dimer to the surface of mitochondria in healthy cells. In cells programmed to die, increased transcriptional expression of EGL-1 leads to the binding of EGL-1 to CED-9, which induces a conformational change of CED-9 and the dissociation of the CED-4 dimer from CED-9 (Chinnaiyan, O'Rourke, Lane, & Dixit, 1997; Conradt & Horvitz, 1998; Parrish, Metters, Chen, & Xue, 2000; Spector, Desnoyers, Hoeppner, & Hengartner, 1997; Wu, Wallen, Inohara, & Nunez, 1997; Yan et al., 2005). The released CED-4 dimer subsequently oligomerizes and translocates to the perinuclear membrane to initiate the activation of the CED-3 caspase (Chen et al., 2000; Yan et al., 2005), whose subcellular localization remains to be determined.

Although the evidence suggesting mitochondria as a docking site for key cell death proteins in *C. elegans* is strong, a direct role of mitochondria in cell death activation in *C. elegans* is lacking. Considering the pivotal position of mitochondria in mammalian apoptosis, the potential contribution of mitochondria to apoptosis in other organisms has been an active area of investigations in the past decade. In this review, we discuss the recent insights into the roles of mitochondria in *C. elegans* apoptosis. We also compare the involvements of mitochondria in apoptosis in mammals and *C. elegans* and related genes in this process.

2. MITOCHONDRIA: A CENTRAL PLAYER IN MAMMALIAN APOPTOSIS

It has been well established that mitochondria play a central role in mammalian apoptosis driven by diverse stress conditions and apoptotic stimuli (Green & Reed, 1998; Wang, 2001). The prominent role of mitochondria in mammalian apoptosis was revealed by the seminal discoveries that cytochrome *c* is released from mitochondria during apoptosis to activate caspases (Kim et al., 2005; Li et al., 1997; Liu, Kim, Yang, Jemmerson, & Wang, 1996; Zou et al., 1997). Soon after this discovery, a string of studies confirm that besides cytochrome *c* the mitochondrial release of several key proapoptotic proteins such as the serine protease HtrA2/Omi (Hegde et al., 2002; Suzuki et al., 2001), apoptosis-inducing factor (AIF) (Susin et al., 1999), Smac/Diablo (Du, Fang, Li, Li, & Wang, 2000; Verhagen et al., 2000), and Endonuclease G (EndoG) (Li, Luo, & Wang, 2001) is also important for apoptosis.

The mitochondrion with its released proapoptotic proteins forms a "life versus death" decision-making point, at both the physiological and the pathological settings in mammalian cells. Mitochondrial outer membrane permeabilization (MOMP) is considered a key event that leads to the release of cytochrome *c* and several other mitochondrial apoptogenic factors, such as Smac/Diablo, AIF, and EndoG (Breckenridge & Xue, 2004; Scorrano & Korsmeyer, 2003).

2.1 Mitochondrial Membrane Permeabilization

Cellular stress conditions such as DNA damages and oxidative stress induce activation of proapoptotic proteins in the cytosol, leading to mitochondrial membrane permeabilization (Bender & Martinou, 2013). The exact mechanism of MOMP is not completely clear. The process is initiated by the opening of a nonspecific large-conductance mitochondrial permeability transition pore (MPTP). MPTP is hypothesized to be a multiprotein channel, which forms at the contact site of the inner and outer mitochondrial membranes (Zamzami & Kroemer, 2001). It was originally proposed to comprise of the voltage-dependent anion channel (VDAC) in the outer membrane and the adenine nucleotide translocator (ANT) in the inner membrane of the mitochondrion along with other proteins from the matrix and the intermembrane space (Zamzami & Kroemer, 2001). However, several latest studies suggest that the c subunit of the mitochondrial F1FO ATP synthase instead constitutes the pore-forming unit of MPTP in the mitochondrial inner membrane (Alavian et al., 2014; Bonora et al., 2013; Giorgio et al., 2013). Proapoptotic proteins, Bcl-2-associated X protein (Bax) and Bcl-2 antagonist/killer-1 (Bak), form a permeability pore in the mitochondrial outer membrane and interact with other MPTP components during permeabilization, inducing the release of apoptogenic factors (Alavian et al., 2014; Karch & Molkentin, 2014). Binding of antiapoptotic members of the Bcl-2 family to MPTP components can prevent the opening of the MPTP and consequently inhibits apoptosis (Bender & Martinou, 2013; Zamzami & Kroemer, 2001).

In this context, we briefly describe the regulation of mitochondrial membrane permeability during cellular insults, directly by the Bcl-2 family of proteins and other regulatory loops involving caspases.

2.1.1 The Roles of Bcl-2 Family Proteins in MOMP and Release of Mitochondrial Factors

Although the most prominent member of this family, Bcl-2, is a robust antiapoptotic protein (Tsujimoto, Cossman, Jaffe, & Croce, 1985), this protein family comprises both antiapoptotic and proapoptotic members. It is a divergent group of proteins that play critical roles in the regulation of programmed cell death by an evolutionarily conserved mechanism found in species as remotely related as humans and nematodes (Adams & Cory, 2001). Most members of the Bcl-2 protein family possess a carboxyl-terminal transmembrane region and contain from one to four Bcl-2 homology (BH) domains, BH1 to BH4 (Adams & Cory, 2001). The antiapoptotic members, including Bcl-2, Bcl-xL, Bcl-w, and Mcl-1, have all four BH domains, while the BH4 domain is absent from the proapoptotic members, such as Bax and Bak.

Most Bcl-2 family proteins are found constitutively localized on the membranes of mitochondria and endoplasmic reticulum. They control MOMP and by this means regulate the release of apoptotic factors from the mitochondrial intermembrane space into the cytoplasm. Upon activation by apoptotic stimuli, including interaction with caspase-8-cleaved

BH3-only Bid (truncated Bid), Bax undergoes a conformational change permitting the uncoupling of its carboxyl-terminal anchor domain from the hydrophobic groove and insertion into the mitochondrial outer membrane (Bender & Martinou, 2013; Gross, Jockel, Wei, & Korsmeyer, 1998; Hsu, Wolter, & Youle, 1997). Bax and Bak oligomers participate in pore formation in the outer mitochondrial membrane, leading to the release of apoptogenic proteins from mitochondria into the cytosol. The exact molecular mechanisms by which the antiapoptotic members of the Bcl-2 family protect against cell death is not entirely clear; a plausible mechanism is dimerization with their proapoptotic counterparts (Adams & Cory, 2001).

Hence, Bcl-2 family proteins regulate the sequestration and release of cytochrome *c* and other mitochondrial apoptogenic factors from mitochondria. The ratio of the proapoptotic to the antiapoptotic Bcl-2 family proteins is crucial in the regulation of apoptosis via the intrinsic pathways.

2.1.2 The Roles of Caspases in Mitochondrial Membrane Permeabilization

Caspase-3, caspase-6, and caspase-7 are effector caspases that execute apoptosis. Activated caspase-3 and caspase-7 can cleave multiple structural and regulatory proteins, which are critical for cell survival and maintenance (Boyce et al., 2004; Crawford & Wells, 2011). Caspase-3 is the most important executioner caspase and is activated by both intrinsic and extrinsic pathways. Mitochondria have also been shown to act as amplifiers of caspase activities. It is already known that upstream caspases, such as caspase-2 and caspase-8, affect mitochondrial events in both intrinsic and extrinsic pathways of apoptosis, either directly or indirectly through interaction with Bcl-2 family members. Moreover, caspase-3 and caspase-7 have been shown to mediate and amplify mitochondrial events, such as cytochrome *c* release, Bax translocation, and AIF release (Lakhani et al., 2006).

2.2 Caspase-Dependent Cell Death Signaling

The caspase-dependent cell death signaling initiates with the release of cytochrome *c* and Smac from the mitochondrial intermembrane space (Du et al., 2000; Liu et al., 1996; Verhagen et al., 2000). About 85% of cytochrome *c* are stored in cristae, partial partitions formed by infolding of the inner mitochondrial membrane, and the rest are located in the intermembrane space (Scorrano et al., 2002). Cytochrome *c* is released from mitochondria during apoptosis, which induces a conformational change of Apaf-1 in the presence of ATP/dATP (Kim et al., 2005; Liu et al., 1996). Apaf-1 recruits procaspase-9 via their respective caspase recruitment domains (CARDs) and oligomerizes into a heptameric scaffold (Li et al., 1997). This complex of cytochrome *c*, dATP, Apaf-1, and procaspase-9, known as the apoptosome, proteolytically activates procaspase-9. The activated caspase-9 stimulates proteolytic activation of procaspase-3, procaspase-7, and DNA fragmentation factor complex, DFF40/DFF45, which causes chromatin fragmentation (Jiang & Wang, 2004; Liu, Zou, Slaughter, & Wang, 1997). Release of two other apoptogenic factors, Smac/Diablo and HtrA2/Omi, from the mitochondrial intermembrane space to the cytosol antagonizes the activities of inhibitors of apoptosis proteins, which further potentiates the activation of caspase-3 and caspase-7.

2.3 The Roles of Mitochondria in Caspase-Independent Cell Death Pathways

EndoG and AIF released from mitochondria have been reported to induce cell death through caspase-independent pathways (Li et al., 2001; Susin et al., 1999). EndoG translocates to the nucleus and induces internucleosomal DNA fragmentation. AIF also translocates to the nucleus and causes condensation and fragmentation of chromosomal DNA. EndoG is a nuclear-encoded mitochondrial protein that belongs to the large family of DNA/RNA nonspecific ββα-Me-finger nucleases (Schafer et al., 2004). Once released from mitochondria, EndoG functions independently of caspases to induce nuclear DNA fragmentation in mammalian cells (Li et al., 2001). AIF, a 67 kDa flavin adenine dinucleotide-binding oxidoreductase, is the first mitochondrial protein reported to be involved in apoptosis through a caspase-independent mechanism (Susin et al., 1999). AIF is cleaved in the mitochondrial matrix by the mitochondrial peptidase and later processed in the intermembrane space of mitochondria, generating a soluble and proapoptotic protein (Otera, Ohsakaya, Nagaura, Ishihara, & Mihara, 2005). Cleaved AIF is released from mitochondria and translocates to the nucleus to promote fragmentation of chromosomal DNA during apoptosis. Both caspase-dependent and -independent pathways may synergize to activate caspases and chromosome fragmentation.

3. PROGRAMMED CELL DEATH IN *C. ELEGANS*

Genetic studies in *C. elegans* in the last 30 years have led to the identification of many genes that function in different aspects of programmed cell death. These genes define a genetic pathway of programmed cell death that is conserved between nematodes and mammals (Horvitz, 2003). Such an extensive and systematic investigation of apoptosis is possible mainly due to the powerful genetic tools available in this model organism.

In *C. elegans*, apoptosis occurs during both embryonic and post-embryonic development in somatic cells and in adult hermaphrodite germ line. Because *C. elegans* is transparent, cell divisions and cell deaths can be observed and tracked *in vivo* using high-magnification Nomarski optics. The cell lineage of *C. elegans* is invariant from animal to animal and allows for analysis of programmed cell death in a single-cell resolution (Sulston & Horvitz, 1977). During the development of hermaphrodites, a total of 1090 somatic cells are generated and 131 of them undergo programmed cell death. Under the Nomarski optics, the dying cells display a decreased refractivity in cytoplasm and nucleus and assume a raised and flattened button-like appearance, termed "cell corpses." As in mammals, these dying cells exhibit distinct morphological and biochemical features such as cell shrinkage, PS externalization, and chromatin fragmentation, which are followed by phagocytosis of cell corpses by neighboring cells (Conradt & Xue, 2005).

Programmed cell death in *C. elegans* can be categorized into three distinguishable phases: (1) the "specification phase" in which a cell is specified by the cell lineage to undergo cell death through regulatory genes such as *ces-1* and *ces-2*; (2) the "killing phase" in which the apoptotic program comprised of genes *ced-3*, *ced-4*, *ced-9*, and *egl-1* is activated in the cell programmed to die and culminates with the activation of the CED-3 caspase; and (3) the "execution phase" in which the activated CED-3 caspase proteolytically activates downstream substrates, which go on to execute different aspects of the cell-killing process, such as chromosome fragmentation (*dcr-1*), PS externalization that triggers clearance of apoptotic cells (*ced-8*), inactivation of the AKT survival signaling (*cnt*-1), and mitochondria elimination (*drp-1*) (Breckenridge et al., 2008; Chen, Mapes, Lee, Skeen-Gaar, & Xue, 2013; Conradt & Xue, 2005; Horvitz, 2003; Nakagawa, Shi, Kage-Nakadai, Mitani, & Xue, 2010; Nakagawa, Sullivan, & Xue, 2014).

The killing phase of programmed cell death is tightly regulated in *C. elegans* by a cascade of protein interactions between proapoptotic (EGL-1, CED-4, and CED-3) and antiapoptotic (CED-9) molecules, largely similar to those observed in mammals (Horvitz, 2003). CED-3 and CED-4 are the key players in all apoptotic deaths in *C. elegans* and loss-of-function (*lf*) mutations in either gene cause survival of almost all 131 cells in soma that normally undergo cell death. CED-3, a homolog of mammalian caspases, contains a CARD similar to those in mammalian caspases (Yuan, Shaham, Ledoux, Ellis, & Horvitz, 1993). Like other caspases, CED-3 is synthesized as a zymogen and is proteolytically cleaved to generate an active protease during apoptosis (Xue et al., 1996). CED-4 is a homolog of mammalian Apaf-1, which activates procaspase-9 by forming an apoptosome in mammalian cells (Zou et al., 1997). Resembling caspase-9 activation in mammalian cells, CED-3 is activated by CED-4 that oligomerizes to form a *C. elegans* apoptosome (Yan et al., 2005). Both CED-4 and Apaf-1 contain a CARD and a nucleotide-binding motif that are crucial for their functions.

As in mammals, both pro- and antiapoptotic members of the Bcl-2 family exist to regulate cell death activation in *C. elegans*. *ced-9* encodes a homolog of human Bcl-2 and Bcl-xL proteins and acts as a cell death inhibitor (Hengartner, Ellis, & Horvitz, 1992; Hengartner & Horvitz, 1994b). Importantly, the human Bcl-2 protein can partially substitute for the function of *ced-9* in *C. elegans*, indicating that they act through a similar mechanism to protect against cell death (Hengartner & Horvitz, 1994b; Vaux, Weissman, & Kim, 1992). Loss-of-function mutations in *ced-9* cause widespread apoptosis and embryonic lethality due to increased ectopic death of normal living cells (Hengartner et al., 1992). EGL-1, a homolog of the human BH3-only proapoptotic proteins such as Bim and Bid, interacts directly with CED-9 to antagonize its death protective function, similar to interactions of the BH3-only proteins with antiapoptotic Bcl-2 proteins (Conradt & Horvitz, 1998). In living cells, CED-4 associates with CED-9 and is tethered to the surface of mitochondria, which prevents CED-4 from activating CED-3 (Chen et al., 2000; Chinnaiyan et al., 1997; Spector et al., 1997; Wu, Wallen, & Nunez, 1997). In cells that are programmed to die, EGL-1 is transcriptionally activated and binds to CED-9, causing a conformational change in CED-9 that results in the dissociation of CED-4 from CED-9 (Conradt & Horvitz, 1998, 1999; del Peso, Gonzalez, & Nunez, 1998; Parrish et al., 2000; Yan et al., 2005). The released CED-4 then oligomerizes to form an apoptosome that recruits and proteolytically activates the CED-3 zymogen (Yan et al., 2005).

4. MITOCHONDRION IS AN IMPORTANT COMPONENT IN *C. ELEGANS* PROGRAMMED CELL DEATH

Mitochondria have been shown to play critical roles in both the activation and the execution of apoptosis in mammals. However, the role of mitochondria in the activation of *C. elegans* cell death remains unclear.

Unlike in mammalian cells, there is no evidence for the involvement of cytochrome *c* in *C. elegans* apoptosis. The protein interaction between EGL-1 and CED-9 that results in the dissociation of the CED-4 dimer from CED-9 and the formation of the CED-4 oligomers appears to be sufficient to activate the CED-3 zymogen *in vitro* without the requirement of additional factors (Yan et al., 2005). However, the mitochondrial membrane localization of CED-9 suggests an association of mitochondria to *C. elegans* cell death (Chen et al., 2000). Moreover, a recent study reports that CED-4 predominantly localizes to the perinuclear membrane in living cells and further accumulates on the perinuclear membrane of dying cells in a manner dependent on EGL-1 (Pourkarimi, Greiss, & Gartner, 2012). This new observation suggests that cell death activation in *C. elegans* may be more complicated than we anticipated and may involve additional factors, including potentially mitochondrial factors.

Interestingly, CED-9 has also been proposed to play a proapoptotic role, because the cell death defect in weak *ced-3* loss-of-function mutants is significantly enhanced by the loss of the *ced-9* function (Hengartner & Horvitz, 1994a). The proapoptotic activity of CED-9 may be explained by its serving as a chaperone/inhibitor in assisting the initial correct folding of CED-4 in the CED-4/CED-9 complex, which is critical for the subsequent formation of the CED-4 oligomer essential for CED-3 activation (Yan et al., 2005). Alternatively, CED-9 might regulate the release of mitochondrial factors. Further analysis of the acting mechanisms of CED-9 is critical in addressing this paradox.

4.1 Release of the Mitochondrial Apoptogenic Factors in *C. elegans*

Although there is no direct evidence that mitochondria play an important role in cell death activation in *C. elegans*, several studies suggest that mitochondrial factors are released from mitochondria and play important roles in cell death execution in *C. elegans* (Parrish et al., 2001; Wang et al., 2007; Wang, Yang, Chai, Shi, & Xue, 2002). WAH-1, a *C. elegans* homolog of human AIF, is released from mitochondria during apoptosis to promote chromosome fragmentation (Wang et al., 2002). WAH-1 acts in the same pathway as CPS-6, another mitochondrial protein and a homolog of human EndoG, to promote apoptotic DNA degradation. WAH-1 directly interacts with CPS-6 to enhance the endonuclease activity of CPS-6. Although CPS-6 normally localizes to mitochondria, the mature form of CPS-6 without the mitochondrial targeting sequence is found in nuclei (Parrish et al.,

2001). How WAH-1 and CPS-6 are released from mitochondria during *C. elegans* apoptosis is not understood. The release of WAH-1 from mitochondria during apoptosis is partially dependent on the CED-3 caspase (Wang et al., 2002). It is also not clear if MOMP or mitochondrial permeability transition (MPT) is involved in releasing CPS-6 and WAH-1. In mammalian cells, adenine nucleotide translocators (ANTs) have been implicated in MPTP formation (Zamzami & Kroemer, 2001). Interestingly, WAN-1, one of the four ANTs in *C. elegans*, has been shown to have a proapoptotic activity. Coimmunoprecipitation assays indicate that WAN-1 interacts with both CED-9 and CED-4 (Shen et al., 2009). It is not known if WAN-1 affects release of WAH-1 and CPS-6 from mitochondria during apoptosis.

4.2 Mitochondrial Proapoptotic Proteins in *C. elegans*

In addition to CED-9, which localizes to mitochondria and is a crucial antiapoptotic regulator, several other mitochondrial proteins play important proapoptotic roles in *C. elegans*.

4.2.1 CPS-6

CPS-6 (CED-3 protease suppressors) was identified through a CED-3 protease suppressor screen and is the first *C. elegans* mitochondrial protein with a specific role in programmed cell death (Parrish et al., 2001). It exhibits remarkable sequence similarity to human EndoG. A CPS-6::GFP fusion, which can rescue the cell death defects of the *cps-6* loss-of-function mutant, specifically colocalizes with MitoTracker Red, a mitochondrion-specific dye, confirming that it is a mitochondrial protein. In mammalian cells, EndoG is synthesized as a propeptide, which is imported into mitochondria through its amino-terminal mitochondrial targeting sequence (Ruiz-Carrillo & Renaud, 1987). During apoptosis, EndoG is released from mitochondria and translocates to the nucleus to promote chromosome fragmentation (Li et al., 2001). Since expression of CPS-6 lacking its mitochondrial targeting sequence in *C. elegans* results in CPS-6 localization in the nucleus (Parrish et al., 2001), this result suggests that mature CPS-6 is also released from mitochondria and translocates to nuclei during apoptosis like human EndoG. How this translocation of CPS-6 from mitochondria to the nucleus occurs during apoptosis is not known.

Once translocated to the nucleus, CPS-6 promotes apoptotic chromosome fragmentation. Loss-of-function mutations in *cps-6*, as well as RNA interference (RNAi) knockdown of *cps-6*, prevent normal chromosome degradation during apoptosis and result in accumulation of chromosomal DNA breaks that can be detected by the TUNEL (TdT-mediated dUTP nick-end labeling) assays (Parrish et al., 2001). *cps-6* mutations and RNAi also cause delayed appearance of cell corpses during embryo development, indicating that CPS-6 promotes normal progression of cell death. Since inactivation of *cps-6* can partially suppress cell death induced by activated CED-3, CPS-6 likely functions downstream of or in parallel to the CED-3 caspase. The proapoptotic activity of CPS-6 is dependent on another *C. elegans* mitochondrial protein WAH-1, as loss of *wah-1* causes identical cell death defects to those of the *cps-6* mutants but cannot enhance the defects of the *cps-6* mutants. WAH-1 directly interacts with CPS-6 to potentiate its endonuclease activity (Wang et al., 2002). CPS-6 also interacts with a nuclear protein CRN-1, a homolog of the human flap endonuclease 1 (FEN-1), and several other cell death-related nucleases (CRN nuclease), to form a multinuclease complex, named degradeosome, which is the workforce in promoting apoptotic DNA degradation (Parrish & Xue, 2003). CRN-1 possesses a 5′–3′ exonuclease activity and a novel gap-dependent endonuclease activity. In normal cells, CRN-1 assists in nuclear DNA replication and repair and is regarded as a genome stabilizer. In apoptotic cells, CRN-1 and other CRN nucleases act as cofactors of CPS-6 to promote stepwise chromosomal DNA degradation (Parrish, Yang, Shen, & Xue, 2003).

4.2.2 WAH-1

AIF was first identified as an apoptogenic factor that is released from mitochondria to mediate caspase-independent apoptosis in mammals (Susin et al., 1999). The release of AIF is stimulated by the BH-3-only proteins such as Bim and Bid. *C. elegans* WAH-1 is a homolog of human AIF and plays a similar role in programmed cell death. WAH-1 localizes to mitochondria in living cells and is released from mitochondria into the cytosol and the nucleus during apoptosis by the BH3-only protein EGL-1 (Wang et al., 2002). The release of WAH-1 during *C. elegans* apoptosis is partially dependent on the CED-3 caspase activity. This observation is in agreement with studies in mammalian cells where activated caspase-3 triggers a feedback regulation on mitochondria that promotes the release of apoptotic factors, such as cytochrome *c* and AIF (Kirsch et al., 1999; Lakhani et al., 2006). Therefore, although AIF can be released through a caspase-independent process in mammals, caspases may still play a role in releasing a significant portion of AIF. It is possible that in *C. elegans* the initial limited release of WAH-1 may be CED-3 independent. But after the activation of the CED-3 caspase, CED-3 could amplify the mitochondrial damage by cleaving downstream CED-3 targets and promote further release of WAH-1.

Once released from mitochondria, WAH-1 acts to promote two important hallmark events of apoptosis: chromosome fragmentation and PS externalization (Wang et al., 2007, 2002). RNAi knockdown of *wah-1* causes delayed appearance of cell corpses during embryo development and accumulation of TUNEL-positive nuclei, both of which are observed in the *cps-6* mutants. Importantly, *wah-1* RNAi does not enhance these two defects in the *cps-6* mutants, indicating that WAH-1 and CPS-6 act in the same pathway to promote apoptotic chromosome fragmentation (Wang et al., 2002). Indeed, WAH-1 physically interacts with CPS-6 *in vitro* and can significantly enhance the endonuclease activity of CPS-6. WAH-1 and CPS-6 hence drive a single conserved apoptotic DNA degradation pathway initiated from the mitochondria (Fig. 1).

Surprisingly, *wah-1* RNAi also affects clearance of apoptotic cells in sensitized genetic backgrounds, such as animals deficient in the engulfment genes *ced-1* and *ced-7* (Wang et al., 2007). Further analysis indicates that *wah-1* RNAi compromises PS externalization on the surface of apoptotic cells, which is a widespread phenomenon of apoptosis in diverse organisms. Externalized PS serves as an "eat me" signal to trigger engulfment of apoptotic cells by phagocytes (Fadok et al., 2001). Through a candidate approach, WAH-1 was found to promote plasma membrane PS externalization in apoptotic germ cells through a downstream factor, phospholipid scramblase 1 (SCRM-1). SCRM-1 is a bidirectional lipid-translocating enzyme that normally is dormant, but upon binding to the released WAH-1 from mitochondria, is activated to catalyze PS externalization (Wang et al., 2007). WAH-1 can activate the lipid-scrambling activity of

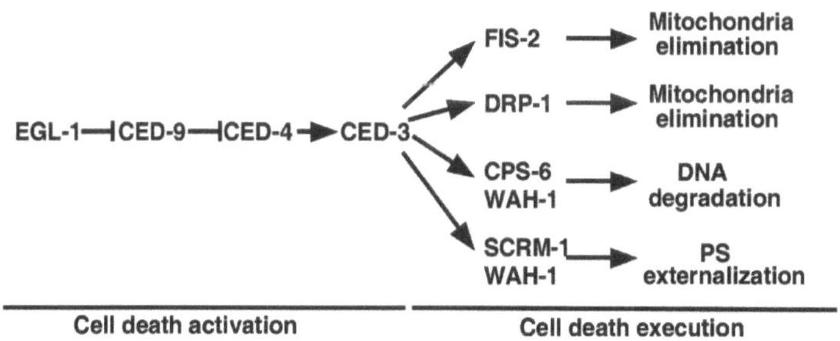

Figure 1 The central cell-killing pathway in *C. elegans* and the mitochondrial proteins involved in multiple different pathways of cell death execution.

SCRM-1 by 10-folds *in vitro* and thus is a strong SCRM-1 activator. Inactivation of the *scrm-1* gene reduces PS externalization on the surface of apoptotic germ cells and causes a defect in the clearance of apoptotic cells. A point mutation on the surface of WAH-1 (K446E) that blocks binding of WAH-1 to SCRM-1 abolishes the activity of WAH-1 to promote PS externalization in apoptotic germ cells and the activity of WAH-1 to simulate the lipid-translocating activity of SCRM-1 *in vitro* (Wang et al., 2007). Hence, WAH-1 and SCRM-1 define a new mitochondrion to cell-surface-signaling pathway important for apoptosis and cell corpse clearance (Fig. 1).

4.2.3 DRP-1 and FIS-2

The double-membrane mitochondria are highly dynamic organelles that constantly undergo morphological changes such as fusion and fission. Numerous stimuli are known to induce mitochondrial morphological changes, which in turn alter the functions of mitochondria (Friedman & Nunnari, 2014; Okamoto & Shaw, 2005). The mitochondrial fusion and fission processes are regulated by the dynamin family of GTPases. The dynamin-related protein 1 (DRP1/DLP1) is required for mitochondrial fission, while other GTPases, the mitofusins (Mfn1 and Mfn2) and optic atrophy 1 (OPA1), are needed for fusion of the outer and inner mitochondrial membrane, respectively (Friedman & Nunnari, 2014; Okamoto & Shaw, 2005). Mitochondrial fragmentation has been shown to increase significantly during cell death in yeast, *C. elegans*, *Drosophila*, and mammals and thus has been proposed to play a role in cell death activation (Fannjiang et al., 2004; Frank et al., 2001; Goyal, Fell, Sarin, Youle, & Sriram, 2007; Jagasia, Grote, Westermann, & Conradt, 2005). In mammals, mitochondrial fragmentation has also been implicated in mitochondrial membrane permeabilization and release of proapoptotic factors (Bender & Martinou, 2013). Whether mitochondrial fragmentation is a cause or a consequence of apoptosis is still under debate (James & Martinou, 2008).

In *C. elegans*, the mitochondrial fusion and fission processes are tightly regulated by similar dynamin-related GTPases. Three *C. elegans* genes, *fzo-1*, *eat-3*, and *drp-1*, encode orthologs of MFN1/FZO1, OPA1, and DRP1, respectively. Two homologs of human Fis1, which may recruit DRP1 to the outer mitochondrial membrane (Okamoto & Shaw, 2005), exist in *C. elegans* and are named FIS-1 and FIS-2. Neither *fis-1* or *fis-2* alone nor these two genes in combination affect mitochondrial dynamics (Breckenridge et al., 2008). The mitochondria undergo excessive fragmentation in *C. elegans* animals deficient in the profusion genes *fzo-1* and *eat-3*. On the other hand, in the *drp-1* loss-of-function mutants, mitochondria fuse to form a long and tubular network (Breckenridge et al., 2008; Jagasia et al., 2005; Kanazawa et al., 2008; Labrousse, Zappaterra, Rube, & van der Bliek, 1999). Despite the dramatic changes in the morphology of mitochondria in animals deficient in *drp-1*, *fzo-1*, or *eat-3*, programmed cell death appears to proceed normally in these mutants (Breckenridge et al., 2008). For example, inactivation of *fzo-1* or *eat-3* does not cause increased or decreased apoptosis either on its own or in sensitized genetic backgrounds, such as mutants defective in the clearance of apoptotic cells or animals partially deficient in cell death activation. Although inactivation of *drp-1* alone does not reduce or delay cell death, it does affect apoptosis in sensitized genetic backgrounds. For instance, inactivation of both *drp-1* and *fis-2* results in reduced embryonic cell death, whereas loss of either *drp-1* or *fis-2* alone has no effect on cell death. Moreover, inactivation of *drp-1* or *fis-2* in animals partially defective in cell death, such as the *ced-3* partial loss-of-function mutants, results in more severe cell death defects, indicating that unlike *fzo-1* or *eat-3*, *drp-1* and *fis-2* have a proapoptotic role in *C. elegans* (Breckenridge et al., 2008).

Genetic epistasis analysis of *drp-1* and *fis-2* reveals that both *drp-1* and *fis-2* act downstream of the activated CED-3 caspase and in distinct cell death pathways to promote apoptosis, as loss of both *drp-1* and *fis-2* inhibits EGL-1 and activated CED-3-induced cell death (Breckenridge et al., 2008). Consistent with the genetic observations, DRP-1 is cleaved by CED-3 *in vitro*. Cleavage of DRP-1 by CED-3 is required for its proapoptotic function *in vivo*, but is dispensable for its mitochondrial fission function, suggesting that the proapoptotic function of DRP-1 can be separated from its mitochondrial fission function. Moreover, expression of various combinations of DRP-1 cleavage products by CED-3 *in vivo* indicates that the carboxyl-terminal cleavage product of DRP-1 needs to work with full-length DRP-1 to promote apoptosis (Breckenridge et al., 2008). FIS-2 is also found to localize to mitochondria, but it does not appear to be a substrate of CED-3 *in vitro*. Electron microscopy analysis of mitochondria in apoptotic cells reveals that mitochondria are reduced or eliminated during apoptosis, which is blocked by loss of either *drp-1* or *fis-2*, indicating that *drp-1* and *fis-2* play a role in the elimination of mitochondria, the major energy source for cell survival and functions, to facilitate execution of apoptosis (Breckenridge et al., 2008). Furthermore, genetic analysis indicates that *drp-1* and *fis-2* act in parallel to the DNA degradation pathway mediated by *wah-1* and *cps-6* to promote apoptosis (Fig. 1). It will be interesting to find out how *drp-1* and *fis-2* promote elimination of mitochondria during apoptosis and if these three mitochondrial pathways coordinate to promote cell death execution.

4.2.4 CED-9 and WAN-1

The aforementioned mitochondrion-related cell death pathways all act downstream of CED-3 to promote efficient cell death execution. There is so far no strong evidence for the release of an apoptogenic factor from mitochondria that is critical for cell death activation in *C. elegans*. However, several mitochondrion-associated proteins are important for cell death activation. The Bcl-2 homolog CED-9 localizes to the outer membrane of mitochondria and has been shown by numerous studies to bind and tether CED-4 to the surface of mitochondria, leading to inhibition of cell death (Chen et al., 2000; Chinnaiyan et al., 1997; Spector et al., 1997; Wu, Wallen, & Nunez, 1997). Transcriptional activation of EGL-1 in cells that are programmed to die results in binding of EGL-1 to CED-9 and a subsequent conformational change of CED-9 that releases the CED-4 dimer from the CED-4/CED-9 complex tethered to the mitochondria (Chen et al., 2000; Conradt & Horvitz, 1998; Parrish et al., 2000; Yan et al., 2005). CED-4 dimers then oligomerize and translocate to the perinuclear region to activate the CED-3 caspase zymogen. Although this proposed CED-3 activation process can be recapitulated *in vitro* using only recombinant EGL-1, CED-9, and CED-4 (Yan et al., 2005), the involvement of additional factors in cell death activation *in vivo* cannot be ruled out. Indeed, two incomplete caspase homologs in *C. elegans*, CSP-2 and CPS-3, have been shown to inhibit the autoactivation of the CED-3 zymogen and thus protect cells from apoptosis in *C. elegans* germ line and soma, respectively (Geng et al., 2008, 2009; Shaham, 1998). Moreover, a recent study reports a predominantly perinuclear localization pattern of CED-4, which is inconsistent with this popular CED-3 activation model, and suggests that alternative mechanisms could be involved in cell death activation in *C. elegans* (Pourkarimi et al., 2012).

One of the potential factors is WAN-1, a *C. elegans* homolog of mammalian ANTs, which have been suggested to be a component of the permeability transition pore complex (Zamzami & Kroemer, 2001). WAN-1 was identified as a protein that associates with CED-4 in coimmunoprecipitation assays in *C. elegans*. It localizes to mitochondria and can associate with both CED-4 and CED-9 in cultured cells. RNAi knockdown of *wan-1* causes reduced cell death in both soma and germ line, supporting a role in apoptosis (Shen et al., 2009). It is not clear how WAN-1 regulates cell death through interacting with CED-4 and CED-9.

It is interesting that the mechanism by which CED-9 inhibits developmental cell death in *C. elegans* appears to differ from those of the antiapoptotic proteins of the Bcl-2 family in mammals, which do not bind Apaf-1 and inhibit apoptosis mainly through regulating the release of mitochondrial apoptogenic factors such as cytochrome *c* and Smac/Diablo (Danial & Korsmeyer, 2004; Wang, 2001). In addition, there is no proapoptotic, multi-BH domain Bcl-2 family protein, like Bax or Bak, in *C. elegans*. A recent study shows that a hepatitis B virus (HBV) protein, HBx, harbors a BH3-like motif and can induce both necrosis and apoptosis in *C. elegans*. HBx interacts with CED-9 through its BH3-like motif and this interaction is critical for its cell-killing activity in *C. elegans* (Geng, Harry, et al., 2012). Unexpectedly, HBx-induced cell death is strictly dependent on the presence of the CED-9 protein, as loss of *ced-9* completely blocks HBx-induced cell death, whereas loss of either *ced-3* or *ced-4* only partially inhibits HBx-induced cell death. This finding suggests that CED-9 acts as a proapoptotic protein to mediate HBx-induced cell killing through a mechanism different from the one used in developmental cell death (Geng, Harry, et al., 2012). Further genetic and cell biological analysis suggests that HBx interacts with CED-9 to trigger elevation of cytosolic calcium, which in turn induces both necrosis and apoptosis (Syntichaki, Xu, Driscoll, & Tavernarakis, 2002; Xu, Tavernarakis, & Driscoll, 2001). Interestingly, cyclosporin A, a peptide that inhibits MPT and cell death in mammalian cells (Zamzami & Kroemer, 2001), significantly reduces HBx-induced cell death and blocks HBx-induced elevation of cytosolic calcium in *C. elegans*, indicating that the MPT also plays a role in cell death in *C. elegans*. This HBx-induced cell-killing mechanism is conserved between *C. elegans* and humans, as HBx similarly interacts with human Bcl-2 and Bcl-xL to trigger increase of cytosolic calcium, which is required for HBV viral replication and induces both apoptosis and necrosis (Geng, Huang, et al., 2012). Therefore, CED-9 is capable of promoting cell death through a CED-3- and CED-4-independent mechanism and a mechanism dependent on the MPT.

4.2.5 ICD-1

In mammals, mitochondria play an essential role in promoting both caspase-dependent and caspase-independent cell death. The classical cell death pathway in *C. elegans* is caspase dependent and requires the CED-3 caspase activity (Horvitz, 2003). Interestingly, another mitochondrial cell death inhibitor, inhibitor of cell death-1 (ICD-1), was found to suppress a noncanonical pathway of apoptosis in *C. elegans* (Bloss, Witze, & Rothman, 2003). The *icd-1* gene encodes a *C. elegans* homologue of the β subunit of the nascent-polypeptide associated complex (NAC), which is proposed to mediate targeting and transport of proteins to mitochondria (Rospert, Dubaquie, & Gautschi, 2002). ICD-1 overexpression inhibits apoptosis in cells that are programmed to die, whereas reduction of the ICD-1 expression by RNAi leads to ectopic apoptosis during development and in differentiated cells of various tissue types, some of which normally do not undergo apoptosis (Bloss et al., 2003). Surprisingly, *icd-1* RNAi-induced apoptosis occurs independent of the CED-3 caspase, but requires CED-4. This finding indicates that the two core proapoptotic proteins, CED-3 and CED-4, could function through discrete apoptotic pathways. It is possible that ICD-1 inhibits a caspase-independent pathway. Alternatively, ICD-1 may inhibit a cell death pathway mediated by another caspase, such as CSP-1, which has been shown to possess a caspase activity (Shaham, 1998). Indeed, ICD-1 contains a putative CARD domain and a potential CED-3 caspase cleavage site that may lead to disruption of the CARD domain upon caspase cleavage, which would inactivate ICD-1 to allow apoptosis to proceed.

5. CONCLUDING REMARKS

Recent studies provide strong evidence regarding the involvement of mitochondria in the execution of programmed cell death downstream of CED-3 in *C. elegans*. There are indications that additional mitochondrial factors or mitochondria-derived mechanisms might be involved in activation and/or execution of cell death in *C. elegans*. Identification of novel mitochondrial cell death components in *C. elegans* using traditional genetic strategies, candidate approaches, and modern cutting-edge techniques such as the CRISPR/Cas9 gene editing method, not only will help resolve several conundrums in *C. elegans* cell death but also may provide significant insights into critical, unaddressed questions in apoptosis, such as the fundamental question of how mitochondrial apoptogenic factors are released from mitochondria during apoptosis.

ACKNOWLEDGMENTS

Due to space limit, we apologize for not being able to cite all the references in this review. Research in D.X.'s laboratory is supported by US National Institutes of Health (grants R01 GM59083, R01 GM79097, and R01 GM088241).

REFERENCES

Adams, J. M. (2003). Ways of dying: Multiple pathways to apoptosis. *Genes & Development, 17*, 2481–2495.

Adams, J. M., & Cory, S. (2001). Life-or-death decisions by the Bcl-2 protein family. *Trends in Biochemical Sciences, 26*, 61–66.

Alavian, K. N., Beutner, G., Lazrove, E., Sacchetti, S., Park, H. A., Licznerski, P., et al. (2014). An uncoupling channel within the c-subunit ring of the F1FO ATP synthase is the mitochondrial permeability transition pore. *Proceedings of the National Academy of Sciences of the United States of America, 111*, 10580–10585.

Bender, T., & Martinou, J. C. (2013). Where killers meet—Permeabilization of the outer mitochondrial membrane during apoptosis. *Cold Spring Harbor Perspectives in Biology, 5*, a011106.

Bloss, T. A., Witze, E. S., & Rothman, J. H. (2003). Suppression of CED-3-independent apoptosis by mitochondrial betaNAC in *Caenorhabditis elegans*. *Nature, 424*, 1066–1071.

Bonora, M., Bononi, A., De Marchi, E., Giorgi, C., Lebiedzinska, M., Marchi, S., et al. (2013). Role of the c subunit of the FO ATP synthase in mitochondrial permeability transition. *Cell Cycle, 12*, 674–683.

Boyce, M., Degterev, A., & Yuan, J. (2004). Caspases: An ancient cellular sword of Damocles. *Cell Death and Differentiation, 11*, 29–37.

Breckenridge, D. G., Kang, B. H., Kokel, D., Mitani, S., Staehelin, L. A., & Xue, D. (2008). Caenorhabditis elegans drp-1 and fis-2 regulate distinct cell-death execution pathways downstream of ced-3 and independent of ced-9. *Molecular Cell, 31*, 586–597.

Breckenridge, D. G., & Xue, D. (2004). Regulation of mitochondrial membrane permeabilization by BCL-2 family proteins and caspases. *Current Opinion in Cell Biology, 16*, 647–652.

Chen, F., Hersh, B. M., Conradt, B., Zhou, Z., Riemer, D., Gruenbaum, Y., et al. (2000). Translocation of C. elegans CED-4 to nuclear membranes during programmed cell death. *Science, 287*, 1485–1489.

Chen, Y. Z., Mapes, J., Lee, E. S., Skeen-Gaar, R. R., & Xue, D. (2013). Caspase-mediated activation of Caenorhabditis elegans CED-8 promotes apoptosis and phosphatidylserine externalization. *Nature Communications, 4*, 2726.

Chinnaiyan, A. M., O'Rourke, K., Lane, B. R., & Dixit, V. M. (1997). Interaction of CED-4 with CED-3 and CED-9: A molecular framework for cell death. *Science, 275*, 1122–1126.

Conradt, B., & Horvitz, H. R. (1998). The C. elegans protein EGL-1 is required for programmed cell death and interacts with the Bcl-2-like protein CED-9. *Cell, 93*, 519–529.

Conradt, B., & Horvitz, H. R. (1999). The TRA-1A sex determination protein of C. elegans regulates sexually dimorphic cell deaths by repressing the egl-1 cell death activator gene. *Cell, 98*, 317–327.

Conradt, B., & Xue, D. (2005). Programmed cell death. *WormBook (Ed.),* The *C. elegans* Research Community, WormBook. http://dx.doi.org/10.1895/wormbook.1.32.1, http://www.wormbook.org.

Crawford, E. D., & Wells, J. A. (2011). Caspase substrates and cellular remodeling. *Annual Review of Biochemistry, 80*, 1055–1087.

Danial, N. N., & Korsmeyer, S. J. (2004). Cell death: Critical control points. *Cell, 116*, 205–219.

del Peso, L., Gonzalez, V. M., & Nunez, G. (1998). Caenorhabditis elegans EGL-1 disrupts the interaction of CED-9 with CED-4 and promotes CED-3 activation. *The Journal of Biological Chemistry, 273*, 33495–33500.

Du, C., Fang, M., Li, Y., Li, L., & Wang, X. (2000). Smac, a mitochondrial protein that promotes cytochrome c-dependent caspase activation by eliminating IAP inhibition. *Cell, 102*, 33–42.

Fadok, V. A., Xue, D., & Henson, P. (2001). If phosphatidylserine is the death knell, a new phosphatidylserine-specific receptor is the bellringer. *Cell Death and Differentiation, 8*, 582–587.

Fannjiang, Y., Cheng, W. C., Lee, S. J., Qi, B., Pevsner, J., McCaffery, J. M., et al. (2004). Mitochondrial fission proteins regulate programmed cell death in yeast. *Genes & Development, 18*, 2785–2797.

Frank, S., Gaume, B., Bergmann-Leitner, E. S., Leitner, W. W., Robert, E. G., Catez, F., et al. (2001). The role of dynamin-related protein 1, a mediator of mitochondrial fission, in apoptosis. *Developmental Cell, 1*, 515–525.

Friedman, J. R., & Nunnari, J. (2014). Mitochondrial form and function. *Nature, 505*, 335–343.

Geng, X., Harry, B. L., Zhou, Q., Skeen-Gaar, R. R., Ge, X., Lee, E. S., et al. (2012). Hepatitis B virus X protein targets the Bcl-2 protein CED-9 to induce intracellular Ca2+ increase and cell death in Caenorhabditis elegans. *Proceedings of the National Academy of Sciences of the United States of America, 109*, 18465–18470.

Geng, X., Huang, C., Qin, Y., McCombs, J. E., Yuan, Q., Harry, B. L., et al. (2012). Hepatitis B virus X protein targets Bcl-2 proteins to increase intracellular calcium, required for virus replication and cell death induction. *Proceedings of the National Academy of Sciences of the United States of America, 109*, 18471–18476.

Geng, X., Shi, Y., Nakagawa, A., Yoshina, S., Mitani, S., & Xue, D. (2008). Inhibition of CED-3 zymogen activation and apoptosis in Caenorhabditis elegans by caspase homolog CSP-3. *Nature Structural & Molecular Biology, 15*, 1094–1101.

Geng, X., Zhou, Q. H., Kage-Nakadai, E., Shi, Y., Yan, N., Mitani, S., et al. (2009). Caenorhabditis elegans caspase homolog CSP-2 inhibits CED-3 autoactivation and apoptosis in germ cells. *Cell Death and Differentiation, 16*, 1385–1394.

Giorgio, V., von Stockum, S., Antoniel, M., Fabbro, A., Fogolari, F., Forte, M., et al. (2013). Dimers of mitochondrial ATP synthase form the permeability transition pore. *Proceedings of the National Academy of Sciences of the United States of America, 110*, 5887–5892.

Goyal, G., Fell, B., Sarin, A., Youle, R. J., & Sriram, V. (2007). Role of mitochondrial remodeling in programmed cell death in Drosophila melanogaster. *Developmental Cell, 12*, 807–816.

Green, D., & Kroemer, G. (1998). The central executioners of apoptosis: Caspases or mitochondria? *Trends in Cell Biology, 8*, 267–271.

Green, D. R., & Reed, J. C. (1998). Mitochondria and apoptosis. *Science, 281*, 1309–1312.

Gross, A., Jockel, J., Wei, M. C., & Korsmeyer, S. J. (1998). Enforced dimerization of BAX results in its translocation, mitochondrial dysfunction and apoptosis. *The EMBO Journal, 17*, 3878–3885.

Hegde, R., Srinivasula, S. M., Zhang, Z., Wassell, R., Mukattash, R., Cilenti, L., et al. (2002). Identification of Omi/HtrA2 as a mitochondrial apoptotic serine protease that disrupts inhibitor of apoptosis protein-caspase interaction. *The Journal of Biological Chemistry, 277*, 432–438.

Hengartner, M. O., Ellis, R. E., & Horvitz, H. R. (1992). Caenorhabditis elegans gene ced-9 protects cells from programmed cell death. *Nature, 356*, 494–499.

Hengartner, M. O., & Horvitz, H. R. (1994a). Activation of C. elegans cell death protein CED-9 by an amino-acid substitution in a domain conserved in Bcl-2. *Nature, 369*, 318–320.

Hengartner, M. O., & Horvitz, H. R. (1994b). C. elegans cell survival gene ced-9 encodes a functional homolog of the mammalian proto-oncogene bcl-2. *Cell, 76*, 665–676.

Horvitz, H. R. (2003). Worms, life, and death (Nobel lecture). *Chembiochem, 4*, 697–711.

Hsu, Y. T., Wolter, K. G., & Youle, R. J. (1997). Cytosol-to-membrane redistribution of Bax and Bcl-X(L) during apoptosis. *Proceedings of the National Academy of Sciences of the United States of America, 94*, 3668–3672.
Jagasia, R., Grote, P., Westermann, B., & Conradt, B. (2005). DRP-1-mediated mitochondrial fragmentation during EGL-1-induced cell death in C. elegans. *Nature, 433*, 754–760.
James, D. I., & Martinou, J. C. (2008). Mitochondrial dynamics and apoptosis: A painful separation. *Developmental Cell, 15*, 341–343.
Jiang, X., & Wang, X. (2004). Cytochrome C-mediated apoptosis. *Annual Review of Biochemistry, 73*, 87–106.
Kanazawa, T., Zappaterra, M. D., Hasegawa, A., Wright, A. P., Newman-Smith, E. D., Buttle, K. F., et al. (2008). The C. elegans Opa1 homologue EAT-3 is essential for resistance to free radicals. *PLoS Genetics, 4*, e1000022.
Karch, J., & Molkentin, J. D. (2014). Identifying the components of the elusive mitochondrial permeability transition pore. *Proceedings of the National Academy of Sciences of the United States of America, 111*, 10396–10397.
Kim, H. E., Du, F., Fang, M., & Wang, X. (2005). Formation of apoptosome is initiated by cytochrome c-induced dATP hydrolysis and subsequent nucleotide exchange on Apaf-1. *Proceedings of the National Academy of Sciences of the United States of America, 102*, 17545–17550.
Kirsch, D. G., Doseff, A., Chau, B. N., Lim, D. S., de Souza-Pinto, N. C., Hansford, R., et al. (1999). Caspase-3-dependent cleavage of Bcl-2 promotes release of cytochrome c. *The Journal of Biological Chemistry, 274*, 21155–21161.
Krammer, P. H. (2000). CD95's deadly mission in the immune system. *Nature, 407*, 789–795.
Labrousse, A. M., Zappaterra, M. D., Rube, D. A., & van der Bliek, A. M. (1999). C. elegans dynamin-related protein DRP-1 controls severing of the mitochondrial outer membrane. *Molecular Cell, 4*, 815–826.
Lakhani, S. A., Masud, A., Kuida, K., Porter, G. A., Jr., Booth, C. J., Mehal, W. Z., et al. (2006). Caspases 3 and 7: Key mediators of mitochondrial events of apoptosis. *Science, 311*, 847–851.
Li, L. Y., Luo, X., & Wang, X. (2001). Endonuclease G is an apoptotic DNase when released from mitochondria. *Nature, 412*, 95–99.
Li, P., Nijhawan, D., Budihardjo, I., Srinivasula, S. M., Ahmad, M., Alnemri, E. S., et al. (1997). Cytochrome c and dATP-dependent formation of Apaf-1/caspase-9 complex initiates an apoptotic protease cascade. *Cell, 91*, 479–489.
Liu, X., Kim, C. N., Yang, J., Jemmerson, R., & Wang, X. (1996). Induction of apoptotic program in cell-free extracts: Requirement for dATP and cytochrome c. *Cell, 86*, 147–157.
Liu, X., Zou, H., Slaughter, C., & Wang, X. (1997). DFF, a heterodimeric protein that functions downstream of caspase-3 to trigger DNA fragmentation during apoptosis. *Cell, 89*, 175–184.
Metzstein, M. M., Stanfield, G. M., & Horvitz, H. R. (1998). Genetics of programmed cell death in C. elegans: Past, present and future. *Trends in Genetics, 14*, 410–416.
Nakagawa, A., Shi, Y., Kage-Nakadai, E., Mitani, S., & Xue, D. (2010). Caspase-dependent conversion of Dicer ribonuclease into a death-promoting deoxyribonuclease. *Science, 328*, 327–334.
Nakagawa, A., Sullivan, K. D., & Xue, D. (2014). Caspase-activated phosphoinositide binding by CNT-1 promotes apoptosis by inhibiting the AKT pathway. *Nature Structural & Molecular Biology, 21*, 1082–1090.
Okamoto, K., & Shaw, J. M. (2005). Mitochondrial morphology and dynamics in yeast and multicellular eukaryotes. *Annual Review of Genetics, 39*, 503–536.
Otera, H., Ohsakaya, S., Nagaura, Z., Ishihara, N., & Mihara, K. (2005). Export of mitochondrial AIF in response to proapoptotic stimuli depends on processing at the intermembrane space. *The EMBO Journal, 24*, 1375–1386.

Parrish, J., Li, L., Klotz, K., Ledwich, D., Wang, X., & Xue, D. (2001). Mitochondrial endonuclease G is important for apoptosis in C. elegans. *Nature, 412*, 90–94.

Parrish, J., Metters, H., Chen, L., & Xue, D. (2000). Demonstration of the in vivo interaction of key cell death regulators by structure-based design of second-site suppressors. *Proceedings of the National Academy of Sciences of the United States of America, 97*, 11916–11921.

Parrish, J. Z., & Xue, D. (2003). Functional genomic analysis of apoptotic DNA degradation in C. elegans. *Molecular Cell, 11*, 987–996.

Parrish, J. Z., Yang, C., Shen, B., & Xue, D. (2003). CRN-1, a Caenorhabditis elegans FEN-1 homologue, cooperates with CPS-6/EndoG to promote apoptotic DNA degradation. *The EMBO Journal, 22*, 3451–3460.

Pennell, R. I., & Lamb, C. (1997). Programmed cell death in plants. *The Plant Cell, 9*, 1157–1168.

Pourkarimi, E., Greiss, S., & Gartner, A. (2012). Evidence that CED-9/Bcl2 and CED-4/Apaf-1 localization is not consistent with the current model for C. elegans apoptosis induction. *Cell Death and Differentiation, 19*, 406–415.

Rospert, S., Dubaquie, Y., & Gautschi, M. (2002). Nascent-polypeptide-associated complex. *Cellular and Molecular Life Sciences, 59*, 1632–1639.

Ruiz-Carrillo, A., & Renaud, J. (1987). Endonuclease G: A (dG)n X (dC)n-specific DNase from higher eukaryotes. *The EMBO Journal, 6*, 401–407.

Schafer, P., Scholz, S. R., Gimadutdinow, O., Cymerman, I. A., Bujnicki, J. M., Ruiz-Carrillo, A., et al. (2004). Structural and functional characterization of mitochondrial EndoG, a sugar non-specific nuclease which plays an important role during apoptosis. *Journal of Molecular Biology, 338*, 217–228.

Scorrano, L., Ashiya, M., Buttle, K., Weiler, S., Oakes, S. A., Mannella, C. A., et al. (2002). A distinct pathway remodels mitochondrial cristae and mobilizes cytochrome c during apoptosis. *Developmental Cell, 2*, 55–67.

Scorrano, L., & Korsmeyer, S. J. (2003). Mechanisms of cytochrome c release by proapoptotic BCL-2 family members. *Biochemical and Biophysical Research Communications, 304*, 437–444.

Shaham, S. (1998). Identification of multiple *Caenorhabditis elegans* caspases and their potential roles in proteolytic cascades. *Journal of Biological Chemistry, 273*, 35109–35117.

Shen, Q., Qin, F., Gao, Z., Cui, J., Xiao, H., Xu, Z., et al. (2009). Adenine nucleotide translocator cooperates with core cell death machinery to promote apoptosis in Caenorhabditis elegans. *Molecular and Cellular Biology, 29*, 3881–3893.

Spector, M. S., Desnoyers, S., Hoeppner, D. J., & Hengartner, M. O. (1997). Interaction between the C. elegans cell-death regulators CED-9 and CED-4. *Nature, 385*, 653–656.

Steller, H. (1995). Mechanisms and genes of cellular suicide. *Science, 267*, 1445–1449.

Strasser, A., O'Connor, L., & Dixit, V. M. (2000). Apoptosis signaling. *Annual Review of Biochemistry, 69*, 217–245.

Sulston, J. E., & Horvitz, H. R. (1977). Post-embryonic cell lineages of the nematode, Caenorhabditis elegans. *Developmental Biology, 56*, 110–156.

Susin, S. A., Lorenzo, H. K., Zamzami, N., Marzo, I., Snow, B. E., Brothers, G. M., et al. (1999). Molecular characterization of mitochondrial apoptosis-inducing factor. *Nature, 397*, 441–446.

Suzuki, Y., Imai, Y., Nakayama, H., Takahashi, K., Takio, K., & Takahashi, R. (2001). A serine protease, HtrA2, is released from the mitochondria and interacts with XIAP, inducing cell death. *Molecular Cell, 8*, 613–621.

Syntichaki, P., Xu, K., Driscoll, M., & Tavernarakis, N. (2002). Specific aspartyl and calpain proteases are required for neurodegeneration in C. elegans. *Nature, 419*, 939–944.

Tsujimoto, Y., Cossman, J., Jaffe, E., & Croce, C. M. (1985). Involvement of the bcl-2 gene in human follicular lymphoma. *Science, 228*, 1440–1443.

Vaux, D. L., Weissman, I. L., & Kim, S. K. (1992). Prevention of programmed cell death in Caenorhabditis elegans by human bcl-2. *Science*, *258*, 1955–1957.

Verhagen, A. M., Ekert, P. G., Pakusch, M., Silke, J., Connolly, L. M., Reid, G. E., et al. (2000). Identification of DIABLO, a mammalian protein that promotes apoptosis by binding to and antagonizing IAP proteins. *Cell*, *102*, 43–53.

Wang, X. (2001). The expanding role of mitochondria in apoptosis. *Genes & Development*, *15*, 2922–2933.

Wang, X., Wang, J., Gengyo-Ando, K., Gu, L., Sun, C. L., Yang, C., et al. (2007). C. elegans mitochondrial factor WAH-1 promotes phosphatidylserine externalization in apoptotic cells through phospholipid scramblase SCRM-1. *Nature Cell Biology*, *9*, 541–549.

Wang, X., Yang, C., Chai, J., Shi, Y., & Xue, D. (2002). Mechanisms of AIF-mediated apoptotic DNA degradation in Caenorhabditis elegans. *Science*, *298*, 1587–1592.

Wu, D., Wallen, H. D., Inohara, N., & Nunez, G. (1997). Interaction and regulation of the Caenorhabditis elegans death protease CED-3 by CED-4 and CED-9. *The Journal of Biological Chemistry*, *272*, 21449–21454.

Wu, D., Wallen, H. D., & Nunez, G. (1997). Interaction and regulation of subcellular localization of CED-4 by CED-9. *Science*, *275*, 1126–1129.

Xu, K., Tavernarakis, N., & Driscoll, M. (2001). Necrotic cell death in C. elegans requires the function of calreticulin and regulators of Ca(2+) release from the endoplasmic reticulum. *Neuron*, *31*, 957–971.

Xue, D., Shaham, S., & Horvitz, H. R. (1996). The Caenorhabditis elegans cell-death protein CED-3 is a cysteine protease with substrate specificities similar to those of the human CPP32 protease. *Genes & Development*, *10*, 1073–1083.

Yan, N., Chai, J., Lee, E. S., Gu, L., Liu, Q., He, J., et al. (2005). Structure of the CED-4-CED-9 complex provides insights into programmed cell death in Caenorhabditis elegans. *Nature*, *437*, 831–837.

Yuan, J., Shaham, S., Ledoux, S., Ellis, H. M., & Horvitz, H. R. (1993). The C. elegans cell death gene ced-3 encodes a protein similar to mammalian interleukin-1 beta-converting enzyme. *Cell*, *75*, 641–652.

Zamzami, N., & Kroemer, G. (2001). The mitochondrion in apoptosis: How Pandora's box opens. *Nature Reviews. Molecular Cell Biology*, *2*, 67–71.

Zou, H., Henzel, W. J., Liu, X., Lutschg, A., & Wang, X. (1997). Apaf-1, a human protein homologous to C. elegans CED-4, participates in cytochrome c-dependent activation of caspase-3. *Cell*, *90*, 405–413.

CHAPTER THREE

Autophagy in Cell Life and Cell Death

Allyson L. Anding, Eric H. Baehrecke[1]
Department of Molecular, Cell, and Cancer Biology, University of Massachusetts Medical School, Worcester, Massachusetts, USA
[1]Corresponding author: e-mail address: Eric.Baehrecke@umassmed.edu

Contents

Abstract

Macroautophagy (hereafter referred to as autophagy) is a process used by the cell to deliver cytoplasmic components to the lysosome for degradation. Autophagy is most often associated with cell survival, as it provides cells with molecular building blocks during periods of nutrient deprivation and also aids in the elimination of damaged organelles and protein aggregates. However, autophagy has also been implicated in cell death. Here, we review what is known about autophagy, its regulation, its role both in cell life and cell death, and what is known about autophagic cell death *in vivo*.

1. INTRODUCTION

Autophagy is an evolutionarily conserved, catabolic process whereby cellular contents are sequestered by a double-membrane vesicle termed the

Current Topics in Developmental Biology, Volume 114
ISSN 0070-2153
http://dx.doi.org/10.1016/bs.ctdb.2015.07.012

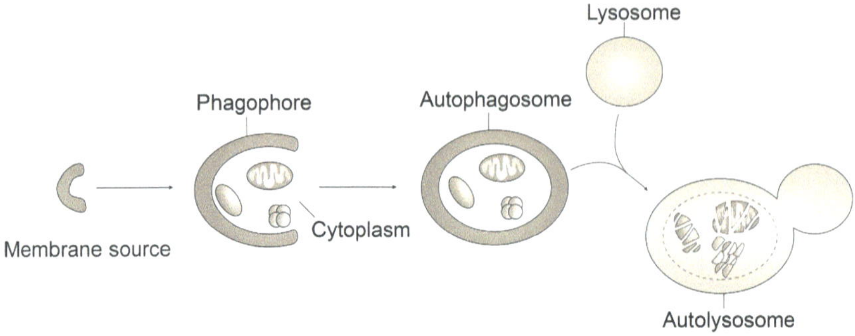

Figure 1 Autophagy delivers cellular contents to the lysosome for degradation. The phagophore is initiated from a membrane source which expands and surrounds cellular cargo, such as organelles and long-lived proteins, to form a double-membrane structure called the autophagosome. The autophagosome then fuses with the lysosome to form an autolysosome in which the cellular cargo is degraded.

autophagosome and brought to the lysosome for degradation (Fig. 1). This multistep process is executed and regulated by autophagy-related (Atg) proteins that control the initiation of autophagy, elongation of the autophagosome, and completion of this process. Traditionally, autophagy is thought of as a cell survival mechanism, acting in response to cellular stressors as a method of cell quality control. Basal autophagy is used by the cell to remove, for example, damaged organelles, long-lived proteins, and protein aggregates. Cellular stressors such as pathogen infection, hypoxia, nutrient limitation, or reactive oxygen species (ROS) can also trigger autophagy.

Most studies have focused on the function of autophagy during cell survival that is triggered by starvation. However, autophagy has also been tightly linked to cell death processes, participating side by side with either apoptosis or other cell death pathways, or independently to kill the cell. Cell death associated with autophagy was termed "autophagic cell death" by the Nomenclature Committee on Cell Death (NCCD), a group consisting of leading researchers in the field of cell death research, which defines it as a process that is blocked by genetic interventions targeting at least two components of the molecular machinery of autophagy (Galluzzi et al., 2012). Additionally, the committee suggests that clonogenic survival assays should be used to demonstrate long-term protection against cell death with genetic inhibition of autophagy, avoiding conclusions based merely on altered kinetics of cell death. Such autophagic cell death was seen in several studies, such as in cell death induced by various chemotherapy drugs and in

Drosophila tissue cell death during the developmental transition from a larva to an adult, also known as metamorphosis. Although there is mounting evidence supporting autophagic cell death to date, the role of this process in mammals *in vivo* is unclear.

2. AUTOPHAGY GENES

Autophagy is a multistep process that is genetically regulated by the *Atg* genes. Pioneering studies in the yeast *Saccharomyces cerevisiae* involving screens for mutants with defects in autophagic structures resulted in the identification of *Atg* genes (Tsukada & Ohsumi, 1993), most of which are conserved in mammals.

Autophagy is initiated with the formation of a preautophagosomal structure (PAS) which serves as the nucleation point for the formation of the isolation membrane (Suzuki et al., 2001). The initiation of autophagy requires the serine/threonine kinase Atg1 (Ulk1 and 2 in mammals; Fig. 2) (Matsuura, Tsukada, Wada, & Ohsumi, 1997) which is sufficient to activate autophagy when misexpressed (Scott, Juhász, & Neufeld, 2007). Upon induction of autophagy, Atg1 forms a complex with Atg13, FAK family kinase-interacting protein of 200 kDa (FIP200; Atg17 in yeast), and ATG101 at the isolation membrane, and this complex is negatively regulated by target of rapamycin complex 1 (TORC1) (Kamada et al., 2000). This Atg1 complex is thought to nucleate the formation of autophagosomal structures. Atg1 kinase activity is also essential for autophagy, as it is known to phosphorylate Atg9, allowing Atg9 to interact with Atg18 (WIPI2 in mammals) which triggers recruitment of this key autophagy factor and Atg8 to the PAS (Papinski et al., 2014).

The Vps34 complex, consisting of Vps34 (class III PI3K), Beclin 1 (Atg6 in *Drosophila* and Atg6/Vps30 in yeast), Vps15 (also called p150 or ird1 in *Drosophila*), and the metazoan-specific components UV irradiation resistance-associated gene (Uvrag) and Barkor (Atg14L in yeast and *Drosophila*), is also essential for the formation of autophagosomes (Juhász et al., 2008). Vps34 phosphorylates phosphatidylinositol (PtdIns) to form PtdIns-3-phosphate (PI3P), a critical step in the initiation of autophagosome biogenesis. Additionally, two ubiquitin-like conjugation pathways involving the ubiquitin-like proteins Atg8 (LC3 in mammals) and Atg12 are required for autophagy. In both the Atg12 and Atg8 conjugation pathways, Atg7 serves as the E1-like activating enzyme (Tanida et al., 1999). In the Atg12 conjugation system, Atg12 is subsequently transferred to the E2

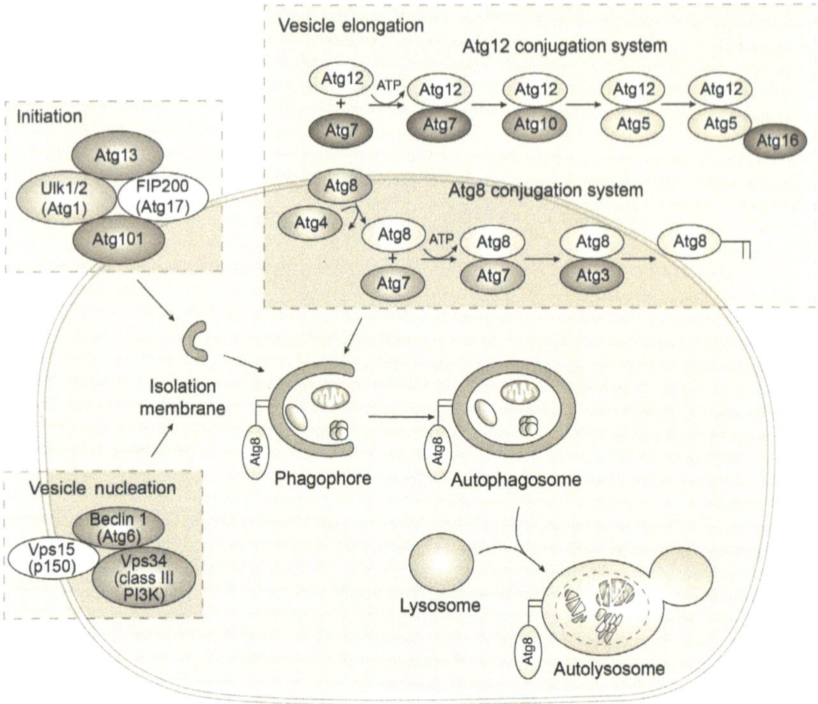

Figure 2 Regulation of autophagy. The initiation of autophagy requires Atg1 and its interacting partners. The Vps34 complex is responsible for vesicle nucleation, and this core complex can also associate with other proteins such as Atg14L and UVRAG. Vesicle elongation is regulated by the Atg12 and Atg8 ubiquitin-like conjugation systems that enable the tethering of Atg8 to the autophagosomal membrane and eventual autophagosome formation. (See the color plate.)

enzyme, Atg10, and then conjugated to Atg5. Finally, the Atg12–Atg5 complex can bind Atg16, forming Atg12–Atg5–Atg16. In contrast, Atg3 acts as the E2 in the Atg8 conjugation system and the Atg12–Atg5 complex acts as the E3 enzyme, bringing the Atg8-carrying E2-like enzyme, Atg3, in proximity to its substrate, the lipid phosphatidylethanolamine (PE) (Hanada et al., 2007). Prior to its interaction with Atg7, Atg8 is cleaved by Atg4, exposing a C-terminal glycine residue to which PE can be coupled. Lipidation of Atg8 facilitates its interaction with the autophagosomal membrane, driving autophagosome maturation steps including cargo capture, growth, and closure of the organelle (Nakatogawa, Ichimura, & Ohsumi, 2007; Weidberg et al., 2010).

3. AUTOPHAGY AND CELL SURVIVAL

3.1 Clearance of Damaged Organelles

Cellular homeostasis is dependent on the maintenance of both organelle integrity and number in the context of varying environments and stressors. Thus, dysfunctional and surplus organelles are cleared from tissues through the process of autophagy (Okamoto, 2014). This clearance of organelles, or organellophagy, varies from the bulk degradation process of starvation-induced autophagy in that it involves the degradation of specific cellular components and is, thus, categorized as selective autophagy (Fig. 3). In selective autophagy, a cellular signal induces downstream events triggering degradation cues for a specific target, molecules tag the target as cargo to be degraded, and autophagy-related components sequester and eliminate the cargo. Thus far, multiple organelles such as peroxisomes, mitochondria, lipid droplets, nuclei, lysosomes, endoplasmic reticulum, and ribosomes have been identified as cargo that can be degraded via autophagy in taxa that are as different as yeast and humans.

Selective sequestration of specific cargo into autophagosomes involves the participation of autophagy receptors. These receptors recognize cargo tagged with degradation signals as well as the autophagosomal membrane through interaction with LC3/Atg8 either directly or indirectly through a consensus sequence called the Atg8-interacting motif or LC3-interacting region (LIR) (Birgisdottir, Lamark, & Johansen, 2013). Pexophagy, or the degradation of peroxisomes, is an example of organellophagy involving

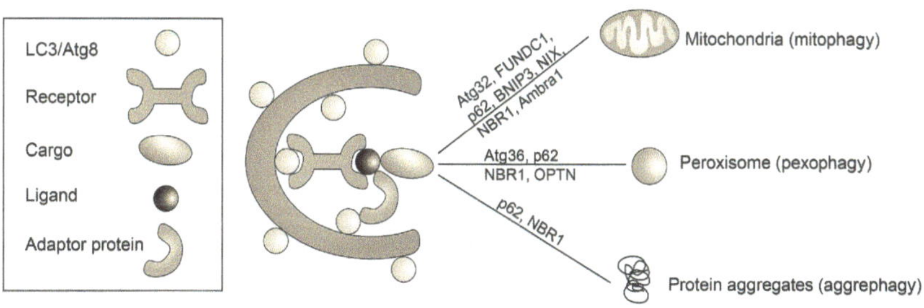

Figure 3 Selective autophagy. In selective autophagy, receptor proteins enable the interaction of the autophagosomal membrane and cargo to be degraded, sometimes with the help of ligands such as ubiquitin or adaptor proteins. Three types of selective autophagy, mitophagy, pexophagy, and aggrephagy, are represented here, along with some of the receptor proteins associated with these processes. (See the color plate.)

receptor-mediated cargo recognition. Peroxisome number increases or decreases in response to the environment in order to maintain appropriate metabolic reactions such as the detoxification of hydrogen peroxide and fatty acid oxidation (Oku & Sakai, 2010). In the yeast *S. cerevisiae*, Atg36 acts as a receptor protein, localizing to the surface of the peroxisome and binding Atg8 to promote autophagy (Motley, Nuttall, & Hettema, 2012). This process was recently shown to be regulated by the kinase, Hrr25, which phosphorylates Atg36, facilitating its interaction with the autophagosomal membrane (Tanaka et al., 2014). In mammals, ubiquitination of peroxisomal proteins promotes recruitment of LC3 to peroxisomes via the receptor proteins p62 (also known as sequestosome 1) and/or NBR1 (neighbor of BRCA1 gene). In COS-7 cells, pexophagy occurs in a p62-dependent manner (Kim, Hailey, Mullen, & Lippincott-Schwartz, 2008), and down-regulation of either NBR1 or p62 has been shown to suppress degradation of peroxisomes in HeLa cells (Deosaran et al., 2013).

Other selective autophagy processes also rely on a ligand such as ubiquitin to facilitate the interaction of cargo with receptors for subsequent degradation. In such cases, proteins that are either associated with or on the surface of the organelle are ubiquitinated by specific E3 ligases and act as degradation signals that are recognized by adaptor proteins, such as p62 and NBR1, that then interact with Atg8 or LC3. It is important to note, however, that others have shown that these two ubiquitin-binding receptors p62 and NBR1 are recruited to autophagosome formation sites independent from LC3, indicating that these receptors may recruit cargo to autophagosomes by binding factors other than LC3/Atg8 (Itakura & Mizushima, 2011). In mammals, cells use the ubiquitin-dependent degradation method to degrade mitochondria, a process known as mitophagy. Mitochondria are extremely important organelles, necessary for various tasks such as the conversion of energy and programmed cell death. ROS are produced by mitochondria as a by-product of cellular respiration, and accumulation of ROS can lead to mitochondrial dysfunction. Thus, in addition to the elimination of surplus mitochondria, elimination of damaged mitochondria is critical for cellular homeostasis. In *Drosophila melanogaster*, for example, clearance of mitochondria seems to be a key step in the degradation of the larval midgut as knockdown of autophagy-related proteins blocks mitochondrial clearance and programmed cell size reduction (Chang et al., 2013).

PINK1, a mitochondrial protein kinase, and Parkin, a cytosolic E3 ubiquitin ligase, are key players in mitophagy. Significantly, mutations in PINK1 and Parkin are associated with Parkinson's disease (Kitada et al.,

1998; Valente et al., 2004), highlighting the potential importance of clearance of mitochondria for human health. In healthy mitochondria, PINK1 is cleaved by presenilin-associated rhomboid-like and rapidly degraded by the proteasome. In dysfunctional mitochondria, PINK1 stays on the outer mitochondrial membrane and recruits Parkin. High-throughput analysis of the Parkin-dependent ubiquitylome has identified several targets of the E3 activity of Parkin in the cell (Sarraf et al., 2013), among these being the proteins previously identified to be targets of Parkin: VDAC1 (Geisler et al., 2010) and mitofusin (MFN1) (Tanaka et al., 2010). Whether ubiquitin/Atg8-binding adaptors such as p62 and NBR1 are necessary for the PINK1/Parkin mitophagy pathway is inconclusive, as the data are conflicting (Geisler et al., 2010; Narendra, Kane, Hauser, Fearnley, & Youle, 2010; Okatsu et al., 2010; Wong & Holzbaur, 2014).

Screens in yeast identified Atg32 to be required for mitophagy (Kanki, Wang, Baba, et al., 2009; Kanki, Wang, Cao, Baba, & Klionsky, 2009; Okamoto, Kondo-Okamoto, & Ohsumi, 2009) and its mammalian counterparts, BNIP3 and NIX, have also been implicated in this process (Hanna et al., 2012; Novak & Dikic, 2011; Novak et al., 2010). FUNDC1 (FUN14 domain containing 1) has also been implicated as a mitophagy receptor protein (Liu et al., 2012). BNIP3, NIX, and FUNDC1 all contain LIR motifs, facilitating their interactions with LC3. The autophagy-promoting protein, Ambra1 (activating molecule in Beclin 1-regulated autophagy), has also been implicated in mitophagy, as it was found to interact with Parkin and to be critically important for mitochondrial clearance (Van Humbeeck et al., 2011). Ambra1 was also recently been shown to interact with LC3 through a LIR motif and this interaction is critical for regulating both canonical Parkin-dependent and -independent mitochondrial clearance (Strappazzon et al., 2015). Importantly, forced localization of Ambra1 to the outer mitochondrial membrane caused the induction of a Parkin- and p62-independent, but LC3-dependent, mitophagy, highlighting a novel role for Ambra1 as a regulator of both canonical and non-canonical pathways of mitophagy.

3.2 Clearance of Protein Aggregates

Protein aggregates can form as a protective mechanism in order to store harmful proteins in an inert form, but they can also be harmful to cells. Aggregate formation is associated with a number of neurodegenerative diseases, such as Alzheimer's and Parkinson's disease (Irvine, El-Agnaf,

Shankar, & Walsh, 2008). Normally, soluble abnormal proteins are either refolded using chaperone proteins or are degraded by the ubiquitin–proteasome system. Protein aggregates that accumulate in cells under conditions of stress, however, can be degraded via a protein aggregate-specific autophagy known as aggrephagy (Lamark & Johansen, 2012). Structures called aggresomes form when misfolded or unsuccessfully refolded proteins pack into larger, insoluble aggregates. In aggrephagy, aggresomes are labeled with ubiquitin and associate with ubiquitin-binding microtubule-associated histone deacetylase 6 (HDAC6) (Kawaguchi et al., 2003). Aggresomes then concentrate in the microtubule-organizing center, and HDAC6 controls the trafficking of aggresomes along the microtubules as well as the fusion of subsequently formed autophagosomes and lysosomes (Kawaguchi et al., 2003; Lee et al., 2010).

As with other types of receptor-mediated autophagy processes, p62 and NBR1 are also essential for aggrephagy. p62/SQSTM1 acts as a cargo receptor in aggrephagy, directly linking ubiquitinated proteins and aggregates to autophagosomal membranes by binding Atg8/LC3 (Bjørkøy et al., 2005; Pankiv et al., 2007). NBR1, a protein that shares similar domain architecture with p62, also binds to ubiquitinated proteins and enhances the formation of aggregates (Kirkin, Lamark, Johansen, & Dikic, 2009; Lamark, Kirkin, Dikic, & Johansen, 2009). Both p62 and NBR1 bind to ubiquitinated proteins via their UBA (ubiquitin-associated) domains and to LC3 via their LIR motifs (Kirkin et al., 2009; Pankiv et al., 2007). p62 is also involved in the clearance of nonubiquitinated substrates, such as in the clearance of an aggregation-prone isoform of STAT5A (Watanabe & Tanaka, 2011).

Another protein associated with aggrephagy is autophagy-linked FYVE-domain-containing protein, or ALFY (Simonsen et al., 2004). ALFY acts as an adaptor protein, facilitating binding between autophagosome machinery and p62 (Clausen et al., 2010; Filimonenko et al., 2010). Optineurin was also shown to associate with protein aggregates and to recruit LC3 to aggregates via its LIR motif (Korac et al., 2013). Ultimately, aggresomes will fuse with the lysosome to form the autolysosome, digesting their contents.

3.3 Starvation-Induced Autophagy

Autophagy plays an important role in the protection of cells in response to stress, such as nutrient limitation. During such periods of starvation, autophagy can be used to degrade cytoplasmic contents in order to yield amino acids and fatty acids that can be used, for example, to produce new proteins. Nutrients are sensed by the conserved kinases AMP-activated protein kinase (AMPK) and target of rapamycin (TOR). AMPK promotes autophagy through the inhibition of TOR by phosphorylating TSC2 and Raptor (Gwinn et al., 2008). Additionally, under starvation conditions, AMPK has been shown to directly phosphorylate and activate Ulk1 (Kim, Kundu, Viollet, & Guan, 2011). TOR is a serine/threonine protein kinase belonging to the phosphatidylinositol kinase-related kinase family that is part of a larger macromolecular complex, TORC1. Under conditions of nutrient deprivation, such as low amino acid or glucose levels, TOR activity is inhibited, relieving repression of autophagy. In fact, treatment with TOR inhibitors such as rapamycin induces autophagy in yeast (Noda & Ohsumi, 1998) and animals (Rusten et al., 2004; Scott, Schuldiner, & Neufeld, 2004). Starvation-induced autophagy has been described in model organisms, such as in the fat body of *D. melanogaster*. The larval fat body, a nutrient storage tissue similar to the liver and adipocytes of mammals, is particularly sensitive to starvation, with high levels of autophagy induced in response to relatively short periods of starvation during larval stages (Scott et al., 2004).

3.4 Developmental Autophagy

Autophagy is important in the development of both invertebrate and vertebrate organisms, and plays a part in the development and differentiation of many tissues. Though autophagy is important for tissue destruction and remodeling in some tissues, such as the salivary glands of *D. melanogaster*, autophagy can also be critical for proper development and survival of organisms. In *Drosophila*, for example, multiple *Atg* genes are crucial for survival to adulthood, with mutations in such genes leading to death at early stages in development (Scott et al., 2004). The same is true in *Caenorhabditis elegans*, with several core autophagy genes, such as *bec-1*, exhibiting a lethal phenotype when mutated (Meléndez et al., 2003; Roggo et al., 2002; Takács-Vellai et al., 2005). In *C. elegans*, conditions of stress, such as starvation or increased temperature, lead to developmental arrest, known as dauer diapause (Riddle, Blumenthal, Meyer, Priess, & Albert, 1997). It has also been shown that autophagy is necessary for this process, acting downstream of the insulin/IGF (insulin-like growth factor) pathway (Meléndez et al., 2003).

Autophagy is also critical throughout development in vertebrates. *Atg5*$^{-/-}$ mice survive embryogenesis, but die shortly after birth due to an inability to feed and overcome the perinatal starvation period (Kuma et al., 2004). Removal of the maternal contribution of *Atg5* leads to the arrest of embryonic development at the eight-cell stage, indicating that *Atg5* and autophagy are critical for both early and later stages in development (Tsukamoto et al., 2008). Other *Atg* knockout neonates, such as *Atg3* (Sou et al., 2008), *Atg7* (Komatsu et al., 2005), *Atg9* (Saitoh et al., 2009), and *Atg16L1* (Saitoh et al., 2008), also die shortly after birth. In contrast with the embryonic survival seen with these knockout mice, knockout of other autophagy genes such as *beclin 1*, *Ambra1*, and *FIP200* produce other phenotypes (Fimia et al., 2007; Gan et al., 2006; Yue, Jin, Yang, Levine, & Heintz, 2003), presumably because of additional roles that these proteins play, such as regulation of endocytosis (Mizushima & Levine, 2010; Ruck et al., 2011).

Autophagy is also important for neuronal health, as *Atg5* and *Atg7* deficiency lead to defects in motor function and results in a cytotoxic accumulation of ubiquitinated proteins, such as is seen in neurodegenerative diseases (Hara et al., 2006). *Ulk1* is also critical for axon outgrowth and differentiation of neurons (Tomoda, Bhatt, Kuroyanagi, Shirasawa, & Hatten, 1999). *Ambra1* knockout causes early lethality in mice and also leads to pronounced defects in the central nervous system, such as neural tube closure defects and exencephaly (Fimia et al., 2007). In the heart, *Atg5* deficiency results in cardiac dysfunction in adult mice (Tomoda et al., 1999). In addition, inactivation of *Atg5* within the first few days after birth leads to a rapid onset of heart failure, suggesting the importance of basal autophagy in cardiac function.

Organisms also rely on autophagy for proper hematopoiesis. During erythroid differentiation, erythroblasts differentiate into anucleated cells called reticulocytes. Reticulocytes then lose their intracellular organelles, becoming erythrocytes. $Atg7^{-/-}$ erythroblasts do not mature, suggesting that autophagy is necessary for this process (Mortensen & Simon, 2010). Specifically, the role of autophagy in mitochondrial clearance has been suggested to be important for this process. Accordingly, loss of the mitophagy receptor, Nix, leads to a loss of mitochondrial clearance in reticulocytes and a lower number of mature erythrocytes (Novak et al., 2010; Sandoval et al., 2008; Schweers et al., 2007). B-lymphocyte- and T-lymphocyte-specific deletion of *Atg* genes such as *Atg5* and *Atg7* has also revealed roles for autophagy in lymphocyte differentiation, though mitochondrial clearance does not appear to be required for this process (Miller et al., 2008; Mortensen & Simon, 2010; Pua, Dzhagalov, Chuck, Mizushima, & He, 2007).

4. AUTOPHAGY AND CELL DEATH

Programmed cell death is an evolutionarily conserved process that occurs in multicellular organisms, enabling the removal of damaged and unwanted cells. Initially, the classification of different types of cell death was based largely on morphological criteria. However, the NCCD has proposed five different modes of programmed cell death, including extrinsic apoptosis, intrinsic apoptosis, regulated necrosis, mitotic catastrophe, and autophagic cell death (Galluzzi et al., 2012). These processes are extremely important and well regulated, as disturbances in the normal activation and regulation of cell death can contribute to developmental problems and numerous diseases, such as cancer and neurodegenerative diseases.

All multicellular organisms undergo programmed cell death both during development and in adulthood. Though most of the programmed cell death in mammals is mediated by apoptosis, a caspase-dependent cell death modality, additional forms of cell death, such as autophagic cell death, have also been described. Autophagic cell death differs depending on the context and can function either independently of or in parallel with apoptotic caspases or other cell death pathways.

4.1 Autophagic Cell Death in Model Organisms

The most convincing evidence in support of autophagic cell death has been from the genetic manipulation of various model organisms. Not only have such studies revealed a direct role for autophagy in programmed cell death, but they have also highlighted a tight connection between autophagy and other cell death processes.

4.1.1 Dictyostelium discoideum

During the formation of its multicellular fruiting body, the amoeboid slime mold, *D. discoideum*, undergoes programmed cell death (Whittingham & Raper, 1960). *Dictyostelium* development begins with the aggregation of amoebae by chemotaxis followed by the formation of a mound of adhering cells. This mound can then form a fruiting body consisting of a spore mass on a thin stalk (Kessin, 2001). Because *Dictyostelium* development only occurs under starvation, this organism has to rely on energy and chemical constituents produced by the turnover of its own macromolecules to supply its developing cells.

Dictyostelium does not encode the machinery for apoptosis, nor does it possess phagocytes, and the programmed cell death that occurs during development is prevented by genetically blocking autophagy by mutating *Atg* genes (Kosta et al., 2004; Otto, Wu, Kazgan, Anderson, & Kessin, 2003). This highlights the important role for autophagy in *Dictyostelium* development. Autophagy-linked programmed cell death in *Dictyostelium* is unique in that it is first activated in response to starvation, but programmed cell death is not activated until an additional signal from the differentiation-inducing factor (DIF-1) is received (Cornillon et al., 1994; de Chastellier & Ryter, 1977). Neither the induction of autophagy nor the DIF-1 signal alone can induce cell death, so autophagy, along with additional signals, is necessary for this process (Levraud et al., 2003; Town, Gross, & Kay, 1976). Although the lack of apoptotic machinery presents an unique opportunity for studying the role of autophagy during cell death in *Dictyostelium*, one must observe autophagy in organisms with intact apoptosis machinery in order to understand the complex relationship between autophagy and cell death that may occur in other higher organisms.

4.1.2 *C. elegans*

Pioneering genetic studies of programmed cell death in *C. elegans* resulted in the identification of the core apoptotic machinery (Hengartner & Horvitz, 1994). Thus, unlike with *Dictyostelium*, this organism can be used to observe autophagic cell death in the context of apoptosis. The *C. elegans* ortholog of mammalian *Beclin 1* is known as *bec-1* and is essential for autophagy in the nematode (Meléndez et al., 2003). Decreased *bec-1* leads to a reduction in autophagy and increased apoptosis during embryonic development as well as in the germline of adults (Takács-Vellai et al., 2005). Though autophagy is dispensable for physiological germline cell death, it can contribute to cell death in the germline in response to genotoxic stress or when apoptotic activity is reduced (Wang, Lu, Cheng, Wang, & Zhang, 2013). Depletion of autophagy gene function reduced germ cell corpses following gamma irradiation or ethylnitrosourea treatment. Autophagy gene depletion also reduced the number of cell corpses in *ced-3* (the *C. elegans* executioner caspase) hypomorphic mutants. Thus, in *C. elegans*, autophagy and apoptosis can cooperate during programmed cell death.

4.1.3 *D. melanogaster*

Similar to other organisms, autophagy is induced in specific tissues upon starvation of *D. melanogaster*, such as the fat body (Barth, Szabad,

Hafen, & Köhler, 2011; Scott et al., 2004). In contrast to this starvation-induced, survival-oriented autophagy, developmentally programmed autophagy that is involved in the removal of tissues has also been observed in *Drosophila* (Berry & Baehrecke, 2007; Lee, Cooksey, & Baehrecke, 2002). During metamorphosis, pulses of the steroid hormone 20-hydroxyecdysone (ecdysone) signal for obsolete tissues, such as the larval salivary glands and midgut, to be degraded using programmed autophagic cell death. During the larval to pupal transition, the *Drosophila* midgut and gastric caeca shrink drastically in size (Chang et al., 2013; Lee et al., 2002). A pulse of ecdysone at the onset of puparium formation triggers this cell size reduction, and within 4 h, the midgut has essentially died (Lee et al., 2002). High levels of autophagy can be detected during this process (Lee et al., 2002), and when autophagy genes are mutated, the midgut cells fail to degrade (Denton et al., 2009). Inhibition of autophagy genes in the dying midgut also disrupts mitochondrial clearance, implicating a role for mitophagy in this process (Chang et al., 2013). Interestingly, this midgut cell death seems to be independent of apoptosis, as the inhibition of caspases had no effect on the induction of autophagy or cell death (Denton et al., 2009). Additionally, no involvement of phagocytes has been implicated during midgut cell death. Thus, in the destruction of the midgut, autophagy is necessary but apoptosis is dispensable.

Destruction of the *Drosophila* larval salivary glands also occurs during the larval to pupal transition. Another pulse of ecdysone approximately 12 h after puparium formation signals for the total degradation of the salivary glands. High levels of autophagy are induced during this process and genetic inhibition of autophagy results in incomplete degradation of the gland (Berry & Baehrecke, 2007; Lee & Baehrecke, 2001). As with the destruction of the midgut, phagocytes do not appear to be involved in this process (Martin & Baehrecke, 2004). However, caspases are involved in the degradation of glands, as inhibition of caspases also results in incomplete gland degradation (Berry & Baehrecke, 2007). When both autophagy and caspases are blocked, the removal of the salivary glands is further delayed compared with when either of these pathways is blocked individually. Thus, autophagy and apoptosis function in parallel in salivary gland degradation.

Though most core autophagy genes regulate both starvation-induced and developmentally programmed autophagy in *Drosophila* (Anding & Baehrecke, 2015), some genes have been shown to be context specific in their regulation. For example, Draper, IP3 kinase 2 (IP3K2), the IP3 receptor, and calmodulin have been shown to specifically regulate autophagy during salivary gland degradation (McPhee & Baehrecke, 2010; Nelson, Ambros, & Baehrecke, 2014). Additionally, the E1 enzyme Atg7 was shown to not be required for midgut cell death, though the E1 ubiquitin ligase, Uba1, was found to be necessary (Chang et al., 2013). Thus, it is important to thoroughly investigate autophagy regulatory mechanisms in a context-specific manner.

4.1.4 Vertebrates

It is clear that autophagy is important in various stages of development in vertebrates. However, our knowledge regarding the role of autophagy in vertebrate programmed cell death is incomplete. Dying cells have been shown to feature the hallmarks of autophagy, but it is unclear whether autophagy is acting in a prodeath fashion or whether autophagy is acting in concert with other prodeath mechanisms. More complete genetic studies are needed to fill the gap in our knowledge of vertebrate autophagic cell death. Unfortunately, thus far, understanding the prodeath role of autophagy through the examination of *Atg* gene knockout mice has been difficult due to lethality or other phenotypes (Cecconi & Levine, 2008).

During mammalian development, there are several processes in which autophagy is observed. In mammals, a transient endocrine gland structure known as the corpus luteum develops from the wall of an ovarian follicle following ovulation. This structure produces progesterone in order to maintain pregnancy. However, if pregnancy is absent or when the corpus luteum is no longer needed to maintain the pregnancy, the structure regresses, leading to its elimination. Though apoptosis seems to be the primary mechanism whereby the corpus luteum regresses (Stocco, Telleria, & Gibori, 2007), studies have also shown that other processes, such as autophagy, are involved in luteolysis (Choi, Jo, Lee, & Choi, 2011; Fraser, Lunn, Harrison, & Kerr, 1999). For example, luteal cell autophagy was found to be directly involved in corpus luteum regression and was correlated with increased apoptosis in rats (Choi et al., 2011). In this study, inhibition of autophagosome formation was found to decrease apoptosis and cell death, indicating that autophagosome accumulation induces luteal cell apoptosis.

Autophagy is also important in the involution of the mammary glands, another pregnancy-related tissue regression process that occurs after completion of lactation (Gajewska, Sobolewska, Kozlowski, & Motyl, 2008; Motyl, Gajewska, Zarzynska, Sobolewska, & Gajkowska, 2007; Zarzyńska, Gajkowska, Wojewódzka, Dymnicki, & Motyl, 2007). Again, the primary mechanism whereby mammary gland involution proceeds is apoptosis (Zarzyńska et al., 2007). However, LC3 is upregulated during the initial stages of involution in rodents (Monks & Henson, 2009), and bovine mammary glands exhibit an increase in *beclin 1* expression and morphological features of autophagy during this process (Zarzyńska et al., 2007). Additionally, murine studies have shown that involution is perturbed by *Atg* gene (*Beclin 1* and *Atg7*) knockout, though this may have to do with the role of apoptosis and autophagy in the clearance of dead cells (Teplova et al., 2013). These studies suggest that autophagy and apoptosis act in parallel during the process of mammary involution.

A role for autophagy in early embryogenesis was suggested by studies of cavitation in a model of early embryonic development (Qu et al., 2007). Two autophagy genes, *Beclin 1* and *Atg5*, were shown to be necessary for the clearance of dying cells during cavitation using an *in vitro* model of embryoid bodies. Again, however, autophagy seems to be acting in parallel with apoptosis during this process, as autophagy seems to be affecting phosphatidylserine exposure leading to a lack of apoptotic cell clearance.

Though no studies to date have shown that autophagy is solely required for cell death in mammalian cells *in vivo*, studies in mammalian cell lines have shown that autophagy is required for cell death in the absence of apoptosis. Autophagic cell death is often revealed *in vitro* when apoptotic pathways have been blocked. For example, in cells deficient in Bax and Bak, cell death induced by etoposide or staurosporine can occur in a manner that is dependent on Beclin-1 and ATG5 (Shimizu et al., 2004). In cells that are competent for apoptosis, high levels of autophagy can also lead to caspase-independent cell death (Elgendy, Sheridan, Brumatti, & Martin, 2011; Gewirtz, 2014; Reef et al., 2006; Sharma, Le, Alotaibi, & Gewirtz, 2014). Additionally, Tat-Beclin 1, a cell-death-inducing peptide, was shown to induce a form of autophagy-dependent cell death known as autosis that was independent from the apoptotic and necroptotic machinery in mammalian cells (Liu et al., 2013). Importantly, autosis was also shown to occur *in vivo* in rats after cerebral hypoxic–ischemic injury, indicating that this phenomenon could be a physiologically relevant form of autophagic cell death.

In the context of cancer, autophagy has been shown to function in either a prosurvival or prodeath capacity, depending on the type and stage of cancer. Autophagy can contribute to cancer cell death, particularly when apoptotic pathways are blocked (Kondo, Kanzawa, Sawaya, & Kondo, 2005; Moretti, Yang, Kim, & Lu, 2007). For example, regression of neuroblastoma cells with upregulated Ras expression by autophagic cell death has been observed both *in vitro* and *in vivo* (Kitanaka et al., 2002). Additionally, RasV12 expression in human ovarian epithelial cancer cells can induce cell death that is dependent on autophagy but caspase-independent (Elgendy et al., 2011). As autophagy can act as either a cell survival or cell death process in cancer cells, knowing exactly how it is acting in a particular cancer type or context is incredibly important for cancer therapy. Ideally, autophagy would enhance the efficacy of cancer therapy treatments by either promoting cell death or cooperating with other cell death pathways. Multiple drug treatments in various cancer cell lines have been shown to induce cell death accompanied by autophagy (Denton, Xu, & Kumar, 2015). However, *in vivo* studies supporting this evidence are incomplete.

4.2 Autophagy and Other Cell Death Mechanisms

It is clear from studies performed in mammalian cell lines and *in vivo* that autophagy cooperates with other cell death modalities, such as apoptosis. There is also evidence of cooperation between autophagy and other cell death mechanisms, such as necroptosis. Necroptosis is a regulated form of necrosis which, like necrosis, is characterized by swelling of the cell and its organelles leading to cell rupture but, like apoptosis, is controlled by a defined molecular cascade (Lalaoui, Lindqvist, Sandow, & Ekert, 2015). Canonically, necroptosis is triggered by signaling through the TNF receptor which then stimulates recruitment of the TNF receptor-associated death domain protein, recruiting the receptor-interacting protein kinase 1 (RIPK1). In the absence of caspase-8 activity, RIPK1 and RIPK3 can interact and form a complex with mixed lineage kinase like (MLKL) called the necrosome, where MLKL is phosphorylated. Phosphorylation of MLKL leads to its oligomerization, allowing MLKL to insert into and permeabilize organelles and the plasma membrane (Su et al., 2014; Wang et al., 2014).

Autophagy has recently been shown to have a role in necroptosis. In mouse L929 fibroblastic cells, the caspase inhibitor zVAD can induce cell death. Reduction of *ATG7* and *Beclin 1* in these cells inhibited this cell death, indicating that autophagy contributes to necroptotic cell death (Yu et al., 2004). Additionally, chemotherapy-induced autophagy resulted in FADD/RIPK1/RIPK3 recruitment to the autophagosomal membranes through interaction with Atg5 in rhabdomyosarcoma cells, suggesting that autophagy may be promoting necroptosis by stimulating assembly of the necrosome on autophagosomal membranes (Basit, Cristofanon, & Fulda, 2013). However, other studies have shown that autophagy suppresses necroptosis. In RCC4 human renal carcinoma cell lines, mTOR inhibition stimulated autophagy and eliminated RIPKs, an effect that was blocked by autophagy inhibition, resulting in RIPK1- and oxidative stress-dependent necroptosis (Bray et al., 2012). This suggests that autophagy may inhibit necroptosis via the degradation of RIPK1 and decreasing ROS. In T-cell receptor-stimulated c-FLIP-deficient T cells, autophagy also has a prosurvival role, inhibiting necroptosis (He & He, 2013). It is clear that the role of autophagy in necroptosis is context specific, and that further studies are needed to truly understand the interplay between autophagy and necroptosis.

5. CONCLUSIONS

Here, we have described the role of autophagy in both cell survival and cell death. Autophagy is most often thought of as a cell survival mechanism, ridding cells of damaged organelles, protein aggregates, or providing energy during periods of stress such as starvation. However, mounting evidence points to the role of autophagy in programmed cell death in varying tissues and organisms.

Cellular homeostasis is crucial for the survival of a cell and is dependent on the maintenance of organelle integrity, removal of toxic substances, and adequate sources of energy. Autophagy has a clear role in the clearance or damaged or surplus organelles, removing peroxisomes and mitochondria, for example, through a non-bulk degradation process known as selective autophagy. The degradation of protein aggregates through aggrephagy is also important for cellular homeostasis, and defects in all of these processes can lead to human diseases, such as neurodegeneration. Autophagy also plays a key role in cell survival under starvation conditions in both developmental and stressed contexts. This could be particularly important in the context of providing proper nutrition to a tumor that may lack adequate vasculature and, thus, energy production. Finally, autophagy plays a key role in development, especially due to its roles in cell survival during starvation and removal of organelles.

Evidence indicates that autophagy acts beyond its role in cell survival and can promote programmed cell death. The best evidence in support of autophagic cell death has been from genetic studies in model organisms, such as *D. melanogaster*. Though some evidence points to autophagic cell death occurring in mammals, further genetic studies are necessary to truly understand this process. This would be particularly useful in the treatment of cancers in which the apoptotic machinery is often disrupted, and autophagy would offer an alternative mechanism for tumor cell death. Finally, the complex relationship between autophagy and other death processes, such as necroptosis, should be considered when developing therapies for human diseases. Though our knowledge of the interplay between autophagy and necroptosis is incomplete, future studies could expand treatment options for cancers particularly sensitive to this type of cell death.

ACKNOWLEDGMENTS

The authors gratefully acknowledge G. Kabachinski and P. Velentzas for critically reading the manuscript. Research on this subject is supported by the National Institutes of Health (GM079431, GM111658, CA159314, AI099708) to E.H.B. and A.L.A. was supported by T32 CA 130807-06 A1. E.H.B. is an Ellison Medical Foundation Scholar.

REFERENCES

Anding, A. L., & Baehrecke, E. H. (2015). Vps15 is required for stress induced and developmentally triggered autophagy and salivary gland protein secretion in Drosophila. *Cell Death and Differentiation*, *22*(3), 457–464. http://dx.doi.org/10.1038/cdd.2014.174.

Barth, J. M. I., Szabad, J., Hafen, E., & Köhler, K. (2011). Autophagy in Drosophila ovaries is induced by starvation and is required for oogenesis. *Cell Death and Differentiation*, *18*(6), 915–924. http://dx.doi.org/10.1038/cdd.2010.157.

Basit, F., Cristofanon, S., & Fulda, S. (2013). Obatoclax (GX15-070) triggers necroptosis by promoting the assembly of the necrosome on autophagosomal membranes. *Cell Death and Differentiation*, *20*(9), 1161–1173. http://dx.doi.org/10.1038/cdd.2013.45.

Berry, D. L., & Baehrecke, E. H. (2007). Growth arrest and autophagy are required for salivary gland cell degradation in Drosophila. *Cell*, *131*(6), 1137–1148. http://dx.doi.org/10.1016/j.cell.2007.10.048.

Birgisdottir, Å. B., Lamark, T., & Johansen, T. (2013). The LIR motif—Crucial for selective autophagy. *Journal of Cell Science*, *126*(Pt. 15), 3237–3247. http://dx.doi.org/10.1242/jcs.126128.

Bjørkøy, G., Lamark, T., Brech, A., Outzen, H., Perander, M., Overvatn, A., et al. (2005). p62/SQSTM1 forms protein aggregates degraded by autophagy and has a protective effect on huntingtin-induced cell death. *The Journal of Cell Biology*, *171*(4), 603–614. http://dx.doi.org/10.1083/jcb.200507002.

Bray, K., Mathew, R., Lau, A., Kamphorst, J. J., Fan, J., Chen, J., et al. (2012). Autophagy suppresses RIP kinase-dependent necrosis enabling survival to mTOR inhibition. *PLoS One*, 7(7), e41831. http://dx.doi.org/10.1371/journal.pone.0041831.

Cecconi, F., & Levine, B. (2008). The role of autophagy in mammalian development: Cell makeover rather than cell death. *Developmental Cell*, *15*(3), 344–357. http://dx.doi.org/10.1016/j.devcel.2008.08.012.

Chang, T.-K., Shravage, B. V., Hayes, S. D., Powers, C. M., Simin, R. T., Wade Harper, J., et al. (2013). Uba1 functions in Atg7- and Atg3-independent autophagy. *Nature Cell Biology*, *15*(9), 1067–1078. http://dx.doi.org/10.1038/ncb2804.

Choi, J., Jo, M., Lee, E., & Choi, D. (2011). The role of autophagy in corpus luteum regression in the rat. *Biology of Reproduction*, *85*(3), 465–472. http://dx.doi.org/10.1095/biolreprod.111.091314.

Clausen, T. H., Lamark, T., Isakson, P., Finley, K., Larsen, K. B., Brech, A., et al. (2010). p62/SQSTM1 and ALFY interact to facilitate the formation of p62 bodies/ALIS and their degradation by autophagy. *Autophagy*, *6*(3), 330–344.

Cornillon, S., Foa, C., Davoust, J., Buonavista, N., Gross, J. D., & Golstein, P. (1994). Programmed cell death in Dictyostelium. *Journal of Cell Science*, *107*(Pt. 10), 2691–2704.

de Chastellier, C., & Ryter, A. (1977). Changes of the cell surface and of the digestive apparatus of Dictyostelium discoideum during the starvation period triggering aggregation. *The Journal of Cell Biology*, *75*(1), 218–236.

Denton, D., Shravage, B., Simin, R., Mills, K., Berry, D. L., Baehrecke, E. H., et al. (2009). Autophagy, not apoptosis, is essential for midgut cell death in Drosophila. *Current Biology: CB*, *19*(20), 1741–1746. http://dx.doi.org/10.1016/j.cub.2009.08.042.

Denton, D., Xu, T., & Kumar, S. (2015). Autophagy as a pro-death pathway. *Immunology and Cell Biology*, *93*(1), 35–42. http://dx.doi.org/10.1038/icb.2014.85.

Deosaran, E., Larsen, K. B., Hua, R., Sargent, G., Wang, Y., Kim, S., et al. (2013). NBR1 acts as an autophagy receptor for peroxisomes. *Journal of Cell Science*, *126*(Pt. 4), 939–952. http://dx.doi.org/10.1242/jcs.114819.

Elgendy, M., Sheridan, C., Brumatti, G., & Martin, S. J. (2011). Oncogenic Ras-induced expression of Noxa and Beclin-1 promotes autophagic cell death and limits clonogenic survival. *Molecular Cell*, *42*(1), 23–35. http://dx.doi.org/10.1016/j.molcel.2011.02.009.

Filimonenko, M., Isakson, P., Finley, K. D., Anderson, M., Jeong, H., Melia, T. J., et al. (2010). The selective macroautophagic degradation of aggregated proteins requires the PI3P-binding protein Alfy. *Molecular Cell*, *38*(2), 265–279. http://dx.doi.org/10.1016/j.molcel.2010.04.007.

Fimia, G. M., Stoykova, A., Romagnoli, A., Giunta, L., Di Bartolomeo, S., Nardacci, R., et al. (2007). Ambra1 regulates autophagy and development of the nervous system. *Nature*, *447*(7148), 1121–1125. http://dx.doi.org/10.1038/nature05925.

Fraser, H. M., Lunn, S. F., Harrison, D. J., & Kerr, J. B. (1999). Luteal regression in the primate: Different forms of cell death during naturaland gonadotropin-releasing hormone antagonist or prostaglandin analogue-induced luteolysis. *Biology of Reproduction*, *61*(6), 1468–1479.

Gajewska, M., Sobolewska, A., Kozlowski, M., & Motyl, T. (2008). Role of autophagy in mammary gland development. *Journal of Physiology and Pharmacology: An Official Journal of the Polish Physiological Society*, *59*(Suppl. 9), 237–249.

Galluzzi, L., Vitale, I., Abrams, J. M., Alnemri, E. S., Baehrecke, E. H., Blagosklonny, M. V., et al. (2012). Molecular definitions of cell death subroutines: Recommendations of the Nomenclature Committee on Cell Death 2012. *Cell Death and Differentiation*, *19*(1), 107–120. http://dx.doi.org/10.1038/cdd.2011.96.

Gan, B., Peng, X., Nagy, T., Alcaraz, A., Gu, H., & Guan, J.-L. (2006). Role of FIP200 in cardiac and liver development and its regulation of TNFalpha and TSC-mTOR signaling pathways. *The Journal of Cell Biology*, *175*(1), 121–133. http://dx.doi.org/10.1083/jcb.200604129.

Geisler, S., Holmström, K. M., Skujat, D., Fiesel, F. C., Rothfuss, O. C., Kahle, P. J., et al. (2010). PINK1/Parkin-mediated mitophagy is dependent on VDAC1 and p62/SQSTM1. *Nature Cell Biology*, *12*(2), 119–131. http://dx.doi.org/10.1038/ncb2012.

Gewirtz, D. A. (2014). When cytoprotective autophagy isn't... and even when it is. *Autophagy*, *10*(3), 391–392. http://dx.doi.org/10.4161/auto.27719.

Gwinn, D. M., Shackelford, D. B., Egan, D. F., Mihaylova, M. M., Mery, A., Vasquez, D. S., et al. (2008). AMPK phosphorylation of raptor mediates a metabolic checkpoint. *Molecular Cell*, *30*(2), 214–226. http://dx.doi.org/10.1016/j.molcel.2008.03.003.

Hanada, T., Noda, N. N., Satomi, Y., Ichimura, Y., Fujioka, Y., Takao, T., et al. (2007). The Atg12-Atg5 conjugate has a novel E3-like activity for protein lipidation in autophagy. *The Journal of Biological Chemistry*, *282*(52), 37298–37302. http://dx.doi.org/10.1074/jbc.C700195200.

Hanna, R. A., Quinsay, M. N., Orogo, A. M., Giang, K., Rikka, S., & Gustafsson, Å. B. (2012). Microtubule-associated protein 1 light chain 3 (LC3) interacts with Bnip3 protein to selectively remove endoplasmic reticulum and mitochondria via autophagy. *The Journal of Biological Chemistry*, *287*(23), 19094–19104. http://dx.doi.org/10.1074/jbc.M111.322933.

Hara, T., Nakamura, K., Matsui, M., Yamamoto, A., Nakahara, Y., Suzuki-Migishima, R., et al. (2006). Suppression of basal autophagy in neural cells causes neurodegenerative disease in mice. *Nature*, *441*(7095), 885–889. http://dx.doi.org/10.1038/nature04724.

He, M.-X., & He, Y.-W. (2013). A role for c-FLIP(L) in the regulation of apoptosis, autophagy, and necroptosis in T lymphocytes. *Cell Death and Differentiation*, *20*(2), 188–197. http://dx.doi.org/10.1038/cdd.2012.148.

Hengartner, M. O., & Horvitz, H. R. (1994). Programmed cell death in Caenorhabditis elegans. *Current Opinion in Genetics & Development*, *4*(4), 581–586.

Irvine, G. B., El-Agnaf, O. M., Shankar, G. M., & Walsh, D. M. (2008). Protein aggregation in the brain: The molecular basis for Alzheimer's and Parkinson's diseases. *Molecular Medicine*, *14*(7–8), 451–464. http://dx.doi.org/10.2119/2007-00100.Irvine.

Itakura, E., & Mizushima, N. (2011). p62 Targeting to the autophagosome formation site requires self-oligomerization but not LC3 binding. *The Journal of Cell Biology*, *192*(1), 17–27. http://dx.doi.org/10.1083/jcb.201009067.

Juhász, G., Hill, J. H., Yan, Y., Sass, M., Baehrecke, E. H., Backer, J. M., et al. (2008). The class III PI(3)K Vps34 promotes autophagy and endocytosis but not TOR signaling in Drosophila. *The Journal of Cell Biology*, *181*(4), 655–666. http://dx.doi.org/10.1083/jcb.200712051.

Kamada, Y., Funakoshi, T., Shintani, T., Nagano, K., Ohsumi, M., & Ohsumi, Y. (2000). Tor-mediated induction of autophagy via an Apg1 protein kinase complex. *The Journal of Cell Biology*, *150*(6), 1507–1513.

Kanki, T., Wang, K., Baba, M., Bartholomew, C. R., Lynch-Day, M. A., Du, Z., et al. (2009). A genomic screen for yeast mutants defective in selective mitochondria autophagy. *Molecular Biology of the Cell*, *20*(22), 4730–4738. http://dx.doi.org/10.1091/mbc.E09-03-0225.

Kanki, T., Wang, K., Cao, Y., Baba, M., & Klionsky, D. J. (2009). Atg32 is a mitochondrial protein that confers selectivity during mitophagy. *Developmental Cell*, *17*(1), 98–109. http://dx.doi.org/10.1016/j.devcel.2009.06.014.

Kawaguchi, Y., Kovacs, J. J., McLaurin, A., Vance, J. M., Ito, A., & Yao, T. P. (2003). The deacetylase HDAC6 regulates aggresome formation and cell viability in response to misfolded protein stress. *Cell*, *115*(6), 727–738.

Kessin, R. H. (2001). *Dictyostelium:* Evolution, cell biology and the development of multicellularity. Cambridge, UK: Cambridge University Press.

Kim, P. K., Hailey, D. W., Mullen, R. T., & Lippincott-Schwartz, J. (2008). Ubiquitin signals autophagic degradation of cytosolic proteins and peroxisomes. *Proceedings of the National Academy of Sciences of the United States of America*, *105*(52), 20567–20574. http://dx.doi.org/10.1073/pnas.0810611105.

Kim, J., Kundu, M., Viollet, B., & Guan, K.-L. (2011). AMPK and mTOR regulate autophagy through direct phosphorylation of Ulk1. *Nature Cell Biology*, *13*(2), 132–141. http://dx.doi.org/10.1038/ncb2152.

Kirkin, V., Lamark, T., Johansen, T., & Dikic, I. (2009). NBR1 cooperates with p62 in selective autophagy of ubiquitinated targets. *Autophagy*, *5*(5), 732–733.

Kitada, T., Asakawa, S., Hattori, N., Matsumine, H., Yamamura, Y., Minoshima, S., et al. (1998). Mutations in the parkin gene cause autosomal recessive juvenile parkinsonism. *Nature*, *392*(6676), 605–608. http://dx.doi.org/10.1038/33416.

Kitanaka, C., Kato, K., Ijiri, R., Sakurada, K., Tomiyama, A., Noguchi, K., et al. (2002). Increased Ras expression and caspase-independent neuroblastoma cell death: Possible

mechanism of spontaneous neuroblastoma regression. *Journal of the National Cancer Institute, 94*(5), 358–368.
Komatsu, M., Waguri, S., Ueno, T., Iwata, J., Murata, S., Tanida, I., et al. (2005). Impairment of starvation-induced and constitutive autophagy in Atg7-deficient mice. *The Journal of Cell Biology, 169*(3), 425–434. http://dx.doi.org/10.1083/jcb.200412022.
Kondo, Y., Kanzawa, T., Sawaya, R., & Kondo, S. (2005). The role of autophagy in cancer development and response to therapy. *Nature Reviews. Cancer, 5*(9), 726–734. http://dx.doi.org/10.1038/nrc1692.
Korac, J., Schaeffer, V., Kovacevic, I., Clement, A. M., Jungblut, B., Behl, C., et al. (2013). Ubiquitin-independent function of optineurin in autophagic clearance of protein aggregates. *Journal of Cell Science, 126*(Pt. 2), 580–592. http://dx.doi.org/10.1242/jcs.114926.
Kosta, A., Roisin-Bouffay, C., Luciani, M.-F., Otto, G. P., Kessin, R. H., & Golstein, P. (2004). Autophagy gene disruption reveals a non-vacuolar cell death pathway in Dictyostelium. *The Journal of Biological Chemistry, 279*(46), 48404–48409. http://dx.doi.org/10.1074/jbc.M408924200.
Kuma, A., Hatano, M., Matsui, M., Yamamoto, A., Nakaya, H., Yoshimori, T., et al. (2004). The role of autophagy during the early neonatal starvation period. *Nature, 432*(7020), 1032–1036. http://dx.doi.org/10.1038/nature03029.
Lalaoui, N., Lindqvist, L. M., Sandow, J. J., & Ekert, P. G. (2015). The molecular relationships between apoptosis, autophagy and necroptosis. *Seminars in Cell & Developmental Biology, 39*, 63–69. http://dx.doi.org/10.1016/j.semcdb.2015.02.003.
Lamark, T., & Johansen, T. (2012). Aggrephagy: Selective disposal of protein aggregates by macroautophagy. *International Journal of Cell Biology, 2012*(12), 736905–736921. http://dx.doi.org/10.1155/2012/736905.
Lamark, T., Kirkin, V., Dikic, I., & Johansen, T. (2009). NBR1 and p62 as cargo receptors for selective autophagy of ubiquitinated targets. *Cell Cycle, 8*(13), 1986–1990.
Lee, C. Y., & Baehrecke, E. H. (2001). Steroid regulation of autophagic programmed cell death during development. *Development (Cambridge, England), 128*(8), 1443–1455.
Lee, C.-Y., Cooksey, B. A. K., & Baehrecke, E. H. (2002). Steroid regulation of midgut cell death during Drosophila development. *Developmental Biology, 250*(1), 101–111.
Lee, J.-Y., Koga, H., Kawaguchi, Y., Tang, W., Wong, E., Gao, Y.-S., et al. (2010). HDAC6 controls autophagosome maturation essential for ubiquitin-selective quality-control autophagy. *The EMBO Journal, 29*(5), 969–980. http://dx.doi.org/10.1038/emboj.2009.405.
Levraud, J.-P., Adam, M., Luciani, M.-F., de Chastellier, C., Blanton, R. L., & Golstein, P. (2003). Dictyostelium cell death: Early emergence and demise of highly polarized paddle cells. *The Journal of Cell Biology, 160*(7), 1105–1114. http://dx.doi.org/10.1083/jcb.200212104.
Liu, L., Feng, D., Chen, G., Chen, M., Zheng, Q., Song, P., et al. (2012). Mitochondrial outer-membrane protein FUNDC1 mediates hypoxia-induced mitophagy in mammalian cells. *Nature Cell Biology, 14*(2), 177–185. http://dx.doi.org/10.1038/ncb2422.
Liu, Y., Shoji-Kawata, S., Sumpter, R. M., Wei, Y., Ginet, V., Zhang, L., et al. (2013). Autosis is a Na+, K+-ATPase-regulated form of cell death triggered by autophagy-inducing peptides, starvation, and hypoxia-ischemia. *Proceedings of the National Academy of Sciences of the United States of America, 110*(51), 20364–20371. http://dx.doi.org/10.1073/pnas.1319661110.
Martin, D. N., & Baehrecke, E. H. (2004). Caspases function in autophagic programmed cell death in Drosophila. *Development (Cambridge, England), 131*(2), 275–284. http://dx.doi.org/10.1242/dev.00933.
Matsuura, A., Tsukada, M., Wada, Y., & Ohsumi, Y. (1997). Apg1p, a novel protein kinase required for the autophagic process in Saccharomyces cerevisiae. *Gene, 192*(2), 245–250.

McPhee, C. K., & Baehrecke, E. H. (2010). The engulfment receptor Draper is required for autophagy during cell death. *Autophagy*, *6*(8), 1192–1193. http://dx.doi.org/10.4161/auto.6.8.13474.

Meléndez, A., Tallóczy, Z., Seaman, M., Eskelinen, E.-L., Hall, D. H., & Levine, B. (2003). Autophagy genes are essential for dauer development and life-span extension in C. elegans. *Science (New York, NY)*, *301*(5638), 1387–1391. http://dx.doi.org/10.1126/science.1087782.

Miller, B. C., Zhao, Z., Stephenson, L. M., Cadwell, K., Pua, H. H., Lee, H. K., et al. (2008). The autophagy gene ATG5 plays an essential role in B lymphocyte development. *Autophagy*, *4*(3), 309–314.

Mizushima, N., & Levine, B. (2010). Autophagy in mammalian development and differentiation. *Nature Cell Biology*, *12*(9), 823–830. http://dx.doi.org/10.1038/ncb0910-823.

Monks, J., & Henson, P. M. (2009). Differentiation of the mammary epithelial cell during involution: Implications for breast cancer. *Journal of Mammary Gland Biology and Neoplasia*, *14*(2), 159–170. http://dx.doi.org/10.1007/s10911-009-9121-0.

Moretti, L., Yang, E. S., Kim, K. W., & Lu, B. (2007). Autophagy signaling in cancer and its potential as novel target to improve anticancer therapy. *Drug Resistance Updates: Reviews and Commentaries in Antimicrobial and Anticancer Chemotherapy*, *10*(4–5), 135–143. http://dx.doi.org/10.1016/j.drup.2007.05.001.

Mortensen, M., & Simon, A. K. (2010). Nonredundant role of Atg7 in mitochondrial clearance during erythroid development. *Autophagy*, *6*(3), 423–425.

Motley, A. M., Nuttall, J. M., & Hettema, E. H. (2012). Pex3-anchored Atg36 tags peroxisomes for degradation in Saccharomyces cerevisiae. *The EMBO Journal*, *31*(13), 2852–2868. http://dx.doi.org/10.1038/emboj.2012.151.

Motyl, T., Gajewska, M., Zarzynska, J., Sobolewska, A., & Gajkowska, B. (2007). Regulation of autophagy in bovine mammary epithelial cells. *Autophagy*, *3*(5), 484–486.

Nakatogawa, H., Ichimura, Y., & Ohsumi, Y. (2007). Atg8, a ubiquitin-like protein required for autophagosome formation, mediates membrane tethering and hemifusion. *Cell*, *130*(1), 165–178. http://dx.doi.org/10.1016/j.cell.2007.05.021.

Narendra, D., Kane, L. A., Hauser, D. N., Fearnley, I. M., & Youle, R. J. (2010). p62/SQSTM1 is required for Parkin-induced mitochondrial clustering but not mitophagy; VDAC1 is dispensable for both. *Autophagy*, *6*(8), 1090–1106.

Nelson, C., Ambros, V., & Baehrecke, E. H. (2014). miR-14 regulates autophagy during developmental cell death by targeting ip3-kinase 2. *Molecular Cell*, *56*(3), 376–388. http://dx.doi.org/10.1016/j.molcel.2014.09.011.

Noda, T., & Ohsumi, Y. (1998). Tor, a phosphatidylinositol kinase homologue, controls autophagy in yeast. *The Journal of Biological Chemistry*, *273*(7), 3963–3966.

Novak, I., & Dikic, I. (2011). Autophagy receptors in developmental clearance of mitochondria. *Autophagy*, 7(3), 301–303.

Novak, I., Kirkin, V., McEwan, D. G., Zhang, J., Wild, P., Rozenknop, A., et al. (2010). Nix is a selective autophagy receptor for mitochondrial clearance. *EMBO Reports*, *11*(1), 45–51. http://dx.doi.org/10.1038/embor.2009.256.

Okamoto, K. (2014). Organellophagy: Eliminating cellular building blocks via selective autophagy. *The Journal of Cell Biology*, *205*(4), 435–445. http://dx.doi.org/10.1083/jcb.201402054.

Okamoto, K., Kondo-Okamoto, N., & Ohsumi, Y. (2009). A landmark protein essential for mitophagy: Atg32 recruits the autophagic machinery to mitochondria. *Autophagy*, *5*(8), 1203–1205.

Okatsu, K., Saisho, K., Shimanuki, M., Nakada, K., Shitara, H., Sou, Y.-S., et al. (2010). p62/SQSTM1 cooperates with Parkin for perinuclear clustering of depolarized mitochondria. *Genes to Cells: Devoted to Molecular & Cellular Mechanisms*, *15*(8), 887–900. http://dx.doi.org/10.1111/j.1365-2443.2010.01426.x.

Oku, M., & Sakai, Y. (2010). Peroxisomes as dynamic organelles: Autophagic degradation. *The FEBS Journal*, *277*(16), 3289–3294. http://dx.doi.org/10.1111/j.1742-4658.2010.07741.x.

Otto, G. P., Wu, M. Y., Kazgan, N., Anderson, O. R., & Kessin, R. H. (2003). Macroautophagy is required for multicellular development of the social amoeba Dictyostelium discoideum. *The Journal of Biological Chemistry*, *278*(20), 17636–17645. http://dx.doi.org/10.1074/jbc.M212467200.

Pankiv, S., Clausen, T. H., Lamark, T., Brech, A., Bruun, J.-A., Outzen, H., et al. (2007). p62/SQSTM1 binds directly to Atg8/LC3 to facilitate degradation of ubiquitinated protein aggregates by autophagy. *The Journal of Biological Chemistry*, *282*(33), 24131–24145. http://dx.doi.org/10.1074/jbc.M702824200.

Papinski, D., Schuschnig, M., Reiter, W., Wilhelm, L., Barnes, C. A., Maiolica, A., et al. (2014). Early steps in autophagy depend on direct phosphorylation of Atg9 by the Atg1 kinase. *Molecular Cell*, *53*(3), 471–483. http://dx.doi.org/10.1016/j.molcel.2013.12.011.

Pua, H. H., Dzhagalov, I., Chuck, M., Mizushima, N., & He, Y.-W. (2007). A critical role for the autophagy gene Atg5 in T cell survival and proliferation. *The Journal of Experimental Medicine*, *204*(1), 25–31. http://dx.doi.org/10.1084/jem.20061303.

Qu, X., Zou, Z., Sun, Q., Luby-Phelps, K., Cheng, P., Hogan, R. N., et al. (2007). Autophagy gene-dependent clearance of apoptotic cells during embryonic development. *Cell*, *128*(5), 931–946. http://dx.doi.org/10.1016/j.cell.2006.12.044.

Reef, S., Zalckvar, E., Shifman, O., Bialik, S., Sabanay, H., Oren, M., et al. (2006). A short mitochondrial form of p19ARF induces autophagy and caspase-independent cell death. *Molecular Cell*, *22*(4), 463–475. http://dx.doi.org/10.1016/j.molcel.2006.04.014.

Riddle, D. L., Blumenthal, T., Meyer, B. J., Priess, J. R., & Albert, P. S. (1997). *Genetic and environmental regulation of dauer larva development* (2nd ed.). Cold Spring Harbor, NY: Cold Spring Harbor Laboratory Press.

Roggo, L., Bernard, V., Kovács, A. L., Rose, A. M., Savoy, F., Zetka, M., et al. (2002). Membrane transport in Caenorhabditis elegans: An essential role for VPS34 at the nuclear membrane. *The EMBO Journal*, *21*(7), 1673–1683. http://dx.doi.org/10.1093/emboj/21.7.1673.

Ruck, A., Attonito, J., Garces, K. T., Núnez, L., Palmisano, N. J., Rubel, Z., et al. (2011). The Atg6/Vps30/Beclin 1 ortholog BEC-1 mediates endocytic retrograde transport in addition to autophagy in C. elegans. *Autophagy*, 7(4), 386–400.

Rusten, T. E., Lindmo, K., Juhász, G., Sass, M., Seglen, P. O., Brech, A., et al. (2004). Programmed autophagy in the Drosophila fat body is induced by ecdysone through regulation of the PI3K pathway. *Developmental Cell*, 7(2), 179–192. http://dx.doi.org/10.1016/j.devcel.2004.07.005.

Saitoh, T., Fujita, N., Hayashi, T., Takahara, K., Satoh, T., Lee, H., et al. (2009). Atg9a controls dsDNA-driven dynamic translocation of STING and the innate immune response. *Proceedings of the National Academy of Sciences of the United States of America*, *106*(49), 20842–20846. http://dx.doi.org/10.1073/pnas.0911267106.

Saitoh, T., Fujita, N., Jang, M. H., Uematsu, S., Yang, B.-G., Satoh, T., et al. (2008). Loss of the autophagy protein Atg16L1 enhances endotoxin-induced IL-1beta production. *Nature*, *456*(7219), 264–268. http://dx.doi.org/10.1038/nature07383.

Sandoval, H., Thiagarajan, P., Dasgupta, S. K., Schumacher, A., Prchal, J. T., Chen, M., et al. (2008). Essential role for Nix in autophagic maturation of erythroid cells. *Nature*, *454*(7201), 232–235. http://dx.doi.org/10.1038/nature07006.

Sarraf, S. A., Raman, M., Guarani-Pereira, V., Sowa, M. E., Huttlin, E. L., Gygi, S. P., et al. (2013). Landscape of the PARKIN-dependent ubiquitylome in response to mitochondrial depolarization. *Nature*, *496*(7445), 372–376. http://dx.doi.org/10.1038/nature12043.

Schweers, R. L., Zhang, J., Randall, M. S., Loyd, M. R., Li, W., Dorsey, F. C., et al. (2007). NIX is required for programmed mitochondrial clearance during reticulocyte maturation. *Proceedings of the National Academy of Sciences of the United States of America*, *104*(49), 19500–19505. http://dx.doi.org/10.1073/pnas.0708818104.

Scott, R. C., Juhász, G., & Neufeld, T. P. (2007). Direct induction of autophagy by Atg1 inhibits cell growth and induces apoptotic cell death. *Current Biology: CB*, *17*(1), 1–11. http://dx.doi.org/10.1016/j.cub.2006.10.053.

Scott, R. C., Schuldiner, O., & Neufeld, T. P. (2004). Role and regulation of starvation-induced autophagy in the Drosophila fat body. *Developmental Cell*, *7*(2), 167–178. http://dx.doi.org/10.1016/j.devcel.2004.07.009.

Sharma, K., Le, N., Alotaibi, M., & Gewirtz, D. A. (2014). Cytotoxic autophagy in cancer therapy. *International Journal of Molecular Sciences*, *15*(6), 10034–10051. http://dx.doi.org/10.3390/ijms150610034.

Shimizu, S., Kanaseki, T., Mizushima, N., Mizuta, T., Arakawa-Kobayashi, S., Thompson, C. B., et al. (2004). Role of Bcl-2 family proteins in a non-apoptotic programmed cell death dependent on autophagy genes. *Nature Cell Biology*, *6*(12), 1221–1228. http://dx.doi.org/10.1038/ncb1192.

Simonsen, A., Birkeland, H. C. G., Gillooly, D. J., Mizushima, N., Kuma, A., Yoshimori, T., et al. (2004). Alfy, a novel FYVE-domain-containing protein associated with protein granules and autophagic membranes. *Journal of Cell Science*, *117*(Pt. 18), 4239–4251. http://dx.doi.org/10.1242/jcs.01287.

Sou, Y.-S., Waguri, S., Iwata, J.-I., Ueno, T., Fujimura, T., Hara, T., et al. (2008). The Atg8 conjugation system is indispensable for proper development of autophagic isolation membranes in mice. *Molecular Biology of the Cell*, *19*(11), 4762–4775. http://dx.doi.org/10.1091/mbc.E08-03-0309.

Stocco, C., Telleria, C., & Gibori, G. (2007). The molecular control of corpus luteum formation, function, and regression. *Endocrine Reviews*, *28*(1), 117–149. http://dx.doi.org/10.1210/er.2006-0022.

Strappazzon, F., Nazio, F., Corrado, M., Cianfanelli, V., Romagnoli, A., Fimia, G. M., et al. (2015). AMBRA1 is able to induce mitophagy via LC3 binding, regardless of PARKIN and p62/SQSTM1. *Cell Death and Differentiation*, *22*(3), 419–432. http://dx.doi.org/10.1038/cdd.2014.139.

Su, L., Quade, B., Wang, H., Sun, L., Wang, X., & Rizo, J. (2014). A plug release mechanism for membrane permeation by MLKL. *Structure*, *22*(10), 1489–1500. http://dx.doi.org/10.1016/j.str.2014.07.014.

Suzuki, K., Kirisako, T., Kamada, Y., Mizushima, N., Noda, T., & Ohsumi, Y. (2001). The pre-autophagosomal structure organized by concerted functions of APG genes is essential for autophagosome formation. *The EMBO Journal*, *20*(21), 5971–5981. http://dx.doi.org/10.1093/emboj/20.21.5971.

Takács-Vellai, K., Vellai, T., Puoti, A., Passannante, M., Wicky, C., Streit, A., et al. (2005). Inactivation of the autophagy gene bec-1 triggers apoptotic cell death in C. elegans. *Current Biology: CB*, *15*(16), 1513–1517. http://dx.doi.org/10.1016/j.cub.2005.07.035.

Tanaka, A., Cleland, M. M., Xu, S., Narendra, D. P., Suen, D.-F., Karbowski, M., et al. (2010). Proteasome and p97 mediate mitophagy and degradation of mitofusins induced by Parkin. *The Journal of Cell Biology*, *191*(7), 1367–1380. http://dx.doi.org/10.1083/jcb.201007013.

Tanaka, C., Tan, L.-J., Mochida, K., Kirisako, H., Koizumi, M., Asai, E., et al. (2014). Hrr25 triggers selective autophagy-related pathways by phosphorylating receptor proteins. *The Journal of Cell Biology*, *207*(1), 91–105. http://dx.doi.org/10.1083/jcb.201402128.

Tanida, I., Mizushima, N., Kiyooka, M., Ohsumi, M., Ueno, T., Ohsumi, Y., et al. (1999). Apg7p/Cvt2p: A novel protein-activating enzyme essential for autophagy. *Molecular Biology of the Cell*, *10*(5), 1367–1379.

Teplova, I., Lozy, F., Price, S., Singh, S., Barnard, N., Cardiff, R. D., et al. (2013). ATG proteins mediate efferocytosis and suppress inflammation in mammary involution. *Autophagy*, *9*(4), 459–475. http://dx.doi.org/10.4161/auto.23164.

Tomoda, T., Bhatt, R. S., Kuroyanagi, H., Shirasawa, T., & Hatten, M. E. (1999). A mouse serine/threonine kinase homologous to C. elegans UNC51 functions in parallel fiber formation of cerebellar granule neurons. *Neuron*, *24*(4), 833–846.

Town, C. D., Gross, J. D., & Kay, R. R. (1976). Cell differentiation without morphogenesis in Dictyostelium discoideum. *Nature*, *262*(5570), 717–719.

Tsukada, M., & Ohsumi, Y. (1993). Isolation and characterization of autophagy-defective mutants of Saccharomyces cerevisiae. *FEBS Letters*, *333*(1–2), 169–174.

Tsukamoto, S., Kuma, A., Murakami, M., Kishi, C., Yamamoto, A., & Mizushima, N. (2008). Autophagy is essential for preimplantation development of mouse embryos. *Science (New York, N.Y.)*, *321*(5885), 117–120. http://dx.doi.org/10.1126/science.1154822.

Valente, E. M., Abou-Sleiman, P. M., Caputo, V., Muqit, M. M. K., Harvey, K., Gispert, S., et al. (2004). Hereditary early-onset Parkinson's disease caused by mutations in PINK1. *Science (New York, N.Y.)*, *304*(5674), 1158–1160. http://dx.doi.org/10.1126/science.1096284.

Van Humbeeck, C., Cornelissen, T., Hofkens, H., Mandemakers, W., Gevaert, K., De Strooper, B., et al. (2011). Parkin interacts with Ambra1 to induce mitophagy. *The Journal of Neuroscience: The Official Journal of the Society for Neuroscience*, *31*(28), 10249–10261. http://dx.doi.org/10.1523/JNEUROSCI.1917-11.2011.

Wang, H., Lu, Q., Cheng, S., Wang, X., & Zhang, H. (2013). Autophagy activity contributes to programmed cell death in Caenorhabditis elegans. *Autophagy*, *9*(12), 1975–1982.

Wang, H., Sun, L., Su, L., Rizo, J., Liu, L., Wang, L.-F., et al. (2014). Mixed lineage kinase domain-like protein MLKL causes necrotic membrane disruption upon phosphorylation by RIP3. *Molecular Cell*, *54*(1), 133–146. http://dx.doi.org/10.1016/j.molcel.2014.03.003.

Watanabe, Y., & Tanaka, M. (2011). p62/SQSTM1 in autophagic clearance of a non-ubiquitylated substrate. *Journal of Cell Science*, *124*(Pt. 16), 2692–2701. http://dx.doi.org/10.1242/jcs.081232.

Weidberg, H., Shvets, E., Shpilka, T., Shimron, F., Shinder, V., & Elazar, Z. (2010). LC3 and GATE-16/GABARAP subfamilies are both essential yet act differently in autophagosome biogenesis. *The EMBO Journal*, *29*(11), 1792–1802. http://dx.doi.org/10.1038/emboj.2010.74.

Whittingham, W. F., & Raper, K. B. (1960). Non-viability of stalk cells in Dictyostelium. *Proceedings of the National Academy of Sciences of the United States of America*, *46*(5), 642–649.

Wong, Y. C., & Holzbaur, E. L. F. (2014). Optineurin is an autophagy receptor for damaged mitochondria in parkin-mediated mitophagy that is disrupted by an ALS-linked mutation. *Proceedings of the National Academy of Sciences of the United States of America*, *111*(42), E4439–E4448. http://dx.doi.org/10.1073/pnas.1405752111.

Yu, L., Alva, A., Su, H., Dutt, P., Freundt, E., Welsh, S., et al. (2004). Regulation of an ATG7-beclin 1 program of autophagic cell death by caspase-8. *Science (New York, N.Y.)*, *304*(5676), 1500–1502. http://dx.doi.org/10.1126/science.1096645.

Yue, Z., Jin, S., Yang, C., Levine, A. J., & Heintz, N. (2003). Beclin 1, an autophagy gene essential for early embryonic development, is a haploinsufficient tumor suppressor. *Proceedings of the National Academy of Sciences of the United States of America*, *100*(25), 15077–15082. http://dx.doi.org/10.1073/pnas.2436255100.

Zarzyńska, J., Gajkowska, B., Wojewódzka, U., Dymnicki, E., & Motyl, T. (2007). Apoptosis and autophagy in involuting bovine mammary gland is accompanied by up-regulation of TGF-beta1 and suppression of somatotropic pathway. *Polish Journal of Veterinary Sciences*, *10*(1), 1–9.

CHAPTER FOUR

The End of the Beginning: Cell Death in the Germline

Jeanne S. Peterson, Allison K. Timmons, Albert A. Mondragon, Kimberly McCall[1]
Department of Biology, Boston University, Boston, Massachusetts, USA
[1]Corresponding author: e-mail address: kmccall@bu.edu

Contents

Abstract

Programmed cell death occurs in the germline of many organisms, both as an essential part of development and throughout adult life. Germline cell death can be apoptotic or nonapoptotic, depending on the stimulus or stage of development. Here, we focus on the *Drosophila* ovary, which is a powerful model for studying diverse types of cell death. In *Drosophila*, the death of primordial germ cells occurs normally during embryonic development, and germline nurse cells are programmed to die during oocyte development in adult flies. Cell death of previtellogenic egg chambers in adults can also be induced by starvation or other environmental cues. Mid-oogenesis seems to be particularly sensitive to such cues and has been proposed to serve as a checkpoint to avoid the energetically expensive cost of egg production. After the germline dies in mid-oogenesis, the remnants are engulfed by an epithelial layer of follicle cells; thus, the fly ovary also serves as a highly tractable model for engulfment by epithelial cells. These examples of cell death in the fly ovary share many similarities to the types of cell death seen in the mammalian germline. Recent progress in elucidating the molecular mechanisms of cell death in the germline is discussed.

Current Topics in Developmental Biology, Volume 114
ISSN 0070-2153
http://dx.doi.org/10.1016/bs.ctdb.2015.07.025

1. OVERVIEW OF CELL DEATH PATHWAYS

Programmed cell death (PCD) refers to any cell death that is controlled or regulated (Fuchs & Steller, 2011; Jacobson, Weil, & Raff, 1997). The cell death process functions to remove excess or damaged cells during the development of an organism or to maintain homeostasis in adult tissues. Importantly, misregulation of PCD has been implicated in human diseases including cancer, neurodegeneration, and autoimmune diseases (Elliott & Ravichandran, 2010a; Fuchs & Steller, 2011). The three most familiar forms of PCD are apoptosis, autophagic cell death, and necrosis. However, there are many more forms of cell death that have been described and that are less well understood (Galluzzi et al., 2012; Kroemer et al., 2009).

The main features of apoptosis are nuclear condensation and fragmentation, a reduction in cellular volume, and plasma membrane blebbing (Kerr, Wyllie, & Currie, 1972; Kroemer et al., 2009). A wide range of physiological and environmental stimuli can induce apoptosis including oxidative stress, DNA damage, starvation, and developmental signals (Galluzzi et al., 2012). Since it was first described by Kerr et al. (1972), apoptosis has been a significant research focus in laboratories worldwide.

An important step in apoptosis is the activation of caspases, a family of cysteine aspartyl proteases that execute apoptosis by cleaving proteins at specific sequences to dismantle and demolish the cell. Caspases are ubiquitously present as proenzymes with little to no protease activity. There are seven caspases in *Drosophila*, and their activity is tightly controlled by the inhibitor of apoptosis protein (IAP), Diap1. In healthy cells, IAPs directly bind and inhibit caspases. Once a cell has committed to apoptosis, IAPs are inactivated by IAP antagonists (such as Reaper, Hid, and Grim (RHG) in *Drosophila*) that are induced in response to death signals. When inhibition of caspases by IAPs is relieved, they become activated via cleavage at specific aspartic residues. Initiator caspases activate effector caspases, leading to the destruction of the cell.

Autophagy is an evolutionarily conserved intracellular degradation system that delivers cellular contents to the lysosomes. Under normal conditions, basal levels of autophagy allow for turnover of old or damaged organelles and proteins. Double membrane structures surround cellular contents, forming an autophagosome. The autophagosome fuses with lysosomes, forming an autolysosome, where its contents are degraded by lysosomal enzymes (Baehrecke, 2005; Kroemer et al., 2009; Nelson & Baehrecke, 2014). In response to stress (such as nutrient deprivation), the activation of autophagy allows cells to consume itself for energy, increasing the likelihood of survival. Therefore, a major role for autophagy is to promote cell survival (Baehrecke, 2005; Mevorach et al., 2010).

Interestingly, autophagy has also been shown to promote cell death. For example, autophagic cell death occurs in the *Drosophila* midgut and salivary gland during the transition from larval to pupal stages of development (Berry & Baehrecke, 2007; Denton et al., 2009). Autophagic cell death and apoptosis function in parallel to promote death of the larval salivary gland (Berry & Baehrecke, 2007; Nelson & Baehrecke, 2014). Removal of cells that die via autophagic cell death is thought to occur independently of phagocytosis (Kroemer et al., 2009; Nelson & Baehrecke, 2014). Compared to autophagy, the mechanisms that promote autophagic cell death are less understood. Overall, autophagy is a dynamic process that promotes cellular survival and cell death during animal development, tissue homeostasis, and disease.

Necrosis, characterized by cytoplasmic and organelle swelling, and plasma membrane rupture, was once thought to be cell death due to damage and not regulated or controlled like apoptosis (de Almagro & Vucic, 2015; Golstein & Kroemer, 2007; Kerr et al., 1972; Kroemer et al., 2009; McCall, 2010; Sun & Wang, 2014). However, recent work has shown that necrosis proceeds through a series of morphological, biochemical, and molecular events, indicating that necrosis can be highly regulated (Golstein & Kroemer, 2007; Kroemer et al., 2009; McCall, 2010).

Cells undergoing necrosis progress through a general series of morphological events. The first indications of necrosis are organelle swelling and the appearance of clumped chromatin in the nucleus, followed by cellular swelling and rupture of the plasma membrane. In addition to the morphological features that define necrosis, there are several intracellular events associated with necrotic cell death including increased reactive oxygen species, ATP depletion, and increased cytosolic Ca^{2+} (Festjens, Vanden Berghe, & Vandenabeele, 2006; Golstein & Kroemer, 2007; McCall, 2010). In mammals, a specific form of necrosis called necroptosis or programmed necrosis has been recently elucidated. This form of necrosis is dependent on the kinases RIP1 and RIP3, and a pseudokinase MLKL (de Almagro & Vucic, 2015; Sun & Wang, 2014). Whether necroptosis is evolutionary conserved in invertebrates remains to be seen, and the chronology of biochemical events occurring in other forms of necrosis remains unclear. Although the word necrosis has been used to describe cell death since the 1800s, necrosis remains poorly understood. Since necrosis is implicated in many human diseases, including heart disease and neurodegenerative disorders, it is important to work toward a better understanding of the characteristics, events, and mechanisms of necrosis.

In addition to cell suicide pathways, cell death can be initiated non-autonomously. For example, Fas-mediated cell death is a well-characterized example of non-autonomous PCD in mammals that is important for cell death related to immunity. Fas ligand (FasL) activates the intrinsic apoptotic machinery of a target cell that expresses the Fas receptor (Ashkenazi & Dixit, 1998; Krammer, 1998). The FasL is a member of the tumor necrosis factor (TNF) superfamily that participates in diverse biological processes. A single TNF family member, Eiger, has been reported in *Drosophila*. Eiger has been shown to act as a long-range death signal released from apoptotic cells to induce apoptosis elsewhere in the same tissue (apoptosis-induced-apoptosis), demonstrating that apoptotic cells themselves can nonautonomously coordinate communal PCD (Perez-Garijo, Fuchs, & Steller, 2013).

Another intriguing example of non-cell-autonomous PCD is entosis (also referred to as "cell in cell" or "cell cannibalism"). During entosis, a viable cell becomes detached from the extracellular matrix and invades another cell, where it is degraded by lysosomes (Overholtzer et al., 2007). Occasionally, internalized cells remain viable and can divide inside the engulfing cell and/or be released (Galluzzi et al., 2012; Overholtzer et al., 2007). Interestingly, cell-in-cell structures are frequently observed in tumors and are proposed to act as tumor suppressors in some contexts and as tumor accelerators in others. Therefore, understanding the role of entosis in tumor biology could have therapeutic implications (Kroemer & Perfettini, 2014). A phenomenon related to entosis is phagoptosis, where a phagocytic cell engulfs an apparently healthy cell (Brown & Neher, 2012). In this case, the death of the cell is dependent on the phagocytic machinery of the engulfing cell. Phagoptosis has been described in *Caenorhabditis elegans* and mammals (Brown & Neher, 2012; Hoeppner, Hengartner, & Schnabel, 2001; Reddien, Cameron, & Horvitz, 2001).

The final step in PCD is removal of cell corpses via phagocytosis, also referred to as engulfment or efferocytosis (Fuchs & Steller, 2011; Taylor, Cullen, & Martin, 2008). Apoptotic cells that fail to be engulfed can progress to secondary necrosis. The release of intracellular contents from secondarily necrotic cells can result in an inflammatory response that leads to autoimmunity. Indeed, defective engulfment has been implicated in diseases including lupus, chronic obstructive pulmonary disease, and macular degeneration (Elliott & Ravichandran, 2010a; Nandrot, Chang, & Finnemann, 2008).

In mammals and *Drosophila*, engulfment of cell corpses is carried out by both professional phagocytes, such as macrophages, and nonprofessional phagocytes (Kinchen, 2010). For example, the *Drosophila* ovary is closed to circulating macrophages and engulfment of the dying germline is completed by neighboring epithelial follicle cells (Giorgi & Deri, 1976; King, 1970). In the mammalian mammary gland, milk-secreting alveolar cells undergo apoptosis following the cessation of lactation and are engulfed by mammary epithelial cells (Monks, Smith-Steinhart, Kruk, Fadok, & Henson, 2008). In another example of nonprofessional phagocytosis in mammals, the retinal pigment epithelial (RPE) cells engulf photoreceptor outer segments (POS) that are shed on a daily basis (Nandrot, 2014). Defects in engulfment of POS by RPE cells have been implicated in diseases such as retinitis pigmentosa and age-related macular degeneration.

Pioneering genetic studies in *C. elegans* identified two partially redundant signaling pathways that control engulfment: the *ced-1, 6, 7* and *ced-2, 5, 12* pathways (Fullard, Kale, & Baker, 2009; Kinchen, 2010; Mangahas & Zhou, 2005; Reddien & Horvitz, 2004). The *ced-1, 6, 7* and *ced-2, 5, 12* pathways act in parallel to promote the activation of CED-10, a Rac GTPase responsible for cytoskeletal rearrangements that allow for internalization of the cell corpse. In *Drosophila*, the roles of the *ced-1, 6, 7* and *ced-2, 5, 12* pathways appear to be generally conserved. The CED-1 ortholog, Draper, is a transmembrane protein that localizes to the surface of engulfing cells and acts as a receptor to recognize dying cells. Draper was first shown to be required for the engulfment of apoptotic neurons in the embryonic central nervous system with mutants displaying lingering cell corpses (Freeman, Delrow, Kim, Johnson, & Doe, 2003). Additionally, Draper has been shown to be important in several other contexts including the engulfment of severed axons, necrotic cells, bacteria, imaginal disc cells, and hemocytes (Cuttell et al., 2008; Li & Baker, 2007; MacDonald et al., 2006; Manaka et al., 2004). Draper activity has been shown to be dependent on the nonreceptor tyrosine kinase Shark, which binds Draper on its intracellular domain. The interaction between Shark and Draper requires phosphorylation of Draper by Src42A, which ultimately leads to downstream phagocytic signaling (Ziegenfuss et al., 2008).

The CED-2, 5, 12 pathway also has a conserved role in engulfment in *C. elegans*, *Drosophila*, and mammals. CED-12 (*Drosophila*: Ced-12, mammals: ELMO) and CED-5 (*Drosophila*: Myoblast City, mammals: Dock180) act as a guanine-nucleotide exchange factor to activate the GTPase, Rac (Kinchen & Ravichandran, 2007; Reddien & Horvitz, 2000). The upstream activators of the CED-2, 5, 12 pathway are largely unknown, although it has been proposed that integrins activate the pathway (Hsu & Wu, 2010).

Here, we describe the mechanisms of germ cell death, focusing on the *Drosophila* ovary. In the fly ovary, there are several distinct forms of germline cell death that occur, depending on the developmental stage and stimulus. The fly ovary also provides a tractable model for characterizing the final step of clearance by nonprofessional phagocytes. We provide a comparison to germ cell death in other species and in the testis. The study of germline cell death has high relevance for fertility in humans but also provides a valuable model for understanding the diversity of cell death mechanisms.

2. DEATH OF PRIMORDIAL GERM CELLS

Cell death of the germline occurs within the first few hours of development during embryogenesis. Germline cells form very early in embryogenesis, when a discrete group of cells at the posterior end of the syncytial embryo are the first to cellularize, forming the pole cells (Coffman, 2003; Starz-Gaiano & Lehmann, 2001). These cells are the primordial germ cells (PGCs). During formation of the hindgut, a gastrulation event that occurs at the posterior of the embryo, the pole cells become internalized into the hindgut. In later stages, the pole cells migrate out of the midgut to form two clusters of cells that associate with mesodermal cells to form the gonads (Fig. 1A–E). Cells that fail to migrate properly are eliminated by PCD (Coffman, 2003; Coffman et al., 2002).

The formation of pole cells and their incorporation into the gonad is under the control of maternally and zygotically expressed genes. Embryos produced by females mutant for *nanos* are able to form the normal number of pole cells, but most of these cells are not incorporated into the embryo and degenerate (Heller & Steinmann-Zwicky, 1998). Nanos normally acts to translationally repress *hid*, so *nanos* mutant pole cells produce Hid and die by apoptosis (Sato et al., 2007). Tao-1, a microtubule-associated serine–threonine protein kinase, and another RHG gene, *sickle*, are also required

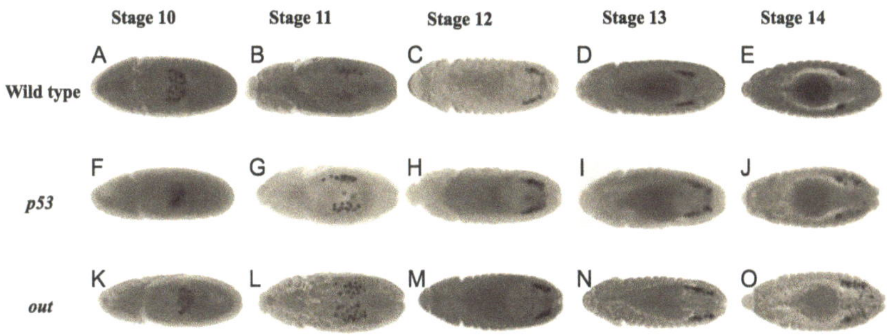

Figure 1 Migration of primordial germ cells in wild-type and *p53* and *outsiders* (*out*) mutant embryos. Anterior is to the left. Stages 10–14 are shown with anti-Vasa antibody staining to highlight the primordial germ cells (PGCs). At stage 10 (A, F, and K), the PGCs have moved out of the midgut and spread laterally. By stage 11 (B, G, and L), the cells have formed two groups on either side of the midline; however, in the mutants, the groups are not as tight as those seen in wild type, with an occasional PGC lingering near the midline. By stage 12 (C, H, and M), the clusters of PGCs appear more compact, but the mutants still have a few cells outside the clusters. In stage 13 in wild type (D), the clusters have become linear, and in the mutants (I and N), the clusters are mostly linear, but some of the cells are in separate groups. By stage 14 (E) in wild type, the clusters are small and tight, but in the mutants (J and O), there are several groups of ectopic cells persisting outside of the main clusters. *Reproduced with permission from Yamada, Davis, and Coffman (2008).*

for PCD in *nanos*-deficient pole cells. The RNAi knockdown of *tao-1* results in disordered migration of PGCs, which disperse within the embryo and undergo cell death (Pflanz, Voigt, Yakulov, & Jackle, 2015).

Interestingly, the death of ectopic PGCs in normal embryonic development occurs independently of the classic cell death genes, *reaper*, *hid*, and *grim*, as well as effector caspases (Sano, Renault, & Lehmann, 2005). However, several novel genes that affect PGC death have been identified. When embryos are mutant for *Tre1* (*scattershot*), the PGCs do not migrate and also do not undergo PCD (Coffman et al., 2002). *Tre1* is required in germ cells and encodes a G protein-coupled receptor (GPCR) (Kamps, Pruitt, Herriges, & Coffman, 2010; Kunwar & Lehmann, 2003). GPCRs have also been implicated in germ cell migration in vertebrates, suggesting evolutionary conservation of *Tre1* function. In embryos mutant for *p53* or *outsiders* (*out*), the PGCs do not undergo PCD and thus persist as ectopic cells in the embryo (Yamada et al., 2008; Fig. 1F–O). *p53* encodes a proapoptotic tumor suppressor gene, and *out* encodes a monocarboxylate transporter. *wunen* and *wunen2* encode lipid phosphate phosphatases that serve to regulate PGC migration and survival by mediating communication between somatic cells and the PGCs

(Hanyu-Nakamura, Kobayashi, & Nakamura, 2004; Sano et al., 2005). Thus, some players in embryonic germ cell death have been uncovered, but much remains to be learned about how this nonapoptotic cell death occurs.

3. GERMLINE CELL DEATH IN THE ADULT OVARY

3.1 Overview of the Ovary

Drosophila females have two ovaries which consist of bundles of 15–20 ovarioles, tubular structures containing the developing egg chambers. The ovarioles from each ovary converge on the lateral oviduct and the two lateral oviducts join the common oviduct (King, 1970; Spradling, 1993; Fig. 2A). Egg chambers originate from a specialized region at the anterior of the ovariole known as the germarium, which harbors germline and somatic stem cells and their niches (Figs. 2 and 3A). Egg chambers contain 16-cell germline cysts, comprised of a single oocyte and 15 nurse cells that support the oocyte throughout 14 stages of oogenesis (Figs. 2 and 3B). The nurse cells remain connected to the oocyte through ring canals that form as a result of incomplete cytokinesis. Each egg chamber also has approximately

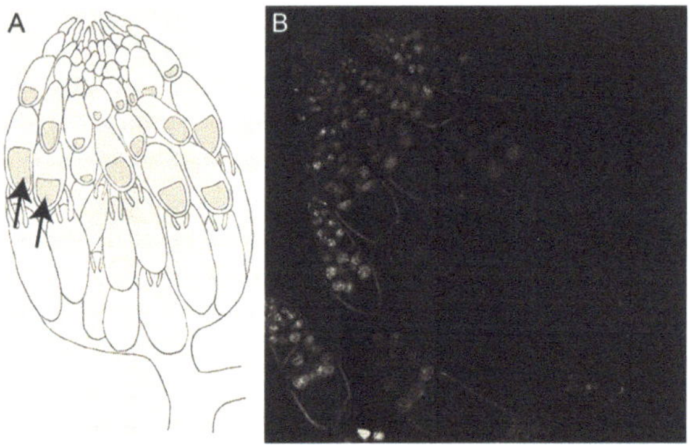

Figure 2 *Drosophila* ovaries. Anterior is toward the upper left in both images. (A) Drawing of a *Drosophila* ovary showing bundles of ovarioles converging on the lateral oviduct that leads into the common oviduct. Prominent in this drawing are the tips of the ovarioles containing the germaria where egg chambers begin development (top). As they continue developing, the oocytes in mid-stage egg chambers accumulate yolk (arrows), shown tinted gray in this drawing. Before exiting the ovary, late-stage egg chambers elongate and acquire dorsal appendages, long filaments at their anterior ends. (B) Fluorescent image of an ovary stained with DAPI to label the large nurse cell nuclei within the egg chambers and the smaller follicle cell nuclei surrounding the egg chambers. In the largest late-stage egg chambers, there are no longer any nurse cell nuclei visible since by then the nurse cells have degenerated.

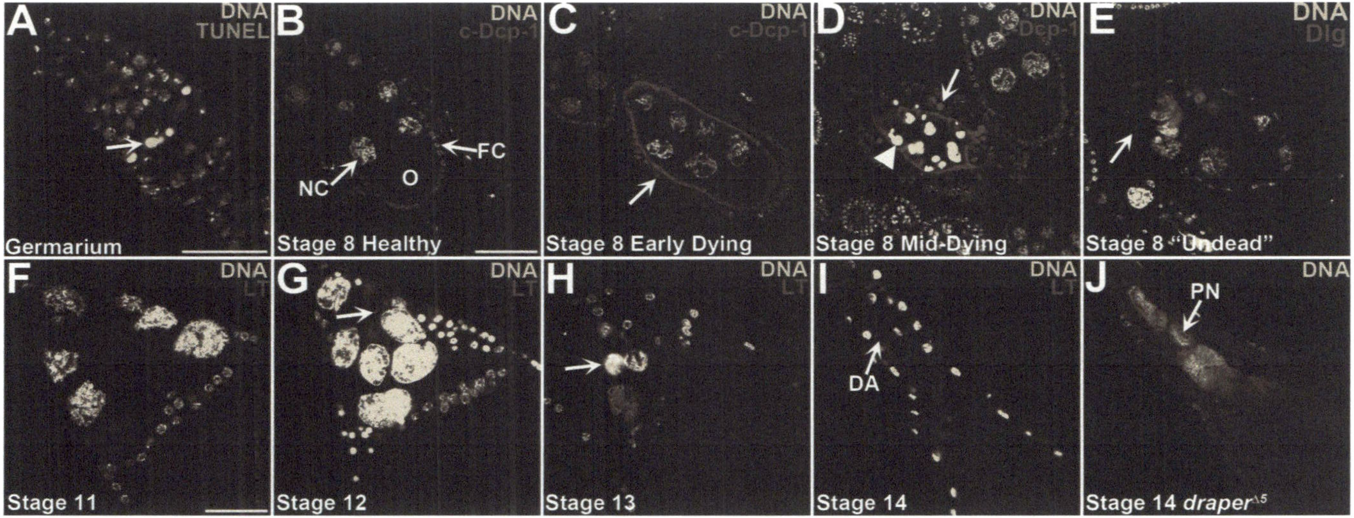

Figure 3 Overview of programmed cell death in the *Drosophila* ovary. (A) Wild-type (w^{1118}) germarium stained with DAPI (cyan) to label DNA and TUNEL (green arrow) to label fragmented DNA. Scale bar = 25 μm. (B–D) Control (*GR1-GAL4 mCD8-GFP/TM3*) stage 8 egg chambers from starved flies were stained with DAPI (cyan) and cleaved (active) Dcp-1 antibody (c-Dcp1, red). (B) Healthy egg chamber has no apparent cleaved Dcp-1 staining. Each egg chamber is comprised of three cell types: nurse cells (NC), follicle cells (FC), and the oocyte (O). (C) Cleaved Dcp-1 staining (red) appears in an early phase dying egg chamber and is present in vesicle structures in the engulfing follicle cells (arrow). (D) A mid-phase dying egg chamber has condensed and fragmented nurse cell nuclei (arrowhead) and stains with cleaved Dcp-1 (arrow). Engulfing follicle cells have enlarged. (E) *Diap1* was overexpressed in the germline (*nanos-GAL4* > *UASp-Diap1*), and ovaries from starved flies were stained with DAPI (cyan) and anti-Discs Large (Dlg, red) to label membranes. The overexpression of *Diap1* in the germline results in an "undead" stage 8 egg chamber, where the nurse cell nuclei have failed to condense or fragment and the follicle cells have largely disappeared (arrow). Scale bar = 50 μm for (B)–(E). (F–I) Wild-type (w^{1118}) stage 11–14 egg chambers were stained with LysoTracker (LT, red) to label acidified structures. (F) Stage 11 egg chamber has no LysoTracker staining. (G) Stage 12 egg chamber has LysoTracker puncta (arrow). (H) Stage 13 egg chamber has LysoTracker-positive nurse cell nuclei (arrow). (I) Stage 14 egg chamber is characterized by the presence of dorsal appendages (DA, arrow) and has no LysoTracker staining. (J) Stage 14 egg chamber from *draper*$^{\Delta 5}$ has persisting nurse cell nuclei (PN, arrow). Scale bar = 50 μm for (F)–(J). (See the color plate.)

800 somatically derived epithelial cells known as the follicle cells that surround each germline cyst.

The germline-derived nurse cells are large polyploid cells that provide nutrients, proteins, mRNAs, and organelles for the developing oocyte. During late oogenesis, the nurse cells "dump" their cytoplasmic contents into the oocyte through the ring canals (Cummings & King, 1970; Spradling, 1993). The somatically derived follicle cells are a dynamic population of cells responsible for yolk, vitelline membrane, and chorion synthesis (Jenkins, Timmons, & McCall, 2013; McCall, 2004; Pritchett, Tanner, & McCall, 2009; Wu, Tanwar, & Raftery, 2008). Furthermore, follicle cells are subdivided into distinct populations that perform specialized tasks throughout oogenesis (Wu et al., 2008). For example, the polar cells are specialized follicle cells that reside on the anterior and posterior poles of the egg chambers. Together with the migratory border cells, the polar cell/border cell cluster migrates from the anterior of the egg chamber to the nurse cell/oocyte interface. The polar and border cells sculpt the micropyle, the sperm entry structure essential for female fertility (Montell, Yoon, & Starz-Gaiano, 2012; Wu et al., 2008). At stage 9 of oogenesis, most of the follicle cells migrate to form a columnar epithelium that surrounds the oocyte. Concurrently, approximately 50 cells at the anterior of the egg chamber remain to cover the nurse cells and are called the stretch follicle cells (Horne-Badovinac & Bilder, 2005; Tran & Berg, 2003; Wu et al., 2008). Furthermore, follicle cells are responsible for the formation of the dorsal appendages (Fig. 2A), structures that facilitate gas exchange in the embryo (Berg, 2005). At the end of oogenesis, mature stage 14 egg chambers pass through the oviducts into the uterus, where they are fertilized.

The *Drosophila* ovary is a powerful model for the study of cell death. Cell death occurs at distinct stages throughout oogenesis: PCD occurs in response to environmental stress such as starvation in the germarium and during mid-oogenesis (Drummond-Barbosa & Spradling, 2001) and in response to developmental signals during late oogenesis (McCall, 2004; Pritchett et al., 2009). There are morphological features of all three major types of cell death (apoptosis, autophagic cell death, and necrosis), indicating that complex cell death mechanisms are at work in the ovary (Bass et al., 2009; Hou, Chittaranjan, Barbosa, McCall, & Gorski, 2008; Nezis et al., 2009; Pritchett et al., 2009). Furthermore, the *Drosophila* ovary is closed to circulating macrophages, thereby providing an ideal system to study engulfment via nonprofessional phagocytes (King, 1970).

3.2 Cell Death in the Germarium

The anterior tip of every ovariole is called the germarium (Fig. 3A), and it contains a population of stem cells, their niches, and early stage egg chambers. Region 1 of the germarium contains two germline stem cells, and between region 2a and 2b are two follicle stem cells (Spradling, 1993; Spradling et al., 2008). Oogenesis begins when a germline stem cell divides, producing a daughter cell that is guided by a somatic escort cell to move posteriorly as it divides 4 times to form a 16-cell cyst. As this cyst reaches region 2b, it is surrounded by new follicle cells produced by the follicle stem cells and thus becomes an egg chamber with a follicle cell layer and 16 germline cells, 1 which becomes the oocyte and 15 which will be nurse cells. When the egg chamber moves into region 3 of the germarium, it is called a stage 1 egg chamber.

In response to protein starvation, egg production is adjusted to conserve resources. This happens partly in mid-stage oogenesis (see Section 3.3) and partly in the germarium by slowing the rate of stem cell proliferation and inducing apoptosis of developing egg chambers (Drummond-Barbosa & Spradling, 2001). Death of a germline cyst occurs primarily at the border between region 2a and 2b, where follicle cells normally proliferate and surround the 16-cell cyst (Fig. 3A). Both of these effects, cell death and proliferation, are mediated by the insulin signaling pathway (Drummond-Barbosa & Spradling, 2001).

Independent of insulin signaling, ecdysone, a steroid hormone, also regulates germline proliferation and survival. Ecdysone functions through the ecdysone receptor (EcR) and its downstream target genes including *E74*, *E75*, and *broad*. Mutations in these members of the pathway result in failure of egg chambers to survive (Ables & Drummond-Barbosa, 2010). In mosaic clones of *E74*, there is increased caspase staining and cell death in regions 2 and 3 of the germarium, indicating that the lack of ecdysone signaling results in cell death (Ables & Drummond-Barbosa, 2010).

Other stimuli can trigger cell death in the germarium. High-dose gamma irradiation induces apoptosis in region 1 of the germarium, shown by TUNEL staining which indicates DNA fragmentation (Shim, Lee, Nguyen, Shim, & Song, 2014). Magnetic field exposure induces cell death in region 2a/2b of the germarium and results in a decline in reproductive capacity (Panagopoulos, Karabarbounis, & Lioliousis, 2013). Cell death has been shown to occur in response to improper chromosome pairing during meiosis in many species (Pritchett et al., 2009). This "pachytene checkpoint" would be expected to occur in regions 2 and 3 and may also be a trigger of cell death in the germarium.

The rate of cell death in the germarium is affected by *Wolbachia*, intracellular bacteria that infect insects and produce various effects on viability and/or fertility. In *Drosophila melanogaster*, infection with *Wolbachia* wMelPop, but not with strain wMel, increases apoptosis in the germarium, shown by acridine orange and TUNEL (Zhukova & Kiseleva, 2012). *Drosophila mauritiana* infected with *Wolbachia w*Mau produce four times as many eggs as uninfected flies due to increased mitosis in germline stem cells and decreased apoptosis in the germarium (Fast et al., 2011). Since *Wolbachia* are strictly maternally inherited like mitochondria, it has been suggested that they modulate host egg production to increase the number of infected progeny in a population.

Apoptosis of developing germline cysts in the germarium is also regulated by somatic cells. The transcription factor Daughterless (Da) plays a role in follicle cell differentiation and is also essential in somatic cells to promote death of germline cysts under certain conditions (Smith, Cummings, & Cronmiller, 2002). In *da* loss-of-function alleles, developing egg chambers are not properly separated from each other by a group of cells called stalk cells and have a merged appearance. In addition, TUNEL staining is reduced compared to wild type, indicating that *da* acts non-cell autonomously to promote germline cell death in the germarium. The BMP family member Dpp also regulates germ cell death nonautonomously. Decreased egg production in aging females is due to a decline in the rate of germline cell division and an increase in cell death in the germarium accompanied by decreased *dpp* signaling in the germ cell niche (Zhao, Xuan, Li, & Xi, 2008).

In addition to apoptotic markers, starved flies show autophagy induction in region 2 of the germarium, as detected by autophagy markers, LC3-GFP (Hou et al., 2008) or mCherry-Dr-Atg8a (Nezis et al., 2009). In well-fed flies, the GFP pattern is diffuse but in starved flies LC3-GFP is punctate, indicating autophagy induction. Interestingly, a decrease in apoptotic markers is seen in autophagy mutants (Nezis et al., 2009). In the germaria of starved *Dcp-1* effector caspase mutants, LysoTracker staining and the number of LC3-GFP puncta are reduced, indicating that Dcp-1 is required for starvation-induced autophagy. In mutants of *dBruce*, encoding a ubiquitin-conjugating enzyme, there is an increase in LysoTracker staining in well-fed flies indicating that Bruce normally suppresses autophagy (Hou et al., 2008). Further discussion of the cross talk between apoptosis and autophagy is described in the section on mid-oogenesis.

3.3 Cell Death in Mid-Oogenesis

During mid-oogenesis (stages 7–9), the nurse cell nuclei continue endoreplicating their DNA, and proteins and RNAs flow from the nurse cells through the ring canals into the oocyte. During these stages, the oocyte begins to accumulate yolk proteins (Fig. 2A, arrows) and grows in size until it fills half of the egg chamber. The onset of vitellogenesis (yolk protein synthesis and uptake) that occurs during mid-oogenesis is an energetically expensive process, so it is thought that this is a checkpoint where environmental conditions are surveyed prior to investing energy into egg development (Giorgi & Deri, 1976; Jenkins et al., 2013). In response to poor conditions such as starvation, the germline cells in some mid-stage egg chambers undergo PCD, which is easily visualized by morphological changes (Fig. 3B–D). Whereas healthy mid-stage egg chambers contain nurse cells with dispersed chromatin surrounded by a thin layer of follicle cells (Fig. 3B), dying egg chambers exhibit features of apoptotic cell death including condensation and fragmentation of the nurse cell nuclei (Fig. 3C–D). Coincident with the progressive degeneration of the germline, the follicle cells synchronously and dramatically enlarge as they engulf the dying germline (Etchegaray et al., 2012; Giorgi & Deri, 1976).

Mid-stage egg chambers seem to be exquisitely sensitive to environmental cues and developmental abnormalities. Even well-fed flies have occasional degenerating mid-stage egg chambers. This property of *Drosophila* oogenesis has been used as a readout for biological responses to environmental hazards. For example, mid-stage egg chambers have been shown to degenerate in response to the chemotherapeutic drug etoposide (Nezis, Stravopodis, Papassideri, Robert-Nicoud, & Margaritis, 2000), cocaine (Willard, Koss, & Cronmiller, 2006), and even mobile phone exposure. Exposure of adult female *Drosophila* to 900-GSM (Global System for Mobile Telecommunications) radiation results in DNA fragmentation of the germline primarily in the germarium and mid-stage egg chambers, and a reduction in the number of offspring produced by the females (Chavdoula, Panagopoulos, & Margaritis, 2010). Additionally, abnormalities in the follicle cells can lead to death of the underlying germline. When apoptosis of follicle cells in developing egg chambers is induced by expression of *reaper* or *hid*, the germline cells of mid-stage egg chambers die by apoptosis, showing nuclear condensation and fragmentation (Chao & Nagoshi, 1999). One notable exception to environmental triggers of mid-stage cell death is ionizing radiation. Germline cells beyond the germarium and follicle cells beyond stage 6 are resistant to ionizing radiation because endocycling cells fail to undergo apoptosis in response to DNA damage (Hassel, Zhang, Dixon, & Calvi, 2014). Recently, it has been found that this is likely due to low levels of p53 protein and epigenetic repression of *reaper* and *hid* (Zhang, Mehrotra, Ng, & Calvi, 2014).

Caspases are essential for the death of the germline during mid-oogenesis. Egg chambers lacking the effector caspase Dcp-1 (death caspase-1), or overexpressing the caspase inhibitors p35 or Diap1, show a complete block to germline death, resulting in "undead" egg chambers (Fig. 3E). Undead egg chambers contain nurse cells that have failed to condense, and the surrounding layer of follicle cells thins and disappears (Jenkins et al., 2013). It remains unknown why or how the follicle cells disappear in undead egg chambers. Downstream of Dcp-1, DNA fragmentation in dying egg chambers is mediated by both caspase-activated DNase (CAD) and DNaseII. In *drep1* (*icad*) mutants, where CAD is inactive, nurse nuclei remain large and fail to be cleared properly (Bass et al., 2009). In DNaseII mutants, although the nurse cell nuclei of dying mid-stage egg chambers appear to condense and fragment normally, the final digestion of DNA fragments is incomplete resulting in an egg chamber that appears opaque (Bass et al., 2009).

Although the requirement for Dcp-1 in mid-stage death is well documented, the upstream signals that regulate cell death during mid-oogenesis are only partially characterized. Unlike most examples of apoptosis in *Drosophila*, cell death during mid-oogenesis does not require the IAP antagonist genes *reaper*, *hid*, and *grim*, indicating that death of the germline is a unique cell death program (Peterson et al., 2007). Interestingly, although mid-stage cell death is completely blocked in mutants of the effector caspase Dcp-1, disruption of initiator caspases leads to a milder phenotype (Baum, Arama, Steller, & McCall, 2007), suggesting that Dcp-1 can be activated independently of initiator caspases. In addition to caspases, mitochondrial remodeling and the Bc1-2 proteins are important regulators of PCD during mid-oogenesis (Tanner, Blute, Brachmann, & McCall, 2011). Further investigations are necessary to determine the upstream signaling pathways that respond to death signals to trigger apoptosis of the germline in mid-oogenesis.

The importance of the steroid hormone ecdysone in the response to nutritional conditions has been investigated through work with EcR isoforms and the ecdysone response gene E75 (Buszczak et al., 1999; Terashima & Bownes, 2006). Under the control of ecdysone and the

BR-C genes acting in the follicle cells, the E75A isoform can induce apoptosis in the nurse cells of mid-stage egg chambers when flies are starved (Terashima & Bownes, 2006). However, when the flies are well fed, E75B suppresses the expression of E75A so that the egg chamber survives. E75 germline clones show dying nurse cells, indicating a cell-autonomous germline requirement for E75 (Buszczak et al., 1999). Ubiquitous knockdown of the ecdysone receptor EcR-B1 in flies results in a high incidence of mid-stage egg chamber death. When knocked down only in follicle cells, reduced EcR-B1 causes the follicle cell layer to delaminate (Romani et al., 2009). Follicle cell clones of EcR-B1 knockdown showed pyknotic nuclei and TUNEL staining, caspase-3 activation, and reduced Diap1 staining, indicating apoptotic death of follicle cells within the mutant clone. The delamination and death of the follicle cells are thought to trigger the death of the egg chamber including the germline cells (Romani et al., 2009; Thomson, Fitzpatrick, & Johnson, 2010).

Autophagic cell death may also contribute to PCD during mid-oogenesis. Dying mid-stage egg chambers stain with LysoTracker, a dye that labels acidified compartments, which may be indicative of autophagic cell death. Dying mid-stage egg chambers also have an increase in LC3-GFP puncta, indicating the formation of autophagosomes (Hou et al., 2008; Nezis et al., 2009). Mid-stage egg chambers that lack the autophagy genes *Atg1* or *Atg7* in nurse cells show reduced LysoTracker staining (Hou et al., 2008). Interestingly, *Atg1* and *Atg7* germline clones show condensation of the nurse cell chromatin but reduced DNA fragmentation, suggesting that autophagy promotes DNA fragmentation (Hou et al., 2008; Nezis et al., 2009). Moreover, autophagic flux is regulated by the effector caspase Dcp-1, through an interaction with mitochondrial adenine nucleotide translocase SesB, which in turn affects ATP levels (DeVorkin et al., 2014). Therefore, autophagy is thought to cooperate with apoptosis to promote PCD during mid-oogenesis.

While insulin is known to mediate cell proliferation and death in the germarium (Drummond-Barbosa & Spradling, 2001), it was found that the normal form of mid-stage death was not induced when attempts were made to mimic starvation by blocking positive components of the insulin signaling pathway in the germline (Pritchett & McCall, 2012). Rather, the follicle cells died, while the nurse cell nuclei resisted fragmentation and the egg chambers failed to show induction of caspases and lysosomes. In this case, a block in insulin signaling in the germline prevents the proper response to starvation. However, germline clones of *tor* did show a phenotype more similar to starvation, although egg chambers degenerated at earlier stages than what normally occurs under starvation (Pritchett & McCall, 2012). Other aspects of the starvation response, such as the formation of RNP aggregates and cortically condensed microtubules in the germline, have been shown to be mediated by decreased insulin signaling in follicle cells (Burn et al., 2015). How the follicle cells communicate the nutritional status to germline cells is unknown.

Treatment with rapamycin to inhibit Tor and simulate mid-stage death by starvation was marked by invasion and phagocytosis of germline cells by somatic follicle cells (Thomson & Johnson, 2010), a phenotype somewhat different than that seen normally during starvation. However, a similar type of cell death caused by follicle cell invasion has been seen in response to flock house virus infection (Thomson, Schneemann, & Johnson, 2012). Thus, somatic cells in this case can respond to external stimuli and destroy healthy oocytes. Also, the polarity genes *discs large*, *warts*, and *merlin* are necessary for the ability of follicle cells to invade and destroy oocytes since flies heterozygous for mutations in these polarity genes are able to resist the effects of rapamycin treatment (Thomson et al., 2010).

Cell death of the germline in mid-oogenesis triggers engulfment by the surrounding epithelial follicle cells (Etchegaray et al., 2012; Giorgi & Deri, 1976; Mazzalupo & Cooley, 2006; Tanner et al., 2011). The follicle cells dramatically change shape, becoming columnar and then appearing rather disorganized as the germline disappears (Etchegaray et al., 2012). The engulfment receptor Draper is essential for proper clearance of the dead germline, and mutants show persisting germline nuclear and cytoplasmic debris (Etchegaray et al., 2012). Engulfment by follicle cells also requires Rac1, Shark, and the JNK pathway, and likely other pathways that remain to be identified. Interestingly, while caspase activity in the germline is required for engulfment, it is not required for enrichment of Draper or activation of JNK, suggesting that there are both caspase-dependent and caspase-independent "eat-me" signals emanating from the dying germline. These eat-me signals also remain to be determined.

3.4 Developmental Cell Death in Late Oogenesis

In stage 11 of oogenesis, the nurse cells (which occupy roughly half the egg chamber by stage 10) transfer their contents to the oocyte rapidly in a process called "dumping" (Spradling, 1993; Fig. 3F–G). Prior to dumping, nurse cell nuclei change shape and actin bundles form in the nurse cells which prevent the nuclei from clogging the ring canals (Pritchett et al., 2009; Spradling, 1993). Following dumping, the nurse cell nuclei condense, the DNA is fragmented (becoming TUNEL-positive), and eventually all remnants of nurse cell organelles and membranes disappear from the end-stage egg chamber (Fig. 3H–I; Jenkins et al., 2013; Pritchett et al., 2009). By the time the chorion is completed with fully developed dorsal appendages, there are typically few traces of nurse cell nuclear remnants (Fig. 3I). As the mature egg leaves the ovariole, the follicle cell layer breaks away from the egg chamber and degenerates.

Much remains to be determined about the mechanism of developmental nurse cell death. Mutants that disrupt nurse cell death have "dumpless" egg chambers, or retain "persisting" nurse cell nuclei at the anterior end of the egg chamber (Fig. 3J). As in mid-oogenesis, the major apoptosis inducers, *reaper*, *hid*, and *grim*, are not required for developmental nurse cell death (Foley & Cooley, 1998). While caspases are known to be essential for mid-stage cell death, the requirement for caspases in late-stage cell death is minor (Baum et al., 2007; Mazzalupo & Cooley, 2006; Peterson, Barkett, & McCall, 2003). Mutations in the initiator caspases *Dronc* and *Dredd* show normal nurse cell death and removal (Baum et al., 2007). Mutations in the initiator caspase *Strica* show a normal phenotype in late oogenesis; however, the combination of *Strica* and *Dronc* produces late-stage egg chambers with a few persisting nurse cell nuclei, indicating redundancy in their function. Redundant function in late-stage cell death is also shown by the effector caspases *Dcp-1* and *Drice* (Baum et al., 2007). The frequency of persisting nurse cell nuclei in both of these genetic combinations is low and although significant, it is far from a complete block in nurse cell death, indicating that there are other pathways contributing to death of the nurse cells in late oogenesis.

In late-stage egg chambers, autophagy contributes to the fragmentation of nurse cell DNA, shown by puncta of the autophagic marker GFP-mCherry-Dr-Atg8a and by mutants of *Atg1*, *Atg13*, and *Vps34*, where some nurse cell nuclei persist in late-stage egg chambers (Bass et al., 2009; Nezis et al., 2010). These autophagy mutants show reduced expression of cleaved caspase-3, and persisting nurse cell nuclei with unfragmented DNA, indicating that autophagy is necessary to activate caspase-3 and to promote DNA fragmentation. The link between autophagy and DNA fragmentation was shown by the accumulation, in autophagy mutants, of undegraded dBruce, an inhibitor of apoptosis, and by the absence of fragmented DNA in autophagy and dBruce double mutants. These results indicated that dBruce controls whether or not DNA is fragmented and that autophagy controls the degradation of dBruce (Nezis et al., 2010).

Follicle cell clones of *Atg1* show abnormalities of dorsal appendages and other anterior structures (Barth, Szabad, Hafen, & Kohler, 2011), and some disruption of nurse cell death and removal. In addition, late-stage egg chambers from flies carrying *Atg1* germline clones, *Atg1* follicle cell clones, or trans-heterozygous mutants of *Atg7* were examined, and a few of the nurse cell nuclei failed to die (Peterson & McCall, 2013). Given the mild phenotypes caused by disrupting autophagy and apoptosis, it was possible that these mechanisms could be functioning redundantly to eliminate the nurse cells. To test this, double mutants affecting both apoptosis and autophagy were examined but were not found to show a significant increase in persisting nurse cells. This indicates that there are yet more factors contributing to the removal of nurse cells from late-stage egg chambers (Peterson & McCall, 2013).

Bass et al. (2009) suggested that the death of nurse cells in late-stage egg chambers has characteristics of programmed necrosis. Nurse cells show increased cytoplasmic calcium and acidification, similar to other examples of necrosis. Several genes associated with lysosomes are defective in nurse cell death and show either dumpless phenotypes or persisting nuclei. Mutants of *CathepsinD*, which encodes a lysosomal cathepsin, show persisting nurse cell nuclei in late stages although this represents only partial disruption of the cell death process since only a few nuclei out of the original 15 are not degraded. Stronger phenotypes are seen in late-stage egg chambers of *deep orange* (*dor*) and *spinster* mutants which show significant numbers of persisting nurse cell nuclei (Bass et al., 2009). Dor is the *Drosophila* ortholog of Vps18, a member of the HOPS (homotypic fusion and protein sorting) complex, which is essential for the fusion of endosomes to lysosomes (Sevrioukov, He, Moghrabi, Sunio, & Kramer, 1999; Solinger & Spang, 2013). Spinster is a transmembrane protein originally identified for its effects on female mating behavior that also functions in lysosomal processing and turnover (Nakano et al., 2001). Lysosomes are seen to cluster around nurse cells prior to their clearance and acidification (Bass et al., 2009), but their exact function in nurse cell death remains to be determined.

As the nurse cells dump their contents into the oocyte and disappear, they are in close contact with a subset of follicle cells called stretch follicle cells (Tran & Berg, 2003). This intimate relationship suggests that the stretch follicle cells could participate actively in the death and/or clearance of the nurse cells. Indeed, mutants of the engulfment receptor Draper show a remarkable persisting nuclei phenotype (Fig. 3F). Nonautonomous control over nurse cell death warrants further investigation.

4. CELL DEATH IN THE TESTIS

Male germline cell death has been reported in many vertebrate and invertebrate species and shows a number of similarities to cell death in the ovary. Like the ovary, entire germline cysts of developing spermatogonia have been shown to undergo PCD. In *Drosophila* spermatogenesis, 20–30% of newly formed cysts die, and similar to developmental cell death in the ovary, the deaths are nonapoptotic (Li, Ma, Cherry, & Matunis, 2014; Yacobi-Sharon, Namdar, & Arama, 2013). Interestingly, this type of germ cell death is associated with acidification as well as a requirement for lysosomal and mitochondrial genes. However, whether these genes act autonomously in the germline or nonautonomously in surrounding somatic cyst cells has not been determined. Loss of the EcR leads to increased cell death of both stem cells and developing cysts, similar to the ovary (Li et al., 2014). Interestingly, expression of EcR in the somatic lineage is sufficient to rescue the germ cell death phenotype observed in EcR mutants, suggesting somatic cyst cell control over germ cell death.

Germline cell death in the mammalian testis occurs during embryogenesis to establish proper germ cell to Sertoli (somatic) cell ratios (Aitken & Baker, 2013; Baum, St. George, & McCall, 2005), similar to the adjustments made to maintain proper follicle cell/nurse cell ratios in the germarium of the fly ovary (Drummond-Barbosa & Spradling, 2001). Cell death also occurs later in adult life to remove damaged germ cells so that they do not develop into spermatozoa (Aitken & Baker, 2013). Male germ cells appear to be very sensitive to environmental toxins, similar to mid-stage egg chambers in *Drosophila*. Also like the *Drosophila* ovary, the vertebrate testis is "immuno-privileged" (insulated from circulating macrophages) and dying germ cells are engulfed by the resident somatic cells, Sertoli cells. Clearance of apoptotic germ cells is mediated by ELMO1, a vertebrate ortholog of CED-12 (Elliott & Ravichandran, 2010b). ELMO1 mutants accumulate germ cell corpses and have a disordered seminiferous epithelium, leading to reduced sperm production. Thus, appropriate levels of cell death and corpse removal are essential for normal function in the testis.

5. COMPARISON OF PCD IN THE *DROSOPHILA* AND MAMMALIAN OVARY

The *Drosophila* ovary shares many similarities with the mammalian ovary and therefore is a relevant model for human fertility. As in *Drosophila*, PCD occurs in the mammalian ovary in response to insults and as a normal part of development. Developmental loss of oocytes during embryogenesis until puberty in mammals is called attrition. In humans, the embryonic ovary has 6.8 million oocytes 5 months post conception, which is reduced to 2 million oocytes within 2 days after birth and 100,000 oocytes by the onset of puberty (Baum et al., 2005; Matova & Cooley, 2001; Thomson et al., 2010). In addition, oocyte loss occurs between puberty and menopause and can be caused by genetic factors, chemical insults, and other physiological factors (Thomson et al., 2010). Degeneration of germ cells encapsulated by somatic granulosa cells (collectively termed a follicle) is called follicular atresia. In the human ovary, 99.9% of oocytes die between birth and menopause. Therefore, death is the ultimate fate of most oocytes in mammals (Morita & Tilly, 1999).

In the embryonic mammalian ovary, oocytes initially develop in clusters, called germ cell cysts, connected via intercellular bridges due to incomplete cytokinesis (Pepling, 2012). These interconnected germ cell cysts are reminiscent of *Drosophila* egg chambers. After birth, germ cell cysts disassemble and the oocytes become individually surrounded by somatic granulosa cells. During germ cell cyst breakdown, approximately two-third of the cells in the cysts undergo cell death (Morita & Tilly, 1999). It is likely that oocyte cell death occurs primarily via apoptosis, but autophagy may also contribute to cell death during cyst breakdown (Pepling, 2012; Rodrigues, Limback, McGinnis, Plancha, & Albertini, 2009). Interestingly, deletion of *Notch2* in the granulosa cells causes defects in the breakdown of germ cell cysts and reduced oocyte death. As mutant mice reach sexual maturity, their ovaries become hemorrhagic, resulting in reduced fertility. Therefore, the granulosa cells likely play a role in the death of the oocytes during development (Xu & Gridley, 2013).

During follicular atresia in adult mammals, both the oocyte and the granulosa cells undergo cell death. The Bcl-2 family member Bok has recently been shown to be important for oocyte attrition (Ke et al., 2013). Similar to the somatic follicle cells in the *Drosophila* ovary, the granulosa cells can act as nonprofessional phagocytes to engulf degenerating oocytes (Inoue, Watanabe, Saito, Hiroi, & Tonosaki, 2000). As described above, somatic follicle cells of the *Drosophila* ovary can be induced to invade and destroy the germline in response to rapamycin treatment. Similarly, in cultures of murine follicles, rapamycin treatment causes granulosa cells to invade and destroy the oocyte (Thomson & Johnson, 2010). Therefore, non-cell-autonomous mechanisms may contribute to cell death in the mammalian ovary.

6. CONCLUSION

Germline cell death in *Drosophila* occurs in response to many stimuli and uses diverse mechanisms, both apoptotic and nonapoptotic. The cell death in mid-oogenesis is caspase dependent but is unusual in that it can occur independently of an initiator caspase. How the Dcp-1 effector caspase becomes activated in mid-oogenesis remains to be determined. Other examples of germline cell death in the fly are nonapoptotic and show some characteristics of autophagic cell death and necrosis. Lysosomal activity seems to be a key component of developmental nurse cell death and cell death in the testis. Somatic follicle cells play an important role in the clearance of the dying germline and may also play a more active role in directing germ cell death. Nonautonomous control over germline cell death has been hinted at in several studies but warrants further investigation. Germline cell death in the fly ovary shares intriguing similarities with the mammalian germline, such as the PCD of germline cysts, and clearance of germline cells by somatic cells in the testis. Thus, the study of germline cell death in *Drosophila* can contribute important insights into human fertility.

ACKNOWLEDGEMENTS

We thank Clark Coffman for permission to use Figure 1, and lab members for helpful discussions and comments. We thank the Developmental Studies Hybridoma Bank and Cell Signaling Technology for antibodies. This work was supported by NIH grants R01 GM060574 and R01 GM094452 (to KM).

REFERENCES

Ables, E. T., & Drummond Barbosa, D. (2010). The steroid hormone ecdysone functions with intrinsic chromatin remodeling factors to control female germline stem cells in *Drosophila*. *Cell Stem Cell*, 7, 581–592.

Aitken, R. J., & Baker, M. A. (2013). Causes and consequences of apoptosis in spermatozoa; contributions to infertility and impacts on development. *The International Journal of Developmental Biology*, 57, 265–272.

Ashkenazi, A., & Dixit, V. M. (1998). Death receptors: Signaling and modulation. *Science*, *281*, 1305–1308.

Baehrecke, E. H. (2005). Autophagy: Dual roles in life and death? *Nature Reviews. Molecular Cell Biology*, *6*, 505–510.

Barth, J. M., Szabad, J., Hafen, E., & Kohler, K. (2011). Autophagy in *Drosophila* ovaries is induced by starvation and is required for oogenesis. *Cell Death and Differentiation, 18*, 915–924.

Bass, B. P., Tanner, E. A., Mateos San Martin, D., Blute, T., Kinser, R. D., Dolph, P. J., et al. (2009). Cell-autonomous requirement for DNaseII in nonapoptotic cell death. *Cell Death and Differentiation, 16*, 1362–1371.

Baum, J. S., Arama, E., Steller, H., & McCall, K. (2007). The *Drosophila* caspases Strica and Dronc function redundantly in programmed cell death during oogenesis. *Cell Death and Differentiation, 14*, 1508–1517.

Baum, J. S., St. George, J. P., & McCall, K. (2005). Programmed cell death in the germline. *Seminars in Cell & Developmental Biology, 16*, 245–259.

Berg, C. A. (2005). The *Drosophila* shell game: Patterning genes and morphological change. *Trends in Genetics, 21*, 346–355.

Berry, D. L., & Baehrecke, E. H. (2007). Growth arrest and autophagy are required for salivary gland cell degradation in *Drosophila. Cell, 131*, 1137–1148.

Brown, G. C., & Neher, J. J. (2012). Eaten alive! Cell death by primary phagocytosis: 'Phagoptosis'. *Trends in Biochemical Sciences, 37*, 325–332.

Burn, K. M., Shimada, Y., Ayers, K., Lu, F., Hudson, A. M., & Cooley, L. (2015). Somatic insulin signaling regulates a germline starvation response in *Drosophila* egg chambers. *Developmental Biology, 398*, 206–217.

Buszczak, M., Freeman, M. R., Carlson, J. R., Bender, M., Cooley, L., & Segraves, W. A. (1999). Ecdysone response genes govern egg chamber development during mid-oogenesis in *Drosophila. Development, 126*, 4581–4589.

Chao, S., & Nagoshi, R. N. (1999). Induction of apoptosis in the germline and follicle layer of *Drosophila* egg chambers. *Mechanisms of Development, 88*, 159–172.

Chavdoula, E. D., Panagopoulos, D. J., & Margaritis, L. H. (2010). Comparison of biological effects between continuous and intermittent exposure to GSM-900-MHz mobile phone radiation: Detection of apoptotic cell-death features. *Mutation Research, 700*, 51–61.

Coffman, C. R. (2003). Cell migration and programmed cell death of *Drosophila* germ cells. *Annals of the New York Academy of Sciences, 995*, 117–126.

Coffman, C. R., Strohm, R. C., Oakley, F. D., Yamada, Y., Przychodzin, D., & Boswell, R. E. (2002). Identification of X-linked genes required for migration and programmed cell death of *Drosophila melanogaster* germ cells. *Genetics, 162*, 273–284.

Cummings, M. R., & King, R. C. (1970). Ultrastructural changes in nurse and follicle cells during late stages of oogenesis in *Drosophila melanogaster. Zeitschrift für Zellforschung und Mikroskopische Anatomie, 110*, 1–8.

Cuttell, L., Vaughan, A., Silva, E., Escaron, C. J., Lavine, M., Van Goethem, E., et al. (2008). Undertaker, a *Drosophila* Junctophilin, links Draper-mediated phagocytosis and calcium homeostasis. *Cell, 135*, 524–534.

de Almagro, M. C., & Vucic, D. (2015). Necroptosis: Pathway diversity and characteristics. *Seminars in Cell & Developmental Biology, 39*, 56–62.

Denton, D., Shravage, B., Simin, R., Mills, K., Berry, D. L., Baehrecke, E. H., et al. (2009). Autophagy, not apoptosis, is essential for midgut cell death in *Drosophila. Current Biology, 19*, 1741–1746.

DeVorkin, L., Go, N. E., Hou, Y. C., Moradian, A., Morin, G. B., & Gorski, S. M. (2014). The *Drosophila* effector caspase Dcp-1 regulates mitochondrial dynamics and autophagic flux via SesB. *The Journal of Cell Biology, 205*, 477–492.

Drummond-Barbosa, D., & Spradling, A. C. (2001). Stem cells and their progeny respond to nutritional changes during *Drosophila* oogenesis. *Developmental Biology, 231*, 265–278.

Elliott, M. R., & Ravichandran, K. S. (2010a). Clearance of apoptotic cells: Implications in health and disease. *The Journal of Cell Biology, 189*, 1059–1070.

Elliott, M. R., & Ravichandran, K. S. (2010b). ELMO1 signaling in apoptotic germ cell clearance and spermatogenesis. *Annals of the New York Academy of Sciences*, *1209*, 30–36.

Etchegaray, J. I., Timmons, A. K., Klein, A. P., Pritchett, T. L., Welch, E., Meehan, T. L., et al. (2012). Draper acts through the JNK pathway to control synchronous engulfment of dying germline cells by follicular epithelial cells. *Development*, *139*, 4029–4039.

Fast, E. M., Toomey, M. E., Panaram, K., Desjardins, D., Kolaczyk, E. D., & Frydman, H. M. (2011). *Wolbachia* enhance *Drosophila* stem cell proliferation and target the germline stem cell niche. *Science*, *334*, 990–992.

Festjens, N., Vanden Berghe, T., & Vandenabeele, P. (2006). Necrosis, a well-orchestrated form of cell demise: Signalling cascades, important mediators and concomitant immune response. *Biochimica et Biophysica Acta*, *1757*, 1371–1387.

Foley, K., & Cooley, L. (1998). Apoptosis in late stage *Drosophila* nurse cells does not require genes within the H99 deficiency. *Development*, *125*, 1075–1082.

Freeman, M. R., Delrow, J., Kim, J., Johnson, E., & Doe, C. Q. (2003). Unwrapping glial biology: Gcm target genes regulating glial development, diversification, and function. *Neuron*, *38*, 567–580.

Fuchs, Y., & Steller, H. (2011). Programmed cell death in animal development and disease. *Cell*, *147*, 742–758.

Fullard, J. F., Kale, A., & Baker, N. E. (2009). Clearance of apoptotic corpses. *Apoptosis*, *14*, 1029–1037.

Galluzzi, L., Vitale, I., Abrams, J. M., Alnemri, E. S., Baehrecke, E. H., Blagosklonny, M. V., et al. (2012). Molecular definitions of cell death subroutines: Recommendations of the Nomenclature Committee on Cell Death 2012. *Cell Death and Differentiation*, *19*, 107–120.

Giorgi, F., & Deri, P. (1976). Cell death in ovarian chambers of *Drosophila melanogaster*. *Journal of Embryology and Experimental Morphology*, *35*, 521–533.

Golstein, P., & Kroemer, G. (2007). Cell death by necrosis: Towards a molecular definition. *Trends in Biochemical Sciences*, *32*, 37–43.

Hanyu-Nakamura, K., Kobayashi, S., & Nakamura, A. (2004). Germ cell-autonomous Wunen2 is required for germline development in *Drosophila* embryos. *Development*, *131*, 4545–4553.

Hassel, C., Zhang, B., Dixon, M., & Calvi, B. R. (2014). Induction of endocycles represses apoptosis independently of differentiation and predisposes cells to genome instability. *Development*, *141*, 112–123.

Heller, A., & Steinmann-Zwicky, M. (1998). No premature gene expression in germ cells of embryos deriving from nos females. *Mechanisms of Development*, *72*, 169–173.

Hoeppner, D. J., Hengartner, M. O., & Schnabel, R. (2001). Engulfment genes cooperate with ced-3 to promote cell death in *Caenorhabditis elegans*. *Nature*, *412*, 202–206.

Horne-Badovinac, S., & Bilder, D. (2005). Mass transit: Epithelial morphogenesis in the *Drosophila* egg chamber. *Developmental Dynamics*, *232*, 559–574.

Hou, Y. C., Chittaranjan, S., Barbosa, S. G., McCall, K., & Gorski, S. M. (2008). Effector caspase Dcp-1 and IAP protein Bruce regulate starvation-induced autophagy during *Drosophila melanogaster* oogenesis. *Journal of Cell Biology*, *182*, 1127–1139.

Hsu, T. Y., & Wu, Y. C. (2010). Engulfment of apoptotic cells in *C. elegans* is mediated by integrin alpha/SRC signaling. *Current Biology*, *20*, 477–486.

Inoue, S., Watanabe, H., Saito, H., Hiroi, M., & Tonosaki, A. (2000). Elimination of atretic follicles from the mouse ovary: A TEM and immunohistochemical study in mice. *Journal of Anatomy*, *196*, 103–110.

Jacobson, M. D., Weil, M., & Raff, M. C. (1997). Programmed cell death in animal development. *Cell*, *88*, 347–354.

Jenkins, V. K., Timmons, A. K., & McCall, K. (2013). Diversity of cell death pathways: Insight from the fly ovary. *Trends in Cell Biology*, *23*, 567–574.

Kamps, A. R., Pruitt, M. M., Herriges, J. C., & Coffman, C. R. (2010). An evolutionarily conserved arginine is essential for Tre1 G protein-coupled receptor function during germ cell migration in *Drosophila melanogaster*. *PLoS One*, *5*, e11839.

Ke, F., Bouillet, P., Kaufmann, T., Strasser, A., Kerr, J., & Voss, A. K. (2013). Consequences of the combined loss of BOK and BAK or BOK and BAX. *Cell Death & Disease*, *4*, e650.

Kerr, J. F. R., Wyllie, A. H., & Currie, A. R. (1972). Apoptosis, a basic biological phenomenon with wide-ranging implications in tissue kinetics. *British Journal of Cancer*, *26*, 239–257.

Kinchen, J. M. (2010). A model to die for: Signaling to apoptotic cell removal in worm, fly and mouse. *Apoptosis*, *15*, 998–1006.

Kinchen, J. M., & Ravichandran, K. S. (2007). Journey to the grave: Signaling events regulating removal of apoptotic cells. *Journal of Cell Science*, *120*, 2143–2149.

King, R. C. (1970). *Ovarian development in Drosophila melanogaster*. New York: Academic Press.

Krammer, P. H. (1998). The CD95(APO-1/Fas)/CD95L system. *Toxicology Letters*, *102–103*, 131–137.

Kroemer, G., Galluzzi, L., Vandenabeele, P., Abrams, J., Alnemri, E. S., Baehrecke, E. H., et al. (2009). Classification of cell death: Recommendations of the Nomenclature Committee on Cell Death 2009. *Cell Death and Differentiation*, *16*, 3–11.

Kroemer, G., & Perfettini, J. L. (2014). Entosis, a key player in cancer cell competition. *Cell Research*, *24*, 1280–1281.

Kunwar, P. S., & Lehmann, R. (2003). Developmental biology: Germ-cell attraction. *Nature*, *421*, 226–227.

Li, W., & Baker, N. E. (2007). Engulfment is required for cell competition. *Cell*, *129*, 1215–1225.

Li, Y., Ma, Q., Cherry, C. M., & Matunis, E. L. (2014). Steroid signaling promotes stem cell maintenance in the *Drosophila* testis. *Developmental Biology*, *394*, 129–141.

MacDonald, J. M., Beach, M. G., Porpiglia, E., Sheehan, A. E., Watts, R. J., & Freeman, M. R. (2006). The *Drosophila* cell corpse engulfment receptor Draper mediates glial clearance of severed axons. *Neuron*, *50*, 869–881.

Manaka, J., Kuraishi, T., Shiratsuchi, A., Nakai, Y., Higashida, H., Henson, P., et al. (2004). Draper-mediated and phosphatidylserine-independent phagocytosis of apoptotic cells by *Drosophila* hemocytes/macrophages. *Journal of Biological Chemistry*, *279*, 48466–48476.

Mangahas, P. M., & Zhou, Z. (2005). Clearance of apoptotic cells in *Caenorhabditis elegans*. *Seminars in Cell & Developmental Biology*, *16*, 295–306.

Matova, N., & Cooley, L. (2001). Comparative aspects of animal oogenesis. *Developmental Biology*, *231*, 291–320.

Mazzalupo, S., & Cooley, L. (2006). Illuminating the role of caspases during *Drosophila* oogenesis. *Cell Death and Differentiation*, *13*, 1950–1959.

McCall, K. (2004). Eggs over easy: Cell death in the *Drosophila* ovary. *Developmental Biology*, *274*, 3–14.

McCall, K. (2010). Genetic control of necrosis—Another type of programmed cell death. *Current Opinion in Cell Biology*, *22*, 882–888.

Mevorach, D., Trahtemberg, U., Krispin, A., Attalah, M., Zazoun, J., Tabib, A., et al. (2010). What do we mean when we write "senescence," "apoptosis," "necrosis," or "clearance of dying cells"? *Annals of the New York Academy of Sciences*, *1209*, 1–9.

Monks, J., Smith-Steinhart, C., Kruk, E. R., Fadok, V. A., & Henson, P. M. (2008). Epithelial cells remove apoptotic epithelial cells during post-lactation involution of the mouse mammary gland. *Biology of Reproduction*, *78*, 586–594.

Montell, D. J., Yoon, W. H., & Starz-Gaiano, M. (2012). Group choreography: Mechanisms orchestrating the collective movement of border cells. *Nature Reviews. Molecular Cell Biology*, *13*, 631–645.

Morita, Y., & Tilly, J. L. (1999). Oocyte apoptosis: Like sand through an hourglass. *Developmental Biology, 213*, 1–17.

Nakano, Y., Fujitani, K., Kurihara, J., Ragan, J., Usui-Aoki, K., Shimoda, L., et al. (2001). Mutations in the novel membrane protein spinster interfere with programmed cell death and cause neural degeneration in *Drosophila melanogaster. Molecular and Cellular Biology, 21*, 3775–3788.

Nandrot, E. F. (2014). Animal models, in "the quest to decipher RPE phagocytosis" *Advances in Experimental Medicine and Biology, 801*, 77–83.

Nandrot, E. F., Chang, Y., & Finnemann, S. C. (2008). Alphavbeta5 integrin receptors at the apical surface of the RPE: One receptor, two functions. *Advances in Experimental Medicine and Biology, 613*, 369–375.

Nelson, C., & Baehrecke, E. H. (2014). Eaten to death. *The FEBS Journal, 281*, 5411–5417.

Nezis, I. P., Lamark, T., Velentzas, A. D., Rusten, T. E., Bjorkoy, G., Johansen, T., et al. (2009). Cell death during *Drosophila melanogaster* early oogenesis is mediated through autophagy. *Autophagy, 5*, 298–302.

Nezis, I. P., Shravage, B. V., Sagona, A. P., Lamark, T., Bjorkoy, G., Johansen, T., et al. (2010). Autophagic degradation of dBruce controls DNA fragmentation in nurse cells during late *Drosophila melanogaster* oogenesis. *The Journal of Cell Biology, 190*, 523–531.

Nezis, I. P., Stravopodis, D. J., Papassideri, I., Robert-Nicoud, M., & Margaritis, L. H. (2000). Stage-specific apoptotic patterns during *Drosophila* oogenesis. *European Journal of Cell Biology, 79*, 610–620.

Overholtzer, M., Mailleux, A. A., Mouneimne, G., Normand, G., Schnitt, S. J., King, R. W., et al. (2007). A nonapoptotic cell death process, entosis, that occurs by cell-in-cell invasion. *Cell, 131*, 966–979.

Panagopoulos, D. J., Karabarbounis, A., & Lioliousis, C. (2013). ELF alternating magnetic field decreases reproduction by DNA damage induction. *Cell Biochemistry and Biophysics, 67*, 703–716.

Pepling, M. E. (2012). Follicular assembly: Mechanisms of action. *Reproduction, 143*, 139–149.

Perez-Garijo, A., Fuchs, Y., & Steller, H. (2013). Apoptotic cells can induce non-autonomous apoptosis through the TNF pathway. *eLife, 2*, e01004.

Peterson, J. S., Barkett, M., & McCall, K. (2003). Stage-specific regulation of caspase activity in *Drosophila* oogenesis. *Developmental Biology, 260*, 113–123.

Peterson, J. S., Bass, B. P., Jue, D., Rodriguez, A., Abrams, J. M., & McCall, K. (2007). Noncanonical cell death pathways act during *Drosophila* oogenesis. *Genesis, 45*, 396–404.

Peterson, J. S., & McCall, K. (2013). Combined inhibition of autophagy and caspases fails to prevent developmental nurse cell death in the *Drosophila melanogaster* ovary. *PLoS One, 8*, e76046.

Pflanz, R., Voigt, A., Yakulov, T., & Jackle, H. (2015). *Drosophila* gene tao-1 encodes proteins with and without a Ste20 kinase domain that affect cytoskeletal architecture and cell migration differently. *Open Biology, 5*, 140161.

Pritchett, T. L., & McCall, K. (2012). Role of the insulin/Tor signaling network in starvation-induced programmed cell death in *Drosophila* oogenesis. *Cell Death and Differentiation, 19*, 1069–1079.

Pritchett, T. L., Tanner, E. A., & McCall, K. (2009). Cracking open cell death in the Drosophila ovary. *Apoptosis, 14*, 969–979.

Reddien, P. W., Cameron, S., & Horvitz, H. R. (2001). Phagocytosis promotes programmed cell death in *C. elegans*. *Nature, 412*, 198–202.

Reddien, P. W., & Horvitz, H. R. (2000). CED-2/CrkII and CED-10/Rac control phagocytosis and cell migration in *Caenorhabditis elegans*. *Nature Cell Biology, 2*, 131–136.

Reddien, P. W., & Horvitz, H. R. (2004). The engulfment process of programmed cell death in *Caenorhabditis elegans*. *Annual Review of Cell and Developmental Biology, 20*, 193–221.

Rodrigues, P., Limback, D., McGinnis, L. K., Plancha, C. E., & Albertini, D. F. (2009). Multiple mechanisms of germ cell loss in the perinatal mouse ovary. *Reproduction, 137*, 709–720.
Romani, P., Bernardi, F., Hackney, J., Dobens, L., Gargiulo, G., & Cavaliere, V. (2009). Cell survival and polarity of *Drosophila* follicle cells require the activity of ecdysone receptor B1 isoform. *Genetics, 181*, 165–175.
Sano, H., Renault, A. D., & Lehmann, R. (2005). Control of lateral migration and germ cell elimination by the *Drosophila melanogaster* lipid phosphate phosphatases Wunen and Wunen 2. *The Journal of Cell Biology, 171*, 675–683.
Sato, K., Hayashi, Y., Ninomiya, Y., Shigenobu, S., Arita, K., Mukai, M., et al. (2007). Maternal Nanos represses *hid/skl*-dependent apoptosis to maintain the germ line in *Drosophila* embryos. *Proceedings of the National Academy of Sciences of the United States of America, 104*, 7455–7460.
Sevrioukov, E. A., He, J. P., Moghrabi, N., Sunio, A., & Kramer, H. (1999). A role for the *deep orange* and *carnation* eye color genes in lysosomal delivery in *Drosophila*. *Molecular Cell, 4*, 479–486.
Shim, H. J., Lee, E. M., Nguyen, L. D., Shim, J., & Song, Y. H. (2014). High-dose irradiation induces cell cycle arrest, apoptosis, and developmental defects during *Drosophila* oogenesis. *PLoS One, 9*, e89009.
Smith, J. E., 3rd, Cummings, C. A., & Cronmiller, C. (2002). Daughterless coordinates somatic cell proliferation, differentiation and germline cyst survival during follicle formation in *Drosophila*. *Development, 129*, 3255–3267.
Solinger, J. A., & Spang, A. (2013). Tethering complexes in the endocytic pathway: CORVET and HOPS. *The FEBS Journal, 280*, 2743–2757.
Spradling, A. C. (1993). Developmental genetics of oogenesis. In M. Bate & A. Martinez Arias (Eds.), *The development of Drosophila melanogaster* (pp. 1–70). Cold Spring Harbor: Cold Spring Harbor Laboratory Press.
Spradling, A. C., Nystul, T., Lighthouse, D., Morris, L., Fox, D., Cox, R., et al. (2008). Stem cells and their niches: Integrated units that maintain *Drosophila* tissues. *Cold Spring Harbor Symposia on Quantitative Biology, 73*, 49–57.
Starz-Gaiano, M., & Lehmann, R. (2001). Moving towards the next generation. *Mechanisms of Development, 105*, 5–18.
Sun, L., & Wang, X. (2014). A new kind of cell suicide: Mechanisms and functions of programmed necrosis. *Trends in Biochemical Sciences, 39*, 587–593.
Tanner, E. A., Blute, T. A., Brachmann, C. B., & McCall, K. (2011). Bcl-2 proteins and autophagy regulate mitochondrial dynamics during programmed cell death in the *Drosophila* ovary. *Development, 138*, 327–338.
Taylor, R. C., Cullen, S. P., & Martin, S. J. (2008). Apoptosis: Controlled demolition at the cellular level. *Nature Reviews. Molecular Cell Biology, 9*, 231–241.
Terashima, J., & Bownes, M. (2006). E75A and E75B have opposite effects on the apoptosis/development choice of the *Drosophila* egg chamber. *Cell Death and Differentiation, 13*, 454–464.
Thomson, T. C., Fitzpatrick, K. E., & Johnson, J. (2010). Intrinsic and extrinsic mechanisms of oocyte loss. *Molecular Human Reproduction, 16*, 916–927.
Thomson, T. C., & Johnson, J. (2010). Inducible somatic oocyte destruction in response to rapamycin requires wild-type regulation of follicle cell epithelial polarity. *Cell Death and Differentiation, 17*, 1717–1727.
Thomson, T. C., Schneemann, A., & Johnson, J. (2012). Oocyte destruction is activated during viral infection. *Genesis, 50*, 453–465.
Tran, D. H., & Berg, C. A. (2003). *bullwinkle* and *shark* regulate dorsal-appendage morphogenesis in *Drosophila* oogenesis. *Development, 130*, 6273–6282.

Willard, S. S., Koss, C. M., & Cronmiller, C. (2006). Chronic cocaine exposure in *Drosophila*: Life, cell death and oogenesis. *Developmental Biology, 296*, 150–163.
Wu, X., Tanwar, P. S., & Raftery, L. A. (2008). *Drosophila* follicle cells: Morphogenesis in an eggshell. *Seminars in Cell & Developmental Biology, 19*, 271–282.
Xu, J., & Gridley, T. (2013). Notch2 is required in somatic cells for breakdown of ovarian germ-cell nests and formation of primordial follicles. *BMC Biology, 11*, 13.
Yacobi-Sharon, K., Namdar, Y., & Arama, E. (2013). Alternative germ cell death pathway in *Drosophila* involves HtrA2/Omi, lysosomes, and a caspase-9 counterpart. *Developmental Cell, 25*, 29–42.
Yamada, Y., Davis, K. D., & Coffman, C. R. (2008). Programmed cell death of primordial germ cells in *Drosophila* is regulated by p53 and the Outsiders monocarboxylate transporter. *Development, 135*, 207–216.
Zhang, B., Mehrotra, S., Ng, W. L., & Calvi, B. R. (2014). Low levels of p53 protein and chromatin silencing of p53 target genes repress apoptosis in *Drosophila* endocycling cells. *PLoS Genetics, 10*, e1004581.
Zhao, R., Xuan, Y., Li, X., & Xi, R. (2008). Age-related changes of germline stem cell activity, niche signaling activity and egg production in *Drosophila. Aging Cell, 7*, 344–354.
Zhukova, M. V., & Kiseleva, E. (2012). The virulent *Wolbachia* strain *wMelPop* increases the frequency of apoptosis in the female germline cells of *Drosophila melanogaster. BMC Microbiology, 12*(Suppl. 1), S15.
Ziegenfuss, J. S., Biswas, R., Avery, M. A., Hong, K., Sheehan, A. E., Yeung, Y. G., et al. (2008). Draper-dependent glial phagocytic activity is mediated by Src and Syk family kinase signalling. *Nature, 453*, 935–939.

CHAPTER FIVE

The HOX–Apoptosis Regulatory Interplay in Development and Disease

Katrin Domsch, Fani Papagiannouli, Ingrid Lohmann[1]
*Centre for Organismal Studies (COS) Heidelberg, Heidelberg University, Heidelberg, Germany
[1]Corresponding author: e-mail address: ingrid.lohmann@cos.uni-heidelberg.de

Contents

Abstract

Apoptosis is a cellular suicide program, which is on the one hand used to remove superfluous cells thereby promoting tissue or organ morphogenesis. On the other hand, the programmed killing of cells is also critical when potentially harmful cells emerge in a developing or adult organism thereby endangering survival. Due to its critical role apoptosis is tightly controlled, however so far, its regulation on the transcriptional level is less studied and understood. *Hox* genes, a highly conserved gene family encoding homeodomain transcription factors, have crucial roles in development. One of their prominent functions is to shape animal body plans by eliciting different developmental programs along the anterior–posterior axis. To this end, Hox proteins transcriptionally regulate numerous processes in a coordinated manner, including cell-type specification, differentiation, motility, proliferation as well as apoptosis. In this review, we will focus on

Current Topics in Developmental Biology, Volume 114
ISSN 0070-2153
http://dx.doi.org/10.1016/bs.ctdb.2015.07.014

how Hox proteins control organismal morphology and function by regulating the apoptotic machinery. We will first focus on well-established paradigms of Hox–apoptosis interactions and summarize how Hox transcription factors control morphological outputs and differentially shape tissues along the anterior–posterior axis by fine-tuning apoptosis in a healthy organism. We will then discuss the consequences when this interaction is disturbed and will conclude with some ideas and concepts emerging from these studies.

1. INTRODUCTION

The discovery of homeotic genes dates back more than 100 years ago when Morgan and Bridges discovered *Drosophila* mutants with characteristic alterations of body structures such as the replacement of antennae with legs (Morgan & Bridges, 1916). We now know that these changes, referred to as homeotic transformations, are caused by the highly conserved family of *Hox* genes, which dictate the identity of different structures along the anterior–posterior (A/P) axis of all bilaterian animals (Cribbs, Pultz, Johnson, Mazzulla, & Kaufman, 1992; Malicki, Schughart, & McGinnis, 1990; McGinnis et al., 1990). On the molecular level, *Hox* genes code for transcription factors (TFs) with a highly conserved DNA-binding domain, the homeodomain (HD), which allows these proteins to bind *cis*-regulatory sequences (CRMs) of close-by genes (Gehring & Hiromi, 1986; Gehring et al., 1994; McGinnis, Levine, Hafen, Kuroiwa, & Gehring, 1984; Scott & Weiner, 1984). Interaction with these CRMs results in activation or repression of target genes (Hueber & Lohmann, 2008) that in turn regulate different processes in a variety of cell types (Foronda, de Navas, Garaulet, & Sanchez-Herrero, 2009). Thus, *Hox* genes have been defined early on as selector genes (Garcia-Bellido, 1975) that regulate realisator genes, defined sets of genes directly instructing morphogenetic processes on the cellular level (Alonso, 2002; Foronda et al., 2009). *Hox* genes are known to control many aspects of development in many different organisms. In vertebrates, formation of digits, mainly regulated by the *Hoxa* and *Hoxd* cluster, is one of the prominent examples showing that *Hox* genes employ the apoptotic machinery to sculpt tissues (Woltering & Duboule, 2010). Furthermore, *Hox* genes often control the development of whole organs, such as *Hoxa5*-mediated skeletal patterning and morphogenesis of the respiration tract (Mandeville et al., 2006). In addition to their role in development, *Hox* genes are also associated with a number of human diseases (Shah & Sukumar, 2010). In several cancer or tumor types, aberrant *Hox* gene function, as a consequence of altered epigenetic, temporal, or spatial gene regulation (Armstrong, Golub, & Korsmeyer, 2003; Faber et al., 2009; Ferrando et al., 2003; Jung et al., 2005; Takahashi et al., 2007), results in a wide spectrum of abnormalities such as defects in cell morphology, differentiation, and proliferation, deregulation of apoptosis and signaling pathways, as well as metastasis and invasion (Argiropoulos & Humphries, 2007; Chen, Chung, & Sukumar, 2004; Chu, Selam, & Taylor, 2004; Hu, Fong, Ferrell, Largman, & Shen, 2009; Lawrence & Largman, 1992; Manohar, Salwen, Furtado, & Cohn, 1996; Zhang et al., 2007). In conclusion, Hox TFs are a major class of developmental regulators that control a diversity of physiological processes throughout the life of an animal. However, since only a few regulatory networks downstream of *Hox* genes have been elucidated, it remains poorly understood how they control and coordinate cellular processes in different tissue contexts. This knowledge is critical to comprehensively grasp their function during normal development as well as in physiopathological processes.

2. *HOX* GENES: HOMEOTIC GENES WITH MASTER REGULATORY FUNCTIONS

2.1 Organization and Regulation of *Hox* Genes

Hox genes are grouped in chromosomal clusters with the number of clusters varying according to organismal complexity (Fig. 1). In the fruit fly *Drosophila melanogaster*, *Hox* genes are assembled in one cluster containing two complexes located on the same chromosome: the *Antennapedia* complex (ANT-C) harbors the *Hox* genes *labial* (*lab*), *proboscipedia* (pb), *Deformed* (*Dfd*), *Sex combs reduced* (*Scr*), and *Antennapedia* (*Antp*), while the *Bithorax* complex (BX-C) includes the *Hox* genes *Ultrabithorax* (*Ubx*), *abdominal-A* (*abd-A*), and *Abdominal-B* (*Abd-B*) (Kaufman, Lewis, & Wakimoto, 1980; Kaufman, Seeger, & Olsen, 1990; Lewis, 1978; Fig. 1B). The nematode *Caenorhabditis elegans* (*C. elegans*) contains six *Hox* genes, *abnormal cell lineage 39* (*lin-39*), *male abnormal 5* (*mab-5*), *homeobox 13* (*ceh-13*), *egg-laying defective 5* (*egl-5*), *posterior Hox gene paralogue 3* (*php-3*), and *knob-like posterior* (*nob-1*) (Burglin & Ruvkun, 1993; Wang et al., 1993). Functional conservation exists between the two anterior *C. elegans Hox* genes *lin-39* and *mab-5* and their *Drosophila* homologs *Scr* and *Antp* (Burglin & Ruvkun, 1993; Hunter & Kenyon, 1995), while *egl-5*, *php-3*, and *nob-1* are all homologs of the *Drosophila Hox* gene *Abd-B* (Aboobaker & Blaxter, 2003; Van Auken et al., 2002; Wang et al., 1993; Fig. 1A). *ceh-13* contains structural

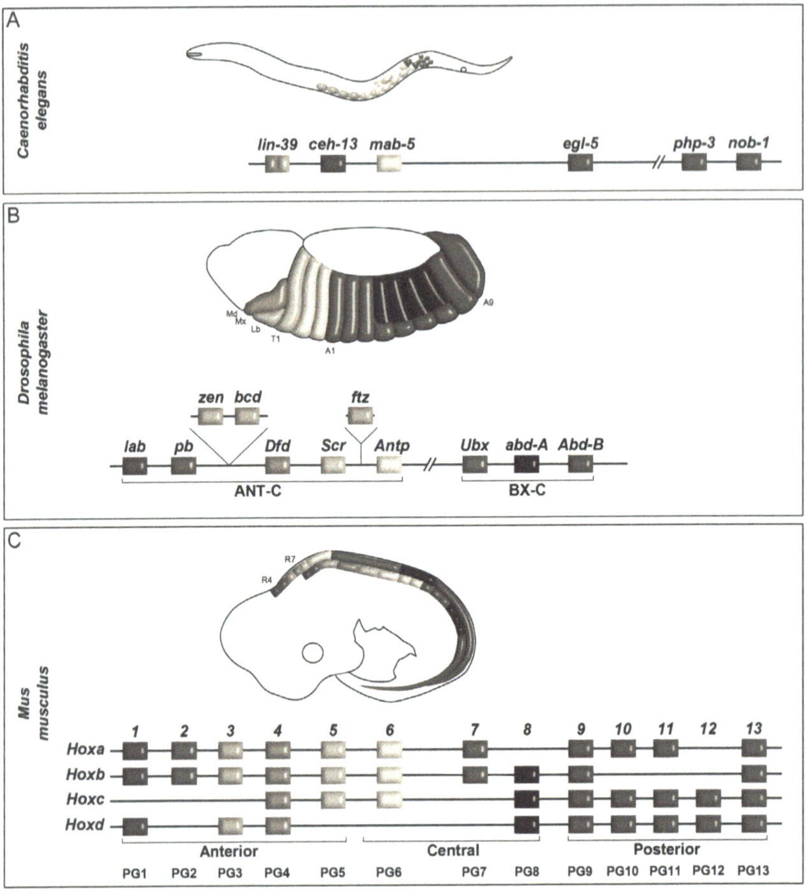

Figure 1 Schematic diagram of the *Hox* gene clusters and their organization in the genome (not in scale). The relative position of *Hox* genes within the clusters reflects their expression pattern along the anterior–posterior axis of the model organisms. *Hox* genes and their corresponding expression domains are always illustrated in the same color. (A) The *Hox* gene cluster in *Caenorhabditis elegans* (*C. elegans*) consists of the *Hox* genes *lin-39* (*abnormal cell lineage 39*), *ceh-13* (*homeobox 13*), *mab-5* (*male abnormal 5*), *egl-5* (*egg-laying defective 5*), *php-3* (*posterior Hox gene paralogue 3*), and *knob-like posterior* (*nob-1*). (B) The *Hox* gene cluster and its segmental expression pattern in the embryo of *Drosophila melanogaster* (stage 14). The *Antennapedia* complex (ANT-C) includes the more anterior *Hox* genes *lab* (*labial*), *pb* (*proboscipedia*), *Dfd* (*Deformed*), *Scr* (*Sex combs reduced*), and *Antp* (*Antennapedia*). The *Bithorax* complex (BX-C) includes the posterior most Hox genes *Ubx* (*Ultrabithorax*), *abd-A* (*abdominal-A*), and *Abd-B* (*Abdominal-B*). (C) *Hox* gene clusters and expression domains in the spinal cord of a mouse embryo (E12.5). Vertebrate *Hox* genes are organized in four different clusters (*Hoxa*, *Hoxb*, *Hoxc*, and *Hoxd*) and consist of up to 13 paralogous groups (PGs). The PGs can be clustered into three classes (anterior, central, and posterior). Orthologous genes and genes of the same PG are indicated in the same color. Md, mandibular; Mx, maxillary; Lb, labial; T, thoracic; A, abdominal. *Panel (B) Picture was adapted from Hartenstein (1993).* (See the color plate.)

and sequence features typical for *Drosophila lab* (Pearson, Lemons, & McGinnis, 2005; Sharkey, Graba, & Scott, 1997). In the vertebrate model organism *Mus musculus*, four *Hox* gene clusters (*Hoxa*, *Hoxb*, *Hoxc*, and *Hoxd*) located on four different chromosomes are described, which contain 13 paralogs (Hox 1–13) (Duboule & Dolle, 1989; Graham, Papalopulu, & Krumlauf, 1989; Maconochie, Nonchev, Morrison, & Krumlauf, 1996). Importantly, Hox proteins are grouped in different paralog groups (anterior, central, and posterior) reflecting the level of conservation and expression domains of the respective proteins (Fig. 1C).

One striking feature of *Hox* genes is their sequential expression along the A/P axis according to their location within the cluster. This means that *Hox* genes active anteriorly in the developing embryo are located normally at the 3′ end of the *Hox* cluster, while those expressed posteriorly are found at 5′ end of the same genomic region, a phenomenon known as spatial colinearity (Gaunt, 1988; Graham et al., 1989; Kmita & Duboule, 2003). Especially in vertebrates, spatial organization of *Hox* genes within each cluster correlates with the temporal sequence of expression, a phenomenon referred to as temporal colinearity (Iimura, Denans, & Pourquie, 2009; Soshnikova & Duboule, 2008). Misexpression of *Hox* genes can lead to severe abnormalities. Thus, their expression is tightly controlled via a network of upstream regulators. In *Drosophila*, as in most other organisms, this is organized by at least three distinct regulatory inputs (Mallo & Alonso, 2013): (1) upstream regulators, such as the segmentation genes, that control *Hox* gene expression at the transcriptional level (Irish, Martinez-Arias, & Akam, 1989); (2) the *Polycomb* (*PcG*) and *Trithorax* (*trxG*) genes, which epigenetically imprint a protein-based cellular memory on *Hox* gene loci (Denell, 1978; Puro & Nygren, 1975); and (3) the cross-regulatory interactions among the *Hox* genes themselves, in which posterior Hox proteins repress the more anteriorly expressed *Hox* genes (Morata & Kerridge, 1982; Struhl, 1982), a phenomenon known as posterior prevalence (Akam, 1987; Duboule & Morata, 1994; Hafen, Levine, & Gehring, 1984; Harding, Wedeen, McGinnis, & Levine, 1985). If one of these mechanisms fails, aberrant *Hox* expression can lead to homeotic transformations (Lewis, 1978) similar to the antenna-to-leg transformation observed in *Antp* mutants (Lewis, 1978; Struhl, 1981, 1982) or to disease situations (Barber & Rastegar, 2010; Procino & Cillo, 2013; Quinonez & Innis, 2014; Shah & Sukumar, 2010). Studies in *Drosophila* revealed that missexpression of *Hox* genes along the A/P axis is intimately linked to *PcG* and *TrxG* proteins (Schuettengruber & Cavalli, 2009). The two Polycomb repressive complexes (PRCs), PRC1 and PRC2, are important for silencing many genomic loci including the *Hox* clusters (Simon & Kingston, 2013). Interestingly, it has been shown that PRC2 components are distributed over large regions of the target genes thereby associating with various RNA classes, including long noncoding RNAs (lncRNAs) (Boyer et al., 2006; Kanhere et al., 2010; Lee et al., 2006; Rinn et al., 2007; Zhao, Sun, Erwin, Song, & Lee, 2008). The interaction of PRC2 with lncRNAs leads to methylation of Histone H3K27 and consequently to gene silencing, as in the case of the *Drosophila* lncRNA *iab-8-ncRNA* (Gummalla, Galetti, Maeda, & Karch, 2014) and the human lncRNA HOTAIR (Kanhere et al., 2010; Rinn et al., 2007; Tsai et al., 2010). Similar to the PcG proteins, the Trx proteins are required for epigenetically controlling *Hox* genes; they were initially characterized as positive regulators of *Hox* genes maintaining expression in the appropriate body segments (Ingham, 1983, 1998; Schuettengruber, Martinez, Iovino, & Cavalli, 2011). The TrxG is a heterogeneous group of proteins that consists of DNA-binding proteins and large multimeric histone-modifying and ATP-dependent chromatin remodeling complexes (Schuettengruber, Chourrout, Vervoort, Leblanc, & Cavalli, 2007; Schuettengruber et al., 2011). The *Drosophila* Trithorax (Trx) and Absent small or homeotic discs 1 (Ash1) proteins as well as the vertebrate mixed-lineage leukemia (Mll) 1–3 proteins and their associated proteins belong to the histone-modifying class. The Switch/Sucrose nonfermentable as well as the Nucleosome-Remodeling Factor complexes also belong to the chromatin remodeling complexes. The vertebrate Trx orthologue Mll1, which was originally discovered as a gene inducing human leukemia (Gu et al., 1992; Schuettengruber et al., 2011; Tkachuk, Kohler, & Cleary, 1992), is required for H3K4 trimethylation at a small subset of mammalian genes including *Hox* genes (Schuettengruber et al., 2011; Wu et al., 2008).

2.2 Hox TFs: Transcriptional Specificity and Target Genes

As TFs, Hox proteins execute their function by interacting with DNA sequences in CRMs thereby regulating the expression of associated genes. However, due to their DNA-binding domain Hox TFs do not display strong sequence specificity as they recognize *in vitro* very similar and frequently encountered consensus motifs, all sharing an –$A^{A}/_{T}$AT– core (Carr & Biggin, 1999; Noyes et al., 2008). This stands in sharp contrast to the highly specific functions Hox proteins execute *in vivo*, a discrepancy referred to as the "Hox paradox" (Mann, 1995; Mann, Lelli, & Joshi, 2009).

A vast amount of work has been invested to solve the long-standing problem of how Hox TFs regulate the precise expression of their target genes and so far several complementary hypotheses have been proposed. Common to all is the idea that regulatory specificity and activity of Hox TFs is not necessarily a matter of Hox proteins alone but the result of their ability to interact with other regulatory proteins. Much attention focused on the pre-B cell (PBC) proteins (Burglin, 1997), which belong to the TALE (three amino acids loop extension) family of HD-containing TFs. Representatives of PBC are the *Drosophila* Extradenticle (Exd) or vertebrate Pbx1–4 proteins and both have been shown to form dimeric complexes with Hox proteins *in vitro*, which is required for their interaction with specific CRMs and regulation of distinct target genes (Chang et al., 1995; Mann et al., 2009; Ryoo, Marty, Casares, Affolter, & Mann, 1999; Shen et al., 1996). Despite the fact that interaction with PBC TFs allow Hox proteins to preferentially recognize different heterodimeric-binding sites, revealing a latent "specificity" (Slattery, Ma, Negre, White, & Mann, 2011), PBC partners are nonetheless widely expressed during embryogenesis and therefore cannot account for all tissue- and/or stage-specific activities of the different Hox proteins. Thus, it is assumed that Hox proteins interact with a diverse set of TFs restricted to particular cell or tissue types to form large regulatory complexes, which depending on their composition can trigger distinct developmental programs in individual cell types. However, so far only a few of the factors fine-tuning the activity of Hox proteins have been identified (Mann & Chan, 1996; Mann et al., 2009).

On the other hand, it has been shown that Hox proteins interact with PBC cofactors through different protein motifs (Hudry, Viala, Graba, & Merabet, 2011). The distinct interaction modes can vary depending on the nature of the surrounding nucleotide sequence. The importance of the enhancer environment is underlined by a recent study showing that a *Drosophila* Ubx/Exd complex regulates *shavenbaby* (*shv*) expression by recognizing clusters of low-affinity binding sites that are strongly divergent from the Hox consensus motif (Crocker et al., 2015). Thus, instead of interacting with different kinds of cofactors specificity in Hox target gene regulation can also be achieved by the same Hox-cofactor complex differentially assembled via different protein domains. Irrespectively of how specificity is achieved *in vivo*, ultimately, Hox proteins regulate target genes to control morphogenesis.

These targets have different functions, many of them coding for signaling molecules and TFs, which instruct cell behavior only indirectly, but a substantial fraction are also realisators, proteins that are directly responsible for the basic cellular functions (Choo, White, & Russell, 2011; Garcia-Bellido, 1975). And as a consequence, morphological adaptations and changes along the A/P axis are put into effect by Hox TFs (together with the appropriate cofactors) by their ability to fine-tune, in a cell-type-specific manner, a multitude of cellular processes, like cell proliferation, cell migration, cell expansion, cell adhesion, and most importantly apoptosis (Foronda et al., 2009).

3. APOPTOSIS: THE CONTROLLED KILLING OF CELLS

The ability of an organism to kill cells in a controlled manner is an important process during normal development as well as in disease-prone situations (Fuchs & Steller, 2011). In a variety of developmental and morphogenetic events, apoptosis is used as a means to sculpt tissue, either by generating a biological scissor that eliminates supernumerous cells, or by triggering a pulling force that induces cellular reorganization of a tissue (Fuchs & Steller, 2011). However, apoptosis is not only critical for normal development but is also essential for the survival of an organism by removing harmful cells and prevent the initiation of a disease processes (Suzanne & Steller, 2013). Cells undergoing apoptosis display a series of characteristic physiological changes, including plasma membrane blebbing, permeabilization of the mitochondria outer membrane, DNA fragmentation, nucleus disintegration, and finally cell disintegration into apoptotic bodies, which are degraded by phagocytes (Suzanne & Steller, 2013). Apart from apoptosis, which is one form of programmed cell death (PCD), additional processes controlling cell death, such as necrosis and autophagy (Kar, Singha, Venkatachalam, & Saikumar, 2009; Yuan & Kroemer, 2010) will not be considered here.

Apoptosis is mediated by a biochemical pathway, which is conserved in multicellular organisms. Molecular analysis revealed the existence of a cell death machinery core that is organized around a family of cysteine proteases, termed caspases (Hengartner, 2000; Thornberry & Lazebnik, 1998). Apoptosis is normally initiated via signaling from the outside (extrinsic or death receptor pathway) or the inside of the cell (intrinsic or mitochondrial pathway). In both cases, it results in the activation of caspases that act in a proteolytic cascade to dismantle the dying cell (Baliga et al., 2003; Cain, Bratton, & Cohen, 2002; Reubold & Eschenburg, 2012). Caspases are normally synthesized as proenzymes or zymogens, which are activated by proteolytic cleavage. In mammals, assembly of the apoptosome, a large cytosolic protein complex, is pivotal for the induction of the mitochondrial pathway of apoptosis, and is prompted by the binding of cytochrome *c* to the apoptotic protease-activating factor 1 (Apaf-1) in the presence of ATP. Once assembled, the apoptosome recruits, processes, and activates Caspase-9 as the initiator caspase to form a holoenzyme complex. This in turn recruits and activates downstream effector caspases, such as Caspase-3. Once initiated, cytochrome *c* and apoptosis antagonists, inhibitors of apoptosis proteins (IAPs), are released from mitochondria, which leads to the activation of Caspase-3 through the association with the adapter protein Apaf-1 (Rodriguez & Lazebnik, 1999) and inhibition of the X-linked inhibitor of apoptosis (Eckelman & Salvesen, 2006; Yang, Fang, Jensen, Weissman, & Ashwell, 2000). Ultimately, Caspase-3 activation drives apoptosis in mammals (Fuchs & Steller, 2011).

The activation of caspases, which is central to apoptotic cell death, and the subsequent mobilization of key elements of this process are highly conserved (Baliga et al., 2003; Cain et al., 2002; Reubold & Eschenburg, 2012). In *C. elegans*, at least four central genes, *ced-3* (*caspase-9 like*), *ced-4* (*Apaf-1 like*), *ced-9* (*Bcl-2 like*), and *egl-1* (*BH3-only like*), are required for the execution of apoptosis. Earlier studies showing that somatic cells normally undergoing apoptosis survive in *egl-1*, *ced-4*, and *ced-3* mutants, uncovered the stepwise activation of apoptosis (Hengartner, 2000). A vast body of work revealed the details of this cascade: uncleaved pro-Ced-3 gets activated through the interaction with Ced-4, which is normally nonfunctional due to its coupling to the mitochondrial membrane via Ced-9. Release of Ced-4 is achieved by the upregulation of the prodeath protein Egl-1 that in turn blocks Ced-9. Ced-4 can now migrate to the perinuclear membrane, where it oligomerizes and "mobilizes" pro-Ced-3 (Baliga et al., 2003; Cain et al., 2002; Reubold & Eschenburg, 2012). Oligomerization of Ced-4 is believed to bring Ced-3 molecules into close proximity, leading to intermolecular autocatalytic cleavage and activation of adjacent Ced-3 molecules (Fuchs & Steller, 2011).

A homologue of Apaf-1 also exists in *Drosophila*, the *Drosophila* Apaf-1-related killer, which shares the same principal domain structure with its mammalian counterpart. In response to different death signals, apoptosis in *Drosophila* is induced by activation of the Reaper-family of proteins that includes Reaper (Rpr), Grim, Head evolution defective (Hid), and Sickle (Skl) (Steller, 2008). Rpr, Grim, and Hid are antagonists of IAPs. They contain a short N-terminal IAP-binding-motif domain that plays an important role in IAP binding and cell death initiation (Shi, 2002). Induction of the Rpr protein family results in inactivation of the *Drosophila* inhibitor of apoptosis protein-1 (IAP-1) (Goyal, McCall, Agapite, Hartwieg, & Steller, 2000; Wang, Hawkins, Yoo, Muller, & Hay, 1999) and the Caspase-9-like protein Dronc, which in turn leads to the maturation of the two Caspase-3-like proteins, DrICE and Dcp-1, thereby initiating apoptosis.

In conclusion, apoptosis is a crucial process which needs to be tightly controlled: its misregulation can lead to inappropriate cell numbers, either too many or too few, leading to developmental errors and/or to the survival of cells able to initiate disease-related processes.

4. HOX–APOPTOSIS REGULATORY INTERACTIONS DURING DEVELOPMENT

Due to their fundamental role during development, it is not surprising that Hox proteins, among many other processes, also control the apoptotic pathway. This regulatory interaction was first described almost 30 years ago when Cynthia Kenyon showed that mutations in the *C. elegans Hox* gene *mab-5* results in the survival of neuronal precursors, which normally undergo apoptosis in the posterior ventral nerve cord (VNC) (Kenyon, 1986; Fig. 2). This finding was ground-breaking, since it demonstrated that *Hox* genes drive morphogenesis, in this case the development of the central nervous system (CNS), along the A/P axis by selectively killing cells in a targeted and controlled manner. Till then, a huge amount of literature has accumulated indicating a regulatory connection between *Hox* genes and apoptosis in different developmental contexts or in disease situations. However, the mechanistic basis of these interactions has been worked out only for a few examples, in most cases only a correlation between *Hox* gene misexpression and apoptosis dysregulation has been reported. In the remaining part of the review, we will first focus on well-established paradigms of Hox–apoptosis interactions and summarize how single Hox TFs or those belonging to the same paralog group control at very restricted locations morphological outputs. This will be followed by examples highlighting how Hox proteins of different paralog groups differentially shape tissues along the A/P axis by controlling apoptosis. The examples chosen do not aim to provide an exhaustive overview, but instead are aimed at offering a sufficient coverage to apprehend common themes and differences emerging within the Hox–apoptosis field.

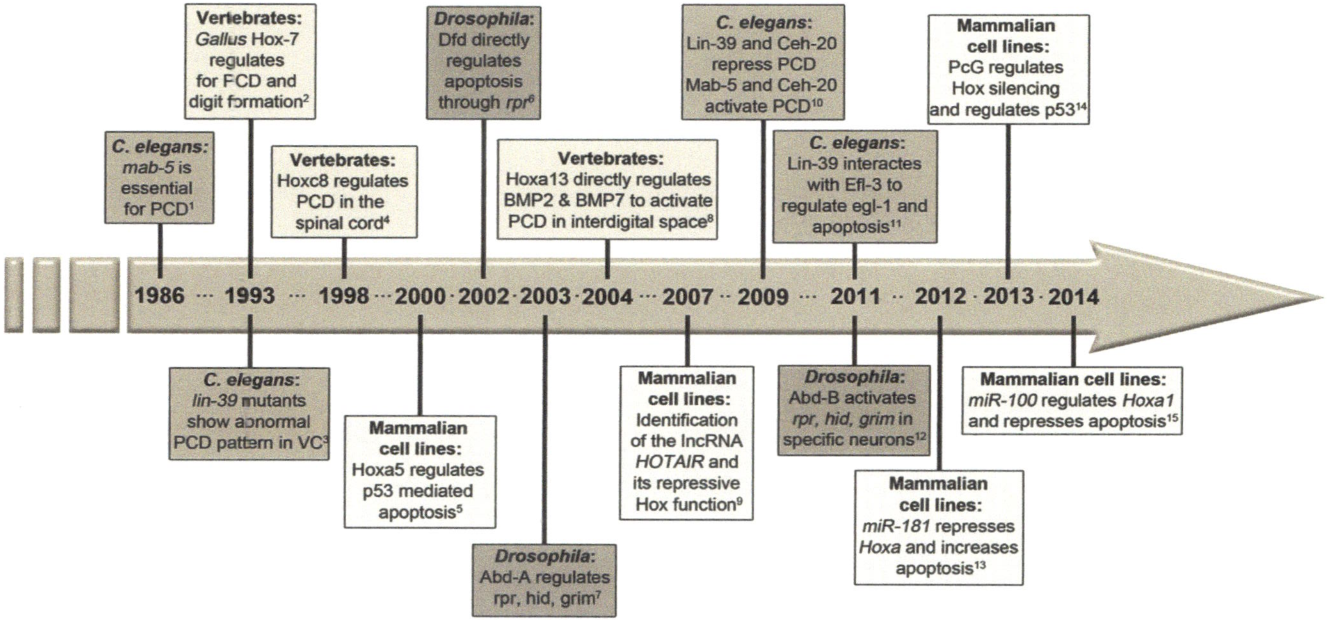

Figure 2 Key discoveries in the field of Hox-related apoptosis described in this review. (1) Kenyon (1986), (2) Coelho, Upholt, and Kosher (1993), (3) Clark, Chisholm, and Horvitz (1993), (4) Tiret, Le Mouellic, Maury, and Brulet (1998), (5) Raman et al. (2000), (6) Lohmann, McGinnis, Bodmer, and McGinnis (2002), (7) Bello, Hirth, and Gould (2003), (8) Knosp, Scott, Bachinger, and Stadler (2004), (9) Rinn et al. (2007), (10) Potts, Wang, and Cameron (2009), (11) Winn, Carter, Avery, and Cameron (2011), (12) Suska, Miguel-Aliaga, and Thor (2011), (13) Li et al. (2012), (14) Yang et al. (2013), and (15) Xiao et al. (2014).

4.1 Restricted Adjustment of Morphologies by Single or Closely Related Hox Proteins

One important example showing that a single Hox TF sculpts a specific morphological aspect of the animal body via the activation of apoptosis is the formation of the maxillary/mandibular segment boundary in *Drosophila* embryos (Lohmann et al., 2002). This morphogenetic event is under the control of the *Hox* gene *Dfd*, which directly activates the expression of the proapoptotic gene *rpr* in a few cells at the segmental border. The subsequent *rpr*-induced death of the cells is both sufficient and required for the boundary to be formed and maintained (Lohmann et al., 2002; Fig. 3B1). In order to do so, apoptosis needs to be tightly regulated otherwise the morphological boundary would be either not formed or too extensive, both outcomes probably detrimental to the organism. Besides being one of the first demonstrations that Hox proteins can carve animal structures using the apoptotic machinery, this study allowed several important conclusions. First of all, it showed that at least in some developmental contexts, Hox proteins directly regulate the expression of critical apoptosis genes. Thereby, this study has confirmed the postulate proposed by Garcia-Bellido 40 years ago that *Hox* genes control the morphogenetic output by controlling genes directly involved in cellular processes (Garcia-Bellido, 1975), in this case the apoptotic machinery. Furthermore, this study highlights that Hox TFs directly influence as micomanagers the cellular behavior of a few cells only, which results in a very characteristic morphological output, in the example discussed the formation of a segmental border. Despite their broad expression, Hox proteins solve this problem by collaborating with other regulatory proteins, and in the case of the maxillary/mandibular boundary, it was shown that Dfd interacts with at least eight other TFs on the *rpr* CRM to achieve localized and timed *rpr* expression (Stobe et al., 2009). Thus, Hox proteins are able to control apoptosis in a very direct, fast, and highly restricted manner via their capacity to interact with a large diversity of regulatory inputs.

Another example of how a single Hox protein shapes a specific feature along the A/P axis using the apoptotic machinery is the acquisition of sex-specific morphological differences characteristic for sexually reproducing animals (Kopp, Duncan, Godt, & Carroll, 2000; Sanchez, Gorfinkiel, & Guerrero, 2001; Wang, Kidd, Carroll, & Yoder, 2011). This sexual dimorphism is relatively well understood in *Drosophila*, where abdominal segments in males are not only darker pigmented (Williams et al., 2008) but also fewer in number in comparison to females. A recent study showed that the

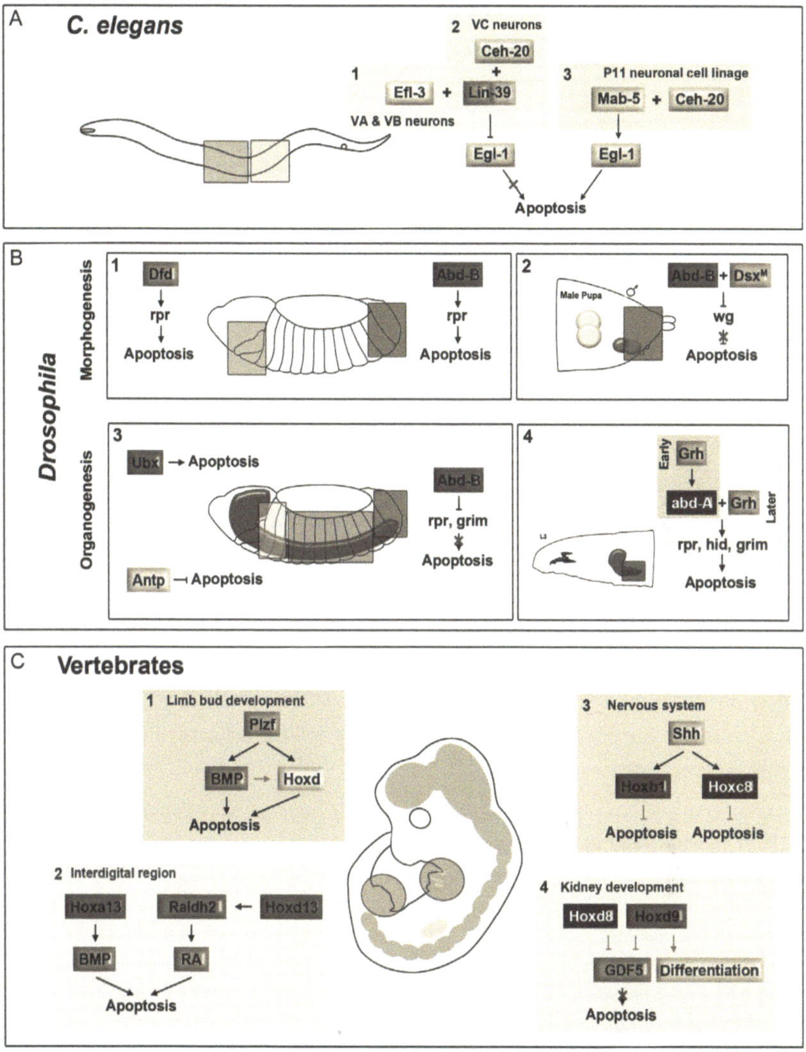

Figure 3 Key paradigms illustrating Hox and apoptosis interplay during the development and morphological diversification of *C. elegans* (A), *Drosophila* (B), and vertebrates (C). Arrows indicate activating and blunt ending arrows indicate inhibitory input: the black-colored arrows represent direct regulations, while the gray ones represent an indirect control or not yet identified mechanism. (A) Cell fate decisions in VA, VB, VC neurons, and P11 neuronal lineage in *C. elegans*. (A1) In the VA and VB neurons, Lin-39 and Efl-3 together inhibit Egl-1 and prevent apoptosis. (A2) In VC neurons, Lin-39 and Ceh-20 form a complex that represses *egl-1* and blocks apoptosis. (A3) In the P11 pioneer neuronal lineage, Mab-5 and Ceh-20 form a complex, which binds the *egl-1* regulatory region and activate apoptosis. (B) Hox and apoptosis interplay during *Drosophila* development. (B1) Shaping the *Drosophila* embryo is achieved through a direct binding of the

(Continued)

male-specific loss of abdominal segments is under the control of the Hox TF Abd-B and the sex-determination TF Doublesex (Dsx) (Keisman, Christiansen, & Baker, 2001; Steller, 2008; Wang et al., 2011). Dsx has male- (Dsx^M) and female (Dsx^F)-specific isoforms, and in order to reduce segment numbers, the male-specific Dsx isoform (Dsx^M) in combination with Abd-B represses *wingless* (*wg*) expression in the abdomen (Wang et al., 2011). This leads to activation of apoptosis and consequently to a loss of segments in males (Fig. 3B2). Accordingly, *Abd-B* and *dsx* mutants show *wg* derepression in females, and a male-like development of abdominal segments. Thus, this study underlines that individual Hox proteins can change and adapt the local morphology of an organism, even for the acquisition of sex-specific features, and that they achieve this goal by their ability to interact with cofactors that can be processed in a context-dependent manner (Wang et al., 2011).

In vertebrates, the regulation of apoptosis by closely related Hox proteins, more precisely those belonging to the same paralog group (Fig. 1C) is also used to shape-specific features along the A/P axis. Without any doubt, the most prominent example is the morphogenesis of the vertebrate limb (Johnson & Tabin, 1997). The development of this tetrapod structure is tightly controlled by various factors. The most important tool to sculpt the limb is the employment of apoptosis, the so-called interdigital cell death

Figure 3—Cont'd Hox TFs Dfd and Abd-B to the *rpr* enhancer, thereby activating apoptosis in specific cells. (B2) During pupal stages, Abd-B interacts with the male-specific form of Dsx (Dsx^M) to repress *wg* expression and activate apoptosis giving rise to male-specific structures. (B3) During embryonic ventral nerve cord development in *Drosophila* (stages 11–15), repression of apoptosis in the thoracic segments by Antp and activation of apoptosis in the abdominal segments by Ubx control the cell fate of defined neurons. In the posterior segments of *Drosophila* embryos (stages 16–17), Abd-B represses *rpr* and *grim* and inhibits apoptosis in the MP1 and dMP2 pioneer neuronal lineages. (B4) In *Drosophila* larvae, Abd-A and its cofactor Grh reduce the number of neuroblasts in CNS by activating apoptosis via *rpr*, *hid*, and *grim* upregulation. (C) Hox-regulated apoptosis in vertebrates. (C1) During limb bud development, Plzf activates BMPs and *Hoxd* cluster gene expression and thereby triggers apoptosis, downstream of this cascade, in the interdigit regions. (C2) During digit formation, apoptosis within the interdigit region is triggered upstream of the cascade by Hoxa13, which activates BMP2 and BMP7, and by Hoxd13 which regulates Raldh2 and activates retinoic acid production. (C3) In vertebrate CNS, Hoxb1 in the hindbrain and Hoxc8 in the spinal cord block apoptosis and at the same time promote specification within the respective CNS regions. (C4) During kidney organogenesis, coordination of differentiation and apoptosis relies primarily on the function of Hoxd8 and Hoxd9, which block GDF expression and prevent the development of polycystic kidneys. Drosophila *pictures were adapted from Hartenstein (1993).* (See the color plate.)

(ICD), for the individualization of the digits (Nakamura, Fujioka, & Mori, 2000; Ros, Macias, Fallon, & Hurle, 1994; Ros, Piedra, Fallon, & Hurle, 1997; Scott, Ritter, & Wilson, 1977, 1980; Wise & Scott, 1982). In brief, limb formation starts in the limb field as undifferentiated mesenchymal cells from the lateral plate mesoderm, the so-called progress zone (PZ), proliferate to the point that they bulge out to create the limb bud. The mesenchymal cells of the PZ are surrounded by an ectodermal cell layer, which is covering the outer part of the growing limb bud, the apical ectodermal ridge (AER). The AER produces fibroblasts growth factors that act on the PZ cells to activate their differentiation along the proximal–distal axis. Furthermore, the AER guides the further development and controls apoptosis, which is necessary to eliminate the webbing between the digits (Chimal-Monroy et al., 2011; Kondo, Zakany, Innis, & Duboule, 1997). Once more, *Hox* genes, in this case *Hoxa* and *Hoxd* cluster genes mostly of the paralog group 13, via regulation of apoptosis, are involved in the shaping of a morphological structure, the limb, by removing undifferentiated mesodermal cells located between the developing digits, therefore, leading to the formation of individual digits from the hand or foot plate (Montero & Hurle, 2010). Consequently, mutations in *Hoxa13* and *Hoxd13* genes result in typical malformations, such as the incomplete digit separation and increased number of digits found in synpolydactyly, or the shortening of the distal limbs accompanied by urogenital tract defects characteristic of the hand-foot-genital syndrome (Goodman, 2002; Sharkey et al., 1997). However, contrary to the formation of the maxillary–mandibular boundary in flies, *Hox* genes of the *Hoxa* and *Hoxd* cluster control ICD in a noncell autonomous manner via the activation of various signaling pathways, mostly the bone morphogenetic protein (BMP) and retinoic acid (RA) pathways (Knosp et al., 2004; Montero & Hurle, 2010; Vitobello et al., 2011). As an example, Hoxa13 directly regulates the expression of BMP2 and BMP7, which in turn induce cell death of neighboring cells located in the interdigital region (Knosp et al., 2004; Fig. 3C2). Thus, in *Hoxa13* mutants supernumerous cells remain in the interdigital space. However, besides controlling apoptosis, Hoxa13-dependent BMP expression is also essential for limb morphogenesis, since it drives digit outgrowth and chondrogenesis (Knosp et al., 2004). This highlights that *Hox* genes often trigger not only one but several critical processes both in a cell and noncell autonomous manner. Limb development is not only disturbed in *Hox* gene mutants but also when upstream regulators of *Hox* genes, like the Promyelocytic leukemia zinc finger (Plzf) protein, are affected (Barna, Hawe, Niswander, & Pandolfi, 2000).

However, there is an additional level of complexity, since Hox TFs have a tendency to regulate downstream processes by interacting with their immediate targets in a feed-forward manner. For example, Hoxd13 directly regulates Raldh2 and affects RA production in the mouse limb (Kuss et al., 2009), and both, RA in combination with Hoxd13 expressed in the interdigit mesenchyme subsequently suppress chondrogenesis in the interdigital space (Kuss et al., 2009; Fig. 3C1). Therefore, it becomes evident that apoptosis induced digit individualization is established through complex regulatory circuits in which *Hox* genes control downstream molecules of limb morphogenesis that function not only on neighboring cells but also feedback and regulate in combination with the Hox proteins subsequent steps of limb formation (Goodman, 2002; Montero & Hurle, 2010; Sharkey et al., 1997; Fig. 3C1 and 2).

4.2 Morphological Diversification of Tissues along the A/P Axis by Different Hox Proteins

The examples discussed above highlight that single Hox proteins or those belonging to the same paralog group are able to control apoptosis in order to shape structures located at specific positions along the A/P axis, like the limb or the maxillary/mandibular boundary. In the following, we will discuss cases in which different Hox proteins regulate apoptosis within one tissue in order to realize its diversification along the A/P axis. The tissue we will focus on mostly in this context is the CNS, which consists of the brain and the nerve cord, termed VNC in invertebrates and spinal cord in vertebrates. The CNS is highly heterogeneous along the A/P axis thereby ensuring the correct matching of neurons and their targets (Lin & Lee, 2012; Suska et al., 2011), which is critical for the execution of region-specific animal behavior such as chewing, breathing, and walking (Baek, Enriquez, & Mann, 2013; Mandeville et al., 2006). *Hox* genes have long been known to play an important role in CNS development (Capecchi, 1997; Dasen, Liu, & Jessell, 2003), and in particular, their role in regulating the segmental variability within this tissue by the differential regulation of apoptosis has been well studied, as described in this section.

The Hox–apoptosis interaction in the CNS was first demonstrated in *C. elegans* when it was recognized that the *Hox* genes *lin-39* and *mab-5* shape different regions of the worm CNS by positively or negatively regulating apoptosis in specific neurons. For example, survival and specification of hermaphrodite-specific neurons VA, VB, VC, VD, and AS generated in the midregion of the VNC (P3–P8) are dependent on the cell-autonomous function of *lin-39* (Clark et al., 1993). In *lin-39* mutants, VC neurons undergo apoptosis, similar to VC-lineal equivalents in anterior and posterior regions (Winn et al., 2011). Therefore, Lin-39 promotes VC survival and provides lineage specification for VC fate under normal conditions (Clark et al., 1993; Potts et al., 2009; Winn et al., 2011). To prevent apoptosis in VC neurons, Lin-39 forms a complex with the TALE-class Pbx homologue Ceh-20 binds to the *egl-1* promoter and repress transcription of this proapoptotic gene (Potts et al., 2009; Fig. 3A2). In contrast to the VC lineage, the VA and VB cell populations require Efl-3 function, an E2F type TF important for cell cycle regulation, in addition to Lin-39 to repress effectively transcription of the *egl-1* cell death gene and thereby promote VA and VB neuronal cell fate (Winn et al., 2011; Fig. 3A1). Thus, Lin-39 provides spatial information to cell lineages developing along the VNC axis, while Efl-3 instructs VA and VB neurons to become different from VC neurons. In the posterior VNC, *mab-5* is essential for the death of two lineally related cells, the P11p and P12p (Kenyon, 1986), as these cells are present in *mab-5* mutants. In the P11 lineage, a complex between Mab-5 and the Pbx homolog Ceh-20 directly regulates transcription of *egl-1* to initiate apoptosis, whereas in the P12 lineage Mab-5 and Ceh-20 only indirectly affect this process (Liu, Strauss, Potts, & Cameron, 2006; Fig. 3A3). Therefore, the shaping of the *C. elegans* nerve cord proceeds through the cooperation of Hox and apoptosis proteins, which in combination with Pbx and Trx family members set up different regulatory networks along the A/P axis of the ventral cord.

The diversification of the CNS along the A/P axis via the Hox-dependent regulation of apoptosis has been extensively studied in *Drosophila*. It starts during embryogenesis when CNS cells are specified and continues during postembryonic stages when the CNS is further trimmed and shaped. For example, the neighboring *Hox* genes *Antp* and *Ubx*, essential for the development of the thoracic and the abdominal segments, respectively (Lewis, 1978; Sanchez-Herrero, Vernos, Marco, & Morata, 1985), control the survival and death of specific motoneurons (Rogulja-Ortmann, Renner, & Technau, 2008). The two Hox TFs control these events in a segment-specific manner, since Antp is responsible for the survival of motoneurons developing from NB7–3 and NB2–4 neuroblasts in the first two thoracic segments, whereas in the third thoracic segment, motoneurons belonging to these neuroblast lineages undergo Ubx-induced apoptosis. In these cells, shortly before they die, Ubx is strongly upregulated and promotes *rpr*-induced death of these cells thereby counteracting the function of

Antp in promoting cell survival (Rogulja-Ortmann et al., 2008; Fig. 3B3). This example demonstrates what is often found in the Hox field: a dual role of the Hox TFs in certain cell lineages (Perrin, Monier, Ponzielli, Astier, & Semeriva, 2004; Rozowski & Akam, 2002). In the case discussed, Ubx first establishes tagma-specific identity of the neuroblasts (which is done by the other *Hox* genes in the other segments), and later it induces apoptosis in individual differentiated motoneurons, most likely to regulate the formation of region-specific neuronal networks. It is this second function that requires Ubx to be upregulated, but it is unclear why Ubx levels need to be increased for apoptosis gene expression to be initiated. However, it is tempting to speculate based on recent findings (Crocker et al., 2015) that this is required in order to achieve specificity in Hox target gene regulation and thereby initiate timely controlled apoptosis induction.

Another interesting feature of this Hox–apoptosis interaction is the activation of apoptosis in postmitotic *Drosophila* neurons, which is rather unusual since cell death in the developing invertebrates nervous system normally occurs soon after neurons have been generated. This has been observed in another *Drosophila* neuronal subtype, the pioneer neurons, which are initially generated throughout the VNC and are required to delineate axon pathways by extending their axons and guiding follower axons (Hidalgo & Brand, 1997; Lin, Auld, & Goodman, 1995). Interestingly, the *Drosophila* pioneer neurons dorsal MP2 (dMP2) and MP1 die in almost all segments at late embryonic stages (stage 16–17), long after they have performed their pioneer function, and survive only in the last three posterior segments through the action of the Hox TF Abd-B (Miguel-Aliaga & Thor, 2004; Fig. 3B3) Thus, this study not only underlines the capacity of Hox TFs to shape the CNS by the selective elimination of differentiated neurons but it also uncovers their antiapoptotic function via the repression of cell death executioners, like *rpr* and *grim*, in selected cells.

The example discussed before shows that one way by which Hox proteins activate apoptosis at a very specific time point is through their pulsed expression (Rogulja-Ortmann et al., 2008). This strategy is also used in a different developmental context, during the postembryonic phase of *Drosophila* CNS development, when more than 90% of the neurons present in the adult CNS are generated by postembryonic neuroblasts (pNBs) (Truman & Bate, 1988). pNBs share the same lineage with their embryonic counterparts. In the early embryo, each hemisegment contains an invariant number of 30 neuroblasts. However, this is no longer the case in larvae. At larval stages, 23 of the 30 embryonic neuroblasts persist in the thoracic segments, whereas in the central abdominal segments only three embryonic neuroblasts survive (Truman & Bate, 1988). The dramatic reduction in the number of abdominal pNBs results from a *rpr*-induced cell death (Abrams, White, Fessler, & Steller, 1993; Peterson, Carney, Taylor, & White, 2002; White et al., 1994). And interestingly, it is a peak of *Hox* expression, in this case of *abd-A*, that mediates cell-autonomously the death of the abdominal pNBs (Bello et al., 2003). Analysis of mosaic clones revealed that the absence of either *abd-A* or the proapoptotic genes *rpr, hid, grim* results in continuous proliferation of pNBs, leading to the production of large pNB progeny clones (Fig. 3B4). This effect is specific to the *Hox* gene *abd-A*, since similar experiments with *Ubx* had no effects on pNB proliferation (Bello et al., 2003). Interestingly, ectopic thoracic expression of *abd-A* results in a massive reduction of the clone size, indicating that Abd-A can selectively activate apoptosis in pNBs. Furthermore, it is not Abd-A alone but the collaboration with the TF Grainyhead (Grh) that allows restriction of pNBs proliferation via the induction of apoptosis (Cenci & Gould, 2005). However, Grh functions not only together with Abd-A in defined pNBs to reduce neuroblast numbers but it is also required to maintain *abd-A* expression during the larval stages, thus acting also upstream of the *abd-A* gene (Cenci & Gould, 2005; Fig. 3B4). Taken together, these examples highlight that *Hox* genes can control the onset of apoptosis due to their upregulation at the time of action. A critical question here is how can elevated Hox protein levels activate the transcription of apoptotic genes while normal levels do not? The answer to this puzzle is likely to be found in the regulatory regions of these genes: it is well known that TFs interact with binding sites of different affinity (Spitz & Furlong, 2012). Thus, CRMs harboring low-affinity sites might be specifically bound by TFs and control target gene expression only when TF levels increase and reach a certain threshold. In the specific examples above, Ubx in the motoneurons and Abd-A in the pNBs would interact with low-affinity Hox-binding sites of CRMs located in the *rpr*-regulatory region and would initiate the apoptotic program in the respective cells when their expression reaches a threshold. However, since these CRMs have not been identified, the mechanistic basis of this regulatory interaction is still unclear. Another interesting question in this context is how the upregulation of Hox expression, which is of uttermost importance for the shaping of the CNS, is achieved only in a few selected cells. Despite the fact that several possibilities have been discussed, such as a transient extrinsic signal or a cell-intrinsic signal that measures the elapsed time, so far the nature of the developmental timers responsible for the Hox expression bursts are unknown.

Hox genes often regulate multiple processes within a cell (Choo et al., 2011), and one emerging view in the field is that Hox TFs impact on two functions: cell fate commitment and cell-type-specific apoptosis. The coordination of these events is frequently observed in the vertebrate CNS and is adapted to pattern and shape the brain and spinal cord. We will discuss here a few prominent cases. For example, the diverse neuronal subtypes in the vertebrate hindbrain arise from progenitor cells specified by the combined actions of signaling molecules along the A/P and dorso-ventral (D/V) axis. In the hindbrain compartment rhombomere 4 (r4), *Hoxb1* shows a defined expression along the entire D/V axis at stage E9.5 controlling the commitment and specification of neuronal progenitor cells along this axis (Gaufo, Flodby, & Capecchi, 2000). Hoxb1 initiates a genetic cascade, which includes major signaling pathways, such as Sonic hedgehog (Shh) signaling involved in early D/V patterning and the Mash1/Ngn signaling which plays a critical role in neural determination. In the absence of *Hoxb1*, neurons that normally express high levels of this *Hox* gene undergo apoptosis within the D/V axis of r4 (Gaufo et al., 2000). A similar link between Hox-mediated specification and apoptosis is observed in the spinal cord segments C_7, C_8, and T_1, where *Hoxc8* is specifically expressed in motorneurons that innervate the forelimb distal muscles essential for the movements of the forepaw (Tiret et al., 1998). In *Hoxc8* mutants, formation of these muscles is unaffected but their innervation is perturbed with a numerical deficit of motoneurons and an irreversible disorganization of motor pools. Interestingly, C7–8 motoneurons lose their selective advantage in growth cone pathfinding and target recognition, the two most essential steps in the establishment and maintenance of functional neuromuscular units. At the same time, apoptosis is specifically enhanced in C7–T1 motoneurons in a segment-specific pattern (Fig. 3C3). These results indicate that Hoxc8 function is essential both for the establishment of the spinal cord patterning and for the survival of motoneurons (Tiret et al., 1998).

Similarly, organogenesis depends on the proper balance between cell-type specification, proliferation, and apoptosis, as in the case of kidney development (Di-Poi, Zakany, & Duboule, 2007). Genes of the *Hoxd* cluster are actively involved in kidney organogenesis and are required both for the growth and the organization of the organ. This is demonstrated when *Hoxd* cluster genes are mutated, which results in reduced kidney size and polycystic kidneys. Within the *Hoxd* cluster, the most posterior genes *Hoxd13* to *Hoxd11* regulate interactions in the metanephric mesenchyme–ureteric bud, while the more anterior *Hox* genes, *Hoxd9* and *Hoxd8*, maintain the integrity of renal tubular epithelia (Di-Poi et al., 2007; Wellik, Hawkes, & Capecchi, 2002). In the absence of *Hoxd9* and *Hoxd8* gene function, animals develop polycystic kidneys associated with increased apoptotic activity in renal tubules, which is likely due to the strong upregulation of the BMP-related *Growth and differentiation factor 5* (*Gdf5*) (Di-Poi et al., 2007; Fig. 3C4). The importance of a coordinated proliferation, differentiation, and apoptosis during kidney development is seen in human and rodent models of polycystic kidney disease whose pathological features include increased apoptosis associated with dedifferentiation, adhesion, and ciliary abnormalities of epithelial cells. Here again, *Hox* genes couple their cell-type specification function to apoptosis, in order to maintain this delicate balance during kidney morphogenesis.

5. INTERPLAY OF *HOX* GENES AND APOPTOSIS IN THE DISEASE CONTEXT

All the paradigms outlined so far show the importance of Hox proteins as cell autonomous and nonautonomous regulators in developmental circuits: from morphogenetic tissue sculpting to segmental patterning and cell fate commitment. Downstream of these cascades, apoptosis is tightly coupled to the Hox-mediated developmental circuits and apoptotic activation or suppression is always developmentally controlled. Within this framework, numerous examples of aberrant Hox expression followed by dramatic effects on organismal survival have been described in cancer malignancies and oncogenic transformations (reviewed in Shah and Sukumar, 2010). Understanding the mechanisms governing the Hox-controlled cancer networks is of critical importance for cancer therapies and the development of medical approaches targeting not only the symptomes but also the cause of these malignancies.

5.1 *Hox* Genes in Leukemic Transformations

All leukemic malignancies are characterized by a high number of abnormal white blood cells; however, on the molecular level this cancer type is extremely multifaceted and in many cases *Hox* genes are involved. One example is the Mll, an acute lymphoblastic leukemia, which is caused by chromosomal translocations affecting the *Mll* gene resulting in an upregulation of the *Hoxa* gene cluster (Dorrance et al., 2006; Golub et al., 1999). In normal tissue, the Mll protein, the human homologue of the *Drosophila* Trithorax (Trx), maintains expression of the *Hoxa* genes by epigenetic means (Dorrance et al., 2006; Milne et al., 2002; Nakamura et al., 2002), suggesting that in the leukemic situation Mll-dependent epigenetic regulation at the *Hoxa* locus is changed. In that context, the finding that several lncRNAs become differentially expressed in Mll samples (Fang et al., 2014) is highly interesting. For example, deregulation of the lncRNA Mistral, which normally activates *Hoxa6* and *Hoxa7* expression by recruiting Mll to the chromatin (Bertani, Sauer, Bolotin, & Sauer, 2011), could result in abnormal chromatin binding of Mll to the *Hoxa* locus thereby causing altered *Hoxa* expression. In line with that hypothesis, deregulation of both *Hoxa* gene cluster and lncRNAs is associated with apoptosis inhibition and increased proliferation in the leukemic condition (Bertani et al., 2011; Fang et al., 2014). Thus, these results highlight the synergism between lncRNAs, Hox function, and apoptosis in Mll pathogenesis (Fang et al., 2014). However, it is unclear so far whether apoptosis inhibition and enhanced proliferation are mediated by direct regulatory inputs from the *Hox* genes, the lncRNAs, or a combination of both and which genes act as the downstream cellular executioners.

One of the key regulators in leukemia development is *Hoxa9*. Elevated expression of Hoxa9 is strongly and causally associated with the development of acute myelogenous leukemia (AML) with Mll rearrangements (Brumatti et al., 2013). *Hoxa9* overexpression maintains the self-renewal capacity of leukemic stem cells, blocks differentiation, and promotes cell survival leading to immortalization of hematopoietic cells. Hoxa9-dependent control of the cell survival pathways is observed in mutant fetal thymus, where *Hoxa9* downregulation promotes apoptosis, suggesting that one key function of Hoxa9 is to repress apoptotic pathways. Several pieces of evidence suggest that on the molecular level, Hoxa9-dependent immortalization in AML relies on Bcl-2: first, downregulation of Hoxa9 induces apoptosis and loss of *Bcl-2* expression, and second, deletion of Bcl-2 in *Hoxa9* expressing cells activates apoptosis. Therefore, Hoxa9-regulated *Bcl-2* expression seems to be important for the survival of myeloid progenitors and contributes to the oncogenic effects of Hoxa9 overexpression (Brumatti et al., 2013).

Another special case of leukemia with Mll rearrangements involves fusion of the Hoxa9 HD to the Nucleoporin 98 (Nup98) (Brumatti et al., 2013). Nup98 belongs to the nuclear pore complex comprised of 30 different proteins termed nucleoporins (D'Angelo & Hetzer, 2008). Genomic translocations affecting the function of nucleoporins lead to several human disorders including AML (Gough, Slape, & Aplan, 2011; Xu & Powers, 2009). Immortalization of the hematopoietic progenitor cells in leukemia cells with a Nup98–Hoxa9 rearrangement no longer requires Bcl-2 and Hox cofactors (Brumatti et al., 2013), suggesting that cancer-promoting apoptosis inhibition is controlled by another mechanism.

Increased expression of Hoxa9 and its cofactor Meis1 are associated with cytogenetically normal AML (CN-AML) development (Mulgrew et al., 2014). In a Hoxa9/Meis model of CN-AML, the hepatocyte growth factor (HGF) is increased in leukemic bone marrow compared to normal controls (Mulgrew et al., 2014). c-Met and its ligand HGF are involved in the development of various cancer types showing increased cell proliferation, cell survival, and metastasis (Ma et al., 2007; Peruzzi & Bottaro, 2006). Interestingly, a drug-screening assay for potential antileukemia-specific activity revealed that an inhibitor of the HGF receptor c-Met (SU11274) could increase apoptosis through caspase activation, decrease cell number, restore differentiation, and alter cell cycle dynamics. The identification of the c-Met inhibitor is of clinical importance for a potential cancer therapy by repressing potential Hoxa9/Meis targets and their function in CN-AML patients (Mulgrew et al., 2014).

So far, we have only discussed the role of *Hoxa* cluster genes and their role in leukemic transformations. However, *Hoxb* cluster genes also play fundamental roles in these leukemic pathologies. In HL60 leukemia cell lines, *Hoxb1* expression is repressed by *Hoxb1* promoter hypermethylation, whereas induced *Hoxb1* reactivation in these cells increases proapoptotic gene expression. This in turn reverts the cell-type-specific hematopoietic differentiation pathway, indicating that Hoxb1 is critical for leukemia development (Petrini et al., 2013). Similar to Hoxb1, Hoxb3 misexpression is also associated with uncontrolled proliferation of hematopoetic cells (Buske & Humphries, 2000), which depends in this case on interleukin (IL)-mediated cytokine signaling (von Freeden-Jeffry et al., 1995). Under normal conditions, Hoxb3 controls B-lymphocyte development in the mouse bone marrow (Ko et al., 2007) by fine-tuning the balance between cell production by proliferation and cell loss by apoptosis. One critical player in this context is the IL-7 receptor IL-7R: in *Hoxb3* depleted B-cells, reduced levels of IL-7R result in enhanced apoptosis and reduced proliferation. Identification of multiple Hox-binding sequences upstream of the IL-7R promoter suggests that Hoxb3 may directly regulate IL-7R expression and thereby activate apoptosis within the mouse bone marrow (Guazzi, Pintonello, Vigano, & Boncinelli, 1998; Ko et al., 2007). Thus, it is plausible to speculate that enhanced proliferation and reduced apoptosis observed in the disease context will depend on Hoxb3-mediated *IL-7R* upregulation.

Interestingly, not only *Hox* genes are deregulated in leukemias but also several miRNAs located within these Hox clusters (Garzon, Garofalo, et al., 2008; Garzon, Volinia, et al., 2008). For example, *miR-10a* is coexpressed with *Hoxb4* and *Hoxb5* in normal developing mouse embryos and all three are upregulated in AML patients. Moreover, miRNA and mRNA expression profiling of bone marrow or peripheral blood samples from AML patients indicates a significant overrepresentation of *Hox* genes from the Hoxa and Hoxb clusters. At the same time, there is a positive correlation of *miR-145*, *miR-181a*, and *miR-181b* with proapoptotic genes in AML cell samples (Havelange et al., 2011). Family members of the miR-181 target the Hoxa cluster (Hoxa7, Hoxa9, and Hoxa11) and the Hox cofactor Pbx3. And ectopic expression of *miR-181b* in leukemic cells promotes apoptosis, reduces cell viability and proliferation (Li et al., 2012). Accordingly, CA-AML patients with an increased expression of *miR-181* show longer overall survival rates when compared to other AML patients (Li et al., 2012). Thus, miRNAs and *Hox* genes seem to control apoptosis in a coordinated manner.

In sum, it becomes clear that Hox genes have a central role in leukemia induction and progression. Although it is not always clear how Hox-regulated apoptosis is mediated, since the mechanisms involved are clarified in only a few cases, it becomes obvious that *Hox* gene deregulation in leukemia has profound effects on apoptosis activation and/or derepression, increased cell survival and overproliferation of malignant leukemia cells.

5.2 Deregulation of *Hox* Genes and Apoptosis in Other Cancer Types

Fundamental insights into how *Hox* genes drive the oncogenic process were first obtained from studies of breast cancer (Raman et al., 2000). Experiments on breast cancer cell lines have shown that Hoxa5 and Hoxa10 regulate p53-mediated apoptosis and that loss of function of any of these genes increases the possibility to develop breast cell carcinomas (Chu et al., 2004; Raman et al., 2000; Reisman & Loging, 1998). Under normal conditions, expression of *Hoxa5* and the tumor suppressor gene *p53* protects the cells against malignant transformation (Raman et al., 2000; Reisman & Loging, 1998). Cell culture experiments showed that this is achieved through the interaction of Hoxa5 with the *p53* promoter region thereby activating *p53* expression. Accordingly, Hoxa5 upregulation increases *p53* expression levels and promotes the activation of apoptosis in cell culture (Raman et al., 2000). Genome-wide analysis of breast cancer cell lines revealed methylation of the *Hoxa5* promoter that results in silencing of its genomic region and thereby loss of *p53* activation (Raman et al., 2000). Therefore, it seems that development of breast cell carcinomas is a consequence of the loss of *Hoxa5* and *p53* expression, which suppresses apoptosis and allows tumor growth. Interestingly, Hoxa5 promotes apoptosis not only through *p53* activation but also by activating the Caspases-2 and -8 (Chen et al., 2004). Similar to Hoxa5, Hoxa10 regulates *p53* and apoptosis. Thus, loss of *Hoxa10* function leads to comparable breast cell abnormalities (Chu et al., 2004). On the other hand, increased metastasis in breast cancers is strongly associated with the *Hoxd* gene cluster. Interestingly, epigenetic silencing of the *Hoxd* cluster is controlled by the lncRNA HOTAIR, which together with PRC2 represses transcription of *Hoxd* genes (Gupta et al., 2010; Ponting, Oliver, & Reik, 2009; Rinn et al., 2007). HOTAIR is overexpressed in primary breast tumors and is associated with increased metastasis risk (Gupta et al., 2010), while reducing HOTAIR expression via isoflavones increases apoptosis and can inhibit breast cancer growth (Chen, Lin, Yong, Ye, & Huang, 2015).

Hox genes belonging to the paralog group 9 are frequently associated with brain tumors. For example, *Hoxa9*, which we have already discussed in the context of leukemic transformations, is also associated with very aggressive adult brain tumors, the so-called glioblastoma. In this pathological situation, Hoxa9 is aberrantly activated, which decreases apoptosis and increases cell proliferation (Costa et al., 2010). Transcriptional activation of *Hoxa9* is controlled by a phosphoinositide 3-kinase (PI3K)-associated epigenetic mechanism that involves histone H3K27 trimethylation. Interestingly, oncogenic *Hoxa9* activation and negative prognosis could be reversed via a PI3K inhibitor, which provides new insights into the role of epigenetic silencing in normal and malignant cells and opens new avenues to PI3K- and mTOR-based therapies (Costa et al., 2010).

However, Hox overexpression does not always result in cancer development; in some cellular contexts, it can reduce the tumor growth. This is the case in neuroblastoma malignancies where *Hoxc9* activation is associated with regression of the tumor. Neuroblastoma is an embryonal malignancy of the sympathetic nervous system, which can be reversed via the so-called spontaneous neuroblastoma regression, a delayed activation of a naturally occurring immune system process that destroys the tumor (Printz, 2001). Although it is not yet understood how this phenomenon works, spontaneous neuroblastoma regression is usually observed in young neuroblastomas and not of advanced stages. However, studies on the role of *Hoxc9* revealed that induced expression of *Hoxc9* in old neuroblastoma leads to a reduced cell proliferation and cell cycle arrest, restores differentiation, and abrogates tumor growth almost completely, "mimicking" the spontaneous neuroblastoma regression (Kocak et al., 2013; Mao et al., 2011). Interestingly, *Hoxc9* growth arrest is associated with the induction of an intrinsic apoptosis pathway that involves the release of cytochrome *c* from the mitochondria to the cytosol and activation of the intrinsic cascade of caspases. This strongly suggests that *Hoxc9* may be actively involved in the process of spontaneous neuroblastoma regression, a finding with a strong prognostic impact (Kocak et al., 2013).

The Hox–apoptosis regulatory interplay can be very different depending on the cellular context and nonetheless produce similar effects. This is the case in oral squamous cell carcinomas where, in contrast to the Hoxa5-mediated breast cancer, Hox expression, in this case expression of Hoxa1, is not lost but increased. High levels of Hoxa1 stimulate cellular proliferation as a result of apoptosis inhibition and subsequently promote epithelial-to-mesenchymal transition (EMT) (Bitu et al., 2012). The oncogenic role of Hoxa1 was also observed in immortal mammary epithelial cells, whereas in immortal human keratinocytes Hoxa1 had no affect on apoptosis, EMT, or invasions. On the other hand, in hematopoietic cells, the Hoxa1 cofactor Meis1 augments the efficiency of Hoxa1 in tumorigenic induction, resulting in even more intense growth.

Interestingly, not only in leukemic malignancies but also in other cancer types, miRNAs control the Hox–apoptosis interplay. A detailed mechanism for miRNAs and Hox regulation has been shown for *miR-196a* in melanoma cells. By binding directly to the *Hoxb8* mRNA, *miR-196a* triggers degradation and silencing of *Hoxc8* that leads to reduced cell survival (Mueller & Bosserhoff, 2011; Yekta, Shih, & Bartel, 2004). Loss of *miR-196a* function in melanoma cells leads to upregulation of *Hoxc8* and activation of Hox target genes involved in melanoma progression. A direct association of miRNAs and *Hox* genes is described for *miR-100*, which downregulates *Hoxa1* leading to survival of lung cancer cells (Xiao et al., 2014). In sum, these examples show that miRNAs, *Hox* genes, and apoptosis are intimately linked as part of tightly coordinated regulatory networks, and thus represent promising targets for cancer therapies.

Taken together, these observations demonstrate that the role of Hox proteins in various cancer types, as in normal development, is very much dependent on the tissue context. The fact that *Hox* genes, promote tumor growth via apoptosis inhibition in some cancers types but not in others, highlights the importance of Hox-cofactor availability in the corresponding cell types and of their stoichiometry compared to the Hox protein levels.

6. CONCLUDING REMARKS

PCD plays a fundamental role in animal development, as it is critically required for tissue homeostasis and remodeling, organogenesis, and the elimination of excessive or abnormal cells (Fuchs & Steller, 2011; Suzanne & Steller, 2013). Altered regulation of any of these processes leads to a variety of developmental disorders and diseases, and therefore, apoptosis needs to be tightly controlled depending on the developmental status and the physiological condition of an organism. However, there is still many open questions about the molecular mechanisms underlying apoptosis regulation *in vivo*—how are cells earmarked for death and how are different signaling pathways, with sometimes opposing functions, involved in the decision of a cell to live or die (Fuchs & Steller, 2011)? In this context, it is may be surprising that very little is known about the control of apoptosis on the transcriptional level, despite its central role shown by the severe consequences on organismal development and survival, which is observed when upstream regulators are dysfunctional.

Very early on, Hox proteins have emerged as critical TFs that control apoptosis induction in a stereotypic and highly precise manner. This is particularly obvious in the developmental context, when Hox TFs due to their function as cell-type-specific micromanagers of gene networks and signaling pathways influence the decision of a cell to die in a very immediate, cell-autonomous as well as in a more indirect, nonautonomous manner (Akam, 1998; Philippidou & Dasen, 2013). One theme emerging from these studies is that *Hox* genes very often regulate cell fate specification and apoptosis in a coordinated manner, and the question is whether this interplay is of functional relevance. A recent study has shown that this is the case for Cut (Ct), another HD-containing TF. Ct controls differentiation and apoptosis repression in *Drosophila*, and the evolutionary rationale for this coupling is that cells unable to follow a specific fate in the absence of Ct, thus potentially harmful to the organism, are removed *in status nascendi* by releasing the transcriptional repression of *rpr* (Zhai et al., 2012). Interestingly, the regulatory wiring of differentiation and apoptosis seems to be conserved, since the human homolog of Ct, Cux-1, activates transcription of the proapoptotic gene *puma* in vertebrate cells (Zhai et al., 2012) and just recently Cux-1 has emerged as a tumor suppressor via its ability to transcriptionally interfere with the PI3K signaling, a pathway known to control apoptosis (Costa et al., 2010). Whether the same regulatory interplay is used by Hox TFs to ensure survival of cells only when they differentiate in the appropriate manner is not clear. However, this seems to be very likely since *Hox* genes are misregulated in many pathological situations, such as cancer, which are often accompanied by a deteriorated control of apoptotic events.

The major challenge in the future will be to elucidate the underlying mechanisms of the Hox–apoptosis interplay. So far, the role of *Hox* genes in cancer development has been primarily addressed by studying gene expression changes in normal and neoplastic tissue. However, very few studies address the complex function of *Hox* genes in the cancer context (Grier et al., 2005; Shah & Sukumar, 2010). For example, hematopoietic progenitors express *Hox* genes in a pattern characteristic of the lineage and stage of differentiation of the cells. Therefore, in order to understand the role of *Hox* genes in malignant transformations, in particular in respect to apoptosis and cell survival, it is crucial to understand Hox function and the Hox–apoptosis interplay under normal physiological conditions and determine the Hox expression signature for different organs (Grier et al., 2005; Shah & Sukumar, 2010). Furthermore, it is also critical to identify Hox downstream genes and cofactors required for target gene regulation, decode their function at the enhancer level and uncover the role of Hox upstream regulators, which are often responsible for Hox deregulation in the various cancer types. Only after fully understanding normal Hox expression and function as part of cell- and tissue-specific regulatory networks, it will be feasible to draw conclusions on how Hox proteins together with their cofactors regulate apoptosis and promote tumorigenesis on the mechanistic level. And only then will scientists be able to design new drugs and personalized treatments for multifactorial diseases such as cancer.

ACKNOWLEDGMENTS

We thank Pedro Pinto, Srividya Tamarisa, and Petra Kaspar for helpful comments on the manuscript. We apologize to colleagues whose work was not included owing to space constraints. Relevant work of I.L. is supported by the DFG.

REFERENCES

Aboobaker, A., & Blaxter, M. (2003). Hox gene evolution in nematodes: Novelty conserved. *Current Opinion in Genetics & Development*, *13*, 593–598.

Abrams, J. M., White, K., Fessler, L. I., & Steller, H. (1993). Programmed cell death during Drosophila embryogenesis. *Development, 117*, 29–43.

Akam, M. (1987). The molecular basis for metameric pattern in the Drosophila embryo. *Development, 101*, 1–22.

Akam, M. (1998). Hox genes: From master genes to micromanagers. *Current Biology, 8*, R676–R678.

Alonso, C. R. (2002). Hox proteins: Sculpting body parts by activating localized cell death. *Current Biology, 12*, R776–R778.

Argiropoulos, B., & Humphries, R. K. (2007). Hox genes in hematopoiesis and leukemogenesis. *Oncogene, 26*, 6766–6776.

Armstrong, S. A., Golub, T. R., & Korsmeyer, S. J. (2003). MLL-rearranged leukemias: Insights from gene expression profiling. *Seminars in Hematology, 40*, 268–273.

Baek, M., Enriquez, J., & Mann, R. S. (2013). Dual role for Hox genes and Hox co-factors in conferring leg motoneuron survival and identity in Drosophila. *Development, 140*, 2027–2038.

Baliga, B. C., Colussi, P. A., Read, S. H., Dias, M. M., Jans, D. A., & Kumar, S. (2003). Role of prodomain in importin-mediated nuclear localization and activation of caspase-2. *The Journal of Biological Chemistry, 278*, 4899–4905.

Barber, B. A., & Rastegar, M. (2010). Epigenetic control of Hox genes during neurogenesis, development, and disease. *Annals of Anatomy = Anatomischer Anzeiger: Official Organ of the Anatomische Gesellschaft, 192*, 261–274.

Barna, M., Hawe, N., Niswander, L., & Pandolfi, P. P. (2000). Plzf regulates limb and axial skeletal patterning. *Nature Genetics, 25*, 166–172.

Bello, B. C., Hirth, F., & Gould, A. P. (2003). A pulse of the Drosophila Hox protein Abdominal-A schedules the end of neural proliferation via neuroblast apoptosis. *Neuron, 37*, 209–219.

Bertani, S., Sauer, S., Bolotin, E., & Sauer, F. (2011). The noncoding RNA Mistral activates Hoxa6 and Hoxa7 expression and stem cell differentiation by recruiting MLL1 to chromatin. *Molecular Cell, 43*, 1040–1046.

Bitu, C. C., Destro, M. F., Carrera, M., da Silva, S. D., Graner, E., Kowalski, L. P., et al. (2012). HOXA1 is overexpressed in oral squamous cell carcinomas and its expression is correlated with poor prognosis. *BMC Cancer, 12*, 146.

Boyer, L. A., Plath, K., Zeitlinger, J., Brambrink, T., Medeiros, L. A., Lee, T. I., et al. (2006). Polycomb complexes repress developmental regulators in murine embryonic stem cells. *Nature, 441*, 349–353.

Brumatti, G., Salmanidis, M., Kok, C. H., Bilardi, R. A., Sandow, J. J., Silke, N., et al. (2013). HoxA9 regulated Bcl-2 expression mediates survival of myeloid progenitors and the severity of HoxA9-dependent leukemia. *Oncotarget, 4*, 1933–1947.

Burglin, T. R. (1997). Analysis of TALE superclass homeobox genes (MEIS, PBC, KNOX, Iroquois, TGIF) reveals a novel domain conserved between plants and animals. *Nucleic Acids Research, 25*, 4173–4180.

Burglin, T. R., & Ruvkun, G. (1993). The Caenorhabditis elegans homeobox gene cluster. *Current Opinion in Genetics & Development, 3*, 615–620.

Buske, C., & Humphries, R. K. (2000). Homeobox genes in leukemogenesis. *International Journal of Hematology, 71*, 301–308.

Cain, K., Bratton, S. B., & Cohen, G. M. (2002). The Apaf-1 apoptosome: A large caspase-activating complex. *Biochimie, 84*, 203–214.

Capecchi, M. R. (1997). Hox genes and mammalian development. *Cold Spring Harbor Symposia on Quantitative Biology, 62*, 273–281.

Carr, A., & Biggin, M. D. (1999). A comparison of *in vivo* and *in vitro* DNA-binding specificities suggests a new model for homeoprotein DNA binding in Drosophila embryos. *The EMBO Journal, 18*, 1598–1608.

Cenci, C., & Gould, A. P. (2005). Drosophila grainyhead specifies late programmes of neural proliferation by regulating the mitotic activity and Hox-dependent apoptosis of neuroblasts. *Development, 132*, 3835–3845.

Chang, C. P., Shen, W. F., Rozenfeld, S., Lawrence, H. J., Largman, C., & Cleary, M. L. (1995). Pbx proteins display hexapeptide-dependent cooperative DNA binding with a subset of Hox proteins. *Genes & Development, 9*, 663–674.

Chen, H., Chung, S., & Sukumar, S. (2004). HOXA5-induced apoptosis in breast cancer cells is mediated by caspases 2 and 8. *Molecular and Cellular Biology, 24*, 924–935.

Chen, J., Lin, C., Yong, W., Ye, Y., & Huang, Z. (2015). Calycosin and genistein induce apoptosis by inactivation of HOTAIR/p-Akt signaling pathway in human breast cancer MCF-7 cells. *Cellular Physiology and Biochemistry: International Journal of Experimental Cellular Physiology, Biochemistry, and Pharmacology, 35*, 722–728.

Chimal-Monroy, J., Abarca-Buis, R. F., Cuervo, R., Diaz-Hernandez, M., Bustamante, M., Rios-Flores, J. A., et al. (2011). Molecular control of cell differentiation and programmed cell death during digit development. *IUBMB Life, 63*, 922–929.

Choo, S. W., White, R., & Russell, S. (2011). Genome-wide analysis of the binding of the Hox protein ultrabithorax and the Hox cofactor homothorax in Drosophila. *PLoS One, 6*, e14778.

Chu, M. C., Selam, F. B., & Taylor, H. S. (2004). HOXA10 regulates p53 expression and matrigel invasion in human breast cancer cells. *Cancer Biology & Therapy, 3*, 568–572.

Clark, S. G., Chisholm, A. D., & Horvitz, H. R. (1993). Control of cell fates in the central body region of *C. elegans* by the homeobox gene lin-39. *Cell, 74*, 43–55.

Coelho, C. N., Upholt, W. B., & Kosher, R. A. (1993). The expression pattern of the chicken homeobox-containing gene GHox-7 in developing polydactylous limb buds suggests its involvement in apical ectodermal ridge-directed outgrowth of limb mesoderm and in programmed cell death. *Differentiation; Research in Biological Diversity, 52*, 129–137.

Costa, B. M., Smith, J. S., Chen, Y., Chen, J., Phillips, H. S., Aldape, K. D., et al. (2010). Reversing HOXA9 oncogene activation by PI3K inhibition: Epigenetic mechanism and prognostic significance in human glioblastoma. *Cancer Research, 70*, 453–462.

Cribbs, D. L., Pultz, M. A., Johnson, D., Mazzulla, M., & Kaufman, T. C. (1992). Structural complexity and evolutionary conservation of the Drosophila homeotic gene proboscipedia. *The EMBO Journal, 11*, 1437–1449.

Crocker, J., Abe, N., Rinaldi, L., McGregor, A. P., Frankel, N., Wang, S., et al. (2015). Low affinity binding site clusters confer hox specificity and regulatory robustness. *Cell, 160*, 191–203.

D'Angelo, M. A., & Hetzer, M. W. (2008). Structure, dynamics and function of nuclear pore complexes. *Trends in Cell Biology, 18*, 456–466.

Dasen, J. S., Liu, J. P., & Jessell, T. M. (2003). Motor neuron columnar fate imposed by sequential phases of Hox-c activity. *Nature, 425*, 926–933.

Denell, R. E. (1978). Homoeosis in Drosophila. II. A genetic analysis of polycomb. *Genetics, 90*, 277–289.

Di-Poi, N., Zakany, J., & Duboule, D. (2007). Distinct roles and regulations for HoxD genes in metanephric kidney development. *PLoS Genetics, 3*, e232.

Dorrance, A. M., Liu, S., Yuan, W., Becknell, B., Arnoczky, K. J., Guimond, M., et al. (2006). Mll partial tandem duplication induces aberrant Hox expression *in vivo* via specific epigenetic alterations. *The Journal of Clinical Investigation, 116*, 2707–2716.

Duboule, D., & Dolle, P. (1989). The structural and functional organization of the murine HOX gene family resembles that of Drosophila homeotic genes. *The EMBO Journal, 8*, 1497–1505.

Duboule, D., & Morata, G. (1994). Colinearity and functional hierarchy among genes of the homeotic complexes. *Trends in Genetics, 10*, 358–364.

Eckelman, B. P., & Salvesen, G. S. (2006). The human anti-apoptotic proteins cIAP1 and cIAP2 bind but do not inhibit caspases. *The Journal of Biological Chemistry*, *281*, 3254–3260.

Faber, J., Krivtsov, A. V., Stubbs, M. C., Wright, R., Davis, T. N., van den Heuvel-Eibrink, M., et al. (2009). HOXA9 is required for survival in human MLL-rearranged acute leukemias. *Blood*, *113*, 2375–2385.

Fang, K., Han, B. W., Chen, Z. H., Lin, K. Y., Zeng, C. W., Li, X. J., et al. (2014). A distinct set of long non-coding RNAs in childhood MLL-rearranged acute lymphoblastic leukemia: Biology and epigenetic target. *Human Molecular Genetics*, *23*, 3278–3288.

Ferrando, A. A., Armstrong, S. A., Neuberg, D. S., Sallan, S. E., Silverman, L. B., Korsmeyer, S. J., et al. (2003). Gene expression signatures in MLL-rearranged T-lineage and B-precursor acute leukemias: Dominance of HOX dysregulation. *Blood*, *102*, 262–268.

Foronda, D., de Navas, L. F., Garaulet, D. L., & Sanchez-Herrero, E. (2009). Function and specificity of Hox genes. *The International Journal of Developmental Biology*, *53*, 1404–1419.

Fuchs, Y., & Steller, H. (2011). Programmed cell death in animal development and disease. *Cell*, *147*, 742–758.

Garcia-Bellido, A. (1975). *Genetic Control of Wing Disc Development in Drosophila. Ciba Foundation Symposium 29* (pp. 161–182). Associated Scientific Publishers.

Garzon, R., Garofalo, M., Martelli, M. P., Briesewitz, R., Wang, L., Fernandez-Cymering, C., et al. (2008). Distinctive microRNA signature of acute myeloid leukemia bearing cytoplasmic mutated nucleophosmin. *Proceedings of the National Academy of Sciences of the United States of America*, *105*, 3945–3950.

Garzon, R., Volinia, S., Liu, C. G., Fernandez-Cymering, C., Palumbo, T., Pichiorri, F., et al. (2008). MicroRNA signatures associated with cytogenetics and prognosis in acute myeloid leukemia. *Blood*, *111*, 3183–3189.

Gaufo, G. O., Flodby, P., & Capecchi, M. R. (2000). Hoxb1 controls effectors of sonic hedgehog and Mash1 signaling pathways. *Development*, *127*, 5343–5354.

Gaunt, S. J. (1988). Mouse homeobox gene transcripts occupy different but overlapping domains in embryonic germ layers and organs: A comparison of Hox-3.1 and Hox-1.5. *Development*, *103*, 135–144.

Gehring, W. J., & Hiromi, Y. (1986). Homeotic genes and the homeobox. *Annual Review of Genetics*, *20*, 147–173.

Gehring, W. J., Qian, Y. Q., Billeter, M., Furukubo-Tokunaga, K., Schier, A. F., Resendez-Perez, D., et al. (1994). Homeodomain-DNA recognition. *Cell*, *78*, 211–223.

Golub, T. R., Slonim, D. K., Tamayo, P., Huard, C., Gaasenbeek, M., Mesirov, J. P., et al. (1999). Molecular classification of cancer: Class discovery and class prediction by gene expression monitoring. *Science*, *286*, 531–537.

Goodman, F. R. (2002). Limb malformations and the human HOX genes. *American Journal of Medical Genetics*, *112*, 256–265.

Gough, S. M., Slape, C. I., & Aplan, P. D. (2011). NUP98 gene fusions and hematopoietic malignancies: Common themes and new biologic insights. *Blood*, *118*, 6247–6257.

Goyal, L., McCall, K., Agapite, J., Hartwieg, E., & Steller, H. (2000). Induction of apoptosis by Drosophila reaper, hid and grim through inhibition of IAP function. *The EMBO Journal*, *19*, 589–597.

Graham, A., Papalopulu, N., & Krumlauf, R. (1989). The murine and Drosophila homeobox gene complexes have common features of organization and expression. *Cell*, *57*, 367–378.

Grier, D. G., Thompson, A., Kwasniewska, A., McGonigle, G. J., Halliday, H. L., & Lappin, T. R. (2005). The pathophysiology of HOX genes and their role in cancer. *The Journal of Pathology*, *205*, 154–171.

Gu, Y., Nakamura, T., Alder, H., Prasad, R., Canaani, O., Cimino, G., et al. (1992). The t(4;11) chromosome translocation of human acute leukemias fuses the ALL-1 gene, related to Drosophila trithorax, to the AF-4 gene. *Cell, 71*, 701–708.
Guazzi, S., Pintonello, M. L., Vigano, A., & Boncinelli, E. (1998). Regulatory interactions between the human HOXB1, HOXB2, and HOXB3 proteins and the upstream sequence of the Otx2 gene in embryonal carcinoma cells. *The Journal of Biological Chemistry, 273*, 11092–11099.
Gummalla, M., Galetti, S., Maeda, R. K., & Karch, F. (2014). Hox gene regulation in the central nervous system of Drosophila. *Frontiers in Cellular Neuroscience, 8*, 96.
Gupta, R. A., Shah, N., Wang, K. C., Kim, J., Horlings, H. M., Wong, D. J., et al. (2010). Long non-coding RNA HOTAIR reprograms chromatin state to promote cancer metastasis. *Nature, 464*, 1071–1076.
Hafen, E., Levine, M., & Gehring, W. J. (1984). Regulation of Antennapedia transcript distribution by the bithorax complex in Drosophila. *Nature, 307*, 287–289.
Harding, K., Wedeen, C., McGinnis, W., & Levine, M. (1985). Spatially regulated expression of homeotic genes in Drosophila. *Science, 229*, 1236–1242.
Hartenstein, V. (1993). *Atlas of drosophila development*. New York: Cold Spring Harbor Laboratory Press.
Havelange, V., Stauffer, N., Heaphy, C. C., Volinia, S., Andreeff, M., Marcucci, G., et al. (2011). Functional implications of microRNAs in acute myeloid leukemia by integrating microRNA and messenger RNA expression profiling. *Cancer, 117*, 4696–4706.
Hengartner, M. O. (2000). The biochemistry of apoptosis. *Nature, 407*, 770–776.
Hidalgo, A., & Brand, A. H. (1997). Targeted neuronal ablation: The role of pioneer neurons in guidance and fasciculation in the CNS of Drosophila. *Development, 124*, 3253–3262.
Hu, Y. L., Fong, S., Ferrell, C., Largman, C., & Shen, W. F. (2009). HOXA9 modulates its oncogenic partner Meis1 to influence normal hematopoiesis. *Molecular and Cellular Biology, 29*, 5181–5192.
Hudry, B., Viala, S., Graba, Y., & Merabet, S. (2011). Visualization of protein interactions in living Drosophila embryos by the bimolecular fluorescence complementation assay. *BMC Biology, 9*, 5.
Hueber, S. D., & Lohmann, I. (2008). Shaping segments: Hox gene function in the genomic age. *BioEssays: News and Reviews in Molecular, Cellular and Developmental Biology, 30*, 965–979.
Hunter, C. P., & Kenyon, C. (1995). Specification of anteroposterior cell fates in Caenorhabditis elegans by Drosophila Hox proteins. *Nature, 377*, 229–232.
Iimura, T., Denans, N., & Pourquie, O. (2009). Establishment of Hox vertebral identities in the embryonic spine precursors. *Current Topics in Developmental Biology, 88*, 201–234.
Ingham, P. W. (1983). Differential expression of bithorax complex genes in the absence of the extra sex combs and trithorax genes. *Nature, 306*, 591–593.
Ingham, P. W. (1998). trithorax and the regulation of homeotic gene expression in Drosophila: A historical perspective. *The International Journal of Developmental Biology, 42*, 423–429.
Irish, V. F., Martinez-Arias, A., & Akam, M. (1989). Spatial regulation of the Antennapedia and Ultrabithorax homeotic genes during Drosophila early development. *The EMBO Journal, 8*, 1527–1537.
Johnson, R. L., & Tabin, C. J. (1997). Molecular models for vertebrate limb development. *Cell, 90*, 979–990.
Jung, C., Kim, R. S., Zhang, H., Lee, S. J., Sheng, H., Loehrer, P. J., et al. (2005). HOXB13 is downregulated in colorectal cancer to confer TCF4-mediated transactivation. *British Journal of Cancer, 92*, 2233–2239.
Kanhere, A., Viiri, K., Araujo, C. C., Rasaiyaah, J., Bouwman, R. D., Whyte, W. A., et al. (2010). Short RNAs are transcribed from repressed polycomb target genes and interact with polycomb repressive complex-2. *Molecular Cell, 38*, 675–688.

Kar, R., Singha, P. K., Venkatachalam, M. A., & Saikumar, P. (2009). A novel role for MAP1 LC3 in nonautophagic cytoplasmic vacuolation death of cancer cells. *Oncogene*, *28*, 2556–2568.
Kaufman, T. C., Lewis, R., & Wakimoto, B. (1980). Cytogenetic analysis of chromosome 3 in Drosophila melanogaster: The homoeotic gene complex in polytene chromosome interval 84a-B. *Genetics*, *94*, 115–133.
Kaufman, T. C., Seeger, M. A., & Olsen, G. (1990). Molecular and genetic organization of the antennapedia gene complex of Drosophila melanogaster. *Advances in Genetics*, *27*, 309–362.
Keisman, E. L., Christiansen, A. E., & Baker, B. S. (2001). The sex determination gene doublesex regulates the A/P organizer to direct sex-specific patterns of growth in the Drosophila genital imaginal disc. *Developmental Cell*, *1*, 215–225.
Kenyon, C. (1986). A gene involved in the development of the posterior body region of *C. elegans*. *Cell*, *46*, 477–487.
Kmita, M., & Duboule, D. (2003). Organizing axes in time and space; 25 years of colinear tinkering. *Science*, *301*, 331–333.
Knosp, W. M., Scott, V., Bachinger, H. P., & Stadler, H. S. (2004). HOXA13 regulates the expression of bone morphogenetic proteins 2 and 7 to control distal limb morphogenesis. *Development*, *131*, 4581–4592.
Ko, K. H., Lam, Q. L., Zhang, M., Wong, C. K., Lo, C. K., Kahmeyer-Gabbe, M., et al. (2007). Hoxb3 deficiency impairs B lymphopoiesis in mouse bone marrow. *Experimental Hematology*, *35*, 465–475.
Kocak, H., Ackermann, S., Hero, B., Kahlert, Y., Oberthuer, A., Juraeva, D., et al. (2013). Hox-C9 activates the intrinsic pathway of apoptosis and is associated with spontaneous regression in neuroblastoma. *Cell Death & Disease*, *4*, e586.
Kondo, T., Zakany, J., Innis, J. W., & Duboule, D. (1997). Of fingers, toes and penises. *Nature*, *390*, 29.
Kopp, A., Duncan, I., Godt, D., & Carroll, S. B. (2000). Genetic control and evolution of sexually dimorphic characters in Drosophila. *Nature*, *408*, 553–559.
Kuss, P., Villavicencio-Lorini, P., Witte, F., Klose, J., Albrecht, A. N., Seemann, P., et al. (2009). Mutant Hoxd13 induces extra digits in a mouse model of synpolydactyly directly and by decreasing retinoic acid synthesis. *The Journal of Clinical Investigation*, *119*, 146–156.
Lawrence, H. J., & Largman, C. (1992). Homeobox genes in normal hematopoiesis and leukemia. *Blood*, *80*, 2445–2453.
Lee, T. I., Jenner, R. G., Boyer, L. A., Guenther, M. G., Levine, S. S., Kumar, R. M., et al. (2006). Control of developmental regulators by Polycomb in human embryonic stem cells. *Cell*, *125*, 301–313.
Lewis, E. B. (1978). A gene complex controlling segmentation in Drosophila. *Nature*, *276*, 565–570.
Li, Z., Huang, H., Li, Y., Jiang, X., Chen, P., Arnovitz, S., et al. (2012). Up-regulation of a HOXA-PBX3 homeobox-gene signature following down-regulation of miR-181 is associated with adverse prognosis in patients with cytogenetically abnormal AML. *Blood*, *119*, 2314–2324.
Lin, D. M., Auld, V. J., & Goodman, C. S. (1995). Targeted neuronal cell ablation in the Drosophila embryo: Pathfinding by follower growth cones in the absence of pioneers. *Neuron*, *14*, 707–715.
Lin, S., & Lee, T. (2012). Generating neuronal diversity in the Drosophila central nervous system. *Developmental Dynamics: An Official Publication of the American Association of Anatomists*, *241*, 57–68.
Liu, H., Strauss, T. J., Potts, M. B., & Cameron, S. (2006). Direct regulation of egl-1 and of programmed cell death by the Hox protein MAB-5 and by CEH-20, a *C. elegans* homolog of Pbx1. *Development*, *133*, 641–650.

Lohmann, I., McGinnis, N., Bodmer, M., & McGinnis, W. (2002). The Drosophila Hox gene deformed sculpts head morphology via direct regulation of the apoptosis activator reaper. *Cell, 110*, 457–466.

Ma, P. C., Tretiakova, M. S., Nallasura, V., Jagadeeswaran, R., Husain, A. N., & Salgia, R. (2007). Downstream signalling and specific inhibition of c-MET/HGF pathway in small cell lung cancer: Implications for tumour invasion. *British Journal of Cancer, 97*, 368–377.

Maconochie, M., Nonchev, S., Morrison, A., & Krumlauf, R. (1996). Paralogous Hox genes: Function and regulation. *Annual Review of Genetics, 30*, 529–556.

Malicki, J., Schughart, K., & McGinnis, W. (1990). Mouse Hox-2.2 specifies thoracic segmental identity in Drosophila embryos and larvae. *Cell, 63*, 961–967.

Mallo, M., & Alonso, C. R. (2013). The regulation of Hox gene expression during animal development. *Development, 140*, 3951–3963.

Mandeville, I., Aubin, J., LeBlanc, M., Lalancette-Hebert, M., Janelle, M. F., Tremblay, G. M., et al. (2006). Impact of the loss of Hoxa5 function on lung alveogenesis. *The American Journal of Pathology, 169*, 1312–1327.

Mann, R. S. (1995). The specificity of homeotic gene function. *BioEssays: News and Reviews in Molecular, Cellular and Developmental Biology, 17*, 855–863.

Mann, R. S., & Chan, S. K. (1996). Extra specificity from extradenticle: The partnership between HOX and PBX/EXD homeodomain proteins. *Trends in Genetics, 12*, 258–262.

Mann, R. S., Lelli, K. M., & Joshi, R. (2009). Hox specificity unique roles for cofactors and collaborators. *Current Topics in Developmental Biology, 88*, 63–101.

Manohar, C. F., Salwen, H. R., Furtado, M. R., & Cohn, S. L. (1996). Up-regulation of HOXC6, HOXD1, and HOXD8 homeobox gene expression in human neuroblastoma cells following chemical induction of differentiation. *Tumour Biology, 17*, 34–47.

Mao, L., Ding, J., Zha, Y., Yang, L., McCarthy, B. A., King, W., et al. (2011). HOXC9 links cell-cycle exit and neuronal differentiation and is a prognostic marker in neuroblastoma. *Cancer Research, 71*, 4314–4324.

McGinnis, W., Jack, T., Chadwick, R., Regulski, M., Bergson, C., McGinnis, N., et al. (1990). Establishment and maintenance of position-specific expression of the Drosophila homeotic selector gene Deformed. *Advances in Genetics, 27*, 363–402.

McGinnis, W., Levine, M. S., Hafen, E., Kuroiwa, A., & Gehring, W. J. (1984). A conserved DNA sequence in homoeotic genes of the Drosophila Antennapedia and bithorax complexes. *Nature, 308*, 428–433.

Miguel-Aliaga, I., & Thor, S. (2004). Segment-specific prevention of pioneer neuron apoptosis by cell-autonomous, postmitotic Hox gene activity. *Development, 131*, 6093–6105.

Milne, T. A., Briggs, S. D., Brock, H. W., Martin, M. E., Gibbs, D., Allis, C. D., et al. (2002). MLL targets SET domain methyltransferase activity to Hox gene promoters. *Molecular Cell, 10*, 1107–1117.

Montero, J. A., & Hurle, J. M. (2010). Sculpturing digit shape by cell death. *Apoptosis, 15*, 365–375.

Morata, G., & Kerridge, S. (1982). The role of position in determining homoeotic gene function in Drosophila. *Nature, 300*, 191–192.

Morgan, T. H., & Bridges, C. B. (1916). Sex-linked inheritance in *Drosophila*. Washington: Carnegie Institution of Washington publication, 237.

Mueller, D. W., & Bosserhoff, A. K. (2011). MicroRNA miR-196a controls melanoma-associated genes by regulating HOX-C8 expression. *International Journal of Cancer, 129*, 1064–1074.

Mulgrew, N. M., Kettyle, L. M., Ramsey, J. M., Cull, S., Smyth, L. J., Mervyn, D. M., et al. (2014). c-Met inhibition in a HOXA9/Meis1 model of CN-AML. *Developmental Dynamics: An Official Publication of the American Association of Anatomists, 243*, 172–181.

Nakamura, N., Fujioka, M., & Mori, C. (2000). Alteration of programmed cell death and gene expression by 5-bromodeoxyuridine during limb development in mice. *Toxicology and Applied Pharmacology*, *167*, 100–106.

Nakamura, T., Mori, T., Tada, S., Krajewski, W., Rozovskaia, T., Wassell, R., et al. (2002). ALL-1 is a histone methyltransferase that assembles a supercomplex of proteins involved in transcriptional regulation. *Molecular Cell*, *10*, 1119–1128.

Noyes, M. B., Christensen, R. G., Wakabayashi, A., Stormo, G. D., Brodsky, M. H., & Wolfe, S. A. (2008). Analysis of homeodomain specificities allows the family-wide prediction of preferred recognition sites. *Cell*, *133*, 1277–1289.

Pearson, J. C., Lemons, D., & McGinnis, W. (2005). Modulating Hox gene functions during animal body patterning. *Nature Reviews. Genetics*, *6*, 893–904.

Perrin, L., Monier, B., Ponzielli, R., Astier, M., & Semeriva, M. (2004). Drosophila cardiac tube organogenesis requires multiple phases of Hox activity. *Developmental Biology*, *272*, 419–431.

Peruzzi, B., & Bottaro, D. P. (2006). Targeting the c-Met signaling pathway in cancer. *Clinical Cancer Research*, *12*, 3657–3660.

Peterson, C., Carney, G. E., Taylor, B. J., & White, K. (2002). Reaper is required for neuroblast apoptosis during Drosophila development. *Development*, *129*, 1467–1476.

Petrini, M., Felicetti, F., Bottero, L., Errico, M. C., Morsilli, O., Boe, A., et al. (2013). HOXB1 restored expression promotes apoptosis and differentiation in the HL60 leukemic cell line. *Cancer Cell International*, *13*, 101.

Philippidou, P., & Dasen, J. S. (2013). Hox genes: Choreographers in neural development, architects of circuit organization. *Neuron*, *80*, 12–34.

Ponting, C. P., Oliver, P. L., & Reik, W. (2009). Evolution and functions of long noncoding RNAs. *Cell*, *136*, 629–641.

Potts, M. B., Wang, D. P., & Cameron, S. (2009). Trithorax, Hox, and TALE-class homeodomain proteins ensure cell survival through repression of the BH3-only gene egl-1. *Developmental Biology*, *329*, 374–385.

Printz, C. (2001). Spontaneous regression of melanoma may offer insight into cancer immunology. *Journal of the National Cancer Institute*, *93*, 1047–1048.

Procino, A., & Cillo, C. (2013). The HOX genes network in metabolic diseases. *Cell Biology International*, *37*, 1145–1148.

Puro, J., & Nygren, T. (1975). Mode of action of a homoeotic gene in Drosophila melanogaster. Localization and dosage effects of Polycomb. *Hereditas*, *81*, 237–248.

Quinonez, S. C., & Innis, J. W. (2014). Human HOX gene disorders. *Molecular Genetics and Metabolism*, *111*, 4–15.

Raman, V., Martensen, S. A., Reisman, D., Evron, E., Odenwald, W. F., Jaffee, E., et al. (2000). Compromised HOXA5 function can limit p53 expression in human breast tumours. *Nature*, *405*, 974–978.

Reisman, D., & Loging, W. T. (1998). Transcriptional regulation of the p53 tumor suppressor gene. *Seminars in Cancer Biology*, *8*, 317–324.

Reubold, T. F., & Eschenburg, S. (2012). A molecular view on signal transduction by the apoptosome. *Cellular Signalling*, *24*, 1420–1425.

Rinn, J. L., Kertesz, M., Wang, J. K., Squazzo, S. L., Xu, X., Brugmann, S. A., et al. (2007). Functional demarcation of active and silent chromatin domains in human HOX loci by noncoding RNAs. *Cell*, *129*, 1311–1323.

Rodriguez, J., & Lazebnik, Y. (1999). Caspase-9 and APAF-1 form an active holoenzyme. *Genes & Development*, *13*, 3179–3184.

Rogulja-Ortmann, A., Renner, S., & Technau, G. M. (2008). Antagonistic roles for Ultrabithorax and Antennapedia in regulating segment-specific apoptosis of differentiated motoneurons in the Drosophila embryonic central nervous system. *Development*, *135*, 3435–3445.

Ros, M. A., Macias, D., Fallon, J. F., & Hurle, J. M. (1994). Formation of extra digits in the interdigital spaces of the chick leg bud is not preceded by changes in the expression of the Msx and Hoxd genes. *Anatomy and Embryology, 190*, 375–382.

Ros, M. A., Piedra, M. E., Fallon, J. F., & Hurle, J. M. (1997). Morphogenetic potential of the chick leg interdigital mesoderm when diverted from the cell death program. *Developmental Dynamics: An Official Publication of the American Association of Anatomists, 208*, 406–419.

Rozowski, M., & Akam, M. (2002). Hox gene control of segment-specific bristle patterns in Drosophila. *Genes & Development, 16*, 1150–1162.

Ryoo, H. D., Marty, T., Casares, F., Affolter, M., & Mann, R. S. (1999). Regulation of Hox target genes by a DNA bound Homothorax/Hox/Extradenticle complex. *Development, 126*, 5137–5148.

Sanchez, L., Gorfinkiel, N., & Guerrero, I. (2001). Sex determination genes control the development of the Drosophila genital disc, modulating the response to Hedgehog, Wingless and Decapentaplegic signals. *Development, 128*, 1033–1043.

Sanchez-Herrero, E., Vernos, I., Marco, R., & Morata, G. (1985). Genetic organization of Drosophila bithorax complex. *Nature, 313*, 108–113.

Schuettengruber, B., & Cavalli, G. (2009). Recruitment of polycomb group complexes and their role in the dynamic regulation of cell fate choice. *Development, 136*, 3531–3542.

Schuettengruber, B., Chourrout, D., Vervoort, M., Leblanc, B., & Cavalli, G. (2007). Genome regulation by polycomb and trithorax proteins. *Cell, 128*, 735–745.

Schuettengruber, B., Martinez, A. M., Iovino, N., & Cavalli, G. (2011). Trithorax group proteins: Switching genes on and keeping them active. *Nature Reviews. Molecular Cell Biology, 12*, 799–814.

Scott, W. J., Ritter, E. J., & Wilson, J. G. (1977). Delayed appearance of ectodermal cell death as a mechanism of polydactyly induction. *Journal of Embryology and Experimental Morphology, 42*, 93–104.

Scott, W. J., Ritter, E. J., & Wilson, J. G. (1980). Ectodermal and mesodermal cell death patterns in 6-mercaptopurine riboside-induced digital deformities. *Teratology, 21*, 271–279.

Scott, M. P., & Weiner, A. J. (1984). Structural relationships among genes that control development: Sequence homology between the Antennapedia, Ultrabithorax, and fushi tarazu loci of Drosophila. *Proceedings of the National Academy of Sciences of the United States of America, 81*, 4115–4119.

Shah, N., & Sukumar, S. (2010). The Hox genes and their roles in oncogenesis. *Nature Reviews. Cancer, 10*, 361–371.

Sharkey, M., Graba, Y., & Scott, M. P. (1997). Hox genes in evolution: Protein surfaces and paralog groups. *Trends in Genetics, 13*, 145–151.

Shen, W. F., Chang, C. P., Rozenfeld, S., Sauvageau, G., Humphries, R. K., Lu, M., et al. (1996). Hox homeodomain proteins exhibit selective complex stabilities with Pbx and DNA. *Nucleic Acids Research, 24*, 898–906.

Shi, Y. (2002). A conserved tetrapeptide motif: Potentiating apoptosis through IAP-binding. *Cell Death and Differentiation, 9*, 93–95.

Simon, J. A., & Kingston, R. E. (2013). Occupying chromatin: Polycomb mechanisms for getting to genomic targets, stopping transcriptional traffic, and staying put. *Molecular Cell, 49*, 808–824.

Slattery, M., Ma, L., Negre, N., White, K. P., & Mann, R. S. (2011). Genome-wide tissue-specific occupancy of the Hox protein Ultrabithorax and Hox cofactor Homothorax in Drosophila. *PLoS One, 6*, e14686.

Soshnikova, N., & Duboule, D. (2008). Epigenetic regulation of Hox gene activation: The waltz of methyls. *BioEssays: News and Reviews in Molecular, Cellular and Developmental Biology, 30*, 199–202.

Spitz, F., & Furlong, E. E. (2012). Transcription factors: From enhancer binding to developmental control. *Nature Reviews. Genetics, 13*, 613–626.
Steller, H. (2008). Regulation of apoptosis in Drosophila. *Cell Death and Differentiation, 15*, 1132–1138.
Stobe, P., Stein, M. A., Habring-Muller, A., Bezdan, D., Fuchs, A. L., Hueber, S. D., et al. (2009). Multifactorial regulation of a hox target gene. *PLoS Genetics, 5*, e1000412.
Struhl, G. (1981). A homoeotic mutation transforming leg to antenna in Drosophila. *Nature, 292*, 635–638.
Struhl, G. (1982). Genes controlling segmental specification in the Drosophila thorax. *Proceedings of the National Academy of Sciences of the United States of America, 79*, 7380–7384.
Suska, A., Miguel-Aliaga, I., & Thor, S. (2011). Segment-specific generation of Drosophila Capability neuropeptide neurons by multi-faceted Hox cues. *Developmental Biology, 353*, 72–80.
Suzanne, M., & Steller, H. (2013). Shaping organisms with apoptosis. *Cell Death and Differentiation, 20*, 669–675.
Takahashi, O., Hamada, J., Abe, M., Hata, S., Asano, T., Takahashi, Y., et al. (2007). Dysregulated expression of HOX and ParaHOX genes in human esophageal squamous cell carcinoma. *Oncology Reports, 17*, 753–760.
Thornberry, N. A., & Lazebnik, Y. (1998). Caspases: Enemies within. *Science, 281*, 1312–1316.
Tiret, L., Le Mouellic, H., Maury, M., & Brulet, P. (1998). Increased apoptosis of motoneurons and altered somatotopic maps in the brachial spinal cord of Hoxc-8-deficient mice. *Development, 125*, 279–291.
Tkachuk, D. C., Kohler, S., & Cleary, M. L. (1992). Involvement of a homolog of Drosophila trithorax by 11q23 chromosomal translocations in acute leukemias. *Cell, 71*, 691–700.
Truman, J. W., & Bate, M. (1988). Spatial and temporal patterns of neurogenesis in the central nervous system of Drosophila melanogaster. *Developmental Biology, 125*, 145–157.
Tsai, M. C., Manor, O., Wan, Y., Mosammaparast, N., Wang, J. K., Lan, F., et al. (2010). Long noncoding RNA as modular scaffold of histone modification complexes. *Science, 329*, 689–693.
Van Auken, K., Weaver, D., Robertson, B., Sundaram, M., Saldi, T., Edgar, L., et al. (2002). Roles of the Homothorax/Meis/Prep homolog UNC-62 and the Exd/Pbx homologs CEH-20 and CEH-40 in *C. elegans* embryogenesis. *Development, 129*, 5255–5268.
Vitobello, A., Ferretti, E., Lampe, X., Vilain, N., Ducret, S., Ori, M., et al. (2011). Hox and Pbx factors control retinoic acid synthesis during hindbrain segmentation. *Developmental Cell, 20*, 469–482.
von Freeden-Jeffry, U., Vieira, P., Lucian, L. A., McNeil, T., Burdach, S. E., & Murray, R. (1995). Lymphopenia in interleukin (IL)-7 gene-deleted mice identifies IL-7 as a nonredundant cytokine. *The Journal of Experimental Medicine, 181*, 1519–1526.
Wang, S. L., Hawkins, C. J., Yoo, S. J., Muller, H. A., & Hay, B. A. (1999). The Drosophila caspase inhibitor DIAP1 is essential for cell survival and is negatively regulated by HID. *Cell, 98*, 453–463.
Wang, W., Kidd, B. J., Carroll, S. B., & Yoder, J. H. (2011). Sexually dimorphic regulation of the Wingless morphogen controls sex-specific segment number in Drosophila. *Proceedings of the National Academy of Sciences of the United States of America, 108*, 11139–11144.
Wang, B. B., Muller-Immergluck, M. M., Austin, J., Robinson, N. T., Chisholm, A., & Kenyon, C. (1993). A homeotic gene cluster patterns the anteroposterior body axis of *C. elegans*. *Cell, 74*, 29–42.
Wellik, D. M., Hawkes, P. J., & Capecchi, M. R. (2002). Hox11 paralogous genes are essential for metanephric kidney induction. *Genes & Development, 16*, 1423–1432.

White, K., Grether, M. E., Abrams, J. M., Young, L., Farrell, K., & Steller, H. (1994). Genetic control of programmed cell death in Drosophila. *Science*, *264*, 677–683.

Williams, T. M., Selegue, J. E., Werner, T., Gompel, N., Kopp, A., & Carroll, S. B. (2008). The regulation and evolution of a genetic switch controlling sexually dimorphic traits in Drosophila. *Cell*, *134*, 610–623.

Winn, J., Carter, M., Avery, L., & Cameron, S. (2011). Hox and a newly identified E2F co-repress cell death in Caenorhabditis elegans. *Genetics*, *188*, 897–905.

Wise, L. D., & Scott, W. J., Jr. (1982). Incorporation of 5-bromo-2′-deoxyuridine into mesenchymal limb-bud cells destined to die: Relationship to polydactyly induction in rats. *Journal of Embryology and Experimental Morphology*, *72*, 125–141.

Woltering, J. M., & Duboule, D. (2010). The origin of digits: Expression patterns versus regulatory mechanisms. *Developmental Cell*, *18*, 526–532.

Wu, M., Wang, P. F., Lee, J. S., Martin-Brown, S., Florens, L., Washburn, M., et al. (2008). Molecular regulation of H3K4 trimethylation by Wdr82, a component of human Set1/COMPASS. *Molecular and Cellular Biology*, *28*, 7337–7344.

Xiao, F., Bai, Y., Chen, Z., Li, Y., Luo, L., Huang, J., et al. (2014). Downregulation of HOXA1 gene affects small cell lung cancer cell survival and chemoresistance under the regulation of miR-100. *European Journal of Cancer*, *50*, 1541–1554.

Xu, S., & Powers, M. A. (2009). Nuclear pore proteins and cancer. *Seminars in Cell & Developmental Biology*, *20*, 620–630.

Yang, Y., Fang, S., Jensen, J. P., Weissman, A. M., & Ashwell, J. D. (2000). Ubiquitin protein ligase activity of IAPs and their degradation in proteasomes in response to apoptotic stimuli. *Science*, *288*, 874–877.

Yang, Y., Wang, C., Zhang, P., Gao, K., Wang, D., Yu, H., et al. (2013). Polycomb group protein PHF1 regulates p53-dependent cell growth arrest and apoptosis. *The Journal of Biological Chemistry*, *288*, 529–539.

Yekta, S., Shih, I. H., & Bartel, D. P. (2004). MicroRNA-directed cleavage of HOXB8 mRNA. *Science*, *304*, 594–596.

Yuan, J., & Kroemer, G. (2010). Alternative cell death mechanisms in development and beyond. *Genes & Development*, *24*, 2592–2602.

Zhai, Z., Ha, N., Papagiannouli, F., Hamacher-Brady, A., Brady, N., Sorge, S., et al. (2012). Antagonistic regulation of apoptosis and differentiation by the Cut transcription factor represents a tumor-suppressing mechanism in Drosophila. *PLoS Genetics*, *8*, e1002582.

Zhang, X., Hamada, J., Nishimoto, A., Takahashi, Y., Murai, T., Tada, M., et al. (2007). HOXC6 and HOXC11 increase transcription of S100beta gene in BrdU-induced *in vitro* differentiation of GOTO neuroblastoma cells into Schwannian cells. *Journal of Cellular and Molecular Medicine*, *11*, 299–306.

Zhao, J., Sun, B. K., Erwin, J. A., Song, J. J., & Lee, J. T. (2008). Polycomb proteins targeted by a short repeat RNA to the mouse X chromosome. *Science*, *322*, 750–756.

CHAPTER SIX

Programmed Cell Death and Caspase Functions During Neural Development

Yoshifumi Yamaguchi[*,†,1], **Masayuki Miura**[*,‡,1]

[*]Department of Genetics, Graduate School of Pharmaceutical Sciences, The University of Tokyo, Bunkyo-ku, Tokyo, Japan

[†]Precursory Research for Embryonic Science and Technology, Japan Science and Technology Agency, Chiyoda-ku, Tokyo, Japan

[‡]Core Research for Evolutional Science and Technology, Japan Agency for Medical Research and Development, Chiyoda-ku, Tokyo, Japan

[1]Corresponding authors: e-mail address: bunbun@mol.f.u-tokyo.ac.jp; miura@mol.f.u-tokyo.ac.jp

Contents

Abstract

Programmed cell death (PCD) is a fundamental component of nervous system development. PCD serves as the mechanism for quantitative matching of the number of projecting neurons and their target cells through direct competition for neurotrophic factors in the vertebrate peripheral nervous system. In addition, PCD plays roles in regulating neural cell numbers, canceling developmental errors or noise, and tissue remodeling processes. These findings are mainly derived from genetic studies that prevent cells from dying by apoptosis, which is a major form of PCD and is executed by

Current Topics in Developmental Biology, Volume 114
ISSN 0070-2153
http://dx.doi.org/10.1016/bs.ctdb.2015.07.016

activation of evolutionarily conserved cysteine protease caspases. Recent studies suggest that caspase activation can be coordinated in time and space at multiple levels, which might underlie nonapoptotic roles of caspases in neural development in addition to apoptotic roles.

1. INTRODUCTION

Programmed cell death (PCD) is widely observed during development in all metazoan organisms. It is widely thought that cell death functions for controlling cell numbers, removal of unwanted or redundant structures, or elimination of harmful or misspecified cells (Buss, Sun, & Oppenheim, 2006; Fuchs & Steller, 2011; Yeo & Gautier, 2004). PCD is an evolutionarily conserved trait and is regarded as an inevitable component of multicellular animal development. It was originally proposed that the pattern and timing of cell death during development are strictly scheduled and tightly regulated following early work in insects and chicken (Lockshin & Williams, 1964; Saunders, 1966), and the concept that PCD depends on cell differentiation programs was confirmed by pioneering works in *Caenorhabditis elegans*. In *C. elegans*, 131 cells undergo PCD, and 105 of them are from the neuronal lineage (Sulston & Horvitz, 1977). It is proposed that as many as 70% of cortical cells die in the developing fetal mouse cortex (Blaschke, Staley, & Chun, 1996). Therefore, the nervous system is one of the major tissues in which cell death can be extensively observed during development in both vertebrates and invertebrates (Buss et al., 2006; Fuchs & Steller, 2011; Oppenheim, 1991; Yamaguchi & Miura, 2015; Yeo & Gautier, 2004).

The major form of PCD in the developing nervous system is considered to be apoptosis, although recent studies have revealed that there are many mechanisms for "regulated cell death" in animal bodies (Galluzzi et al., 2012, 2014). Apoptosis was originally defined by morphological criteria such as nuclear condensation and cell shrinkage (Kerr, Wyllie, & Currie, 1972) and is now known to be executed by members of an evolutionarily conserved family of cysteine proteases named caspases (Degterev & Yuan, 2008). Caspases are activated by extrinsic death ligands or intrinsic stimuli including DNA damage, excessive reactive oxygen, metabolic stress, and Endoplasmic reticulum stress (Green, Galluzzi, & Kroemer, 2014). Activated caspases cleave intracellular substrates including cytoskeletal proteins, nuclear lamina, and inhibitors of caspase-dependent deoxyribonuclease (DNase), leading not only to destruction of cellular structures but also granting specific features of apoptosis to the dying cell. These features include "eat-me" signals or "find-me" signals that facilitate engulfment by phagocytes, regulation of immune responses, and generation of forces that promote smooth movements of surrounding cells (Grimsley & Ravichandran, 2003; Kuranaga, 2011; Monier et al., 2015). Furthermore, recent studies have revealed that nonapoptotic caspase activation, which refers to the activation of caspases without causing cell death, contributes to several aspects of development and physiology (Hyman & Yuan, 2012). In this review, we will focus on the roles of apoptosis and non-apoptotic caspase activation in nervous system development.

2. NEUROTROPHIC SUPPORT FOR NEURONAL OR GLIAL SURVIVAL

During neural development in vertebrates, excess neural cells are eliminated by PCD in both the central nervous system (CNS) and peripheral nervous system (PNS) (Oppenheim, 1991). The neurotrophic theory, first demonstrated using vertebrate models in the 1940s, states that limiting amounts of nerve growth factor (NGF) released by target tissues determine the fate of innervating neurons and enable quantitative matching between them, while excess neuronal cells in the PNS are eliminated (Cowan, 2001; Levi-Montalcini, 1987; Purves, Snider, & Voyvodic, 1988; Raff, 1992). In the PNS, tropomyosin receptor kinase (also known as tryosine receptor kinase) A (TrkA) and TrkC act as dependence receptors, allowing cells to survive when stimulated by trophic factors and other prosurvival signals, and inducing apoptosis in the absence of these signals through the loss of antiapoptotic signaling and the activation of proapoptotic signaling (Mehlen & Bredesen, 2011). Neurotrophins (NTs) serve as survival signals for developing sympathetic and sensory neurons, inhibiting cell death mediated by TrkA or TrkC signaling (Kristiansen & Ham, 2014; Nikoletopoulou et al., 2010). Glial cell line derived neurotrophic factor (GDNF) also acts as survival factor for developing motor neurons (Gould & Oppenheim, 2011; Oppenheim et al., 1995). In the mammalian CNS, brain-derived neurotrophic factor (BDNF) and its receptor TrkB are widely expressed. However, TrkB does not function as a dependence receptor, and therefore the exact action of BDNF/TrkB in the developmental death of CNS neurons remains to be elucidated (Dekkers, Nikoletopoulou, & Barde, 2013; Nikoletopoulou et al., 2010).

Neurotrophic support for neural survival also operates in invertebrates. *Drosophila melanogaster* neurotrophins (DNTs) were first identified by using 28 vertebrate full-length NT sequences or Cystine-knot (Cysknot) domains to query the *Drosophila* genome using TBLASTN and PSI-BLAST. This analysis identified DNT1 (spätzle 2). Further analysis revealed that there are three NT superfamily members in *Drosophila*: Spätzle, DNT1, and DNT2 (Spätzle 5) (Zhu et al., 2008). Loss of DNTs induces neuronal cell death. DNTs are expressed in target tissues, for example, the optic lobe for retinal neurons, muscle for motor neurons, and the ventral midline for embryonic CNS interneurons. DNT1 knockdown in the embryonic midline induces CNS apoptosis, and overexpression of the DNT1 Cysknot or Cysknot3-tail domains (the putative active form of DNT1) reduces naturally occurring cell death in the embryonic CNS (Zhu et al., 2008). Toll is a receptor for Spätzle, and Toll family members Toll-6 and Toll-7 function as NT receptors for DNT1 and DNT2, respectively (McIlroy et al., 2013). Like Toll, Toll-6 and Toll-7 receptors function upstream of NF-κB activation.

Another neurotrophic factor, *Drosophila* MANF, a homolog of mammalian MANF (mesencephalic astrocyte-derived neurotrophic factor)/CDNF (conserved dopamine neurotrophic factor), is produced by glia in the embryo and is essential to support survival of dopaminergic neurons (Palgi et al., 2009). Survival signals from glial cells to neurons are also observed in retinal development (Xiong & Montell, 1995). In *Drosophila*, survival of retinal neurons depends on their target optic lobes. When mosaic mutant clones of the transcription factor *disconnected* (*disc*) were generated in optic lobes, there was a strong correlation between defects in the connection of retinal neurons to the optic lobe and retinal degeneration (Campos, Fischbach, & Steller, 1992). In mutants for the transcription factor *sine oculis* (*so*), innervation of photoreceptor neurons from the retina into the optic lobe failed, and excess neuronal cell death in the optic lobe was observed (Dearborn & Kunes, 2004; Fischbach & Technau, 1984). Glial cell migration from retina to optic lobe is also affected in *so* mutants, suggesting migrating glia may have a trophic function in the optic lobe. Glial cell migration also occurs from the optic lobe into the retina. *Reversed polarity* (*repo*) encodes a transcription factor required for glial cell differentiation. In *repo* mutants, retinal neurons die in excess, which correlates with a lack of glia in the retina, implying trophic support of retinal cell survival by glia (Xiong & Montell, 1995).

Conversely, survival of glia often depends on trophic factors from neurons. In *Drosophila*, longitudinal glial survival relies on the trophic factor Vein, a neuregulin homolog produced by pioneer neurons (Hidalgo, Kinrade, & Georgiou, 2001). Midline glia (MG) are required for the separation of commissural axons (Jacobs, 2000). In each segment, 12 MG are produced and nine undergo cell death during normal development. Survival of MG is supported by TGF-β Spitz, a ligand of the EGF receptor (EGFR), and EGFR-mediated MAPK activation (Dong & Jacobs, 1997; Lanoue, Gordon, Battye, & Jacobs, 2000; Stemerdink & Jacobs, 1997). Activated MAPK phosphorylates head involution defective (Hid) to inhibit its apoptosis-inducing activity (Bergmann, Tugentman, Shilo, & Steller, 2002). In vertebrates, survival of oligodendrocytes is supported by axon-dependent survival signals. This mechanism seems to adjust the number of oligodendrocytes (Barres, Schmid, Sendnter, & Raff, 1993). Likewise, Schwann cell survival depends on axons that secrete glial growth factor (Winseck & Oppenheim, 2006).

3. INVOLVEMENT IN MORPHOGENESIS

PCD is observed at the stages before neural circuit formation, especially in the CNS. Significant amounts of dying cells in the CNS have been observed during embryonic development in both vertebrates and invertebrates. In the early phase of mouse brain development, many apoptotic cells are observed in the boundary region between the neural plate and the non-neural ectoderm (Fig. 1A). Deficiency of intrinsic apoptotic pathway genes (*caspase-3*, *caspase-9*, or *apaf-1*) causes brain malformations including brain ventricle compression, indented neuroepithelial sheets, and exencephaly, a condition in which the cranial neural tube fails to close and as a result, brain neuroepithelium protrudes outside the skull. However, total embryonic brain cell numbers are not affected at a stage when morphological abnormalities have begun to appear, indicating that the initial cause of embryonic brain malformations in apoptosis-deficient embryos can be explained simply by defects in neural tube closure (NTC) (Nonomura et al., 2013).

NTC is a dynamic process in which the neural plate bends and fuses to form the neural tube and brain (Hensey & Gautier, 1998; Nonomura et al., 2013; Yamaguchi & Miura, 2013; Yamaguchi et al., 2011). The boundary regions, where many apoptotic cells are observed, are ultimately fused to close the neural tube. Apoptosis in the boundary region seems to contribute to the normal progression of cranial NTC (Yamaguchi et al., 2011). This phenomenon resembles the role of apoptosis in leg disc development and dorsal closure in *Drosophila* embryos (Monier et al., 2015; Toyama, Peralta, Wells, Kiehart, & Edwards, 2008).

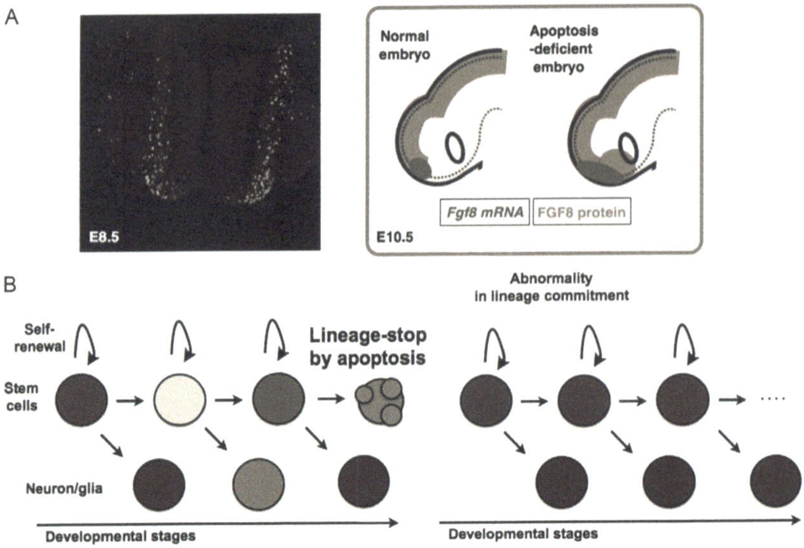

Figure 1 Examples of early neural cell death. (A) Elimination of morphogen-producing cells. (Left) A lot of apoptotic (TUNEL+) cells (green) appear in the boundary regions between the neural plate and the nonneural ectoderm. Purple: nucleus. (Right) Expansions of fgf8-expressing domain and distribution of FGF8 protein in apoptosis-deficient (apaf-1 KO or Caspase-9 KO) embryos (Nonomura et al., 2013). (B) Size regulation of specific cell lineage. Each colored cell expresses different combination of a series of transcriptional factors, which may constitute an intrinsic developmental timer. (See the color plate.)

4. ELIMINATION OF MORPHOGEN-PRODUCING CELLS

Apoptosis in the boundary region also seems to be involved in the proper size regulation of neural and nonneural domains at the neural plate stages in vertebrates including *Xenopus laevis* and mouse (Fig. 1A) (Nonomura et al., 2013; Offner, Duval, Jamrich, & Durand, 2005). In the developing mouse embryo, the anterior neural ridge (ANR) is the most rostral boundary region between the neural plate and nonneural ectoderm. The ANR acts as an organizing center by secreting morphogen Fibroblast growth factor (Fgf) 8 (Shimamura & Rubenstein, 1997). Fgf8-expressing cells undergo massive cell death in the ANR before and after NTC, and apoptosis deficiency results in abnormally prolonged persistence of Fgf8-expressing cells in the rostral ANR (Nonomura et al., 2013). This leads to abnormal distribution of Fgf8 protein and perturbed ventral forebrain development (Fig. 1A). It is not yet known whether a similar scheme can be applied to other signaling centers where substantial apoptosis is observed (Gibson, Robinson, Streit, Sheng, & Stern, 2011; Homma, Yaginuma, & Oppenheim, 1994; Miller & Briglin, 1996; Sanz-Ezquerro & Tickle, 2000).

5. INTRINSIC REGULATION OF THE NUMBER OF PROGENY FROM A SPECIFIC CELL LINEAGE

In *Drosophila*, apoptosis is evident at embryonic stage 11 and peaks in later embryonic stages (from stage 14) (Abrams, White, Fessler, & Steller, 1993). The majority of neuroblasts, which resembles neural stem cells in vertebrates in that they change their identity along development and produce neurons and glia, in the abdominal segments of the developing ventral nerve cord (VNC) become quiescent and undergo cell death at the end of embryogenesis (Fig. 1B; Rogulja-Ortmann, Luer, Seibert, Rickert, & Technau, 2007; Truman & Bate, 1988; White et al., 1994). Developmental apoptosis is blocked in embryos homozygous for the chromosomal deficiency Df(3L) *H99* (White et al., 1994). This region contains head involution defective (*hid*)/*Wrinkled*/*W*, *grim*, and *reaper* (*rpr*), collectively known as the RHG genes, that function as inhibitor of apoptosis (IAP) antagonists to promote caspase activation and cell death. In the *H99* strain, the total number of cells and BrdU uptake is increased, with high variability between individual flies (Rogulja-Ortmann et al., 2007). In a mutant lacking the *reaper* and *grim* region (*MM2* deletion), a fourfold increase in abdominal neuroblasts in early-mid third instar larvae is observed. In 1- to 2-day-old adults carrying the *MM2* deletion, the VNC is more than twice as long as wild type (Tan et al., 2011). This suggests that apoptosis has a role in regulating cell numbers during development.

Recent studies in flies have suggested that endogenous apoptosis of neuroblasts and their descendants are under the regulation of intrinsic developmental programs consisting of transcription factor cascades (Yamaguchi & Miura, 2015). Termination of neuroblast division can be achieved by cell cycle exit or cell death (Fig. 1B). Detailed analysis of the abdominal dorsolateral (dl) neuroblast lineage indicates that the death of dl neuroblasts plays a role in terminating the neuroblast lineage and controls the number of its progeny (Fig. 1B; Maurange, Cheng, & Gould, 2008). Interestingly, when the expression of several transcription factors that can specify or determine cell identity is altered by genetic manipulation, the pattern and timing of

PCD are also changed to match the new cell identity acquired by the manipulation (Fig. 1B; Homem & Knoblich, 2012; Kohwi & Doe, 2013). Intrinsic developmental program also seems to determine the timing of cell death in interneurons of mammalian brain (Southwell et al., 2012). These lines of evidence highlight the concept that the occurrence of PCD depends strongly on cell identity determined by intrinsic developmental programs (Yamaguchi & Miura, 2015).

6. CANCELING DEVELOPMENTAL ERRORS

Apoptosis also plays a role in canceling developmental errors or noise. These are stochastic events and inevitable by-products of developmental processes. During mouse embryonic cerebral cortex development, a significant number of aneuploid cells are generated. Some of them are removed in an apoptosis-dependent manner, although the reason and mechanisms of such spontaneous generation of aneuploidy are not clear (Bushman & Chun, 2013; Peterson et al., 2012; Rehen et al., 2001). Likewise, there are many examples wherein aberrant neural projection that occurs either spontaneously or by genetic mutation is eliminated by PCD in both vertebrates and invertebrates (Baek, Enriquez, & Mann, 2013; Berry & Baehrecke, 2007; Buss et al., 2006; Clarke, 1992; Jiang & Reichert, 2012; Rogulja-Ortmann et al., 2007). Therefore, PCD may serve as a quality control mechanism to eliminate abnormal cells that are spontaneously or accidentally generated during nervous system development.

There are many studies suggesting a role for apoptosis in canceling developmental noise in both vertebrates and invertebrates. One example is observed in the developing *Drosophila* sensory organ, a model for the study of the neural cell fate determination process, including life-or-death decisions. The molecular mechanisms by which sensory organ precursors (SOPs) develop from proneural cell clusters have been well studied (Heitzler & Simpson, 1991; Simpson, 1990). Notch/Delta-mediated lateral inhibition functions to generate a pattern of uniformly spaced SOPs. SOPs are identified by expression of the neurogenic gene *neuralized*. About 20% of neuralized-positive cells die accompanied by high caspase activation and nuclear fragmentation and are ultimately eliminated (Gho, Bellaiche, & Schweisguth, 1999; Koto, Kuranaga, & Miura, 2011). Cells termed as SOP-like cells show high Notch activation, which is not observed in normal SOPs. In *Notch* heterozygous mutant flies, the proportion of SOP-like cells is decreased, and the number of adult bristles increased by 20%, indicating that the Notch activation level is a factor in determining whether neuralized-positive cells become SOPs. If a neuralized-positive cell fails to complete its neural differentiation, accompanied by inappropriate Notch activation, it is specifically eliminated via caspase-dependent cell death (Koto et al., 2011).

7. NEURAL CELL DEATH IN SENSORY SYSTEM DEVELOPMENT: THE VISUAL SYSTEM AS A MODEL

Apoptosis is an essential player in functional eye development in *Drosophila*. The *Drosophila* compound eye is composed of ~750 ommatidia, each containing eight photoreceptor neurons, four cone cells, and two primary pigment cells. Between each ommatidium, secondary and tertiary pigment cells and mechanosensory bristles form the interommatidial lattice. An excess number of secondary and tertiary pigment cells are produced and surplus cells (approximately one-third of the total) are eliminated by apoptosis to form a precise hexagonal lattice. Interommatidial lattice cell death is spatially regulated to ensure it is executed in the correct positions. Tracing cell death in the pupal retina has revealed that 85% of lattice cell death occurs in two specific regions termed "death zones" between ommatidial units and adjacent to bristle groups (Monserrate & Brachmann, 2007).

Lattice cell death is regulated by cell–cell interactions. Cell ablation experiments have shown that cone cells and primary pigment cells are required for the survival of surrounding lattice cells (Miller & Cagan, 1998). Mutants for irregular chiasm-C-roughest (irreC-rst), a member of the transmembrane immunoglobulin superfamily, have excess lattice cells (Ramos et al., 1993). IrreC-rst is thought to be required for aligning lattice cells end-to-end prior to apoptosis (Reiter, Schimansky, Nie, & Fischbach, 1996). Notch is expressed in retinal lattice cells (Kooh, Fehon, & Muskavitch, 1993), and reduction of Notch activity during the period of cell death using a temperature sensitive *Notch* allele prevents lattice cell death (Cagan & Ready, 1989; Wolff & Ready, 1991), indicating that Notch signaling is required for lattice cell death. A connection between IrreC-rst and Notch signaling was uncovered because *Delta* was identified as a genetic modifier of a hypomorphic allele of *irreC-rst* (Tanenbaum, Gorski, Rusconi, & Cagan, 2000). Furthermore, irreC-rst mislocalization is observed in *Delta* and *Notch* mutants (Gorski, Brachmann, Tanenbaum, & Cagan, 2000).

Argos is a soluble antagonist of the *Drosophila* EGF receptor (dEGFR) (Freeman, Klambt, Goodman, & Rubin, 1992; Sawamoto et al., 1994;

Schweitzer, Howes, Smith, Shilo, & Freeman, 1995). Overexpression of *argos* or dominant negative dEGFR induces excess apoptosis of lattice cells, indicating the survival signal for lattice cells acts through dEGFR/dRas1 (Freeman, 1996; Sawamoto et al., 1994). Spitz, a ligand for dEGFR, is expressed in cone and primary pigment cells, and overexpression of the activated forms of dEGFR or dRas1 rescues lattice cell death (Miller & Cagan, 1998), further supporting the importance of dEGFR/dRas1 signaling for lattice cell survival. One of the mechanisms by which dEGFR/dRas1 regulates cell survival could be phosphorylation-mediated inactivation of Drosophila Inhibitor of apoptosis protein 1 (DIAP1) antagonist Hid (Bergmann, Agapite, McCall, & Steller, 1998; Kurada & White, 1998). Initiator caspase Dronc and its downstream effector caspase Drice are activated within the pupal lattice (Yu et al., 2002). DIAP1 and its regulator (UBCD-1, Morgue: proteasomal degradation pathway for DIAP1, DIAP1 antagonist: Reaper, Hid, Grim, Sickle, Jafrac 2) play central role for apoptosis and regulate apoptosis of interommatidial cells (Lisi, Mazzon, & White, 2000; Ryoo, Bergmann, Gonen, Ciechanover, & Steller, 2002; Tenev, Zachariou, Wilson, Paul, & Meier, 2002; Wing et al., 2002). As described above, Notch signaling is proapoptotic for interommatidial lattice cells. Notch signaling works to antagonize the dEGFR survival signal in lattice cells (Yu et al., 2002).

Apoptosis also participates in functional eye development in mammals. During the development of intrinsically photosensitive retinal ganglion cells (ipRGCs), which contain melanopsin and function in circadian rhythms and papillary light responses, Bax-dependent apoptosis occurs in the population of ipRGCs. Bax-deficient mutants show disrupted ipRGC spacing, dendritic stratification, and ectopic synapses. The disruption of ipRGC spacing in Bax mutants does not affect melanopsin-driven circadian photoentrainment but impairs rod/cone-driven photoentrainment (Chen et al., 2013).

8. THE ROLE OF NEURAL CELL DEATH IN TISSUE REMODELING

Cell death is essential for tissue remodeling, the process during which juvenile or larval tissues and structures are destroyed and replaced by adult tissues and structures. A typical example of tissue remodeling is metamorphosis in insects. During insect metamorphosis, adult-specific imaginal neurons remain alive, but prune their dendrites and axons for remodeling of neural circuit. However, other larval neurons including larval motor neurons, mushroom body neurons, and FMRFamide-expressing thoracic ventral neurons are eliminated during the pupal stage. RP2 larval motor neurons in abdominal neuromeres are eliminated after the prepupal pulse of ecdysteroids. RP2 neuronal cell death requires ecdysone receptor B (*EcR-B*; *B1* and/or *B2* isoforms), *reaper* (but not *hid*), and *dark*, suggesting that EcR-B-mediated *reaper* induction and subsequent apoptosome activation cause RP2 death (Winbush & Weeks, 2011). Peptidergic neurons expressing the neuropeptide Corazonin (Crz) in the VNC undergo PCD by 6 h after puparium formation (APF). Similar to RP2 neurons, Crz neuron death is *reaper*- and *dark*-dependent, and is suppressed by *p35* but not by *diap1* (Choi, Lee, & Park, 2006). The engulfment of dying Crz neurons by glia has been suggested to promote apoptosis of Crz neurons (Tasdemir-Yilmaz & Freeman, 2014). Many neurons of the optic lobe die during the pupal stage (Togane et al., 2012). EcR-B1 is required for most neuronal death in the optic lobe from 24 h APF; however, none of the EcR isoforms are required for optic lobe cell death at 0 h APF (Hara et al., 2012).

Each type I neuroblast generates a ganglion mother cell (GMC), which divides once to produce two postmitotic cells. Type II neuroblasts generate transit amplifying cells (intermediate neural progenitors, INPs) each of which produces a further INP and one GMC; thus, type II neuroblasts generate larger populations of neural cells in the central brain (Bello, Izergina, Caussinus, & Reichert, 2008). A study of cell death of dorsomedial (DM) amplifying type II neuroblasts found that one-quarter of the neuronal (but not glial) cells undergo apoptosis during the pupal stage. When DM apoptosis was blocked by expression of *p35*, abnormal innervation was observed in the major central complex (ellipsoid body, fan-shaped body, nodule, and protocerebral bridge) neuropil (Jiang & Reichert, 2012). However, flies with blocked DM apoptosis are viable and fertile, and do not show any abnormalities in behavior tests (walking, climbing, simple negative geotaxis, and positive phototaxis assays).

In vertebrates, several studies have reported the involvement of apoptosis in tissue remodeling. Apoptosis is an essential player in tissue remodeling of the auditory system in mice. Mice deficient for caspase-3 in a C57BL/6 background can survive into adulthood, but they exhibit severe hearing loss (Takahashi et al., 2001). The cause of deafness in caspase-3 mice can be attributed to lack of apoptosis in the greater epithelial ridge (GER), a transient structure that is formed during prenatal development and regresses to form the inner sulcus after birth (Takahashi et al., 2001). Regression of the

GER leads to the proper formation of the organ of Corti, which transduces auditory stimuli via hair cells to the brain. This regression is mediated by apoptosis occurring around postnatal day 8 in a thyroid hormone-dependent manner (Ng, Kelley, & Forrest, 2013; Peeters, Ng, Ma, & Forrest, 2015). Premature exposure to excessive T3 around birth accelerates the onset of apoptosis and results in deafness. These data highlight the importance of timing control in apoptosis under systemic regulation for functional auditory system development (Peeters et al., 2015).

9. GLIAL FUNCTION IN CONTROLLING NEURAL ARCHITECTURE BY ELIMINATION OF DEAD CELLS

Drosophila larval neurons undergo remodeling to generate the adult neural circuit during metamorphosis. Before metamorphosis, myoglianin (myo), a TGF-β ligand, is induced in glia of larval brain. Expression of myo activates EcR-B1 in mushroom body γ neurons. EcR-B1 expression enables γ neurons to receive hormone signals that initiate axon pruning (Awasaki, Huang, O'Connor, & Lee, 2011). The perpendicular axonal branches are pruned during metamorphosis and axons are reextended horizontally in adult brain. Pan-glial-specific knockdown of *myo* prevents the pruning. In the pupal stage, glia infiltrate the mushroom body neuropil and axon fragmentation also occurs. These axon fragments are cleared by glia through mechanisms resembling the engulfment of dying cells in *C. elegans* (Awasaki et al., 2006; Tasdemir-Yilmaz & Freeman, 2014). Genetic studies of *C. elegans* have revealed the mechanisms of dying cell engulfment (Venegas & Zhou, 2007). CED-1 (*Drosophila* Draper/mouse Jedi/human mEGF10 and mEGF11) is a phagocytic receptor for apoptotic cells that is stimulated by phosphatidylserine exposed on the cell surface of apoptotic cells. CED-6, a PTB-domain protein, is a candidate adaptor for CED-1. CED-5 (*Drosophila* Mbc/mammalian Dock180) and CED-12 (*Drosophila* dCed-12/mammalian ELMO1) form a guanine nucleotide exchange factor that activates CED-10/Rac1 GTPase to promote cytoskeleton reorganization during cell corpse internalization. An SH2–SH3 domain protein CED-2 (*Drosophila* Crk/mammalian CrkII) is proposed to connect a phagocytic receptor with the CED-5/CED-12 complex. Draper and the Crk/Mbc/dCed-12 complex are used in the engulfment events of axon pruning of γ neurons (Awasaki et al., 2006; Tasdemir-Yilmaz & Freeman, 2014).

In the adult mouse hippocampus, the majority of newborn cells undergo apoptosis during postnatal days 1–4. In this time period, amplifying neuroprogenitors in the subgranular zone of the dentate gyrus transit to neuroblasts. These apoptotic cells are eliminated by microglia (Sierra et al., 2010). CD11b- and DAP12-positive microglia contact with apoptotic neurons in the developing hippocampus. Mice deficient for CD11b or DAP12 suppress developmental death, suggesting active roles for these receptors in neuronal death. DAP12 and CD11b control the production of microglial superoxide ions; thus, neuronal death is facilitated by microglia (Wakselman et al., 2008). Microglia in the cerebellum also promote the Purkinje cell death that occurs in the cerebellum of postnatal day 3 mice (Marin-Teva et al., 2004) and may control cerebellar cortex compartmentation (Jankowski, Miething, Schilling, Oberdick, & Baader, 2011).

10. THE NONAPOPTOTIC FUNCTIONS OF CASPASES IN NEURAL DEVELOPMENT

In addition to causing cell death, caspases have nonapoptotic roles in multicellular organisms. For example, caspase-1, an inflammatory caspase that was identified as the first mammalian ortholog of CED-3 in *C. elegans*, plays an important role in processing inflammatory cytokine IL-1β from the precursor to the mature form (Franchi, Eigenbrod, Munoz-Planillo, & Nunez, 2009; Miura, Zhu, Rotello, Hartwieg, & Yuan, 1993). There is now accumulating evidence that caspases participate in biological events in a nonapoptotic manner. In vertebrates, there are several studies that propose nonapoptotic caspase activation. For example, mice defective in the mitochondrial apoptotic pathway (*caspase-9* or *apaf-1* knockout mice) exhibit misrouted axons, impaired synaptic formation, and defects in olfactory sensory neuron maturation that are difficult to explain by lack of apoptosis (Ohsawa et al., 2010). Inhibition of caspase-3 in *Xenopus* leads to an increase in diencephalic cell number and accumulation of the active form of β-catenin, an intracellular transducer of the canonical Wnt signaling pathway (Juraver-Geslin, Ausseil, Wassef, & Durand, 2011). Caspase-3 can directly cleave β-catenin and the overgrowth phenotype caused by inhibition of caspase-3 is canceled by β-catenin depletion, suggesting that caspase-3 restricts Wnt signaling activity by decreasing β-catenin activity in diencephalic development (Juraver-Geslin et al., 2011). It is also worth noting, from an evolutionary perspective, that metacaspases in unicellular organisms function in fundamental processes including cell proliferation and quality control of proteins (Hill, Hao, Liu, & Nystrom, 2014; Jung et al., 2008). Questions arise as to when such nonapoptotic roles are required, and how they are regulated.

11. NEURITE PRUNING

Elimination of preformed structures in the developing nervous system can be performed without cell death. Massive elimination of synapses (~90%) at neuromuscular junctions occurs after birth in mice (Sanes & Lichtman, 1999; Tapia et al., 2012). This synapse loss is explained as a terminal branch loss, like neurite pruning (Fig. 2A). In contrast to cell death, pruning can be flexible and used for remodeling and plasticity of neuronal circuitry. Recent studies have suggested that apoptotic machineries are involved in some, if not all, cases of pruning.

Pruning of axons and dendrites occurs in development and in the adult nervous system remodeling stage (Schuldiner & Yaron, 2015). In rat,

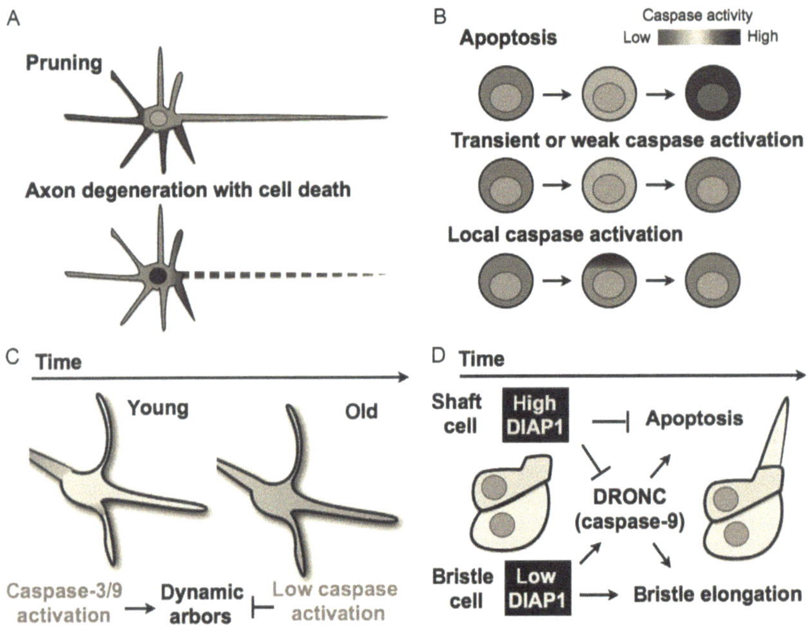

Figure 2 Nonapoptotic caspase activation in neural development. (A) Pruning of neurites without causing cell death and axon degeneration accompanying cell death. Blue indicates caspase activation. (B) Schematic illustration of distinct modes of caspase activation. (C) Local caspase activation observed in real time in young axonal arbors of developing zebrafish embryos (Campbell & Okamoto, 2013). (D) Nonapoptotic caspase activation through the regulation of DIAP1, a caspase inhibitor, in sensory bristle formation of fly in a cell-type-specific manner. DRONC activity is required for bristle elongation without causing cell death (Koto, Kuranaga, & Miura, 2009). (See the color plate.)

neurons from the motor and visual regions of cortical layer 5 send identical projections to the spinal cord and the superior colliculus initially, and then selective elimination of axons occurs at the second to third postnatal week (Luo & O'Leary, 2005; O'Leary & Koester, 1993; Stanfield, O'Leary, & Fricks, 1982). It was also shown that the elimination of axons of retinal ganglion cells (RGCs) occurs at the superior colliculus (Nakamura & O'Leary, 1989). Small-scale pruning of axons by retraction is observed in infrapyramidal bundles of hippocampal granule cells and Cajal–Retzius neurons (Bagri, Cheng, Yaron, Pleasure, & Tessier-Lavigne, 2003; Portera-Cailliau, Weimer, De Paola, Caroni, & Svoboda, 2005).

During metamorphosis in *Drosophila*, dendrites of dendritic arborization (da) sensory neurons are eliminated (Parrish, Emoto, Kim, & Jan, 2007; Williams, Kondo, Krzyzanowska, Hiromi, & Truman, 2006). Down-regulation of cell adhesion molecule Neuroglian by the Rab5/ESCRT endocytic pathway triggers dendrite pruning of da neurons (Zhang et al., 2014). Dendrite pruning of da neurons depends on the ubiquitin–proteasome system (UPS) (Kuo, Jan, & Jan, 2005; Kuo, Zhu, Younger, Jan, & Jan, 2006; Wong et al., 2013). Valosin-containing protein, a ubiquitin-dependent ATPase, is required for dendritic pruning of da neurons through control of mRNA metabolism (Portera-Cailliau et al., 2005; Rumpf et al., 2014).

Dendrites and dorsal and medial axonal branches are eliminated from larval mushroom body γ neurons (Lee, Lee, & Luo, 1999; Technau & Heisenberg, 1982; Watts, Hoopfer, & Luo, 2003). As well as being required for dendrite pruning, UPS also regulates mushroom body γ neuron axon pruning (Watts et al., 2003). Involvement of caspases in neurite pruning has been studied *in vivo* and *in vitro*. Local caspase activities have been observed in fragmented da dendrites and partly contribute to elimination in *Drosophila* (Kuo et al., 2006; Williams et al., 2006). The DIAP1-degrading kinase DmIKKε/IK2 is required for dendritic pruning of da neurons (Lee, Jan, & Jan, 2009).

12. MECHANISMS THAT CONTROL NONAPOPTOTIC CASPASE ACTIVATION

In apoptotic cells, caspase activation occurs in a positive feedback manner. Since executioner caspase-3 contains self-cleavage sites, a low level of caspase activation could lead to activation of other caspases and be propagated to whole-cell regions. Such propagation would ultimately destroy cell structures (Dekkers et al., 2013). In nervous system development, two possible strategies for nonapoptotic caspase activation exist: transient or weak activation and local activation (Fig. 2B).

13. TRANSIENT OR WEAK CASPASE ACTIVATION

Several studies have also suggested the involvement of transient or weak caspase activation in neural differentiation in differentiating PC12 cells or neurospheres (Fernando, Brunette, & Megeney, 2005; Rohn, Cusack, Kessinger, & Oxford, 2004). However, these studies did not demonstrate nonapoptotic cell-autonomous caspase activation directly within a single cell in real time, therefore, not excluding the possibility that a few surrounding apoptotic cells send signals and exert noncell-autonomous effects on those processes, as in the case of compensatory proliferation, regeneration, and terminal differentiation (Bergmann & Steller, 2010; Hochreiter-Hufford et al., 2013). Live imaging analysis is a possible solution for demonstrating nonapoptotic caspase activation. Recent studies using reporters for caspase activity or apoptotic signals have successfully demonstrated nonapoptotic caspase activation occurring within a single cell in real time in the nervous system (Hyman & Yuan, 2012; Spires-Jones et al., 2008). Using a genetically encoded fluorescent reporter for caspase-3 activation, it has been shown that caspase-9 and caspase-3 are locally and temporally activated, possibly downstream of Slit–Robo signaling, at the branch points of young RGCs in zebrafish embryos (Fig. 2C; Campbell & Okamoto, 2013).

DmIKKε is a homolog of the noncanonical members of the IκB kinases (IKKε/IKKi or NAK/T2K/TBK1), which regulate the activation of NF-κB or interferon regulatory factor (IRF)-3 and -7 in mammals (Kawai & Akira, 2006). DmIKKε determines the endogenous level of DIAP1 through DIAP1 phosphorylation. Using a sensitive fluorescence resonance energy transfer-based caspase activity-monitoring probe, SCAT3 (Takemoto, Nagai, Miyawaki, & Miura, 2003), a weak or low-level activation of caspase was detected in the *scabrous*-expressing proneural clusters in wing discs (Kuranaga et al., 2006). However, the knockdown of *DmIKKε* suppressed this low-level caspase activity. Similar to the phenotype of *p35* or *Dronc-DN* overexpression in *scabrous*-expressing proneural clusters, the knockdown of *DmIKKε* leads to the generation of an extra SOP in *scabrous*-expressing proneural clusters of the scutellum. Thus, DmIKKε appears to regulate the amount of DIAP1 protein, which determines the nonapoptotic caspase activity required to define the number of SOPs (i.e., two) in the scutellar proneural cluster (Kuranaga et al., 2006).

Signals regulating caspase activity have also been visualized using the reporter protein PRAP (PRe-Apoptosis signal-detecting probe, based on DIAP1 degradation), which was developed for monitoring the turnover of DIAP1, an endogenous caspase inhibitor. DIAP1 degradation is regulated in a cell lineage- and stage-specific manner during sensory organ (microchaetae) development in notum (Fig. 2D). Genetic and live imaging analysis of DIAP1 degradation in shaft cells revealed that DIAP1 executes two distinct functions: one that supports the survival of the shaft cell just after the final division and one in the morphogenesis of the shaft cell during its maturation. Both DIAP1 functions are exerted through the regulation of DRONC (a fly ortholog of caspase-9) activity. Thus, the precise temporal control of DIAP1 degradation is critical for sensory bristle differentiation by maintaining the balance between cell viability and the execution of the non-apoptotic functions of caspase (Koto et al., 2009).

14. LOCAL CASPASE ACTIVATION

As in developing neurites of zebrafish embryos (Campbell & Okamoto, 2013), local caspase activation is observed also in dendrite pruning of pupal multidendritic neurons in *Drosophila* (Fig. 2A) (Kuo et al., 2006; Williams et al., 2006). Caspase activation is restricted to areas distant from the soma, preventing the effective execution of cytoplasmic death. The IKK-related kinase DmIKKε/IK2 controls DIAP1 degradation as described previously and regulates dendrite pruning (Lee et al., 2009). In the developing *Xenopus* retina, caspases are activated in axons and dendrites as part of the chemotropic response in growth cones (Campbell & Holt, 2003). In mouse, it has been shown that activation of caspase-6 via amyloid precursor protein (APP) cleavage and activation of DR6 induces axon degeneration after NGF deprivation. APP and amyloid precursor-like protein signals through DR6 may be involved in pruning at the neuromuscular junction in the diaphragm (Nikolaev, McLaughlin, O'Leary, & Tessier-Lavigne, 2009). Interestingly, the requirement of caspase-6 for axon degeneration is specific to NGF deprivation, indicating that caspase activation is not universal to axon degeneration. The mechanisms that regulate local caspase activation can vary in a context-dependent manner and operate at multiple levels, including transcription, transport, translation, and degradation of apoptotic machinery (Conforti, Gilley, & Coleman, 2014; Cosker, Pazyra-Murphy, Fenstermacher, & Segal, 2013; Cusack, Swahari, Hampton Henley, Michael Ramsey, & Deshmukh, 2013; Dekkers et al., 2013; Schoenmann et al., 2010; Unsain, Higgins, Parker, Johnstone, & Barker, 2013; Vohra et al., 2010).

15. CONCLUSION

Recent studies have revealed the significance and regulation of PCD, particularly apoptosis, in neural development. Apoptosis is involved not only in the morphological aspects of development but also in functional maturation of the nervous system. Furthermore, nonapoptotic roles of caspases in neural development and physiology have emerged (Galluzzi et al., 2015; Hyman & Yuan, 2012). Elucidating the regulatory mechanisms and functions of apoptosis and nonapoptotic caspase activation will lead to a deeper understanding of the blueprint of neural development.

ACKNOWLEDGMENTS

We apologize to the colleagues whose works have not been cited or cited only indirectly because of space limitations. This work was supported by grants from the Japanese Ministry of Education, Science, Sports, Culture, and Technology (to Y.Y. and M.M.).

REFERENCES

Abrams, J. M., White, K., Fessler, L. I., & Steller, H. (1993). Programmed cell death during Drosophila embryogenesis. *Development, 117*, 29–43.

Awasaki, T., Huang, Y., O'Connor, M. B., & Lee, T. (2011). Glia instruct developmental neuronal remodeling through TGF-beta signaling. *Nature Neuroscience, 14*, 821–823.

Awasaki, T., Tatsumi, R., Takahashi, K., Arai, K., Nakanishi, Y., Ueda, R., et al. (2006). Essential role of the apoptotic cell engulfment genes draper and ced-6 in programmed axon pruning during Drosophila metamorphosis. *Neuron, 50*, 855–867.

Baek, M., Enriquez, J., & Mann, R. S. (2013). Dual role for Hox genes and Hox co-factors in conferring leg motoneuron survival and identity in Drosophila. *Development, 140*, 2027–2038.

Bagri, A., Cheng, H. J., Yaron, A., Pleasure, S. J., & Tessier-Lavigne, M. (2003). Stereotyped pruning of long hippocampal axon branches triggered by retraction inducers of the semaphorin family. *Cell, 113*, 285–299.

Barres, B. A., Schmid, R., Sendnter, M., & Raff, M. C. (1993). Multiple extracellular signals are required for long-term oligodendrocyte survival. *Development, 118*, 283–295.

Bello, B. C., Izergina, N., Caussinus, E., & Reichert, H. (2008). Amplification of neural stem cell proliferation by intermediate progenitor cells in Drosophila brain development. *Neural Development, 3*, 5.

Bergmann, A., Agapite, J., McCall, K., & Steller, H. (1998). The Drosophila gene hid is a direct molecular target of Ras-dependent survival signaling. *Cell, 95*, 331–341.

Bergmann, A., & Steller, H. (2010). Apoptosis, stem cells, and tissue regeneration. *Science Signaling, 3*, re8.

Bergmann, A., Tugentman, M., Shilo, B. Z., & Steller, H. (2002). Regulation of cell number by MAPK-dependent control of apoptosis: A mechanism for trophic survival signaling. *Developmental Cell, 2*, 159–170.

Berry, D. L., & Baehrecke, E. H. (2007). Growth arrest and autophagy are required for salivary gland cell degradation in Drosophila. *Cell, 131*, 1137–1148.

Blaschke, A. J., Staley, K., & Chun, J. (1996). Widespread programmed cell death in proliferative and postmitotic regions of the fetal cerebral cortex. *Development, 122*, 1165–1174.

Bushman, D. M., & Chun, J. (2013). The genomically mosaic brain: Aneuploidy and more in neural diversity and disease. *Seminars in Cell & Developmental Biology, 24*, 357–369.

Buss, R. R., Sun, W., & Oppenheim, R. W. (2006). Adaptive roles of programmed cell death during nervous system development. *Annual Review of Neuroscience, 29*, 1–35.

Cagan, R. L., & Ready, D. F. (1989). Notch is required for successive cell decisions in the developing Drosophila retina. *Genes & Development, 3*, 1099–1112.

Campbell, D. S., & Holt, C. E. (2003). Apoptotic pathway and MAPKs differentially regulate chemotropic responses of retinal growth cones. *Neuron, 37*, 939–952.

Campbell, D. S., & Okamoto, H. (2013). Local caspase activation interacts with Slit-Robo signaling to restrict axonal arborization. *The Journal of Cell Biology, 203*, 657–672.

Campos, A. R., Fischbach, K. F., & Steller, H. (1992). Survival of photoreceptor neurons in the compound eye of Drosophila depends on connections with the optic ganglia. *Development, 114*, 355–366.

Chen, S. K., Chew, K. S., McNeill, D. S., Keeley, P. W., Ecker, J. L., Mao, B. Q., et al. (2013). Apoptosis regulates ipRGC spacing necessary for rods and cones to drive circadian photoentrainment. *Neuron, 77*, 503–515.

Choi, Y. J., Lee, G., & Park, J. H. (2006). Programmed cell death mechanisms of identifiable peptidergic neurons in *Drosophila melanogaster. Development, 133*, 2223–2232.

Clarke, P. G. (1992). Neuron death in the developing avian isthmo-optic nucleus, and its relation to the establishment of functional circuitry. *Journal of Neurobiology, 23*, 1140–1158.

Conforti, L., Gilley, J., & Coleman, M. P. (2014). Wallerian degeneration: An emerging axon death pathway linking injury and disease. *Nature Reviews. Neuroscience, 15*, 394–409.

Cosker, K. E., Pazyra-Murphy, M. F., Fenstermacher, S. J., & Segal, R. A. (2013). Target-derived neurotrophins coordinate transcription and transport of bclw to prevent axonal degeneration. *The Journal of Neuroscience: The Official Journal of the Society for Neuroscience, 33*, 5195–5207.

Cowan, W. M. (2001). Viktor Hamburger and Rita Levi-Montalcini: The path to the discovery of nerve growth factor. *Annual Review of Neuroscience, 24*, 551–600.

Cusack, C. L., Swahari, V., Hampton Henley, W., Michael Ramsey, J., & Deshmukh, M. (2013). Distinct pathways mediate axon degeneration during apoptosis and axon-specific pruning. *Nature Communications, 4*, 1876.

Dearborn, R., Jr., & Kunes, S. (2004). An axon scaffold induced by retinal axons directs glia to destinations in the Drosophila optic lobe. *Development, 131*, 2291–2303.

Degterev, A., & Yuan, J. (2008). Expansion and evolution of cell death programmes. *Nature Reviews. Molecular Cell Biology, 9*, 378–390.

Dekkers, M. P., Nikoletopoulou, V., & Barde, Y. A. (2013). Cell biology in neuroscience: Death of developing neurons: New insights and implications for connectivity. *The Journal of Cell Biology, 203*, 385–393.

Dong, R., & Jacobs, J. R. (1997). Origin and differentiation of supernumerary midline glia in Drosophila embryos deficient for apoptosis. *Developmental Biology, 190*, 165–177.

Fernando, P., Brunette, S., & Megeney, L. A. (2005). Neural stem cell differentiation is dependent upon endogenous caspase 3 activity. *FASEB Journal, 19*, 1671–1673.

Fischbach, K. F., & Technau, G. (1984). Cell degeneration in the developing optic lobes of the sine oculis and small-optic-lobes mutants of *Drosophila melanogaster. Developmental Biology, 104*, 219–239.

Franchi, L., Eigenbrod, T., Munoz-Planillo, R., & Nunez, G. (2009). The inflammasome: A caspase-1-activation platform that regulates immune responses and disease pathogenesis. *Nature Immunology, 10*, 241–247.

Freeman, M. (1996). Reiterative use of the EGF receptor triggers differentiation of all cell types in the Drosophila eye. *Cell, 87*, 651–660.

Freeman, M., Klambt, C., Goodman, C. S., & Rubin, G. M. (1992). The argos gene encodes a diffusible factor that regulates cell fate decisions in the Drosophila eye. *Cell, 69*, 963–975.

Fuchs, Y., & Steller, H. (2011). Programmed cell death in animal development and disease. *Cell, 147*, 742–758.

Galluzzi, L., Bravo-San Pedro, J. M., Vitale, I., Aaronson, S. A., Abrams, J. M., Adam, D., et al. (2014). Essential versus accessory aspects of cell death: Recommendations of the NCCD 2015. *Cell Death and Differentiation, 22*, 58–73.

Galluzzi, L., Bravo-San Pedro, J. M., Vitale, I., Aaronson, S. A., Abrams, J. M., Adam, D., et al. (2015). Essential versus accessory aspects of cell death: Recommendations of the NCCD 2015. *Cell Death and Differentiation, 22*, 58–73.

Galluzzi, L., Vitale, I., Abrams, J. M., Alnemri, E. S., Baehrecke, E. H., Blagosklonny, M. V., et al. (2012). Molecular definitions of cell death subroutines: Recommendations of the Nomenclature Committee on Cell Death 2012. *Cell Death and Differentiation, 19*, 107–120.

Gho, M., Bellaiche, Y., & Schweisguth, F. (1999). Revisiting the Drosophila microchaete lineage: A novel intrinsically asymmetric cell division generates a glial cell. *Development, 126*, 3573–3584.

Gibson, A., Robinson, N., Streit, A., Sheng, G., & Stern, C. D. (2011). Regulation of programmed cell death during neural induction in the chick embryo. *The International Journal of Developmental Biology, 55*, 33–43.

Gorski, S. M., Brachmann, C. B., Tanenbaum, S. B., & Cagan, R. L. (2000). Delta and notch promote correct localization of irreC-rst. *Cell Death and Differentiation, 7*, 1011–1013.

Gould, T. W., & Oppenheim, R. W. (2011). Motor neuron trophic factors: Therapeutic use in ALS? *Brain Research Reviews, 67*, 1–39.

Green, D. R., Galluzzi, L., & Kroemer, G. (2014). Cell biology. Metabolic control of cell death. *Science, 345*, 1250256.

Grimsley, C., & Ravichandran, K. S. (2003). Cues for apoptotic cell engulfment: Eat-me, don't eat-me and come-get-me signals. *Trends in Cell Biology, 13*, 648–656.

Hara, Y., Hirai, K., Togane, Y., Akagawa, H., Iwabuchi, K., & Tsujimura, H. (2012). Ecdysone-dependent and ecdysone-independent programmed cell death in the developing optic lobe of Drosophila. *Developmental Biology, 374*, 127–141.

Heitzler, P., & Simpson, P. (1991). The choice of cell fate in the epidermis of Drosophila. *Cell, 64*, 1083–1092.

Hensey, C., & Gautier, J. (1998). Programmed cell death during Xenopus development: A spatio-temporal analysis. *Developmental Biology, 203*, 36–48.

Hidalgo, A., Kinrade, E. F., & Georgiou, M. (2001). The Drosophila neuregulin vein maintains glial survival during axon guidance in the CNS. *Developmental Cell, 1*, 679–690.

Hill, S. M., Hao, X., Liu, B., & Nystrom, T. (2014). Life-span extension by a metacaspase in the yeast Saccharomyces cerevisiae. *Science, 344*, 1389–1392.

Hochreiter-Hufford, A. E., Lee, C. S., Kinchen, J. M., Sokolowski, J. D., Arandjelovic, S., Call, J. A., et al. (2013). Phosphatidylserine receptor BAI1 and apoptotic cells as new promoters of myoblast fusion. *Nature, 497*, 263–267.

Homem, C. C., & Knoblich, J. A. (2012). Drosophila neuroblasts: A model for stem cell biology. *Development, 139*, 4297–4310.

Homma, S., Yaginuma, H., & Oppenheim, R. W. (1994). Programmed cell death during the earliest stages of spinal cord development in the chick embryo: A possible means of early phenotypic selection. *The Journal of Comparative Neurology, 345*, 377–395.

Hyman, B. T., & Yuan, J. (2012). Apoptotic and non-apoptotic roles of caspases in neuronal physiology and pathophysiology. *Nature Reviews. Neuroscience, 13*, 395–406.

Jacobs, J. R. (2000). The midline glia of Drosophila: A molecular genetic model for the developmental functions of glia. *Progress in Neurobiology, 62*, 475–508.

Jankowski, J., Miething, A., Schilling, K., Oberdick, J., & Baader, S. (2011). Cell death as a regulator of cerebellar histogenesis and compartmentation. *Cerebellum, 10*, 373–392.

Jiang, Y., & Reichert, H. (2012). Programmed cell death in type II neuroblast lineages is required for central complex development in the Drosophila brain. *Neural Development, 7*, 3.

Jung, A. R., Kim, T. W., Rhyu, I. J., Kim, H., Lee, Y. D., Vinsant, S., et al. (2008). Misplacement of Purkinje cells during postnatal development in Bax knock-out mice: A novel role for programmed cell death in the nervous system? *The Journal of Neuroscience: The Official Journal of the Society for Neuroscience, 28*, 2941–2948.

Juraver-Geslin, H. A., Ausseil, J. J., Wassef, M., & Durand, B. C. (2011). Barhl2 limits growth of the diencephalic primordium through Caspase3 inhibition of beta-catenin activation. *Proceedings of the National Academy of Sciences of the United States of America, 108*, 2288–2293.

Kawai, T., & Akira, S. (2006). Innate immune recognition of viral infection. *Nature Immunology, 7*, 131–137.

Kerr, J. F., Wyllie, A. H., & Currie, A. R. (1972). Apoptosis: A basic biological phenomenon with wide-ranging implications in tissue kinetics. *British Journal of Cancer, 26*, 239–257.

Kohwi, M., & Doe, C. Q. (2013). Temporal fate specification and neural progenitor competence during development. *Nature Reviews. Neuroscience, 14*, 823–838.

Kooh, P. J., Fehon, R. G., & Muskavitch, M. A. (1993). Implications of dynamic patterns of Delta and Notch expression for cellular interactions during Drosophila development. *Development, 117*, 493–507.

Koto, A., Kuranaga, E., & Miura, M. (2009). Temporal regulation of Drosophila IAP1 determines caspase functions in sensory organ development. *The Journal of Cell Biology, 187*, 219–231.

Koto, A., Kuranaga, E., & Miura, M. (2011). Apoptosis ensures spacing pattern formation of Drosophila sensory organs. *Current Biology, 21*, 278–287.

Kristiansen, M., & Ham, J. (2014). Programmed cell death during neuronal development: The sympathetic neuron model. *Cell Death and Differentiation, 21*, 1025–1035.

Kuo, C. T., Jan, L. Y., & Jan, Y. N. (2005). Dendrite-specific remodeling of Drosophila sensory neurons requires matrix metalloproteases, ubiquitin-proteasome, and ecdysone signaling. *Proceedings of the National Academy of Sciences of the United States of America, 102*, 15230–15235.

Kuo, C. T., Zhu, S., Younger, S., Jan, L. Y., & Jan, Y. N. (2006). Identification of E2/E3 ubiquitinating enzymes and caspase activity regulating Drosophila sensory neuron dendrite pruning. *Neuron, 51*, 283–290.

Kurada, P., & White, K. (1998). Ras promotes cell survival in Drosophila by downregulating hid expression. *Cell, 95*, 319–329.

Kuranaga, E. (2011). Caspase signaling in animal development. *Development, Growth & Differentiation, 53*, 137–148.

Kuranaga, E., Kanuka, H., Tonoki, A., Takemoto, K., Tomioka, T., Kobayashi, M., et al. (2006). Drosophila IKK-related kinase regulates nonapoptotic function of caspases via degradation of IAPs. *Cell, 126*, 583–596.

Lanoue, B. R., Gordon, M. D., Battye, R., & Jacobs, J. R. (2000). Genetic analysis of vein function in the Drosophila embryonic nervous system. *Genome, 43*, 564–573.

Lee, H. H., Jan, L. Y., & Jan, Y. N. (2009). Drosophila IKK-related kinase Ik2 and Katanin p60-like 1 regulate dendrite pruning of sensory neuron during metamorphosis. *Proceedings of the National Academy of Sciences of the United States of America, 106*, 6363–6368.

Lee, T., Lee, A., & Luo, L. (1999). Development of the Drosophila mushroom bodies: Sequential generation of three distinct types of neurons from a neuroblast. *Development, 126*, 4065–4076.

Levi-Montalcini, R. (1987). The nerve growth factor 35 years later. *Science, 237*, 1154–1162.

Lisi, S., Mazzon, I., & White, K. (2000). Diverse domains of THREAD/DIAP1 are required to inhibit apoptosis induced by REAPER and HID in Drosophila. *Genetics, 154*, 669–678.

Lockshin, R. A., & Williams, C. M. (1964). Programmed cell death. II. Endocrine potentiation of the breakdown of the intersegmental muscles of silkmoths. *Journal of Insect Physiology*, *10*, 643–649.

Luo, L., & O'Leary, D. D. (2005). Axon retraction and degeneration in development and disease. *Annual Review of Neuroscience*, *28*, 127–156.

Marin-Teva, J. L., Dusart, I., Colin, C., Gervais, A., van Rooijen, N., & Mallat, M. (2004). Microglia promote the death of developing Purkinje cells. *Neuron*, *41*, 535–547.

Maurange, C., Cheng, L., & Gould, A. P. (2008). Temporal transcription factors and their targets schedule the end of neural proliferation in Drosophila. *Cell*, *133*, 891–902.

McIlroy, G., Foldi, I., Aurikko, J., Wentzell, J. S., Lim, M. A., Fenton, J. C., et al. (2013). Toll-6 and Toll-7 function as neurotrophin receptors in the *Drosophila melanogaster* CNS. *Nature Neuroscience*, *16*, 1248–1256.

Mehlen, P., & Bredesen, D. E. (2011). Dependence receptors: From basic research to drug development. *Science Signaling*, *4*, mr2.

Miller, S. A., & Briglin, A. (1996). Apoptosis removes chick embryo tail gut and remnant of the primitive streak. *Developmental Dynamics*, *206*, 212–218.

Miller, D. T., & Cagan, R. L. (1998). Local induction of patterning and programmed cell death in the developing Drosophila retina. *Development*, *125*, 2327–2335.

Miura, M., Zhu, H., Rotello, R., Hartwieg, E. A., & Yuan, J. (1993). Induction of apoptosis in fibroblasts by IL-1 beta-converting enzyme, a mammalian homolog of the C. elegans cell death gene ced-3. *Cell*, *75*, 653–660.

Monier, B., Gettings, M., Gay, G., Mangeat, T., Schott, S., Guarner, A., et al. (2015). Apico-basal forces exerted by apoptotic cells drive epithelium folding. *Nature*, *518*, 245–248.

Monserrate, J. P., & Brachmann, C. B. (2007). Identification of the death zone: A spatially restricted region for programmed cell death that sculpts the fly eye. *Cell Death and Differentiation*, *14*, 209–217.

Nakamura, H., & O'Leary, D. D. (1989). Inaccuracies in initial growth and arborization of chick retinotectal axons followed by course corrections and axon remodeling to develop topographic order. *The Journal of Neuroscience: The Official Journal of the Society for Neuroscience*, *9*, 3776–3795.

Ng, L., Kelley, M. W., & Forrest, D. (2013). Making sense with thyroid hormone—The role of T(3) in auditory development. *Nature Reviews. Endocrinology*, *9*, 296–307.

Nikolaev, A., McLaughlin, T., O'Leary, D. D., & Tessier-Lavigne, M. (2009). APP binds DR6 to trigger axon pruning and neuron death via distinct caspases. *Nature*, *457*, 981–989.

Nikoletopoulou, V., Lickert, H., Frade, J. M., Rencurel, C., Giallonardo, P., Zhang, L., et al. (2010). Neurotrophin receptors TrkA and TrkC cause neuronal death whereas TrkB does not. *Nature*, *467*, 59–63.

Nonomura, K., Yamaguchi, Y., Hamachi, M., Koike, M., Uchiyama, Y., Nakazato, K., et al. (2013). Local apoptosis modulates early mammalian brain development through the elimination of morphogen-producing cells. *Developmental Cell*, *27*, 621–634.

Offner, N., Duval, N., Jamrich, M., & Durand, B. (2005). The pro-apoptotic activity of a vertebrate Bar-like homeobox gene plays a key role in patterning the Xenopus neural plate by limiting the number of chordin- and shh-expressing cells. *Development*, *132*, 1807–1818.

Ohsawa, S., Hamada, S., Kuida, K., Yoshida, H., Igaki, T., & Miura, M. (2010). Maturation of the olfactory sensory neurons by Apaf-1/caspase-9-mediated caspase activity. *Proceedings of the National Academy of Sciences of the United States of America*, *107*, 13366–13371.

O'Leary, D. D., & Koester, S. E. (1993). Development of projection neuron types, axon pathways, and patterned connections of the mammalian cortex. *Neuron*, *10*, 991–1006.

Oppenheim, R. W. (1991). Cell death during development of the nervous system. *Annual Review of Neuroscience*, *14*, 453–501.

Oppenheim, R. W., Houenou, L. J., Johnson, J. E., Lin, L. F., Li, L., Lo, A. C., et al. (1995). Developing motor neurons rescued from programmed and axotomy-induced cell death by GDNF. *Nature, 373*, 344–346.

Palgi, M., Lindstrom, R., Peranen, J., Piepponen, T. P., Saarma, M., & Heino, T. I. (2009). Evidence that DmMANF is an invertebrate neurotrophic factor supporting dopaminergic neurons. *Proceedings of the National Academy of Sciences of the United States of America, 106*, 2429–2434.

Parrish, J. Z., Emoto, K., Kim, M. D., & Jan, Y. N. (2007). Mechanisms that regulate establishment, maintenance, and remodeling of dendritic fields. *Annual Review of Neuroscience, 30*, 399–423.

Peeters, R. P., Ng, L., Ma, M., & Forrest, D. (2015). The timecourse of apoptotic cell death during postnatal remodeling of the mouse cochlea and its premature onset by triiodothyronine (T3). *Molecular and Cellular Endocrinology, 407*, 1–8.

Peterson, S. E., Yang, A. H., Bushman, D. M., Westra, J. W., Yung, Y. C., Barral, S., et al. (2012). Aneuploid cells are differentially susceptible to caspase-mediated death during embryonic cerebral cortical development. *The Journal of Neuroscience: The Official Journal of the Society for Neuroscience, 32*, 16213–16222.

Portera-Cailliau, C., Weimer, R. M., De Paola, V., Caroni, P., & Svoboda, K. (2005). Diverse modes of axon elaboration in the developing neocortex. *PLoS Biology, 3*, e272.

Purves, D., Snider, W. D., & Voyvodic, J. T. (1988). Trophic regulation of nerve cell morphology and innervation in the autonomic nervous system. *Nature, 336*, 123–128.

Raff, M. C. (1992). Social controls on cell survival and cell death. *Nature, 356*, 397–400.

Ramos, R. G., Igloi, G. L., Lichte, B., Baumann, U., Maier, D., Schneider, T., et al. (1993). The irregular chiasm C-roughest locus of Drosophila, which affects axonal projections and programmed cell death, encodes a novel immunoglobulin-like protein. *Genes & Development, 7*, 2533–2547.

Rehen, S. K., McConnell, M. J., Kaushal, D., Kingsbury, M. A., Yang, A. H., & Chun, J. (2001). Chromosomal variation in neurons of the developing and adult mammalian nervous system. *Proceedings of the National Academy of Sciences of the United States of America, 98*, 13361–13366.

Reiter, C., Schimansky, T., Nie, Z., & Fischbach, K. F. (1996). Reorganization of membrane contacts prior to apoptosis in the Drosophila retina: The role of the IrreC-rst protein. *Development, 122*, 1931–1940.

Rogulja-Ortmann, A., Luer, K., Seibert, J., Rickert, C., & Technau, G. M. (2007). Programmed cell death in the embryonic central nervous system of *Drosophila melanogaster*. *Development, 134*, 105–116.

Rohn, T. T., Cusack, S. M., Kessinger, S. R., & Oxford, J. T. (2004). Caspase activation independent of cell death is required for proper cell dispersal and correct morphology in PC12 cells. *Experimental Cell Research, 295*, 215–225.

Rumpf, S., Bagley, J. A., Thompson-Peer, K. L., Zhu, S., Gorczyca, D., Beckstead, R. B., et al. (2014). Drosophila Valosin-Containing Protein is required for dendrite pruning through a regulatory role in mRNA metabolism. *Proceedings of the National Academy of Sciences of the United States of America, 111*, 7331–7336.

Ryoo, H. D., Bergmann, A., Gonen, H., Ciechanover, A., & Steller, H. (2002). Regulation of Drosophila IAP1 degradation and apoptosis by reaper and ubcD1. *Nature Cell Biology, 4*, 432–438.

Sanes, J. R., & Lichtman, J. W. (1999). Development of the vertebrate neuromuscular junction. *Annual Review of Neuroscience, 22*, 389–442.

Sanz-Ezquerro, J. J., & Tickle, C. (2000). Autoregulation of Shh expression and Shh induction of cell death suggest a mechanism for modulating polarising activity during chick limb development. *Development, 127*, 4811–4823.

Saunders, J. W., Jr. (1966). Death in embryonic systems. *Science, 154*, 604–612.

Sawamoto, K., Okano, H., Kobayakawa, Y., Hayashi, S., Mikoshiba, K., & Tanimura, T. (1994). The function of argos in regulating cell fate decisions during Drosophila eye and wing vein development. *Developmental Biology*, *164*, 267–276.

Schoenmann, Z., Assa-Kunik, E., Tiomny, S., Minis, A., Haklai-Topper, L., Arama, E., et al. (2010). Axonal degeneration is regulated by the apoptotic machinery or a NAD+−sensitive pathway in insects and mammals. *The Journal of Neuroscience: The Official Journal of the Society for Neuroscience*, *30*, 6375–6386.

Schuldiner, O., & Yaron, A. (2015). Mechanisms of developmental neurite pruning. *Cellular and Molecular Life Sciences*, *72*, 101–119.

Schweitzer, R., Howes, R., Smith, R., Shilo, B. Z., & Freeman, M. (1995). Inhibition of Drosophila EGF receptor activation by the secreted protein Argos. *Nature*, *376*, 699–702.

Shimamura, K., & Rubenstein, J. L. (1997). Inductive interactions direct early regionalization of the mouse forebrain. *Development*, *124*, 2709–2718.

Sierra, A., Encinas, J. M., Deudero, J. J., Chancey, J. H., Enikolopov, G., Overstreet-Wadiche, L. S., et al. (2010). Microglia shape adult hippocampal neurogenesis through apoptosis-coupled phagocytosis. *Cell Stem Cell*, *7*, 483–495.

Simpson, P. (1990). Lateral inhibition and the development of the sensory bristles of the adult peripheral nervous system of Drosophila. *Development*, *109*, 509–519.

Southwell, D. G., Paredes, M. F., Galvao, R. P., Jones, D. L., Froemke, R. C., Sebe, J. Y., et al. (2012). Intrinsically determined cell death of developing cortical interneurons. *Nature*, *491*, 109–113.

Spires-Jones, T. L., de Calignon, A., Matsui, T., Zehr, C., Pitstick, R., Wu, H. Y., et al. (2008). *In vivo* imaging reveals dissociation between caspase activation and acute neuronal death in tangle-bearing neurons. *The Journal of Neuroscience: The Official Journal of the Society for Neuroscience*, *28*, 862–867.

Stanfield, B. B., O'Leary, D. D., & Fricks, C. (1982). Selective collateral elimination in early postnatal development restricts cortical distribution of rat pyramidal tract neurones. *Nature*, *298*, 371–373.

Stemerdink, C., & Jacobs, J. R. (1997). Argos and Spitz group genes function to regulate midline glial cell number in Drosophila embryos. *Development*, *124*, 3787–3796.

Sulston, J. E., & Horvitz, H. R. (1977). Post-embryonic cell lineages of the nematode, Caenorhabditis elegans. *Developmental Biology*, *56*, 110–156.

Takahashi, K., Kamiya, K., Urase, K., Suga, M., Takizawa, T., Mori, H., et al. (2001). Caspase-3-deficiency induces hyperplasia of supporting cells and degeneration of sensory cells resulting in the hearing loss. *Brain Research*, *894*, 359–367.

Takemoto, K., Nagai, T., Miyawaki, A., & Miura, M. (2003). Spatio-temporal activation of caspase revealed by indicator that is insensitive to environmental effects. *The Journal of Cell Biology*, *160*, 235–243.

Tan, Y., Yamada-Mabuchi, M., Arya, R., St Pierre, S., Tang, W., Tosa, M., et al. (2011). Coordinated expression of cell death genes regulates neuroblast apoptosis. *Development*, *138*, 2197–2206.

Tanenbaum, S. B., Gorski, S. M., Rusconi, J. C., & Cagan, R. L. (2000). A screen for dominant modifiers of the irreC-rst cell death phenotype in the developing Drosophila retina. *Genetics*, *156*, 205–217.

Tapia, J. C., Wylie, J. D., Kasthuri, N., Hayworth, K. J., Schalek, R., Berger, D. R., et al. (2012). Pervasive synaptic branch removal in the mammalian neuromuscular system at birth. *Neuron*, *74*, 816–829.

Tasdemir-Yilmaz, O. E., & Freeman, M. R. (2014). Astrocytes engage unique molecular programs to engulf pruned neuronal debris from distinct subsets of neurons. *Genes & Development*, *28*, 20–33.

Technau, G., & Heisenberg, M. (1982). Neural reorganization during metamorphosis of the corpora pedunculata in *Drosophila melanogaster. Nature, 295*, 405–407.

Tenev, T., Zachariou, A., Wilson, R., Paul, A., & Meier, P. (2002). Jafrac2 is an IAP antagonist that promotes cell death by liberating Dronc from DIAP1. *The EMBO Journal, 21*, 5118–5129.

Togane, Y., Ayukawa, R., Hara, Y., Akagawa, H., Iwabuchi, K., & Tsujimura, H. (2012). Spatio-temporal pattern of programmed cell death in the developing Drosophila optic lobe. *Development, Growth & Differentiation, 54*, 503–518.

Toyama, Y., Peralta, X. G., Wells, A. R., Kiehart, D. P., & Edwards, G. S. (2008). Apoptotic force and tissue dynamics during Drosophila embryogenesis. *Science, 321*, 1683–1686.

Truman, J. W., & Bate, M. (1988). Spatial and temporal patterns of neurogenesis in the central nervous system of *Drosophila melanogaster. Developmental Biology, 125*, 145–157.

Unsain, N., Higgins, J. M., Parker, K. N., Johnstone, A. D., & Barker, P. A. (2013). XIAP regulates caspase activity in degenerating axons. *Cell Reports, 4*, 751–763.

Venegas, V., & Zhou, Z. (2007). Two alternative mechanisms that regulate the presentation of apoptotic cell engulfment signal in Caenorhabditis elegans. *Molecular Biology of the Cell, 18*, 3180–3192.

Vohra, B. P., Sasaki, Y., Miller, B. R., Chang, J., DiAntonio, A., & Milbrandt, J. (2010). Amyloid precursor protein cleavage-dependent and -independent axonal degeneration programs share a common nicotinamide mononucleotide adenylyltransferase 1-sensitive pathway. *The Journal of Neuroscience: The Official Journal of the Society for Neuroscience, 30*, 13729–13738.

Wakselman, S., Bechade, C., Roumier, A., Bernard, D., Triller, A., & Bessis, A. (2008). Developmental neuronal death in hippocampus requires the microglial CD11b integrin and DAP12 immunoreceptor. *The Journal of Neuroscience: The Official Journal of the Society for Neuroscience, 28*, 8138–8143.

Watts, R. J., Hoopfer, E. D., & Luo, L. (2003). Axon pruning during Drosophila metamorphosis: Evidence for local degeneration and requirement of the ubiquitin-proteasome system. *Neuron, 38*, 871–885.

White, K., Grether, M. E., Abrams, J. M., Young, L., Farrell, K., & Steller, H. (1994). Genetic control of programmed cell death in Drosophila. *Science, 264*, 677–683.

Williams, D. W., Kondo, S., Krzyzanowska, A., Hiromi, Y., & Truman, J. W. (2006). Local caspase activity directs engulfment of dendrites during pruning. *Nature Neuroscience, 9*, 1234–1236.

Winbush, A., & Weeks, J. C. (2011). Steroid-triggered, cell-autonomous death of a Drosophila motoneuron during metamorphosis. *Neural Development, 6*, 15.

Wing, J. P., Schreader, B. A., Yokokura, T., Wang, Y., Andrews, P. S., Huseinovic, N., et al. (2002). Drosophila Morgue is an F box/ubiquitin conjugase domain protein important for grim-reaper mediated apoptosis. *Nature Cell Biology, 4*, 451–456.

Winseck, A. K., & Oppenheim, R. W. (2006). An *in vivo* analysis of Schwann cell programmed cell death in embryonic mice: The role of axons, glial growth factor, and the pro-apoptotic gene Bax. *The European Journal of Neuroscience, 24*, 2105–2117.

Wolff, T., & Ready, D. F. (1991). Cell death in normal and rough eye mutants of Drosophila. *Development, 113*, 825–839.

Wong, J. J., Li, S., Lim, E. K., Wang, Y., Wang, C., Zhang, H., et al. (2013). A Cullin1-based SCF E3 ubiquitin ligase targets the InR/PI3K/TOR pathway to regulate neuronal pruning. *PLoS Biology, 11*, e1001657.

Xiong, W. C., & Montell, C. (1995). Defective glia induce neuronal apoptosis in the repo visual system of Drosophila. *Neuron, 14*, 581–590.

Yamaguchi, Y., & Miura, M. (2013). How to form and close the brain: Insight into the mechanism of cranial neural tube closure in mammals. *Cellular and Molecular Life Sciences, 70*, 3171–3186.

Yamaguchi, Y., & Miura, M. (2015). Programmed cell death in neurodevelopment. *Developmental Cell, 32*, 478–490.

Yamaguchi, Y., Shinotsuka, N., Nonomura, K., Takemoto, K., Kuida, K., Yosida, H., et al. (2011). Live imaging of apoptosis in a novel transgenic mouse highlights its role in neural tube closure. *The Journal of Cell Biology, 195*, 1047–1060.

Yeo, W., & Gautier, J. (2004). Early neural cell death: Dying to become neurons. *Developmental Biology, 274*, 233–244.

Yu, S. Y., Yoo, S. J., Yang, L., Zapata, C., Srinivasan, A., Hay, B. A., et al. (2002). A pathway of signals regulating effector and initiator caspases in the developing Drosophila eye. *Development, 129*, 3269–3278.

Zhang, H., Wang, Y., Wong, J. J., Lim, K. L., Liou, Y. C., Wang, H., et al. (2014). Endocytic pathways downregulate the L1-type cell adhesion molecule neuroglian to promote dendrite pruning in Drosophila. *Developmental Cell, 30*, 463–478.

Zhu, B., Pennack, J. A., McQuilton, P., Forero, M. G., Mizuguchi, K., Sutcliffe, B., et al. (2008). Drosophila neurotrophins reveal a common mechanism for nervous system formation. *PLoS Biology, 6*, e284.

CHAPTER SEVEN

Regulation of Cell Death by IAPs and Their Antagonists

Deepika Vasudevan, Hyung Don Ryoo[1]
Department of Cell Biology, New York University School of Medicine, New York, New York, USA
[1]Corresponding author: e-mail address: hyungdon.ryoo@nyumc.org

Contents

Abstract

Inhibitors of apoptosis (IAPs) family of genes encode baculovirus IAP-repeat domain-containing proteins with antiapoptotic function. These proteins also contain RING or UBC domains and act by binding to major proapoptotic factors and ubiquitylating them. High levels of IAPs inhibit caspase-mediated apoptosis. For these cells to undergo apoptosis, IAP function must be neutralized by IAP-antagonists. Mammalian IAP knock-outs do not exhibit obvious developmental phenotypes, but the cells are more sensitized to apoptosis in response to injury. Loss of the mammalian IAP-antagonist ARTS results in reduced stem cell apoptosis. In addition to the antiapoptotic properties, IAPs regulate the innate immune response, and the loss of IAP function in humans is associated with immunodeficiency. The roles of IAPs in *Drosophila* apoptosis regulation are more apparent, where the loss of IAP1, or the expression of IAP-antagonists in *Drosophila* cells, is sufficient to trigger apoptosis. In this organism, apoptosis as a fate is conferred by the transcriptional induction of the IAP-antagonists. Many signaling pathways often converge on shared enhancer regions of IAP-antagonists. Cell death sensitivity is further regulated by posttranscriptional mechanisms, including those regulated by kinases, miRs, and ubiquitin ligases. These mechanisms are employed to eliminate damaged

Current Topics in Developmental Biology, Volume 114
ISSN 0070-2153
http://dx.doi.org/10.1016/bs.ctdb.2015.07.026

or virus-infected cells, limit neuroblast (neural stem cell) numbers, generate neuronal diversity, and sculpt tissue morphogenesis.

1. INTRODUCTION

Apoptosis is one of the best-understood forms of cell death that is regulated through a combination of positive and negative factors. Among the negative regulators are antiapoptotic proteins that share the baculovirus IAP-repeat (BIR) domains, which are now widely referred to as inhibitors of apoptosis (IAPs) family of proteins.

The name of the BIR domain originates from the shared sequence of viral IAPs found in the functional homologs of baculovirus p35, Cp-IAP, and Op-IAP (Birnbaum, Clem, & Miller, 1994; Clem, Fechneimer, & Miller, 1991, Clem & Miller, 1994; Crook, Clem, & Miller, 1993). The family now includes X-linked IAP (XIAP), c-IAP1 and 2, *Drosophila* IAP1 and 2 (DIAP1 and 2), and BRUCE (BIR domain-containing ubiquitin conjugating enzyme) (Fig. 1). Not all BIR domain-containing proteins regulate cell death, and certain BIR domain proteins are dedicated to the regulation of mitosis (Silke & Vaux, 2001). The antiapoptotic BIR domain proteins found in *Drosophila* and vertebrates mostly have C-terminal RING domains that have ubiquitin ligase activities (Yang, Fang, Jensen, Weissman, & Ashwell, 2000). One exception to this is BRUCE, a potent antiapoptotic protein that contains an ubiquitin conjugating enzyme (UBC) motif instead of RING. These IAPs bind and ubiquitylate major proapoptotic proteins to exert their antiapoptotic function. In addition, they are actively regulated in cells by their inhibitory molecules, referred to as IAP-antagonists. In this review, we will discuss the latest advances in the field, focusing on the roles of IAPs and their antagonists during animal development.

2. IAP/ANTAGONIST INTERACTION

In many cells, IAPs bind and inhibit active caspases to exert their antiapoptotic function (Deveraux, Takahashi, Salvesen, & Reed, 1997; Goyal, McCall, Agapite, Hartwieg, & Steller, 2000; Wang, Hawkins, Yoo, Muller, & Hay, 1999). Caspases gain full catalytic activity after being proteolytically cleaved, so that the resulting small and large subunits of caspases

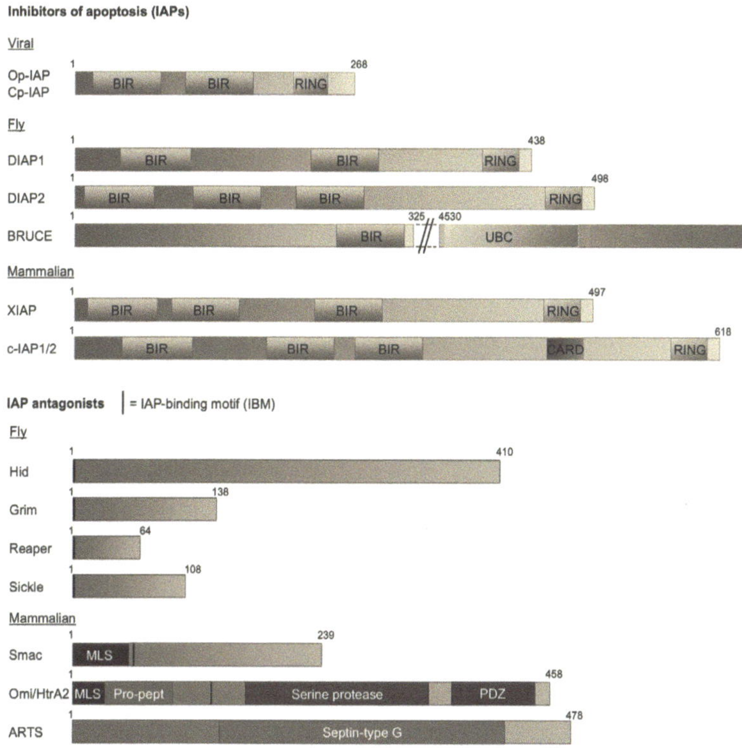

Figure 1 Domain maps of IAPs and their antagonists from various model systems. All known IAPs contain at least one baculovirus IAP-repeat (BIR) domain. In addition, most have RING domains. BRUCE is the largest IAP (as indicated by the break in sequence in figure) and an exception in that it does not contain a RING domain but instead has an ubiquitin conjugation domain (UBC). Most IAP-antagonists contain a short 5–10 amino acid IAP-binding motif (IBM) at their N-terminii, usually immediately after the methionine, which is cleaved to expose the IBM. The mammalian IAP-antagonists, Smac, ARTS, and Omi/HtrA2, localize to the mitochondria for their function and hence contain a mitochondria localization sequence (MLS) among other domains. ARTS belongs is a noncanonical IAP-antagonist that does not have an N-terminal IBM and instead uses the C-terminal sequences to bind IAPs. *Domain maps to scale, source: www.uniprot.org.* (See the color plate.)

can assemble to form active catalytic sites. IAPs can inhibit such proteolytically activated caspases (Muro, Hay, & Clem, 2002; Shapiro, Hsu, Jung, Robbins, & Ryoo, 2008; Srinivasula et al., 2001), and therefore, high levels of IAPs can block apoptosis at the last stage. However, cells with high levels of IAPs can undergo caspase-mediated apoptosis, if IAP-antagonizing molecules are around to neutralize IAP function (Fig. 2). The so-called IAP-antagonists were first discovered in *Drosophila* and remain best characterized in this organism. Reflecting their important role in cell killing, these

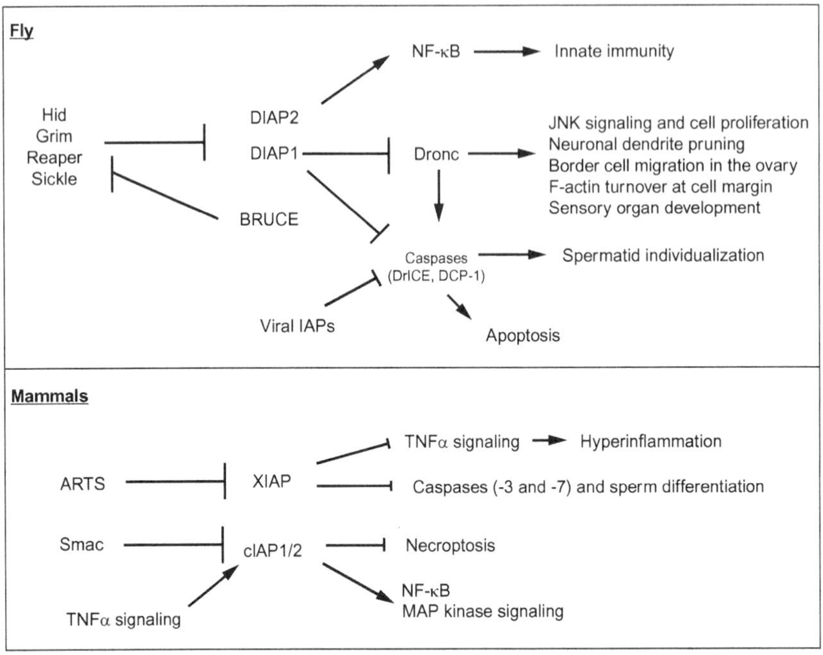

Figure 2 A schematic diagram of IAP-antagonists and their targets. Upper panel shows the relationships between *Drosophila* genes, whereas the lower panel shows mammalian IAPs and their antagonists.

IAP-antagonists were named *grim, reaper, hid,* and *sickle* (Chen, Nordstrom, Gish, & Abrams, 1996; Christich et al., 2002; Grether, Abrams, Agapite, White, & Steller, 1995; Srinivasula et al., 2002; White et al., 1994; Wing et al., 2002).

IAP-antagonists play particularly visible roles in *Drosophila* apoptosis regulation: Virtually all apoptosis is abolished in the absence of these genes, whereas their overexpression is sufficient to kill cells (Chen et al., 1996; Grether et al., 1995; White et al., 1994; White, Tahaoglu, & Steller, 1996). Genetic interaction screens have identified DIAP1, DIAP2, and BRUCE as downstream targets (Arama, Agapite, & Steller, 2003; Goyal et al., 2000; Hay, Wassarman, & Rubin, 1995; Lisi, Mazzon, & White, 2000; Vernooy et al., 2002; Wang et al., 1999). In living cells of *Drosophila*, DIAP1 normally inhibits both, initiator and effector caspases (Hawkins, Wang, & Hay, 1999; Meier, Silke, Leevers, & Evan, 2000; Tenev, Zachariou, Wilson, Ditzel, & Meier, 2005; Yan, Wu, Chai, Li, & Shi, 2004). DIAP1 uses its ubiquitin ligase activity to directly ubiquitylate the initiator caspase DRONC (Lee et al., 2011; Ryoo, Gorenc, & Steller, 2004; Wilson et al., 2002) and also helps to destabilize the upstream adaptor, the apoptosome holoenzyme, which is a protein complex that serves to activate DRONC (Akdemir et al., 2006; Shapiro et al., 2008). DIAP1 is a key player as demonstrated by the observation that virtually all somatic cells undergo apoptosis in *diap1* mutant embryos (Goyal et al., 2000; Lisi et al., 2000; Wang et al., 1999). DIAP2 has a more confined role in inhibiting a specific effector caspase (Ribeiro et al., 2007), and while overexpression of DIAP2 can inhibit IAP-antagonist-induced apoptosis (Hay et al., 1995), the loss of this gene does not show the dramatic apoptosis phenotype as seen in *diap1* mutants (Huh et al., 2007; Ribeiro et al., 2007). BRUCE is also a potent antiapoptotic gene, and this protein exerts its effect by using its UBC domain to ubiquitylate IAP-antagonists for proteasomal degradation (Arama et al., 2003; Bartke, Pohl, Pyrowolakis, & Jentsch, 2004; Domingues & Ryoo, 2012; Hao et al., 2004; Vernooy et al., 2002). Mammalian IAP-antagonists, Smac and Omi/HtrA2, were also identified based on its ability to physically bind to XIAP (Du, Fang, Li, Li, & Wang, 2000; Verhagen et al., 2000). However, mouse genetics studies indicate that IAP-antagonists primarily target c-IAP1 *in vivo* (Vince et al., 2007; Fig. 2).

IAP-antagonists share a conserved N-terminal 4–8 residues that directly bind to a groove within the IAP BIR domain, allowing caspases to be liberated from IAPs (Wu et al., 2000; Wu, Cocina, Chai, Hay, & Shi, 2001). Furthermore, they promote the autoubiquitination and degradation of IAPs (Li, Wang, & Shi, 2011; Ryoo, Bergmann, Gonen, Ciechanover, & Steller, 2002; Yoo et al., 2002). Notable in this interaction is the fact that the first methionine of the N-terminal IAP-binding motif must be lost, and the new N-terminus must start with an alanine residue, in order to fit into an IAP BIR groove (Wu et al., 2000).

3. MITOCHONDRIAL ASSOCIATION OF IAP-ANTAGONISTS

How can cells make peptides that do not start with an N-terminal methionine residue? In mammals, the IAP-antagonist Smac encodes an N-terminal mitochondrial localization motif followed by an IAP-binding motif that is similar to the N-terminal residues of GRIM, REAPER, HID, and SICKLE (Du et al., 2000; Verhagen et al., 2000; Fig. 1). Omi/HtrA2 is a mitochondrial protease that has a similar IAP-binding motif (Hegde et al., 2002; Martins et al., 2002; van Loo et al., 2002)—although in *Drosophila*, Omi/HtrA2 does not appear to regulate IAPs and downstream caspases (Yacobi-Sharon, Namdar, & Arama, 2013). The N-terminal mitochondrial localization sequences of Smac and Omi/HtrA2 are cleaved off while being trafficked into the mitochondrial intermembrane space, generating new N-terminal motifs that start with alanine residues and with the ability to bind to BIR domain grooves. Because mammalian IAP-antagonists are sequestered into the mitochondrial intermembrane space, they do not inhibit IAPs that reside in the cytoplasm during nonapoptotic conditions (Du et al., 2000; Hegde et al., 2002; Martins et al., 2002; van Loo et al., 2002; Verhagen et al., 2000). Only when Bcl-2 family proteins help release the mitochondrial proteins into the cytoplasm during apoptosis do these proteins get into contact with IAPs in the cytoplasm, neutralizing their target IAPs.

In case of the *Drosophila* IAP-antagonists, there are no N-terminal mitochondrial localization sequences, and it remains unclear how these *Drosophila* proteins lose their N-terminal methionine residues. *Drosophila* IAP-antagonists do not enter the mitochondrial intermembrane space but localize to the mitochondrial outer membrane. Hid contains a C-terminal tail anchor sequence that inserts into the mitochondrial outer membrane, with the IAP-binding motif facing the cytoplasm (Abdelwahid et al., 2007; Haining, Carboy-Newcomb, Wei, & Steller, 1999). GRIM and REAPER each contain an amphipathic helix that are required for their mitochondrial outer membrane localization (Claveria, Caminero, Martinez-A, Campuzano, & Torres, 2002; Olson, Holley, Gan, et al., 2003; Sandu, Ryoo, & Steller, 2010). These proteins form multimers with each other, and such association is important for their mitochondrial localization (Sandu et al., 2010). Mutating their mitochondrial localization sequences disrupt their proapoptotic function (Abdelwahid et al., 2007; Claveria et al., 2002; Olson, Holley, Gan, et al., 2003). More recently, it was found that *cdk7* mutants block the mitochondrial localization of Hid, and such conditions abolished Hid's cell killing activity (Morishita, Kang, Fidelin, & Ryoo, 2013). Why should these proteins localize to the mitochondrial outer membrane to trigger apoptosis? Certain studies have implicated proapoptotic roles of IAP-antagonists that are independent of DIAP1 (Abdelwahid et al., 2007; Thomenius et al., 2011; Thress, Evans, & Kornbluth, 1999). On the other hand, DIAP1 overexpression almost completely blocks IAP-antagonist-induced apoptosis *in vivo* (Hay et al., 1995), suggesting that DIAP1-independent effects of IAP-antagonists are likely to be subtle, at best. A different explanation was proposed recently, suggesting that the IAP-binding activity of the antagonists is linked to the mitochondrial localization. Specifically, subcellular fractionation studies indicate that only the mitochondria-associated pool of Hid, but not the cytoplasmic pool, has the ability to bind to recombinant DIAP1 proteins (Morishita et al., 2013).

4. THE ROLE OF MAMMALIAN IAP-ANTAGONISTS

Genetic analysis of mammalian IAP-antagonists has shown varying outcomes. Perhaps, Smac has drawn the most attention, but a knockout study brings into question its biological significance: The Smac-deficient mice grow normally, and the knockout cell lines respond normally to the apoptotic stimuli that were tested (Okada et al., 2002). Omi/HtrA2 mutations are associated with Parkinson's disease (Strauss et al., 2005), which is more consistent with its role in mitochondrial homeostasis, but not with IAP-antagonist function. However, there is another mammalian IAP-antagonist, ARTS (Fig. 2). This is a splice isoform of Septin4, originally identified in a retroviral insertion screen (Larisch et al., 2000). Subsequent studies have revealed that this protein binds to XIAP, but not through an N-terminal sequence as found in other IAP-antagonists. Instead, the nine C-terminal end residues of ARTS serve as the XIAP1-binding motif (Reingewertz et al., 2011). Upon binding, ARTS promotes the ubiquitin-mediated degradation of XIAP, similar to the *Drosophila* IAP-antagonists (Gottfried, Rotem, Lotan, Steller, & Larisch, 2004). ARTS knockout mice have elevated levels of XIAP, resulting in enhanced cell death resistance. Perhaps as a result, these animals have increased numbers of hematopoietic stem cells and hair follicle stem cells. On the negative side, these mice are more prone to develop tumors (Garcia-Fernandez et al., 2010). On the positive side, they display marked improvement in wound healing and regeneration (Fuchs et al., 2013).

5. TRANSCRIPTIONAL REGULATION OF *DROSOPHILA* IAP-ANTAGONISTS

Unlike mammalian IAP-antagonists, which are initially segregated into the mitochondrial intermembrane space, transcriptional induction of IAP-antagonists in *Drosophila* is sufficient to trigger apoptosis. In fact, the transcription of *grim, reaper*, and *sickle* foreshadows apoptosis induction in this organism (White et al., 1994; Chen et al., 1996). As a result, there is much interest in understanding the transcriptional regulation mechanisms of *Drosophila* IAP-antagonists.

Stress-activated pathways mediated by the tumor suppressor gene p53 respond to a number of distinct stress conditions, including DNA damage (Brodsky et al., 2000; Ollmann et al., 2000) and viral infection (Liu, Becnel, Zhang, & Zhou, 2011; Liu et al., 2013). p53 directly binds to a regulatory sequence that lies between *reaper* and *sickle* (Brodsky et al., 2000; Fig. 3). This site is part of a broader irradiation-responsive element that controls the induction of multiple IAP-antagonists, including *hid* that lies more than 250 kbp away (Zhang et al., 2008). Interestingly, this locus is active only during early embryogenesis, and subsequently becomes silenced through epigenetic regulation, thereby making cells insensitive to irradiation-induced apoptosis (Zhang et al., 2008).

Jun N-terminal kinase (JNK) signaling is also tightly associated with IAP-antagonist induction in *Drosophila* (Fig. 3). The heterodimeric

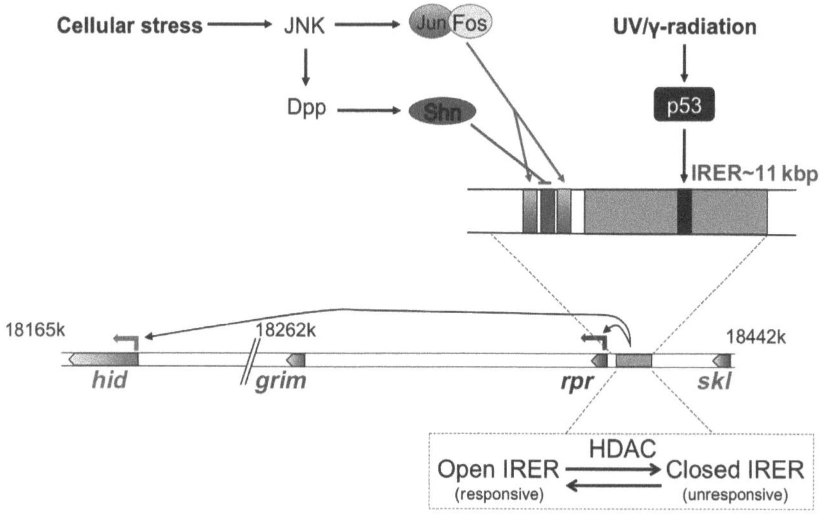

Figure 3 A schematic showing the *Drosophila* H99 locus and its regulation. The 3L chromosome arm contains all four of the IAP-antagonists, *hid, grim, reaper* (*rpr*), and *sickle* (*skl*), clustered in the H99 locus. A ~12 kbp region upstream of *rpr* is the hub of most transcriptional regulation and controls transcription of *rpr* and *hid*, although the *hid* locus is more than 250 kbp away. It contains a ~11 kbp irradiation-responsive enhancer region (IRER) containing a 20 bp p53-binding site critical for radiation-induced apoptosis. The IRER region is heavily epigenetically regulated during development by histone deacetylases (HDAC) and methylases. The *rpr* upstream promoter region contains sites for binding several transcription factors, notably Jun/Fos and Schnurri (Shn), which mediate context-dependent JNK-driven apoptosis. (See the color plate.)

transcription factors Jun and Fos form the AP-1 complex to mediate the transcriptional response to this pathway. In fact, IAP-antagonist expression triggers a transcriptional feedback loop that induces other IAP-antagonists to augment their proapoptotic effects, and this is mediated by JNK signaling and p53 (Kuranaga et al., 2002; Shlevkov & Morata, 2012). Analysis of the *reaper* upstream sequence in fact shows functionally significant binding sites. However, JNK is also involved in cell migration and other physiological events without the induction of apoptosis. Why certain cells evade JNK-induced apoptosis has been a mystery up until recently. It has now been found that JNK-mediated apoptosis signal is integrated at the *reaper* regulatory sequences, together with other antiapoptotic signals. DPP signaling, mediated by the transcription factor Schnurri, represses JNK-mediated *reaper* expression through a binding site located in between those of AP-1 (Beira et al., 2014; Fig. 3). During embryonic development, dorsal closure is regulated by JNK signaling, but these cells do not undergo apoptosis, due to Dpp-Schnurri-mediated repression of IAP-antagonist gene expression.

The steroid hormone ecdysone, which triggers the onset of metamorphosis, is also well known to induce massive cell death during metamorphosis. The effect is, in part, due to the transcriptional induction of *grim, reaper*, and *hid* by the ecdysone receptor, which binds to a *reaper*-upstream enhancer sequence (Jiang, Lamblin, Steller, & Thummel, 2000). In addition to the control of IAP-antagonist, a recent study shows that ecdysone signaling is required for the proper induction of downstream caspases. That study noted that the resulting increase in caspase expression renders cells sensitive to *reaper* and *hid*-induced apoptosis during the late third instar stage of *Drosophila* development, as ecdysone signaling becomes active at this stage (Kang & Bashirullah, 2014). In addition, loss of EGF receptor/MAP kinase survival signaling pathway induces *hid* transcription (Kurada & White, 1998).

6. POSTTRANSCRIPTIONAL REGULATION OF IAP-ANTAGONISTS

When IAP-antagonists were first discovered, it had been noted that the transcripts of *hid* are distributed more broadly than the actual pattern of apoptosis, suggestive of posttranscriptional regulatory mechanisms (Grether et al., 1995). Ever since, distinct types of regulatory mechanisms have been elucidated.

As in many other cellular mechanisms, kinases play roles in IAP-antagonist regulation. A well-characterized example is MAP Kinase, which phosphorylates Hid to inhibit its cell-killing activity (Bergmann, Agapite, McCall, & Steller, 1998; Bergmann, Tugentman, Shilo, & Steller, 2002). This EGF receptor/MAP kinase survival signaling pathway also regulates *hid* transcription (Kurada & White, 1998). More recently, *cdk7* was identified as a gene required for IAP-antagonist-induced cell death. However, it remains unclear whether CDK7 directly phosphorylates IAP-antagonists (Morishita et al., 2013).

Many IAPs and their antagonists are regulated through ubiquitylation. As introduced earlier, both mammalian and insect IAPs undergo autoubiquitylation when bound by IAP-antagonists (Ryoo et al., 2002; Yang et al., 2000; Yoo et al., 2002). IAP-antagonists are ubiquitylated by IAPs (Olson, Holley, Yoo, et al., 2003). Lysine-deficient REAPER is more stable, but it was recently demonstrated that BRUCE can add ubiquitins on REAPER even at nonlysine residues and target it for degradation (Domingues & Ryoo, 2012). In the developing eye disks of *Drosophila*, sensitivity to apoptosis changes as cells transition from an unspecified state to differentiated photoreceptors. Unspecified cells are more vulnerable to apoptosis, as DIAP1 is kept low through ubiquitylation by the Cullin-3 complex in these cells. Differentiated photoreceptors accumulate DIAP1 to gain apoptotic resistance (Fan & Bergmann, 2014).

The role of micro RNAs (miRs) in various cellular processes has been intensely studied in the past decade. Not surprisingly, IAP-antagonists are also targets of miRs in *Drosophila*. One of the first to be discovered was the miR, bantam, which targets *hid* for translational suppression (Brennecke, Hipfner, Stark, Russell, & Cohen, 2003). *bantam* mutants are homozygous viable but are smaller in size due to reduced tissue growth. Consistently, overexpression leads to overgrowth of tissue due to an increase in cell numbers. GFP-reporter studies show that *bantam* binds to at least five sites in the *hid* 3′UTR region and target it for degradation by RNAi. Similar reporter studies show miR-2 to target sites in *reaper, grim*, and *sickle* transcripts (Stark, Brennecke, Russell, & Cohen, 2003) and miR-14 to target *reaper* (Xu, Vernooy, Guo, & Hay, 2003). More recent studies have found miR-6 and -11 as additional regulators of all four IAP-antagonists (Ge et al., 2012; Truscott, Islam, Lopez-Bigas, & Frolov, 2011). Though single mutants are viable, miR-6/miR-11 double mutants are embryonic lethal and show defects in the CNS, suggesting that they may have overlapping roles. Broadly, spatial expression patterns of miRs and their targets contribute to regulation of cell death, but an additional layer of complexity is added by competition amongst miRs for the same targets.

7. THE ROLE OF IAP-ANTAGONISTS IN NERVOUS SYSTEM DEVELOPMENT

Arguably, a majority of apoptosis observed during development occurs in the nervous system. Reinforcing this view, recent studies on the *Drosophila* mutants for *grim*, *reaper*, and *hid* revealed their intricate roles in neuronal development. In the ventral nerve cord of the central nervous system, most neuroblasts (neural stem cells) in the abdominal segments undergo apoptosis during late embryogenesis, and a few additional dying cells are observed in the late third instar larval stage. *grim* and *reaper* double mutants show a dramatic impairment of neuroblast cell death in this tissue, leading to a significantly enlarged ventral nerve cord (Peterson, Carney, Taylor, & White, 2002; Tan et al., 2011). This indicates that, in wild type animals, developmental cues confer death as a fate to the abdominal segment neuroblasts. Apoptosis is induced specifically in the abdominal segment neuroblast due to the regulation of *grim, reaper*, and *hid* by the Hox gene expressed in that region, AbdA (Bello, Hirth, & Gould, 2003). A temporal series of transcription factors that are expressed in these cells make up a combinatorial code to determine the timing of apoptosis (Maurange, Cheng, & Gould, 2008). In addition, there appears to be a cell death signal originating from the progeny: Those progeny express the Notch ligand, delta, and the resulting Notch signaling contributes to AbdA induction to express IAP-antagonists (Arya, Sarkissian, Tan, & White, 2015).

Analogous to the example of cell death induction in the ventral nerve cord, a recent study has found that a combinatorial transcription code induces *grim* and *reaper* to confer death as a fate to certain differentiating cells. In the region of the optic lobe outer proliferation center, neuroblasts temporally express a series of transcription factors to confer different fates to the differentiating neurons. In one of the early lineages, a specific transcription factor helps the cells interpret Notch signal as an apoptotic signal. In a later lineage, a different transcription factor helps the cells to perceive Notch signaling in an opposite way—as a survival signal (Bertet et al., 2014). Such a strategy allows a diverse array of neuronal subtypes to be established in this tissue.

8. IAPs AND THEIR ANTAGONISTS IN SCULPTING MORPHOGENESIS

In *Drosophila*, mutant alleles of DIAP1 are referred to as *thread*, as one of the hypomorphic alleles, *thread*^{1} causes the fly antenna tip to appear as thin as a thread. Normally, that region of the antenna, which is called the arista, has many branches, which disappear in *thread*[1] mutants due to excessive death of cells in the larval antennal disks. Conversely, loss of the IAP-antagonist *hid* causes excessive branches to appear in the arista (Cullen & McCall, 2004). These observations indicate that apoptosis is regulated in antennal disks to regulate the morphogenesis of the arista during development.

One of the important functions of apoptosis is to sculpt body structures. An example is the role of the *Drosophila* Hox gene Deformed (Dfd) in inducing *reaper* to kill cells along the segment boundary (Lohmann, McGinnis, Bodmer, & McGinnis, 2002). A more recent study has found that *Drosophila* joint boundaries also undergo apoptosis. How does the tissue know where to make the joint boundary? The data seem to indicate that the Dpp morphogenic gradient is a key determinant. Dpp is secreted from the organizers to form a gradient, but when there is a sharp discontinuity in the gradient, it is known to activate JNK signaling and the induction of IAP-antagonists. This is what happens at the joints, leading to the death of those boundary cells (Manjon, Sanchez-Herrero, & Suzanne, 2007). Notch pathway also contributes to tarsal joint development. This pathway induces the transcription factor Dys, which in turn induce *reaper* and *hid* to help sculpt joints (Cordoba & Estella, 2014).

Recent work on *Drosophila* IAP-antagonists has revealed a surprising role of apoptosis in unexpected morphogenic processes, such as tissue rotation. It had been noted that *hid* mutants can survive to adulthood, and the surviving males have their genitalia rotated in abnormal angles. A live imaging study of the developing male genitalia in the pupal stage helped elucidate this rotation process in detail. Two distinct domains, each rotating 180 °C, have the incremental effect of rotating 360 °C. These two domains are initially part of the same epithelial layer, and the investigators found that *hid*-induced cell death allows the two domains to separate and rotate away from each other. A failure to separate the two domains leads to a rotational defect (Suzanne et al., 2010).

9. NONAPOPTOTIC ROLES OF IAPs IN MORPHOGENESIS, CELL MIGRATION, AND PROLIFERATION

Caspases were primarily studied as proteins dedicated to apoptosis induction, but now, there are increasing numbers of studies implicating caspases in diverse nonapoptotic roles. Since caspases in *Drosophila* are tightly regulated by DIAP1, IAP-antagonists and DIAP1 are also involved in those nonapoptotic processes. The examples include their roles in regulating non-autonomous cell proliferation, cell migration, and dendritic morphology regulation.

The loss of DIAP1, or activation of IAP-antagonists, not only induces caspase activation but also activates a pathway that triggers mitogen expression and the proliferation of neighboring cells. This phenotype is dramatically augmented to cause tissue overgrowth if effector caspases are blocked with p35. It turns out p35 does not inhibit the initiator caspase DRONC, and latter has an apoptosis-independent function in activating the JNK pathway to promote the expression of mitogenic genes such as *wingless* and *dpp* (Fan & Bergmann, 2008; Huh, Guo, & Hay, 2004; Kondo, Senoo-Matsuda, Hiromi, & Miura, 2006; Perez-Garijo, Martin, & Morata, 2004; Ryoo et al., 2004).

DIAP1's target DRONC is involved in a number of other nonapoptotic roles. One of the nonapoptotic cellular processes that resemble apoptosis is neuronal dendrite pruning. In order to make, or break, proper synaptic connections, certain dendrites have to undergo dramatic morphological changes. Certain neurons have adopted the caspases regulatory network to eliminate, not the entire cell, but specific dendrites. For example, the *Drosophila* initiator caspase DRONC can promote dendrite pruning during development, and this process is inhibited by DIAP1 (Kuo, Zhu, Younger, Jan, & Jan, 2006). Other regulators of dendrite pruning have been discovered, and in one example, it was discovered that the AAA ATPase, VCP, regulates dendrite pruning by binding to DIAP1 and facilitating its degradation (Rumpf, Lee, Jan, & Jan, 2011).

Sensory organ development of *Drosophila* also involves caspases in a non-apoptotic mechanism. During the formation of the precursor cells, caspases cleave Shaggy, a negative regulator of Wnt signaling. The cleavage by caspase converts the substrate to an active kinase, promoting the formation of sensory organ precursors (Kanuka et al., 2005). Not surprisingly, such caspase activity is under the control of DIAP1. DIAP1 is regulated in these cells, not by IAP-antagonists, but through phosphorylation by IKK epsilon. Loss of this kinase results in the stabilization of DIAP1, which in turn, blocks caspase-mediated shaggy cleavage and activation (Kuranaga et al., 2006). The IKK epsilon/DIAP1/caspase cascade is also involved in F-actin turnover at the cellular margin and contributes to the morphogenic changes of cultured cells (Oshima et al., 2006).

Such effects of DIAP1 on the cytoskeleton are not limited to morphological changes but can also affect cell migration. In the *Drosophila* ovary, border cells migrate during ovary development to a specific position, and this process requires the GTPase Rac that helps to rearrange the cytoskeleton during this process. Evidence indicates that DIAP1 and DRONC regulate Rac, and without DIAP1, border cells fail to migrate properly (Geisbrecht & Montell, 2004). In an analogous mechanism, XIAP1 and c-IAP-1/2 regulate mammalian cell motility by ubiquitylating C-RAF (Dogan et al., 2008). The nonapoptotic roles of DIAP1 bring up an important question. How can some cells regulate DIAP1 without triggering apoptosis? Live imaging of DIAP1 in the sensory organ precursor cells of *Drosophila* indicates that the turnover rate of DIAP1 varies between cell types, and such temporal regulation of DIAP1 may determine whether the downstream caspases are utilized for apoptotic or nonapoptotic roles (Koto, Kuranaga, & Miura, 2009).

Spermatid differentiation is a process that involves dramatic morphological changes, including the removal of the bulk of cytoplasm along the elongating spermatids through a process termed "spermatid individualization." Caspases have been adopted in these cells to mediate the massive cytoplasmic removal in this differentiation process (Arama et al., 2003). In *Drosophila*, caspases form a gradient in their activity to regulate spermatid differentiation, so that the regions of the spermatids that are the last to individualize have the lowest caspase activity. This gradient is formed by a counter gradient of the IAP protein, BRUCE (Kaplan, Gibbs-Bar, Kalifa, Feinstein-Rotkopf, & Arama, 2010). Consistently, *Drosophila bruce* mutants show male sterility (Arama et al., 2003). An analogous regulation of caspases occurs during mammalian spermatid differentiation through XIAP1 and its antagonist, ARTS. In fact, mice lacking the ARTS/Septin 4 locus show defects in sperm cell maturation (Kissel et al., 2005).

10. THE ROLES OF IAPs IN THE INNATE IMMUNE RESPONSE

Early studies of mammalian IAPs, XIAP and c-IAP-1/2 focused on their ability to regulate effector caspases and apoptosis, but *in vivo* studies of these IAPs increasingly point to their important roles in TNFα signaling and the innate immune response. Upon infection by virus or pathogenic bacteria, various cells in our body produce TNFα to initiate immune response signaling. There are at least three distinct pathways that can be activated downstream of TNFα receptors: The extrinsic cell death pathway mediated by caspase-8 and -3, NF-κB signaling that leads to cytokine production, and RIP (receptor-interacting protein)-1 and -3/MLKL-mediated "necroptotic cell death" (Silke, 2011). As introduced earlier, c-IAP1 and 2 are mammalian IAP proteins initially identified based on their physical association with TNFα receptor 2 (Rothe, Pan, Henzel, Ayres, & Goeddel, 1995; Uren, Pakusch, Hawkins, Puls, & Vaux, 1996). c-IAP1/2 specifically bind to RIP1 while in a complex with TNFα receptor. Upon TNFα stimulation, c-IAP1 and 2 ubiquitylate RIP1 to activate NK-κB and MAP kinase signaling (Bertrand et al., 2008). Loss of c-IAP1/2 or deubiquitination of RIP1 triggers the formation of a different TNF receptor signaling complex that, instead of promoting NF-κB signaling, activates the caspase-8-mediated apoptosis or RIP3/MLKL-dependent necroptosis (Tenev et al., 2011; Vince et al., 2007). c-IAP1 and 2 also ubiquitylate a different, yet related protein, RIP2, and such ubiquitination promotes NF-κB signaling and cytokine production (Bertrand et al., 2009). In *Drosophila*, DIAP1 and 2 are the closest homologs of c-IAP1/2, and while DIAP1 primarily regulates apoptosis, DIAP2 promotes the activation of the *Drosophila* NF-κB homolog, Relish, as part of an innate immune response to Gram-positive bacteria infection (Huh et al., 2007; Fig. 2).

XIAP was originally characterized as an antiapoptotic protein that primarily inhibits at least two effector caspases, caspase-3 and -7 (Deveraux et al., 1997). There was a slight disappointment to the field when it was first reported that XIAP1$^{-/-}$ mice do not show obvious developmental abnormalities (Olayioye et al., 2005). Subsequent studies revealed subtle cell death phenotypes: XIAP1-deficient sympathetic neurons are more vulnerable to apoptosis after cytochrome c injection (Potts, Singh, Knezek, Thompson, & Deshmukh, 2003), and the mutant fibroblasts are sensitized to TNFα-induced apoptosis (Schile, Garcia-Fernandez, & Steller, 2008).

Interestingly, more recent studies also implicate XIAP in TNFα signaling and immune response. Mutations in human XIAP (also referred to as BIRC4) have been found to underlie immunodeficiency with aberrant activation of macrophages and dendritic cells, and the accumulation of activated T lymphocytes after viral infection (Damgaard et al., 2013; Marsh et al., 2010; Pachlopnik Schmid et al., 2011; Rigaud et al., 2006). Similarly, XIAP knockout mice show reduced ability to clear infectious pathogens (Bauler, Duckett, & O'Riordan, 2008; Prakash, Albrecht, Becker, Kuhlmann, & Rudel, 2010). These immunodeficiency phenotypes are difficult to explain through XIAP's ability to inhibit effector caspases. Recent studies have found that XIAP1 has an inhibitory effect on TNFα signaling, a process that is also regulated by c-IAP1/2. However, XIAP1 and c-IAP1/2 have different mechanisms of action: Whereas c-IAP1/2 ubiquitylate RIP1 while bound with TNF receptor to activate NF-κB signaling, XIAP1 appears to ubiquitylate RIP1 at a later stage of signaling, within a distinct complex. The loss of such XIAP activity results in abnormally high inflammasome activity, caspase-1 activation, and IL-1β secretion from dendritic cells (Yabal et al., 2014). Human patients with XIAP mutation suffer from hyper-inflammation, and XIAP's effect on inflammasome provides a molecular explanation.

Invertebrate IAPs and their antagonists also respond to viral infection as part of an innate immune response. Mosquitos and *Drosophila* induce IAP-antagonists when infected with DNA or RNA viruses, which helps to kill cells infected with virus and block their propagation (Liu et al., 2013). Certain virus have evolved to inhibit such innate immune response by evolving IAPs in their genome (Clem, 2005), and the best-characterized examples of viral IAPs include *Orgyia pseudotsugata* Op-IAP, which primarily inhibit initiator caspase activity (Birnbaum et al., 1994; LaCount, Hanson, Schneider, & Friesen, 2000). It appears that these viral IAPs have a more stable antiapoptotic activity than their cellular homologs: Whereas XIAP, c-IAP1/2, and DIAP1 have short half-lives and undergo autoubiquitination and degradation upon binding to IAP-antagonists (Ryoo et al., 2002; Yang et al., 2000), or after cleavage by caspases near the N-terminus (Ditzel et al., 2003; Yokokura et al., 2004), Op-IAPs lack the N-terminal degrons found in cellular IAPs and exhibit more stability (Cerio, Vandergaast, & Friesen, 2010; Vandergaast, Mitchell, Byers, & Friesen, 2015). Restoring apoptosis by introducing *Drosophila reaper* into Sindbis virus impaired their ability to infect mosquitoes and a gradual negative selection against *reaper* expression in the recovered virus (O'Neill, Olson, Huang, Unis, & Clem, 2015). Together, these observations indicate that IAPs and IAP-antagonists regulate the degree of viral propagation in insect hosts.

11. CONCLUDING REMARKS

It has been more than two decades since IAPs and their antagonists were first discovered, but dramatic new discoveries continue to be made in this field. A number of them are particularly notable: For example, although it had been thought that IAP-antagonists play no obvious roles in mammalian development, knockout of ARTS revealed defects in stem cell death. Exciting biological roles of IAPs and IAP-antagonists in innate immune response, in contributing to neuronal diversity and numbers and playing unexpected roles in morphogenesis, have been discovered only recently. There are still many unanswered questions. We still do not fully understand why *Drosophila* IAP-antagonists must localize to the mitochondrial outer membrane, and how it is trafficked to that site. The intricate regulatory mechanisms that converge on transcription, translation, and posttranslational levels are only beginning to be understood. We hope to see major advances in these areas in coming years.

ACKNOWLEDGMENTS

This work was supported by the NIH Grant R01 EY020866 to H.D.R.

REFERENCES

Abdelwahid, E., Yokokura, T., Krieser, R. J., Balasundaram, S., Fowle, W. H., & White, K. (2007). Mitochondrial disruption in Drosophila apoptosis. *Developmental Cell, 12*, 793–806.

Akdemir, F., Farkas, R., Chen, P., Juhasz, G., Medved'ova, L., Sass, M., et al. (2006). Autophagy occurs upstream or parallel to the apoptosome during histolytic cell death. *Development, 113*, 1457–1465.

Arama, E., Agapite, J., & Steller, H. (2003). Caspase activity and a specific cytochrome C are required for sperm differentiation in Drosophila. *Developmental Cell, 4*, 687–697.

Arya, R., Sarkissian, T., Tan, Y., & White, K. (2015). Neural stem cell progeny regulate stem cell death in a Notch and Hox dependent manner. *Cell Death and Differentiation, 22*, 1378–1387.

Bartke, T., Pohl, C., Pyrowolakis, G., & Jentsch, S. (2004). Dual role of BRUCE as an antiapoptotic IAP and a chimeric E2/E3 ubiquitin ligase. *Molecular Cell, 14*, 801–811.

Bauler, L. D., Duckett, C. S., & O'Riordan, M. X. (2008). XIAP regulates cytosol-specific innate immunity to Listeria infection. *PLoS Pathogens, 4*, e1000142.

Beira, J. V., Spinghorn, A., Gunther, S., Hufnagel, L., Pyrowolakis, G., & Vincent, J. P. (2014). The Dpp/TGFβ-dependent corepressor Schnurri protects epithelial cells from JNK-induced apoptosis in Drosophila embryos. *Developmental Cell, 31*, 240–247.

Bello, B. C., Hirth, F., & Gould, A. P. (2003). A pulse of the Drosophila Hox protein Abdominal-A schedules the end of neural proliferation via neuroblast apoptosis. *Neuron, 37*, 209–219.

Bergmann, A., Agapite, J., McCall, K., & Steller, H. (1998). The Drosophila gene hid is a direct molecular target of Ras-dependent survival signaling. *Cell, 95*, 331–341.

Bergmann, A., Tugentman, M., Shilo, B. Z., & Steller, H. (2002). Regulation of cell number by MAPK-dependent control of apoptosis: A mechanism for trophic survival signaling. *Developmental Cell, 2*, 159–170.

Bertet, C., Li, X., Erclik, T., Cavey, M., Wells, B., & Desplan, C. (2014). Temporal patterning of neuroblasts controls Notch-mediated cell survival through regulation of Hid or Reaper. *Cell, 158*, 1173–1186.

Bertrand, M. J., Doiron, K., Labbe, K., Korneluk, R. G., Barker, P. A., & Saleh, M. (2009). Cellular inhibitors of apoptosis cIAP1 and cIAP2 are required for innate immunity signaling by the pattern recognition receptors NOD1 and NOD2. *Immunity, 30*, 789–801.

Bertrand, M. J., Milutinovic, S., Dickson, K. M., Ho, W. C., Boudreault, A., Durkin, J., et al. (2008). cIAP1 and cIAP2 facilitate cancer cell survival by functioning as E3 ligases that promote RIP1 ubiquitination. *Molecular Cell, 30*, 689–700.

Birnbaum, M. J., Clem, R. J., & Miller, L. K. (1994). An apoptosis-inhibiting gene from a nuclear polyhedrosis virus encoding a polypeptide with Cys/His sequence motifs. *Journal of Virology*, *68*, 2521–2528.

Brennecke, J., Hipfner, D. R., Stark, A., Russell, R. B., & Cohen, S. M. (2003). bantam encodes a developmentally regulated microRNA that controls cell proliferation and regulates the proapoptotic gene hid in Drosophila. *Cell*, *113*, 25–36.

Brodsky, M. H., Nordstrom, W., Tsang, G., Kwan, E., Rubin, G. M., & Abrams, J. M. (2000). Drosophila p53 binds a damage response element at the reaper locus. *Cell*, *101*, 103–113.

Cerio, R. J., Vandergaast, R., & Friesen, P. D. (2010). Host insect inhibitor-of-apoptosis SfIAP functionally replaces baculovirus IAP but is differentially regulated by its N-terminal leader. *Journal of Virology*, *84*, 11448–11460.

Chen, P., Nordstrom, W., Gish, B., & Abrams, J. M. (1996). grim, a novel cell death gene in Drosophila. *Genes & Development*, *10*, 1773–1782.

Christich, A., Kauppila, S., Chen, P., Sogame, N., Ho, S.-I., & Abrams, J. M. (2002). The damage-responsive Drosophila gene sickle encodes a novel IAP binding protein similar to but distinct from reaper. *Current Biology*, *12*, 137–140.

Claveria, C., Caminero, E., Martinez-A, C., Campuzano, S., & Torres, M. (2002). GH3, a novel proapoptotic domain in Drosophila Grim, promotes a mitochondrial death pathway. *The EMBO Journal*, *21*, 3327–3336.

Clem, R. J. (2005). The role of apoptosis in defense against baculovirus infection in insects. *Current Topics in Microbiology and Immunology*, *289*, 113–129.

Clem, R. J., Fechneimer, M., & Miller, L. K. (1991). Prevention of apoptosis by a baculovirus gene during infection of insect cells. *Science*, *254*, 1388–1390.

Clem, R. J., & Miller, L. K. (1994). Control of programmed cell death by the baculovirus genes p35 and iap. *Molecular and Cellular Biology*, *14*, 5212–5222.

Cordoba, S., & Estella, C. (2014). The bHLH-PAS transcription factor dysfusion regulates tarsal joint formation in response to Notch activity during drosophila leg development. *PLoS Genetics*, *10*, e1004621.

Crook, N. E., Clem, R. J., & Miller, L. K. (1993). An apoptosis-inhibiting baculovirus gene with a zinc finger-like motif. *Journal of Virology*, *67*, 2168–2174.

Cullen, K., & McCall, K. (2004). Role of programmed cell death in patterning the Drosophila antennal arista. *Developmental Biology*, *275*, 82–92.

Damgaard, R. B., Fiil, B. K., Speckmann, C., Yabal, M., zur Stadt, U., Bekker-Jensen, S., et al. (2013). Disease-causing mutations in the XIAP BIR2 domain impair NOD2-dependent immune signaling. *EMBO Molecular Medicine*, *5*, 1278–1295.

Deveraux, Q. L., Takahashi, R., Salvesen, G. S., & Reed, J. C. (1997). X-linked IAP is a direct inhibitor of cell death proteases. *Nature*, *388*, 300–304.

Ditzel, M., Wilson, R., Tenev, T., Zachariou, A., Paul, A., Deas, E., et al. (2003). Degradation of DIAP1 by the N-end rule pathway is essential for regulating apoptosis. *Nature Cell Biology*, *5*, 467–473.

Dogan, T., Harms, G. S., Hekman, M., Karreman, C., Oberoi, T. K., Alnemri, E. S., et al. (2008). X-linked and cellular IAPs modulate the stability of C-RAF kinase and cell motility. *Nature Cell Biology*, *10*, 1447–1455.

Domingues, C., & Ryoo, H. D. (2012). Drosophila BRUCE inhibits apoptosis through non-lysine ubiquitination of the IAP-antagonist REAPER. *Cell Death and Differentiation*, *19*, 470–477.

Du, C., Fang, M., Li, Y., Li, L., & Wang, X. (2000). Smac, a mitochondrial protein that promotes cytochrome c-dependent caspase activation by eliminating IAP proteins. *Cell*, *102*, 33–42.

Fan, Y., & Bergmann, A. (2008). Distinct mechanisms of apoptosis-induced compensatory proliferation in proliferating and differentiating tissues in the Drosophila eye. *Developmental Cell*, *14*, 399–410.

Fan, Y., & Bergmann, A. (2014). Multiple mechanisms modulate distinct cellular susceptibilities toward apoptosis in the developing Drosophila. *Developmental Cell, 30*, 48–60.

Fuchs, Y., Brown, S., Gorenc, T., Rodriguez, J., Fuchs, E., & Steller, H. (2013). Sept4/ARTS regulates stem cell apoptosis and skin regeneration. *Science, 341*, 286–289.

Garcia-Fernandez, M., Kissel, H., Brown, S., Gorenc, T., Schile, A. J., Rafii, S., et al. (2010). Sept4/ARTS is required for stem cell apoptosis and tumor suppression. *Genes & Development, 24*, 2282–2293.

Ge, W., Chen, Y. W., Weng, R., Lim, S. F., Buescher, M., Zhang, R., et al. (2012). Overlapping functions of microRNAs in control of apoptosis during Drosophila embryogenesis. *Cell Death and Differentiation, 19*, 839–846.

Geisbrecht, E. R., & Montell, D. J. (2004). A role for Drosophila IAP1-mediated caspase inhibition in Rac-dependent cell migration. *Cell, 118*, 111–125.

Gottfried, Y., Rotem, A., Lotan, R., Steller, H., & Larisch, S. (2004). The mitochondrial ARTS protein promotes apoptosis through targeting XIAP. *The EMBO Journal, 23*, 1627–1635.

Goyal, L., McCall, K., Agapite, J., Hartwieg, E., & Steller, H. (2000). Induction of apoptosis by Drosophila reaper, hid and grim through inhibition of IAP function. *The EMBO Journal, 19*, 589–597.

Grether, M. E., Abrams, J. M., Agapite, J., White, K., & Steller, H. (1995). The head involution defective gene of Drosophila melanogaster functions in programmed cell death. *Genes & Development, 9*, 1694–1708.

Haining, W. N., Carboy-Newcomb, C., Wei, C. L., & Steller, H. (1999). The proapoptotic function of Drosophila Hid is conserved in mammalian cells. *Proceedings of the National Academy of Sciences of the United States of America, 96*, 4936–4941.

Hao, Y., Sekine, K., Kawabata, A., Nakamura, H., Ishioka, T., Ohata, H., et al. (2004). Apollon ubiquitinates SMAC and caspase-9, and has an essential cytoprotection function. *Nature Cell Biology, 6*, 849–860.

Hawkins, C. J., Wang, S. L., & Hay, B. A. (1999). A cloning method to identify caspases and their regulators in yeast: Identification of Drosophila IAP1 as an inhibitor of the Drosophila caspase DCP-1. *Proceedings of the National Academy of Sciences of the United States of America, 96*, 2885–2890.

Hay, B. A., Wassarman, D. A., & Rubin, G. M. (1995). Drosophila homologs of baculovirus inhibitor of apoptosis proteins function to block cell death. *Cell, 83*, 1253–1262.

Hegde, R., Srinivasula, S. M., Zhang, Z., Wassell, R., Mukattash, R., Cilenti, L., et al. (2002). Identification of Omi/HtrA2 as a mitochondrial apoptotic serine protease that disrupts IAP-caspase interaction. *The Journal of Biological Chemistry, 277*, 432–438.

Huh, J. R., Foe, I., Muro, I., Chen, C. H., Seol, J. H., Yoo, S. J., et al. (2007). The Drosophila inhibitor of apoptosis (IAP) DIAP2 is dispensable for cell survival, required for the innate immune response to gram-negative bacterial infection, and can be negatively regulated by the reaper/hid/grim family of IAP-binding apoptosis inducers. *The Journal of Biological Chemistry, 282*, 2056–2068.

Huh, J. R., Guo, M., & Hay, B. A. (2004). Compensatory proliferation induced by cell death in the Drosophila wing disc requires activity of the apical cell death caspase Dronc in a nonapoptotic role. *Current Biology, 14*, 1262–1266.

Jiang, C., Lamblin, A. F., Steller, H., & Thummel, C. S. (2000). A steroid-triggered transcriptional hierarchy controls salivary gland cell death during Drosophila metamorphosis. *Molecular Cell, 5*, 445–455.

Kang, Y., & Bashirullah, A. (2014). A steroid-controlled global switch in sensitivity to apoptosis during Drosophila development. *Developmental Biology, 386*, 34–41.

Kanuka, H., Kuranaga, E., Takemoto, K., Hiratou, T., Okano, H., & Miura, M. (2005). Drosophila caspase transduces Shaggy/GSK-3beta kinase activity in neural precursor development. *The EMBO Journal, 24*(21), 3793–3806.

Kaplan, Y., Gibbs-Bar, L., Kalifa, Y., Feinstein-Rotkopf, Y., & Arama, E. (2010). Gradients of a ubiquitin E3 ligase inhibitor and a caspase inhibitor determine differentiation or death in spermatids. *Developmental Cell, 19*, 160–173.

Kissel, H., Georgescu, M., Larisch, S., Manova, K., Hunnicutt, G. R., & Steller, H. (2005). The Sept4 septin locus is required for sperm terminal differentiation in mice. *Developmental Cell, 8*, 353–364.

Kondo, S., Senoo-Matsuda, N., Hiromi, Y., & Miura, M. (2006). DRONC coordinates cell death and compensatory proliferation. *Molecular and Cellular Biology, 26*, 7258–7268.

Koto, A., Kuranaga, E., & Miura, M. (2009). Temporal regulation of Drosophila IAP1 determines caspase functions in sensory organ development. *The Journal of Cell Biology, 182*, 219–231.

Kuo, C. T., Zhu, S., Younger, S., Jan, L. Y., & Jan, Y. N. (2006). Identification of E2/E3 ubiquitinating enzymes and caspase activity regulating Drosophila sensory neuron dendrite pruning. *Neuron, 51*, 283–290.

Kurada, P., & White, K. (1998). Ras promotes cell survival in Drosophila by downregulating hid expression. *Cell, 95*, 319–329.

Kuranaga, E., Kanuka, H., Igaki, T., Sawamoto, K., Ichijo, H., Okano, H., et al. (2002). Reaper-mediated inhibition of DIAP1-induced DTRAF1 degradation results in activation of JNK in Drosophila. *Nature Cell Biology, 4*, 705–710.

Kuranaga, E., Kanuka, H., Tonoki, A., Takemoto, K., Tomioka, T., Kobayashi, M., et al. (2006). Drosophila IKK-related kinase regulates nonapoptotic function of caspases via degradation of IAPs. *Cell, 126*(3), 583–596.

LaCount, D. J., Hanson, S. F., Schneider, C. L., & Friesen, P. D. (2000). Caspase inhibitor p35 and inhibitor of apoptosis Op-IAP block in vivo proteolytic activation of an effector caspase at different steps. *The Journal of Biological Chemistry, 275*, 15657–15664.

Larisch, S., Yi, Y., Lotan, R., Kerner, H., Eimerl, S., Tony Parks, W., et al. (2000). A novel mitochondrial septin-like protein, ARTS, mediates apoptosis dependent on its P-loop motif. *Nature Cell Biology, 2*, 915–921.

Lee, T. V., Fan, Y., Wang, S., Srivastava, M., Broemer, M., Meier, P., et al. (2011). Drosophila IAP1-mediated ubiquitylation controls activation of the initiator caspase DRONC independent of protein degradation. *PLoS Genetics*, 7, e1002261.

Li, X., Wang, J., & Shi, Y. (2011). Structural mechanisms of DIAP1 auto-inhibition and DIAP1-mediated inhibition of drICE. *Nature Communications, 2*, 408.

Lisi, S., Mazzon, I., & White, K. (2000). Diverse domains of THREAD/DIAP1 are required to inhibit apoptosis induced by REAPER and HID in Drosophila. *Genetics, 154*, 669–678.

Liu, B., Becnel, J. J., Zhang, Y., & Zhou, L. (2011). Induction of reaper ortholog mx in mosquito midgut cells following baculovirus infection. *Cell Death and Differentiation, 18*, 1337–1345.

Liu, B., Behura, S. K., Clem, R. J., Schneemann, A., Becnel, J., Severson, D. W., et al. (2013). p53-mediated rapid induction of apoptosis conveys resistance to viral infection in Drosophila melanogaster. *PLoS Pathogens, 9*, e1003137.

Lohmann, I., McGinnis, N., Bodmer, M., & McGinnis, W. (2002). The Drosophila Hox gene deformed sculpts head morphology via direct regulation of the apoptosis activator reaper. *Cell, 110*, 457–466.

Manjon, C., Sanchez-Herrero, E., & Suzanne, M. (2007). Sharp boundaries of Dpp signalling trigger local cell death required for Drosophila leg morphogenesis. *Nature Cell Biology, 9*, 57–63.

Marsh, R. A., Madden, L., Kitchen, B. J., Mody, R., McClimon, B., Jordan, M. B., et al. (2010). XIAP1 deficiency: A unique primary immunodeficiency best classified as X-linked familial hemophagocytic lymphohistiocytosis and not as X-linked lymphoproliferative disease. *Blood, 116*, 1079–1082.

Martins, L. M., Iaccarino, I., Tenev, T., Gschmeissner, S., Totty, N. F., Lemoine, N. R., et al. (2002). The serine protease Omi/HtrA2 regulates apoptosis by binding XIAP through a Reaper-like motif. *The Journal of Biological Chemistry*, *277*, 439–444.

Maurange, C., Cheng, L., & Gould, A. P. (2008). Temporal transcription factors and their targets schedule the end of neural proliferation in Drosophila. *Cell*, *133*, 891–902.

Meier, P., Silke, J., Leevers, S. J., & Evan, G. I. (2000). The Drosophila caspase DRONC is regulated by DIAP1. *The EMBO Journal*, *19*, 598–611.

Morishita, J., Kang, M. J., Fidelin, K., & Ryoo, H. D. (2013). CDK7 regulates the mitochondrial localization of a tail-anchored proapoptotic protein, Hid. *Cell Reports*, *5*, 1481–1488.

Muro, I., Hay, B. A., & Clem, R. J. (2002). The Drosophila DIAP1 protein is required to prevent accumulation of a continuously generated processed form of the apical caspase DRONC. *The Journal of Biological Chemistry*, *20*, 49644–49650.

O'Neill, K., Olson, B. J., Huang, N., Unis, D., & Clem, R. J. (2015). Rapid selection against arbovirus-induced apoptosis during infection of a mosquito vector. *Proceedings of the National Academy of Sciences of the United States of America*, *112*, E1152–E1161.

Okada, H., Suh, W. K., Jin, J., Woo, M., Du, C., Elia, A., et al. (2002). Generation and characterization of Smac/DIABLO-deficient mice. *Molecular and Cellular Biology*, *22*, 3509–3517.

Olayioye, M. A., Kaufmann, H., Pakusch, M., Vaux, D. L., Lindeman, G. J., & Visvader, J. E. (2005). XIAP-deficiency leads to delayed lobuloalveolar development in the mammary gland. *Cell Death and Differentiation*, *12*, 87–90.

Ollmann, M., Young, L. M., Di Como, C. J., Karim, F., Belvin, M., Robertson, S., et al. (2000). Drosophila p53 is a structural and functional homolog of the tumor suppressor p53. *Cell*, *101*, 91–101.

Olson, M. R., Holley, C. L., Gan, E. C., Colon-Ramos, D. A., Kaplan, B., & Kornbluth, S. (2003). A GH3-like domain in reaper is required for mitochondrial localization and induction of IAP degradation. *The Journal of Biological Chemistry*, *278*, 44758–44768.

Olson, M. R., Holley, C. L., Yoo, S. L., Huh, J. R., Hay, B. A., & Kornbluth, S. (2003). Reaper is regulated by IAP-mediated ubiquitination. *The Journal of Biological Chemistry*, *278*, 4028–4034.

Oshima, K., Takeda, M., Kuranaga, E., Ueda, R., Aigaki, T., Miura, M., et al. (2006). IKK epsilon regulates F actin assembly and interacts with Drosophila IAP1 in cellular morphogenesis. *Current Biology*, *16*(15), 1531–1537.

Pachlopnik Schmid, J., Canioni, D., Moshous, D., Touzot, F., Mahlaoui, N., Hauck, F., et al. (2011). Clinical similarities and differences of patients with X-linked lymphoproliferative syndrome type 1 (XLP-1/SAP deficiency) versus type 2 (XLP-2/XIAP deficiency). *Blood*, *117*, 1522–1529.

Perez-Garijo, A., Martin, F. A., & Morata, G. (2004). Caspase inhibition during apoptosis causes abnormal signaling and developmental aberrations in Drosophila. *Development*, *131*, 5591–5598.

Peterson, C., Carney, G. E., Taylor, B. J., & White, K. (2002). Reaper is required for neuroblast apoptosis during Drosophila development. *Development*, *129*, 1467–1476.

Potts, P. R., Singh, S., Knezek, M., Thompson, C. B., & Deshmukh, M. (2003). Critical function of endogenous XIAP in regulating caspase activation during sympathetic neuronal apoptosis. *The Journal of Cell Biology*, *163*, 789–799.

Prakash, H., Albrecht, M., Becker, D., Kuhlmann, T., & Rudel, T. (2010). Deficiency of XIAP leads to sensitization for Chlamydophila pneumoniae pulmonary infection and dysregulation of innate immune response in mice. *The Journal of Biological Chemistry*, *285*, 20291–20302.

Reingewertz, T. H., Shalev, D. E., Sukenik, S., Blatt, O., Rotem-Bamberger, S., Lebendiker, M., et al. (2011). Mechanism of the interaction between the intrinsically disordered C-terminus of the pro-apoptotic ARTS protein and the Bir3 domain of XIAP. *PLoS One*, *6*, e24655.

Ribeiro, P. S., Kuranaga, E., Tenev, T., Leulier, F., Miura, M., & Meier, P. (2007). Diap2 functions as a mechanism-based regulator of drICE that contributes to the caspase activity threshold in living cells. *The Journal of Cell Biology*, *31*, 7.

Rigaud, S., Fondaneche, M. C., Lambert, N., Pasquier, B., Mateo, V., Soulas, P., et al. (2006). XIAP deficiency in humans causes an X-linked lymphoproliferative syndrome. *Nature*, *444*, 110–114.

Rothe, M., Pan, M. G., Henzel, W. J., Ayres, T. M., & Goeddel, D. V. (1995). The TNF-R2-TRAF signaling complex contains two novel proteins related to baculoviral-inhibitor of apoptosis proteins. *Cell*, *83*, 1243–1252.

Rumpf, S., Lee, S. B., Jan, L. Y., & Jan, Y. N. (2011). Neuronal remodeling and apoptosis require VCP-dependent degradation of the apoptosis inhibitor DIAP1. *Development*, *138*, 1153–1160.

Ryoo, H. D., Bergmann, A., Gonen, H., Ciechanover, A., & Steller, H. (2002). Regulation of Drosophila IAP1 degradation and apoptosis by reaper and ubcD1. *Nature Cell Biology*, *4*, 432–438.

Ryoo, H. D., Gorenc, T., & Steller, H. (2004). Apoptotic cells can induce compensatory cell proliferation through the JNK and the Wingless signaling pathways. *Developmental Cell*, *7*, 491–501.

Sandu, C., Ryoo, H. D., & Steller, H. (2010). Drosophila IAP antagonists form multimeric complexes to promote cell death. *The Journal of Cell Biology*, *190*, 1039–1052.

Schile, A. J., Garcia-Fernandez, M., & Steller, H. (2008). Regulation of apoptosis by XIAP ubiquitin-ligase activity. *Genes & Development*, *22*, 2256–2266.

Shapiro, P. J., Hsu, H. H., Jung, H., Robbins, E. S., & Ryoo, H. D. (2008). Regulation of the Drosophila apoptosome through feedback inhibition. *Nature Cell Biology*, *10*, 1440–1446.

Shlevkov, E., & Morata, G. (2012). A dp53/JNK-dependant feedback amplification loop is essential for the apoptotic response to stress in Drosophila. *Cell Death and Differentiation*, *19*, 451–460.

Silke, J. (2011). The regulation of TNF signalling: What a tangled web we weave. *Current Opinion in Immunology*, *23*, 620–626.

Silke, J., & Vaux, D. L. (2001). Two kinds of BIR-containing protein—Inhibitors of apoptosis, or required for mitosis. *Journal of Cell Science*, *114*, 1821–1827.

Srinivasula, S., Datta, P., Kobayashi, M., Wu, J.-W., Fujioka, M., hedge, R., et al. (2002). Sickle, a novel Drosophila death gene in the reaper/hid/grim region, encodes an IAP-inhibitory protein. *Current Biology*, *12*, 125–130.

Srinivasula, S. M., Hegde, R., Saleh, A., Datta, P., Shiozaki, E., Chai, J., et al. (2001). A conserved XIAP-interaction motif in caspase-9 and Smac/DIABLO regulates caspase activity and apoptosis. *Nature*, *410*, 112–116.

Stark, A., Brennecke, J., Russell, R. B., & Cohen, S. M. (2003). Identification of Drosophila MicroRNA targets. *PLoS Biology*, *1*, e60.

Strauss, K. M., Martins, L. M., Plun-Favreau, H., Marx, F. P., Kautzmann, S., Berg, D., et al. (2005). Loss of function mutations in the gene encoding Omi/HtrA2 in Parkinson's disease. *Human Molecular Genetics*, *14*, 2099–2111.

Suzanne, M., Petzoldt, A. G., Speder, P., Coutelis, J. B., Steller, H., & Noselli, S. (2010). Coupling of apoptosis and L/R patterning controls stepwise organ looping. *Current Biology*, *20*, 1773–1779.

Tan, Y., Yamada-Mabuchi, M., Arya, R., St. Pierre, S., Tang, W., Tosa, M., et al. (2011). Coordinated expression of cell death genes regulates neuroblast apoptosis. *Development*, *138*, 2197–2206.

Tenev, T., Bianchi, K., Darding, M., Broemer, M., Langlais, C., Wallberg, F., et al. (2011). The ripoptosome, a signaling platform that assembles in response to genotyoxic stress and loss of IAPs. *Molecular Cell, 43*, 432–448.

Tenev, T., Zachariou, A., Wilson, R., Ditzel, M., & Meier, P. (2005). IAPs are functionally non-equivalent and regulate effector caspases through distinct mechanisms. *Nature Cell Biology, 7*, 70–77.

Thomenius, M., Freel, C. D., Horn, S., Krieser, R., Abdelwahid, E., Cannon, R., et al. (2011). Mitochondrial fusion is regulated by Reaper to modulate Drosophila programmed cell death. *Cell Death and Differentiation, 18*, 1640–1650.

Thress, K., Evans, E. K., & Kornbluth, S. (1999). Reaper-induced dissociation of a Scythe-sequestered cytochrome c-releasing activity. *The EMBO Journal, 18*, 5486–5493.

Truscott, M., Islam, A. B., Lopez-Bigas, N., & Frolov, M. V. (2011). Mir-11 limits the proapoptotic function of its host gene, dE2f1. *Genes & Development, 25*, 1820–1834.

Uren, A. G., Pakusch, M., Hawkins, C. J., Puls, K. L., & Vaux, D. L. (1996). Cloning and expression of apoptosis inhibitory protein homologs that function to inhibit apoptosis and/or bind tumor necrosis factor receptor-associated factors. *Proceedings of the National Academy of Sciences of the United States of America, 1996*, 4974–4978.

van Loo, G., van Gurp, M., Depuydt, B., Srinivasula, S. M., Rodriguez, I., Alnemri, E. S., et al. (2002). The serine protease Omi/HtrA2 is released from mitochondria during apoptosis. Omi interacts with caspase-inhibitor XIAP and induces enhanced caspase activity. *Cell Death and Differentiation, 9*, 20–26.

Vandergaast, R., Mitchell, J. K., Byers, N. M., & Friesen, P. D. (2015). Insect inhibitor-of-Apoptosis (IAP) proteins are negatively regulated by signal-induced N-terminal degrons absent within viral IAP proteins. *Journal of Virology, 89*, 4481–4493.

Verhagen, A., Ekert, P. G., Pakusch, M., Silke, J., Connolly, L. M., Reid, G. E., et al. (2000). Identification of DIABLO, a mammalian protein that promotes apoptosis by binding to and antagonizing IAP proteins. *Cell, 102*, 43–54.

Vernooy, S. Y., Chow, V., Su, J., Verbrugghe, K., Yang, J., Cole, S., et al. (2002). Drosophila Bruce can potently suppress Rpr- and Grim-dependent but not Hid-dependent cell death. *Current Biology, 12*, 1164–1168.

Vince, J. E., Wong, W. W., Khan, N., Feltham, R., Chau, D., Ahmed, A. U., et al. (2007). IAP antagonists target cIAP1 to induce TNFalpha-dependent apoptosis. *Cell, 131*, 682–693.

Wang, S. L., Hawkins, C. J., Yoo, S. J., Muller, H. A., & Hay, B. A. (1999). The Drosophila caspase inhibitor DIAP1 is essential for cell survival and is negatively regulated by HID. *Cell, 98*, 453–463.

White, K., Grether, M. E., Abrams, J. M., Young, L., Farrell, K., & Steller, H. (1994). Genetic control of programmed cell death in Drosophila. *Science, 264*, 677–683.

White, K., Tahaoglu, E., & Steller, H. (1996). Cell killing by the Drosophila gene reaper. *Science, 271*, 805–807.

Wilson, R., Goyal, L., Ditzel, M., Zachariou, A., Baker, D. A., Agapite, J., et al. (2002). The DIAP1 RING finger mediates ubiquitination of Dronc and is indispensable for regulating apoptosis. *Nature Cell Biology, 4*, 445–450.

Wing, J. P., Karres, J. S., Ogdahl, J. L., Zhou, L., Schwartz, L. M., & Nambu, J. R. (2002). Drosophila sickle is a novel grim-reaper cell death activator. *Current Biology, 12*, 131–135.

Wu, G., Chai, J., Suber, T. L., Wu, J. W., Du, C., Wang, X., et al. (2000). Structural basis of IAP recognition by Smac/DIABLO. *Nature, 408*, 1008–1012.

Wu, J. W., Cocina, A. E., Chai, J., Hay, B. A., & Shi, Y. (2001). Structural analysis of a functional DIAP1 fragment bound to grim and hid peptides. *Molecular Cell, 8*, 95–104.

Xu, P., Vernooy, S. Y., Guo, M., & Hay, B. A. (2003). The Drosophila microRNA Mir-14 suppresses cell death and is required for normal fat metabolism. *Current Biology, 13*, 790–795.

Yabal, M., Muller, N., Adler, H., Knies, N., Bross, C. J., Damgaard, R. B., et al. (2014). XIAP restricts TNF- and RIP3-dependent cell death and inflammasome activation. *Cell Reports*, 7, 1796–1808.

Yacobi-Sharon, K., Namdar, Y., & Arama, E. (2013). Alternative germ cell death pathway in Drosophila involves HtrA2/Omi, lysosomes, and a caspase-9 counterpart. *Developmental Cell*, *25*, 29–42.

Yan, N., Wu, J. W., Chai, J., Li, W., & Shi, Y. (2004). Molecular mechanisms of DrICE inhibition by DIAP1 and removal of inhibition by Reaper, Hid and Grim. *Nature Structural & Molecular Biology*, *11*, 420–428.

Yang, Y., Fang, S., Jensen, J. P., Weissman, A. M., & Ashwell, J. D. (2000). Ubiquitin protein ligase activity of IAPs and their degradation in proteasomes in response to apoptotic stimuli. *Science*, *288*, 874–877.

Yokokura, T., Dresnek, D., Huseinovic, N., Lisi, S., Abdelwahid, E., Bangs, P., et al. (2004). Dissection of DIAP1 functional domains via a mutant replacement strategy. *The Journal of Biological Chemistry*, *279*, 52603–52612.

Yoo, S. J., Huh, J. R., Muro, I., Yu, H., Wang, L., Wang, S., et al. (2002). Hid, Rpr and Grim negatively regulate DIAP1 levels through distinct mechanisms. *Nature Cell Biology*, *4*, 416–424.

Zhang, Y., Lin, N., Carroll, P. M., Chan, G., Guan, B., Xiao, H., et al. (2008). Epigenetic blocking of an enhancer region controls irradiation-induced proapoptotic gene expression. *Developmental Cell*, *14*, 481–493.

CHAPTER EIGHT

Ubiquitin-Mediated Regulation of Cell Death, Inflammation, and Defense of Homeostasis

Pascal Meier*,[1], Otto Morris*, Meike Broemer†,[1]

*The Breakthrough Toby Robins Breast Cancer Research Centre, Chester Beatty Laboratories, Institute of Cancer Research, London, UK
†German Center for Neurodegenerative Diseases (DZNE), Bonn, Germany
[1]Corresponding authors: e-mail address: pmeier@icr.ac.uk; Meike.Broemer@dzne.de

Contents

Abstract

Cell death and inflammation are ancient processes of fundamental biological importance in both normal physiology and human disease pathologies. The recent observation that apoptosis regulatory components have dual roles in cell death and inflammation suggests that these proteins function, not primarily to kill, but to coordinate tissue repair and remodeling. This perspective unifies cell death components as positive regulators of tissue repair that replaces malfunctioning or damaged tissues and enhances the resilience of epithelia to insult. It is now recognized that cells that die by apoptosis do not do so silently, but release a variety of paracrine signals to communicate with their cellular environment to ensure tissue regeneration, and wound healing. Moreover, inflammatory signaling pathways, such as those emanating from the TNF receptor or Toll-related receptors, take part in cell competition to eliminate developmentally aberrant clones. Ubiquitylation has emerged as crucial mediator of signal transduction in cell death and inflammation. Here, we focus on recent advances on ubiquitin-mediated regulation of cell death and inflammation, and how this is used to regulate the defense of homeostasis.

Current Topics in Developmental Biology, Volume 114
ISSN 0070-2153
http://dx.doi.org/10.1016/bs.ctdb.2015.07.015

1. INTRODUCTION

The remarkable capacity to correct for tissue stress and malfunction is one of the most fascinating hallmarks of multicellular organisms. While we marvel at the high level of tissue plasticity during animal development, we struggle with the incredible ability of cancer cells to escape growth inhibitory signals or develop resistance to treatments. For the maintenance of tissue homeostasis and tissue fitness, it is essential that damaged or malfunctioning cells are detected, eliminated, and replaced (Jacobson, Weil, & Raff, 1997). The decision as to whether a cell lives or dies relies not just on the regulation of the intrinsic apoptotic program, but critically depends on paracrine interactions between damaged cells and cells of the surrounding tissue (Neves, Demaria, Campisi, & Jasper, 2015). It is now clear that cellular malfunction, and this includes activation of oncogenes or loss of tumor suppressor proteins, results in the production of cytokines, chemokines, and mitogens, which stimulate an adaptive response for the defense of homeostasis (Medzhitov, 2008). Activation of the cell death program is, thereby, critically important to eliminate and replace malfunctioning or damaged cells (Thompson, 1995).

Studies in flies show that tissue homeostasis in epithelia is governed by "collective" decision mechanisms that determine cell death and proliferation across tissues (Vincent, Fletcher, & Baena-Lopez, 2013). These mechanisms include cell competition and apoptosis-induced compensatory proliferation (Fig. 1). Cell competition is a process in which fast-growing fitter cells (winners) kill neighboring slow-growing, "less fit" (losers) cells, even when the weaker cells are fully viable in a nonmosaic tissue. While cell competition and compensatory proliferation have been studied extensively in flies, recent studies reveal the existence and importance of similar processes in mammals (Bondar & Medzhitov, 2010; Oertel, Menthena, Dabeva, & Shafritz, 2006; Oliver, Saunders, Tarle, & Glaser, 2004). For example, cell competition selects the fittest stem cell in the epiblast that gives rise to the entire organism (Claveria, Giovinazzo, Sierra, & Torres, 2013), while in young individuals, cell competition is used to ensure tissue fitness (Amoyel & Bach, 2014). Due to chronic engagement of the tissue repair program in older individuals, age-related decline of tissue homeostasis can occur, resulting in degeneration, metabolic dysfunction, and cancer (Martins et al., 2014; Neves et al., 2015). Consistently, loss of homeostasis is one of the hallmarks of aging (Chovatiya & Medzhitov, 2014).

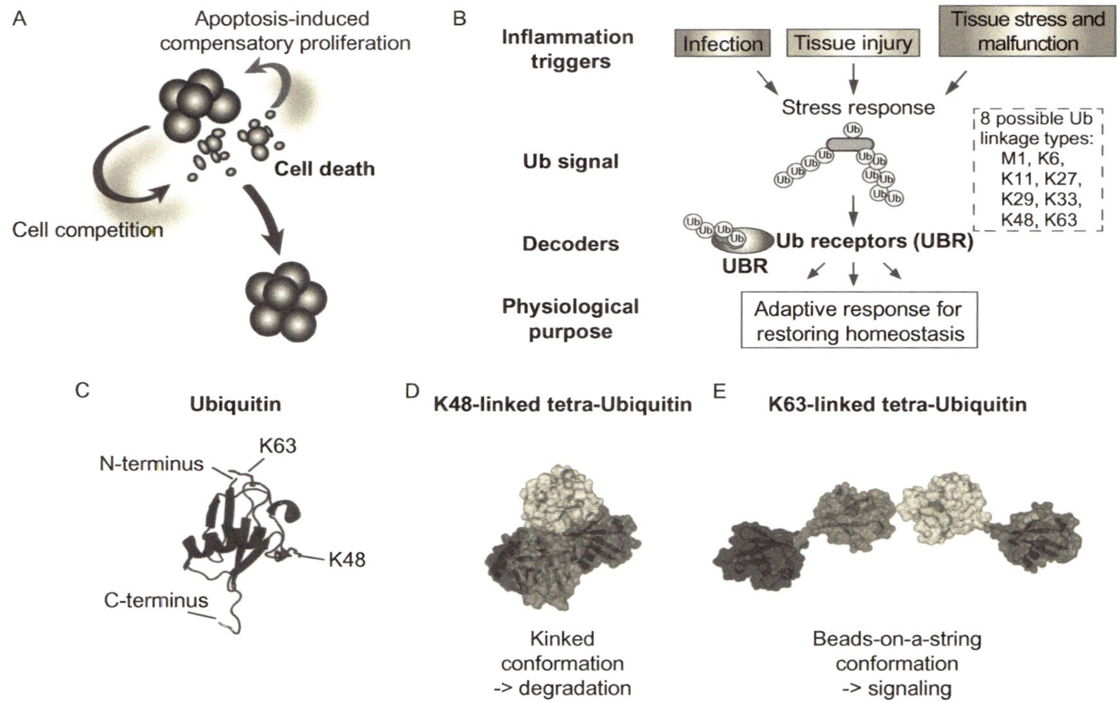

Figure 1 Ubiquitylation as mediator and regulator of signal transduction in cell death, inflammation, and defense of homeostasis. (A) Tissue malfunction results in a secretory program and inflammatory response whose purpose is to restore homeostasis. Tissue homeostasis is regulated by a collective decision mechanism that influences cell death and proliferation across tissues. These include cell competition and apoptosis-induced compensatory proliferation. (B) Tissue stress response and inflammation underlie a common principle in which the conjugation of typical Ub chains produces robust networks that are decoded by Ub receptors whose actions serve to coordinate adaptation to tissue stress. (C) Ribbon structure of Ub. Lysines at position 48 and 63 of Ub are highlighted. Topology of K48- (D) and K63-linked (E) tetra-Ub chains. (See the color plate.)

2. UBIQUITIN AS MEDIATOR OF SIGNALING EVENTS

While superficially tissue stress, inflammation, and cell death appear as separate phenomena, recent advances in our understanding of these stress responses strongly indicate that the underlying signaling events are in fact linked. Ubiquitylation, thereby, has emerged as crucial mediator and regulator of signal transduction in cell death and inflammation (Dikic, Wakatsuki, & Walters, 2009; Fig. 1).

The covalent attachment of ubiquitin (Ub) to target proteins can alter the protein's conformation or binding properties and thus influences protein activities, localization, or stability. Ub is a small protein modifier that is covalently attached to proteins in a stepwise process that involves Ub-activating enzymes (E1), Ub-conjugating (UBC) enzymes (E2), and Ub protein ligases (E3) (Hershko, Ciechanover, & Varshavsky, 2000). E3s confer substrate specificity by bringing Ub-loaded E2 to target substrates and promoting the formation of an isopeptide linkage between the carboxyl terminus of Ub (glycine (G)76) and the amino group of a lysine (K) residue of the substrate.

Ub can be conjugated either as a single moiety or as chains of variable length (Komander, 2009; Fig. 1). Different linkage types provide further complexity, as Ub moieties can be conjugated to one another via each of the seven K residues within Ub, or via Ub's N-terminal methionine. This allows the formation of homotypic chains linked via Ub's K6, K11, K27, K29, K33, K48, K63, or M1 (Kirisako et al., 2006). In addition to homotypic chains that are sequentially linked through the same successive linkage type, mixed-linkage chain types also exist in which several distinct K residues are used to connect consecutive Ub moieties (Meyer & Rape, 2014). The complexity and versatility of Ub-dependent modifications are further increased through the generation of heterologous chain types where Ub is connected with other Ub-like modifiers, such as SUMO (Tatham et al., 2008).

The eight different types of homotypic Ub chains exert distinct effects on cellular processes (Bhoj & Chen, 2009). This is because the differently linked poly-Ub chains adopt distinct structures. For instance, K48-linked poly-Ub chains take up a kinked topology while K63- and M1-linked chains adopt an open configuration that resembles "beads-on-a-string" (Komander, 2009). While it is well established that K48-linked modifications can promote degradation through recognition by the 26S proteasome, recent evidence indicate that other linkage types, such as M1, K29, K33, and K63, can regulate biological processes in a degradation-independent manner (Komander & Rape, 2012). Whether ubiquitylation targets proteins for degradation or mediates nondegradative signaling depends on protein interactions between the ubiquitylated protein and Ub-binding proteins—often termed Ub "receptors" (Hoeller, Hecker, & Dikic, 2006). Ub receptors (UBRs) carry small Ub-binding domains (UBDs) that bind to the Ub modification via low-affinity, noncovalent interactions. Currently, more than 20 different types of UBDs are known that detect overlapping as well as distinct Ub modifications. UBRs that selectively recognize K48-linked poly-Ub chains, such as the proteasome subunit Rpn13 (Lundgren, Masson, Realini, & Young, 2003), recruit modified proteins to the proteasome for degradation. In contrast, UBRs that bind to mono-Ub, K63-linkages, or linear (M1) Ub allow Ub-dependent association with signaling molecules (Hoeller et al., 2006). In particular, K63-linked Ub chains, and their respective UBRs, are critical for tumor necrosis factor (TNF)-mediated NF-κB activation and cell survival (Stickle et al., 2004).

3. UB AS ARBITER OF LIFE AND DEATH—TNF SIGNALING AS PARADIGM

The role of Ub as crucial mediator and regulator of signal transduction in the defense of homeostasis is best illustrated by the signaling pathways emanating from the TNF receptor (Fig. 2). TNF is a major inflammatory cytokine that was first identified for its ability to induce rapid hemorrhagic necrosis of experimental cancers (Carswell et al., 1975). Now it is clear that TNF functions as a master regulator of the cytokine network that coordinates defense of homeostasis via controlling inflammation, cell proliferation, differentiation, survival, and death (Balkwill, 2009).

In mammals, binding of TNF to its extracellular receptor TNFR1 triggers either prosurvival/inflammatory or prodeath signaling pathways in a strictly Ub-dependent manner (Walczak, 2013; Fig. 2). TNF can regulate tissue homeostasis in at least three different ways: through (1) activation of NF-κB-dependent and MAPK/JNK-dependent transcriptional programs, (2) induction of caspase-8-dependent apoptosis, or (3) stimulation of receptor-interacting protein kinase (RIPK)-mediated necrosis (necroptosis) (Declercq, Vanden Berghe, & Vandenabeele, 2009).

Binding of TNF to TNFR1 results in the assembly of a protein complex at the receptor's cytoplasmic tail. This complex, which is frequently referred

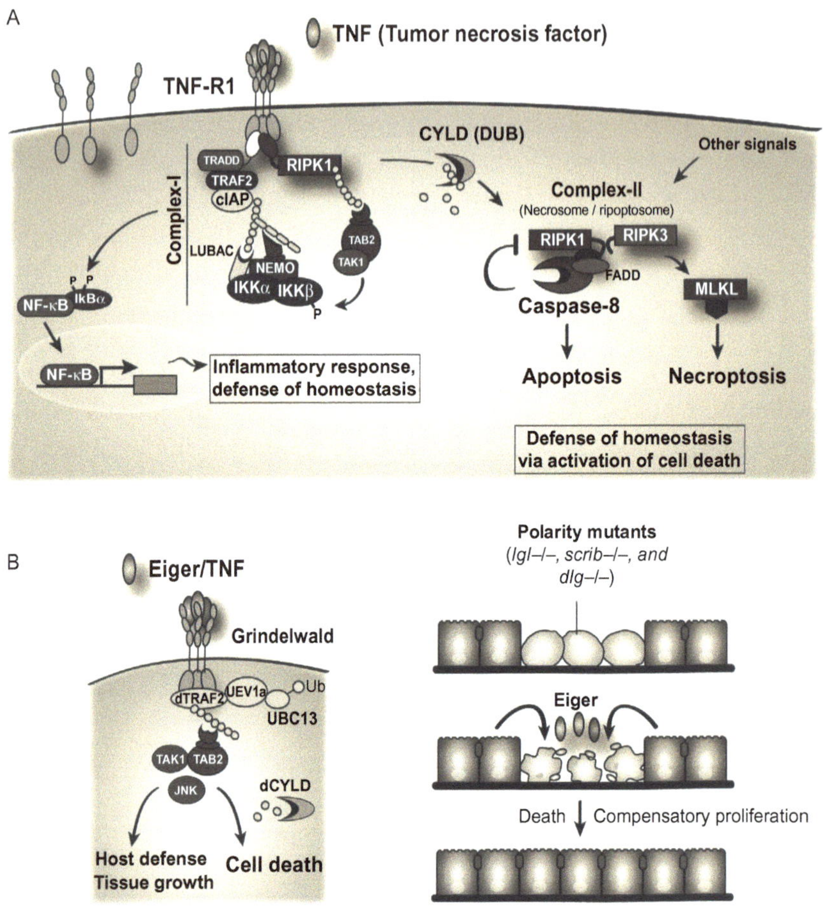

Figure 2 Evolutionary conservation of TNF-induced cell death and the defense of homeostasis. (A) TNF signaling and the transition of complex-I to complex-II. Upon TNF binding, cIAPs are recruited to the TNF-R1 signaling complex (complex-I) via TRADD/TRAF2. cIAPs ubiquitylate several molecules within the complex. RIPK1 ubiquitylation is the most readily observed. Ubiquitylation of components of complex-I, such as RIPK1, drives the recruitment of HOIL-1/HOIP/Sharpin that together form the Linear Ubiquitin Assembly Complex (LUBAC). LUBAC generates linear Ub chains on NEMO and RIPK1 that in turn recruits more NEMO molecules via its linear Ub-binding UBAN domain. NEMO is probably constitutively associated with IKKα/IKKβ and IKKβ is phosphorylated and activated by TAK1 that is independently recruited to ubiquitylated complex-I via its Ub receptors TAB2 and TAB3 that bind only to K63-linked Ub chains. Phosphorylated and activated IKKβ in turn phosphorylates IκBα, which leads to recruitment of a HECT E3 ligase. This E3 ligase promotes K48-linked ubiquitylation

to as complex-I (Micheau & Tschopp, 2003), consists of TNFR1, the adaptors TRADD, TRAF2, the kinase RIPK1, and the E3 Ub-ligases cellular Inhibitor of Apoptosis cIAP1 and cIAP2. Within this complex, RIPK1 is rapidly conjugated with poly-Ub chains by cIAPs. cIAP-mediated conjugation of Ub to RIPK1 allows subsequent recruitment of the Linear Ubiquitin chain Assembly Complex (LUBAC, composed of HOIL/HOIP/Sharpin), the kinase complexes TAK1/TAB2/TAB3, and IKK (composed of NEMO/IKKα/IKKβ) (Silke, 2011). Ub-dependent recruitment of LUBAC, TAK1/TAB2/TAB3, and IKKs is mediated by UBDs present in TAB2, NEMO, and HOIP. Once recruited, LUBAC then modifies NEMO and RIPK1 with M1-linked Ub chains, resulting in increased stability of the TNF signaling complex. Additionally, the binding of NEMO to M1-linked Ub chains causes a conformational change of the IKK complex that is thought to facilitate its activation (Rahighi et al., 2009). Complex-I then signals inflammation and cell survival through TAK1 and IkB kinase (IKK)-dependent activation of NF-κB. cIAPs are also required for JNK signaling (Gardam et al., 2011; Matsuzawa et al., 2008). This has been most clearly demonstrated for signaling that emanates from CD40, a TNF super family receptor, but similar concepts likely hold true for TNFR1 signaling too. Ultimately, this drives expression of a transcriptional program that restores homeostasis and lowers the apoptotic threshold through the induction of antiapoptotic molecules (Fig. 2).

Figure 2—Cont'd and proteasomal degradation of IκBα, allowing translocation of NF-κB subunits p50/p65 to drive production of cytokines. p50/p65 also promote expression of *IκBα*, to cause feedback inhibition, as well as genes such as *cFLIP* that are required to protect cells from complex-II-induced cell death. The numbered arrows provide a tentative indication of temporal sequence. Complex-II is most likely generated from complex-I, in an as yet undefined manner, and comprises RIPK1, FADD, and caspase-8. Deubiquitylation by CYLD is thereby a decisive step in the transition of complex-I to complex-II. Caspase-8 limits complex-II formation by cleaving and inactivating RIPK1. Consequently, loss of IAPs, LUBAC, or caspase-8 activity results in formation of complex-II that is able to drive necroptosis. Formation of complex-II, necrosome, or ripoptosome can also occur following stimulation of pattern recognition receptors or genotoxic stress. (B) Eiger-mediated signaling that regulates a variety of cellular and tissue processes, including the elimination of polarity mutant cells. Eiger mediates its effect through binding to its cognate receptor Grindelwald. This results in activation of JNK in a DTRAF2/Bendless/dUev1A-dependent manner. The *Drosophila* homologue of TAB2/3 (dTAB2) links TAK1 to the presumptive Ub chains conjugated by DTRAF2. dCYLD influences the decision as to whether JNK drives cell death or noncell death processes (see text for further details). (See the color plate.)

As its name suggests, TNF can also potently induce cell death. This is mediated by a RIPK1-based secondary complex that is frequently referred to as complex-II or necrosome (Pasparakis & Vandenabeele, 2015; Wang, Du, & Wang, 2008). This RIPK1-based complex can either kill via caspase-8 leading to apoptosis, or through RIPK3 and MLKL, which results in necroptosis. Under normal conditions, a small fraction of RIPK1 dissociates from complex-I within 30 min to 3 h, and together with TRADD associates with the adaptor protein FADD and procaspase-8 to form complex-II (Micheau & Tschopp, 2003) or necrosome (Pasparakis & Vandenabeele, 2015). Whether lethal levels of complex-II forms critically depend on the ubiquitylation status of RIPK1 in complex-I, RIPK1 deubiquitylation by deubiquitylating enzymes such as CYLD, or NF-κB-dependent expression of antiapoptotic genes, such as cFLIP or cIAP2. The formation and activity of complex-II is tightly regulated by cIAPs, LUBAC, NEMO, and TAK1. cIAPs, NEMO, and TAK1 not only suppress formation complex-II via activation of NF-κB but also thwart transition of complex-I to complex-II through mechanisms that are NF-κB-independent. It is believed that the Ub chains conjugated to RIPK1 by cIAP1/2 and LUBAC in complex-I constitute the decisive factor preventing RIPK1 from forming complex-II, and limiting its killing potential. The ability of IAPs to ubiquitylate RIPK1 and suppress its killing potential is antagonist by the deubiquitylating enzyme CYLD, which can cleave M1, and K63-linked Ub chains on RIPK1 (Komander et al., 2008). It is currently thought that deubiquitylation of RIPK1 drives complex-II formation and favors RIPK1-dependent apoptosis. However, when the levels of RIPK3 and MLKL are sufficiently high and caspase-8 activity is reduced, blocked or absent, complex-II can recruit and activate RIPK3, which in turn can drive MLKL-dependent necroptosis (Pasparakis & Vandenabeele, 2015). Caspase-8 together with $cFLIP_L$ reportedly inhibits necroptosis through cleaving RIPK1 and RIPK3 (Feng et al., 2007; Lin, Devin, Rodriguez, & Liu, 1999). In addition, caspase-8 can also cleave CYLD (O'Donnell et al., 2011), which removes Ub chains from RIPK1 and contributes to necroptosis *in vitro* and *in vivo*.

Ub-dependent regulation of TNF-induced cell death is evolutionary conserved, and is critically important for the maintenance of tissue fitness and the elimination of developmentally aberrant cells in *Drosophila* (Igaki & Miura, 2014). The *Drosophila* genome encodes a single member of the TNF family, named Eiger (Igaki et al., 2002; Moreno, Yan, & Basler, 2002; Fig. 2). Eiger is required for inducing the death of cells mutant for apico-basal polarity genes, such as *scribbled* (*scrib*), *discs large* (*dlg*), and *lethal giant larvae* (*lgl*). In the absence of Eiger, such mutant clones grow aggressively and develop into neoplastic tumors. This suggests that in *Drosophila* Eiger functions as part of a surveillance program that actively eliminates oncogenic polarity-deficient cells from the tissue (Igaki & Miura, 2014). Intriguingly, the Eiger-dependent elimination system is the result of cell competition that only operates when mutant cells are confronted with wild-type cells in a mosaic tissue. While *Drosophila* imaginal epithelium entirely mutant for *scribbled* or *discs large* results in tumorous overgrowth, such polarity-deficient oncogenic mutant cells, in an otherwise wild-type tissue, do not overgrow but are instead eliminated from the tissue by cell death.

Eiger-dependent cell elimination critically depends on Ub-dependent activation of TAK1 and JNK signaling, which in turn drives a cell death program that relies on inputs from *Drosophila* initiator caspase DRONC and the metabolic state of a cell (Igaki & Miura, 2014). Eiger-mediated cell death requires the E3 ligase TRAF2 and the K63-selective E2-heteromeric complex Bendless (Ubc13)/dUev1a (Ma et al., 2013, 2014). In addition, the E3 ligase NOPO (no poles), which is the *Drosophila* orthologue of mammalian TRAF-interacting protein (TRIP), also contributes to Eiger-induced cell death (Ma et al., 2012). Most likely, TRAF2 in conjunction with Bendless/dUev1a promotes the conjugation of K63-linked Ub chains that in turn allows recruitment and activation of TAK1 via its UBR TAB2. Activation of TAK1 subsequently results in activation of Hemipterous (also known as JNKK or MKK7) and the *Drosophila* JNK orthologue Basket (Igaki & Miura, 2014). Through a process that is molecularly ill defined, Eiger-mediated activation of Basket results in the induction of cell death signaling.

Intriguingly, Eiger-mediated cell death is strictly dependent on deubiquitylation. In the absence of the deubiquitylating enzyme CYLD, Eiger-mediated cell death and elimination of neoplastic tumors is blocked (Xue et al., 2007). Given that mammalian CYLD is an M1 and K63-selective deubiquitylating enzyme (Komander et al., 2008), and that the catalytic domains of *Drosophila* and mammalian CYLD are 53% identical, it is highly likely that the removal of nondegradative, K63-linked Ub chains is critical for the execution of cell death. The current literature suggests that CYLD promotes Eiger-induced cell death via the removal of degradative K48-poly-Ub chains on dTRAF2, thereby stabilizing TRAF2 and allowing efficient TRAF2-dependent activation of TAK1, Hep, and JNK (Xue et al.,

2007). Such a model is, however, inconsistent with the notion that TRAF2 and Bendless/Uev1a exclusively promote the formation of K63-linked non-degradative Ub chains, and that CYLD, at least in mammals, lacks affinity for K48 di-Ub chains and instead preferentially cleaves K63- and M1-Ub linkages (Komander, Clague, & Urbe, 2009).

Taken together, in both *Drosophila* and mammals, ligation of TNF and Eiger to their cognate receptors results in the transient formation of Ub-dependent signaling hubs that allow recruitment and activation of kinases via specialized adaptor molecules with UBDs. While assembly of these Ub-dependent signaling centers can mediate various cellular phenotypes ranging from cytokine production, proliferation, canalization, pain sensitization to host defense, TNF/Eiger-dependent cell death appears to be achieved through specific deubiquitylation events that tip the balance of TNF/Eiger signaling in favor of death.

4. UB-DEPENDENT REGULATION OF RIPK1 AND THE RIPOPTOSOME

In mammals, RIPK1-dependent cell death not only occurs in response to TNF but also operates downstream of many other cytokine receptors, damage-associated molecular pattern receptors, pathogen-associated molecular pattern receptors, or in response to genotoxic stress (Pasparakis & Vandenabeele, 2015; Fig. 2). This suggests that RIPK1 functions as a more generic stress sentinel. Under these conditions, RIPK1 assembles a protein complex containing the core components RIPK1, FADD, and caspase-8 (Feoktistova et al., 2011; Tenev et al., 2011). While these are the same components as the ones of complex-II (Micheau & Tschopp, 2003; Wang et al., 2008), the fact that this complex forms independently of TNFR1 indicates that it cannot constitute complex-II, which, per definition, originates from complex-I (Micheau & Tschopp, 2003; Wang et al., 2008). To uncouple it from the TNF-dependent complex-II, TNFR1-independent assembly of the RIPK1/FADD/caspase-8 complex is, therefore, better referred to as "ripoptosome" (Feoktistova et al., 2011; Tenev et al., 2011). Although the core of this complex consists of RIPK1, FADD, and caspase-8, the ripoptosome can also include additional proteins such as FLIP and RIPK3, depending on cell type and stimulus (Green, Oberst, Dillon, Weinlich, & Salvesen, 2011).

Proper regulation of RIPK1, and the ripoptosome, is critically important for normal development and physiology (Bonnet et al., 2011; Dannappel et al., 2014; Declercq et al., 2009; Dondelinger et al., 2013; Duprez et al., 2011; Ermolaeva et al., 2008; Green et al., 2011; Oberst et al., 2011; O'Donnell et al., 2011; Pasparakis & Vandenabeele, 2015; Polykratis et al., 2014; Vandenabeele, Galluzzi, Vanden Berghe, & Kroemer, 2010; Welz et al., 2011). Consistently, genetic deletion of RIPK1 in mouse models results in lethality due to systemic inflammation. Although cell death is frequently considered to be the result of inflammation, recent evidence suggest that cell death may in fact precede, trigger, or amplify the inflammatory response. Indeed, neither apoptosis nor necroptosis are "silent" processes. Although caspase-mediated apoptosis is widely thought to be immunologically inert or even tolerogenic, it is now clear that activation of caspases not only contributes to apoptosis but also leads to the generation of paracrine signals that ensure tissue homeostasis and coordinate tissue repair (Fuchs & Steller, 2011; Martin, Henry, & Cullen, 2012). Further, a complex relationship exists between caspase activation, apoptosis, and necroptosis (Green et al., 2011). Accordingly, necroptosis typically occurs under circumstances in which caspase activation is absent or blocked. Because necroptotic death results in the spillage of cytosolic components and alarmins, this form of death triggers secondary inflammation. The important physiological role of necroptosis was highlighted by a number of genetic studies showing that caspase-8 or FADD deficiency causes embryonic lethality and triggers inflammation *in vivo* by sensitizing cells to RIPK3-mediated necroptosis (Bonnet et al., 2011; Dannappel et al., 2014; Declercq et al., 2009; Dondelinger et al., 2013; Duprez et al., 2011; Ermolaeva et al., 2008; Green et al., 2011; Oberst et al., 2011; O'Donnell et al., 2011; Pasparakis & Vandenabeele, 2015; Polykratis et al., 2014; Vandenabeele et al., 2010; Welz et al., 2011). The activity of RIPK3 is subject to modulation by RIPK1. This is achieved through the RIP homotypic interaction motif present in both RIPK1 and RIPK3 that allows the formation of RIPK1:RIPK3 protein complexes. The recruitment of RIPK1 to RIPK3 not only allows activation of RIPK3 but also permits RIPK1-dependent negative regulation of RIPK3, most likely via recruitment of caspase-8/FLIP heterodimers, which cleave and inactivate RIPK3 (Feng et al., 2007).

Formation and activity of the ripoptosome is also subject to tight regulation by multiple members of the IAP protein family. Of the IAPs, cIAP1 and cIAP2 are the most critical regulators of ripoptosome assembly (Feoktistova et al., 2011; Geserick et al., 2009; Tenev et al., 2011). Nevertheless, XIAP also contributes to the regulation of this RIPK1-based platform because in the absence of XIAP, depletion of cIAPs results in increased assembly of this complex (Tenev et al., 2011). The extent to which individual IAPs contribute to the inhibition of ripoptosome assembly most likely depends upon cell type and stimulus. IAP-mediated inactivation of RIPK1 and/or ripoptosome occurs in an Ub-dependent manner, most likely by targeting RIPK1, and other components of the ripoptosome, for proteasomal degradation. Caspase-8-mediated cleavage of cFLIP generates cFLIP(p43), permitting its binding to TRAF2 and the formation of cFLIP(p43)–caspase-8–TRAF2 tertiary complex (Kataoka & Tschopp, 2004; Micheau et al., 2002). TRAF2 then recruits cIAPs which target cFLIP(p43) and caspase-8 for ubiquitylation (Tenev et al., 2011). This indicates that cIAP1 and cIAP2 target "active" cFLIP–caspase-8 complexes for ubiquitylation and inactivation. The importance of IAP-mediated regulation of RIPK1-based death complexes is illustrated by the notion that $Xiap^{-/-}Ciap1^{-/-}$ and $Ciap1^{-/-}Ciap2^{-/-}$ animals are embryonic lethal, and that this lethality is rescued by crossing the $Xiap^{-/-}Ciap1^{-/-}$ and $Ciap1^{-/-}Ciap2^{-/-}$ mice to $Ripk1^{-/-}$ and $Ripk3^{-/-}$ mice (Moulin et al., 2012). This demonstrates that multiple IAPs function together to regulate an embryonic decision point involving RIP kinase activity.

5. IAP-MEDIATED REGULATION OF CASPASES

In both *Drosophila* and mammals, members of the IAP protein family are the most prominent E3 ligases that modulate caspases and apoptosis. Apoptosis regulatory IAPs carry either two or three NH_2-terminal BIR domains and a C-terminal RING finger that provides them with Ub E3 ligase activity (Silke & Meier, 2013). The BIR domain mediates protein interactions, and in most cases, binds to IAP-binding motifs (IBMs) present in active caspases and IAP antagonists such as mammalian Smac/DIABlO and Omi/HtrA2 or *Drosophila* REAPER, GRIM, and Head Involution Defective (Shi, 2002a). The main feature of an IBM is the presence of an NH_2-terminal alanine (Fig. 3). However, in some cases, IBMs can also harbor a serine at the first position (Verhagen et al., 2007). The NH_2-terminal alanine or serine, which must be exposed and unblocked (devoid of NH_2-terminal acetylation), inserts into the extensive hydrophobic cleft on the surface of BIRs and forms hydrogen bonds with neighboring residues, thereby anchoring the IBM-carrying protein to IAPs (Wu et al., 2000). Subtle changes in the peptide-binding groove of BIR domains alter their preference for particular client proteins with IBMs. Therefore, proteins with IBMs

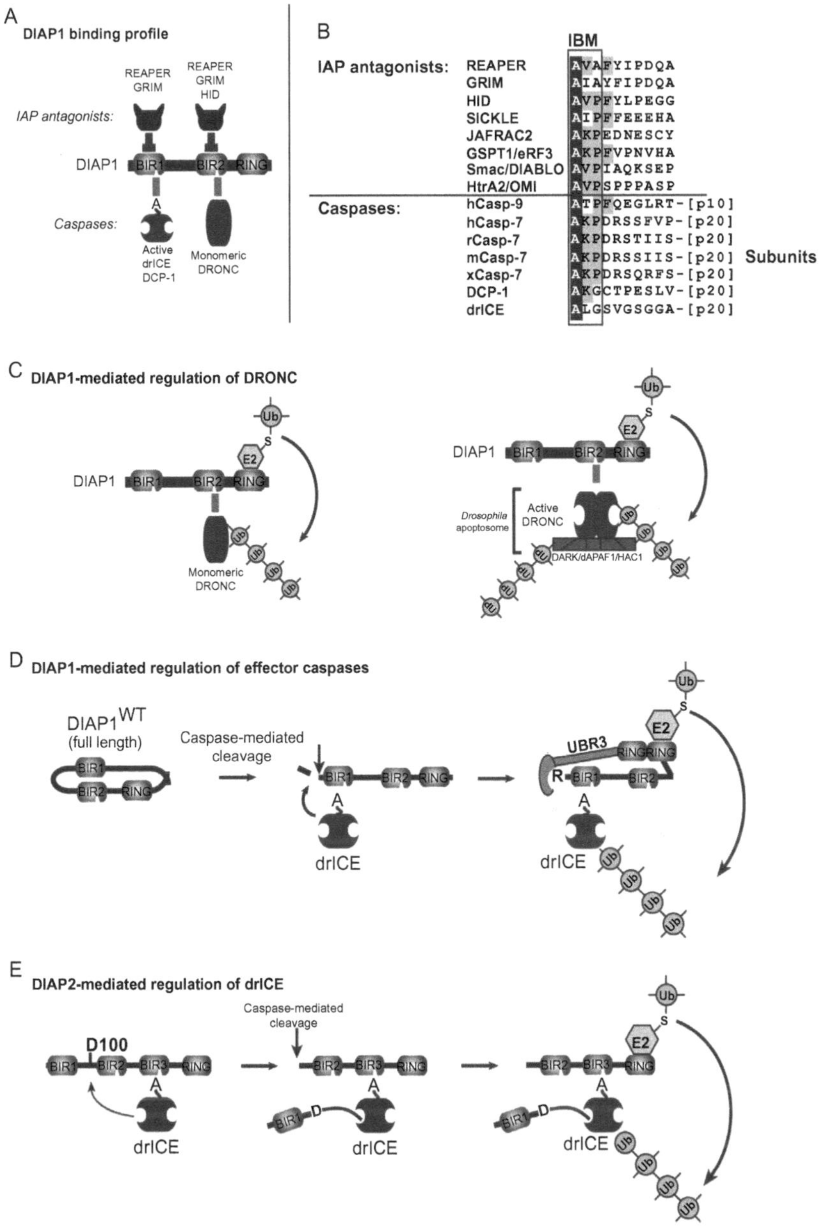

Figure 3 IAP-mediated regulation of caspases in *Drosophila*. (A) Binding profile of DIAP1 with caspases and IAP antagonists. Direct physical interaction with the effector caspases drICE or DCP-1 and the initiator caspase DRONC is mediated through DIAP1's BIR1 and BIR2 domains, respectively. Following their activation, drICE and DCP-1 expose an NH_2-terminal IBM (depicted as A), which allows their binding to BIR1. (B) Sequence *(Continued)*

display differential and selective binding to specific BIR domains. Apoptosis-regulatory IAPs such as XIAP, cIAP1, cIAP2, and *Drosophila* IAP1 (DIAP1) and DIAP2 carry two BIR domains capable of binding IBMs. The tandem arrangement (i) increases the repertoire of proteins with which they can interact and (ii) potentially enhances the binding affinity to particular IBM-containing target proteins, particularly when they are dimeric or oligomeric in nature.

In *Drosophila*, DIAP1, encoded by the *thread* (*th*) gene, was the first BIR-containing IAP identified. DIAP1-mediated inhibition of caspases is essential for cell survival as loss of DIAP1 function instigates spontaneous caspase-mediated cell death (Goyal, McCall, Agapite, Hartwieg, & Steller, 2000; Lisi, Mazzon, & White, 2000; Wang, Hawkins, Yoo, Muller, & Hay, 1999). Conversely, gain-of-function mutations significantly suppress cell death, and lead to overgrowth phenotypes due to supernumerary cells (Goyal et al., 2000; Lisi et al., 2000; Wang et al., 1999).

The BIR domain, in combination with its flanking regions, functions as a protein interaction module that, for DIAP1, mediates binding to both initiator (DRONC) and effector caspases (drICE and DCP-1; Fig. 3). Importantly, different caspases bind to distinct BIRs: while the BIR1 region of DIAP1 is essential for binding to the effector caspases drICE and DCP-1 (Hawkins, Wang, & Hay, 1999; Kaiser, Vucic, & Miller, 1998; Zachariou et al., 2003), the BIR2 region directly associates with the initiator caspase DRONC (Meier, Silke, Leevers, & Evan, 2000). As a consequence of this differential binding, one molecule of DIAP1 can bind simultaneously

Figure 3—Cont'd alignment of IBM-bearing proteins. Identical residues are highlighted in black. Residues conserved in four or more IBM proteins are indicated in gray. (C) DIAP1's BIR2-DRONC association is essential for DIAP1 to neutralize DRONC. Following binding, DIAP1's RING finger promotes Ub conjugation of DRONC, leading to its inactivation through nondegradative ubiquitylation of monomeric DRONC (left panel), and by targeting apoptosome-associated active DRONC for degradation (right panel). Mechanism of effector caspase (drICE) inactivation by DIAP1 (D) and DIAP2 (E). (D) Full-length wild-type DIAP1 is held in an inactive conformation and requires caspase-mediated proteolytic cleavage at residue 20 for its activation. After cleavage, BIR-mediated caspase binding occurs more efficiently. Cleavage also facilitates recruitment of N-end rule UBR E3 ligases, which together with DIAP1's RING domain promote ubiquitylation and inactivation of drICE and DCP-1. (E) drICE is also subject to regulation by DIAP2. drICE binds to the BIR3 of DIAP2 in an IBM-dependent manner, and following binding cleaves DIAP2 at D100. DIAP2 cleavage results in a covalent adduct between D100 and the catalytic machinery of drICE, trapping the caspase. Full inactivation of drICE is achieved through RING-mediated ubiquitylation. (See the color plate.)

to DRONC and drICE or DCP-1 (Zachariou et al., 2003). Physical association between DIAP1 and effector caspases is essential for cell survival. Embryos homozygous for *diap1* loss-of-function mutations that completely abrogate binding to effector or initiator caspases die during embryogenesis due to inappropriate cell death (Goyal et al., 2000; Lisi et al., 2000; Rodriguez, Chen, Oliver, & Abrams, 2002; Wang et al., 1999; Zachariou et al., 2003). On the other hand, mutations that enhance DIAP1's ability to associate with activated effector caspases result in a gain-of-function phenotype (Goyal et al., 2000; Zachariou et al., 2003). Therefore, DIAP1:caspase association represents a pivotal step in the regulation of the apoptotic caspase cascade.

The mechanism of caspase binding differs greatly depending on the caspase involved (Fig. 3). While zymogenic DRONC readily binds to the BIR2 region of DIAP1, DIAP1 only associates with proteolytically cleavage forms of drICE/DCP-1. As proteolytic cleavage and removal of the prodomain of drICE or DCP-1 results in activation of the effector caspases, this indicates that DIAP1 only regulates active versions of drICE and DCP-1. The mechanism behind this selectivity resides in the exposure of an IBM at the neo-NH_2 terminus of drICE and DCP-1, which is uncovered following cleavage and activation of these effector caspases (Tenev, Zachariou, Wilson, Ditzel, & Meier, 2005). Although IAP:caspase association is the decisive step in the regulation of apoptosis in *Drosophila*, physical interaction between DIAP1 and caspases alone is insufficient to regulate caspases. This is evident because DIAP1-bound effector caspases remain catalytically active under *in vitro* conditions (Tenev et al., 2005). Moreover, DIAP1 mutants with a dysfunctional RING finger completely fail to suppress caspase-mediated cell death, even though these proteins bind to caspases with the same affinity as their wild-type counterparts. Ultimately, suppression of caspases and apoptosis results from DIAP1-mediated ubiquitylation of the zymogenic form of DRONC and active drICE or DCP-1 (Chai et al., 2003; Ditzel et al., 2008; Lisi et al., 2000; Wilson et al., 2002).

The mechanism by which ubiquitylation of DRONC causes its inactivation appears to be context dependent, involving degradative as well as nondegradative ubiquitylation. Outside of the apoptosome, DIAP1-mediated ubiquitylation of DRONC neutralizes it through an unknown mechanism that operates independent of the proteasome (Lee et al., 2011; Fig. 3). However, when part of the apoptosome, DIAP1 conjugates K48-linked poly-Ub chains to DRONC, targeting it for proteasomal destruction (Shapiro, Hsu, Jung, Robbins, & Ryoo, 2008). Hence, only apoptosome-associated DRONC (as well as DARK itself), but not free monomeric DRONC, is targeted for proteasomal degradation. Interestingly, DRONC-mediated cleavage of DARK is required for proteasomal degradation of the DRONC/DARK complex. This suggests that the cleavage event recruits the E3 ligase (Shapiro et al., 2008).

DIAP1-mediated regulation of effector caspases is also dependent on the conjugation of Ub. Attachment of nondegradative (K63-linked) poly-Ub chains to the effector caspase drICE (homologue of caspase-3/-7) directly reduces its proteolytic potency, affecting kinetic parameters of the enzyme (Ditzel et al., 2008). Computational modeling of an ubiquitylated effector caspase suggests that the Ub chains sterically occlude the catalytic pocket of the caspase and would interfere with substrate entry.

In addition to Ub, DIAP1 can also inactivate effector caspases via the covalent attachment of the Ub-like modifier NEDD8. NEDD8-mediated suppression of drICE occurs via a mechanism that relies on noncompetitive inhibition, most likely through a NEDD8-induced conformational change of the caspase. Disruption of drICE ubiquitylation or NEDDylation, either by loss of DIAP1's E3 activity or generation of a nonmodifyable form of drICE, renders this effector caspase resistant to DIAP1-mediated inactivation (Broemer et al., 2010; Ditzel et al., 2008).

Surprisingly, DIAP1 in its full-length form is incapable of binding and regulating caspases, or acting as an E3 ligase. To activate its antiapoptotic potential, it requires proteolytic cleavage. Removal of the first 20 amino acid residues of DIAP1 radically changes its properties: the cleaved form interacts with caspases far better than full-length, noncleaved DIAP1. It seems that full-length DIAP1 resides in an inactive, "closed" configuration that precludes caspase binding (Yan, Wu, Chai, Li, & Shi, 2004). Only when it is cleaved can it bind tightly to DRONC or effector caspases, and function as an E3 (Ditzel et al., 2008). In this respect, caspases activate their own inhibition in a regulatory feedback loop.

Cleavage of DIAP1 not only removes the presumptive inhibitory NH_2-terminal portion of the protein, but it also exposes a new docking site for UBR-containing E3 ligases of the NH_2-end rule pathway (Ditzel et al., 2003; Herman-Bachinsky, Ryoo, Ciechanover, & Gonen, 2007; Fig. 3). DIAP1's two ubiquitylation-associated activities—UBR-E3 recruitment and DIAP1's own RING finger—are both required for its full antiapoptotic activity. The presence of a functional RING alone seems not to be sufficient, since a DIAP1 mutant that retains a functional RING but fails to bind UBRs also fails to protect from cell death induced by ectopic expression of proapoptotic proteins. Likewise, DIAP1 that has a defective RING but is otherwise fully competent in recruiting UBR-E3s also fails to regulate apoptosis properly. Among the different UBRs, UBR3 appears to be particularly important for the regulation of caspases (Huang et al., 2014), as RNAi-mediated knockdown of *ubr3* leads to caspase activation and cell death. Intriguingly, in the absence of *ubr3*, overexpression of DIAP1 fails to suppress caspases, corroborating the notion that DIAP1 relies on recruitment of the NH_2-end rule system to suppress caspase activity. Therefore, UBR recruitment and DIAP1's RING are both required for DIAP1's full anti-apoptotic activity (Ditzel et al., 2008, 2003; Muro, Means, & Clem, 2005).

drICE, but not DRONC or DCP-1, is also regulated by the second *Drosophila* IAP, DIAP2. Intriguingly, DIAP2 functions as a mechanism-based regulator of drICE, acting as a pseudosubstrate, which, following cleavage, traps the active caspase via a covalent linkage between DIAP2 and the catalytic machinery of drICE (Ribeiro et al., 2007; Fig. 3). Despite direct inhibition of drICE's catalytic cysteine, DIAP2's E3 ligase activity also contributes to proper drICE inhibition (Ribeiro et al., 2007). Despite the ability of DIAP2 to regulate drICE, this function seems not to be essential for cell viability as *diap2* mutant animals are fertile and fully viable. A cell death phenotype in *diap2*-mutant animals is only revealed when flies are subjected to ionizing radiation (Ribeiro et al., 2007).

While DIAP2-mediated regulation of caspases is less important for the regulation of cell death, recent evidence indicate that the ability of DIAP2 to regulate caspases is essential for proper regulation of the *Drosophila* innate immune response to infection with Gram-negative bacteria. DIAP2-mediated conjugation of K63-linked Ub chains on DREDD, the *Drosophila* orthologue of caspase-8, thereby allows Ub-mediated aggregation and activation of DREDD (Meinander et al., 2012; Fig. 4). Active DREDD subsequently cleaves IMD (*immune deficiency*; Paquette et al., 2010). Upon cleavage, IMD exposes an IBM at its neo-NH_2 terminus, which binds to the BIR2/3 of DIAP2. This provides DIAP2 with an additional docking site, reinforcing complex stability and allowing DIAP2-mediated ubiquitylation of IMD, and quite possibly other components of the signaling complex (Paquette et al., 2010). The Ub chains on IMD and DREDD appear to serve as scaffolds for the recruitment of dTAK1, IKK, and the precursor form of the NF-κB transcription factor RELISH (Ferrandon, Imler, Hetru, & Hoffmann, 2007; Kanayama et al., 2004; Kleino et al., 2005; Lu, Wu, & Anderson, 2001; Rutschmann et al., 2000; Rutschmann, Kilinc, &

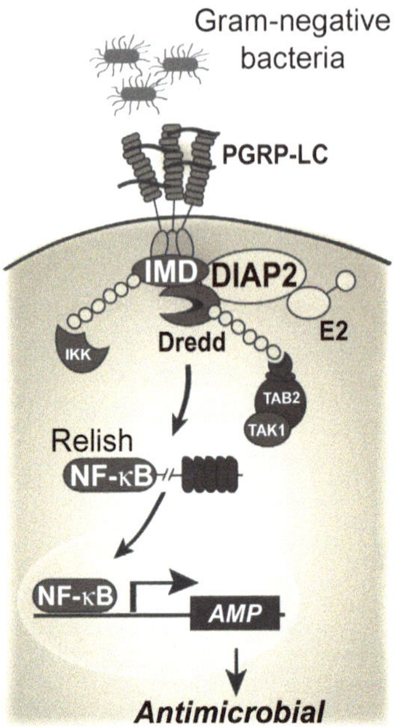

Figure 4 Model depicting Ub-dependent activation of NF-κB during *Imd* signaling. Ligand binding induces recruitment of Imd and dFADD to PGRP-LC. This allows binding of DREDD to dFADD, which leads to dimerization-induced activation of DREDD. Active DREDD subsequently cleaves IMD, which leads to the exposure of an IBM at the neo-amino terminus of cleaved IMD. This in turn allows recruitment of DIAP2 into the signaling complex, followed by DIAP2-mediated ubiquitylation of various components of this complex, such as IMD and DREDD. Ubiquitylation of DREDD is key for DREDD-mediated cleavage of RELISH. The Ub chains of DREDD may serve as a scaffold for the recruitment of IKK. Since IKK binds to RELISH, as evidenced by its ability to phosphorylate RELISH, IKK might bring RELISH into close proximity of DREDD for proteolysis. Ubiquitylation of components of the IMD pathway also leads to activation of TAK1 and IKK, which in turn phosphorylates RELISH. Phosphorylated and cleaved RELISH subsequently translocates to the nucleus where it drives expression of RELISH target genes.

Ferrandon, 2002; Silverman et al., 2003, 2000; Vidal et al., 2001; Zhuang et al., 2006). This brings RELISH into close proximity of ubiquitylated and active DREDD, allowing DREDD-mediated proteolysis of Relish. The proximity to the signaling complex also allows phosphomediated activation of RELISH (Erturk-Hasdemir et al., 2009). Subsequently, cleaved and phosphoryated RELISH translocates to the nucleus where it drives expression of antimicrobial peptide genes.

While ubiquitylation of DREDD and IMD is essential for activation of RELISH, several deubiquitylating enzymes have been identified that negatively regulate the ability of IMD and DREDD to drive induction of RELISH-dependent target genes (Engel et al., 2014; Thevenon et al., 2009). Among these are USP36, USP2, and USP34, which prevent inappropriate activation of IMD-dependent immune signaling in unchallenged conditions.

Like DIAP2, mammalian XIAP similarly functions both as mechanism-based caspase inhibitor and potent regulator of Ub-dependent activation of NF-κB (Fig. 5). XIAP-mediated inactivation of caspase-3, -7, and -9 does not require a functional RING finger under *in vitro* conditions or when overexpressed (Shi, 2002b). Residues located immediately upstream of XIAP's BIR2 domain directly bind to the active site pocket of caspase-3 and -7 and obstruct substrate entry (Chai et al., 2001; Huang et al., 2001; Riedl et al., 2001; Silke et al., 2001; Suzuki, Nakabayashi, Nakata, Reed, & Takahashi, 2001). XIAP-mediated inactivation of caspase-9 occurs differently, namely by keeping caspase-9 in a monomeric, inactive state (Shiozaki et al., 2003).

Even though the RING domain of XIAP is not required for caspase inhibition *in vitro*, it contributes to XIAP's function *in vivo*. A recent report now finds that XIAP's RING indeed is critical for its antiapoptotic function (Schile, Garcia-Fernandez, & Steller, 2008). Using gene targeting of endogenous XIAP, the authors show that deletion of the RING finger sensitized fibroblasts to TNFα-induced cell death and led to increased rates of apoptosis in an Eμ-Myc mouse lymphoma model. In both these systems, $XIAP^{\Delta RING}$ mutant cells responded in the same way as XIAP null cells, even though $XIAP^{\Delta RING}$ protein was abundantly expressed. Moreover, following apoptosis induction caspase activity was significantly higher in $XIAP^{\Delta RING}$ cells compared to wild-type controls. This indicates that the BIR domains are not sufficient on their own to block caspase activity *in vivo*. Consistent with a requirement of the RING finger, caspase-3 poly-ubiquitylation was reduced in mutant cells. Remarkably, no increased levels of caspase-3 were observed, suggesting a nondegradative mode of caspase ubiquitylation, a finding that contradicts an earlier study (Suzuki, Nakabayashi, & Takahashi, 2001). In addition, XIAP reportedly also ubiquitylates caspase-9 (Morizane, Honda, Fukami, & Yasuda, 2005), though the functional outcome of this is yet to be determined.

As indicated above, XIAP not only regulates cellular processes by controlling caspases but also by mediating activation of NF-κB. In particular, XIAP is indispensable for innate immune signaling triggered by NOD1 and NOD2 (Bauler, Duckett, & O'Riordan, 2008; Bertrand et al., 2009; Krieg et al., 2009). NODs belong to the nucleotide-binding oligomerization

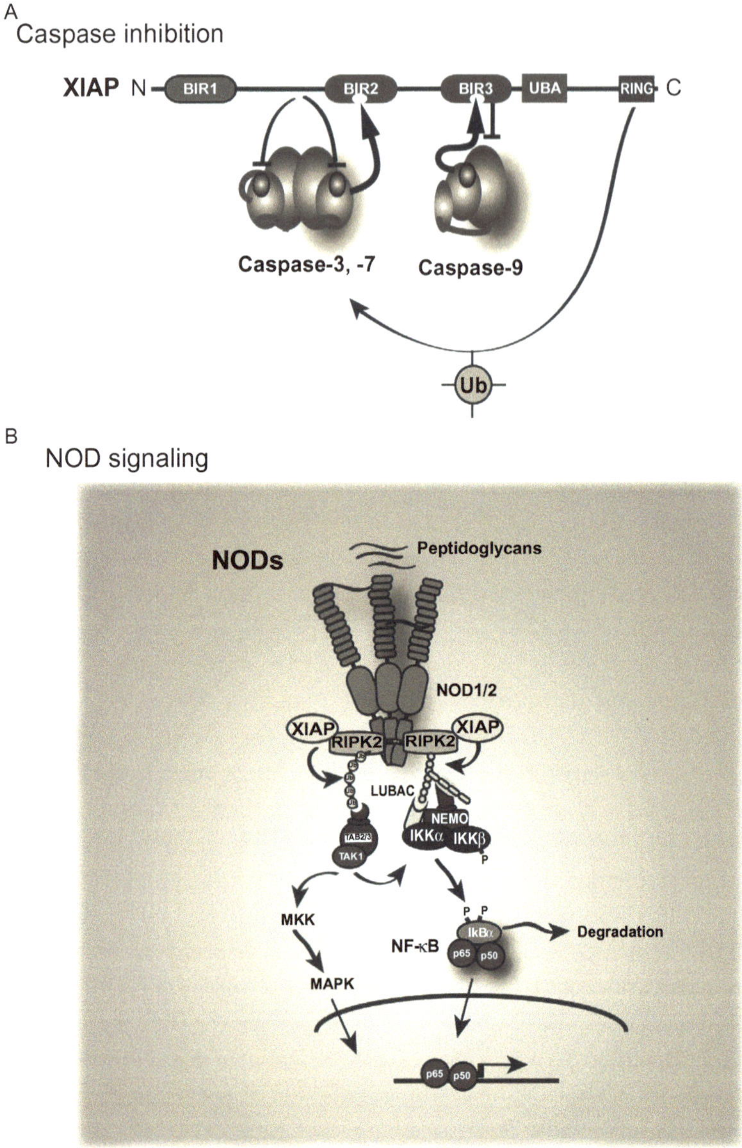

Figure 5 XIAP-mediated regulation of caspases and NF-κB. (A) XIAP directly inhibits the effector caspase-3 and -7, and the initiator caspase-9. The sequence preceding the BIR2 domain of XIAP occupies the catalytic pocket of caspase-3 or -7, thereby blocking substrate entry. In addition, the BIR2 domain interacts with the IBM of caspase-3 or -7 that is exposed following their proteolytic activation (shown as an arrow). XIAP-mediated inhibition of caspase-9 requires proteolytic cleavage of caspase-9, which exposes an IBM that binds to the BIR3 of XIAP. Caspase-9 activity is blocked because XIAP prevents caspase-9 dimerization, a prerequisite for initiator caspase activity. The RING domain of XIAP also contributes to caspase inhibition. (B) NOD-mediated activation of NF-κB and MAPK signaling. Detection of bacterial peptidoglycans by NOD1 and NOD2 results in the formation of an oligomeric signaling complex that recruits RIPK2 and XIAP. XIAP-mediated ubiquitylation of RIPK2 allows the recruitment of TAB2/TAB3/TAK1 and IKKs, thereby triggering NF-κB and MAPK signaling. (See the color plate.)

domain family that contain an NH_2-terminal caspase recruitment domain, a centrally located nucleotide-binding and oligomerization domain, and 10 tandem leucine-rich repeats (LRRs) in their C-terminus through which they detect bacterial peptidoglycans shed during bacterial growth (Chen, Shaw, Kim, & Nunez, 2009).

The pathways activated by NODs are highly reminiscent to the Ub-dependent signal transduction cascade downstream of TNF-R1 (Fig. 5). Like TNF-R1, NODs stimulate Ub-mediated activation of NF-κB and the MAP kinases p38 and JNK. Stimulation of NOD1 and NOD2 receptors triggers the formation of a NOD-dependent multiprotein complex that recruits RIPK2, a close homologue of RIPK1 (Inohara et al., 2000). XIAP in conjunction with LUBAC then conjugates RIPK2 with M1- and K63-linked Ub chains (Bertrand et al., 2009; Hasegawa et al., 2008; Yang et al., 2007), which subsequently result in TAK1/TAB2/3-mediated activation of MAPK and NF-κB pathways leading to production of cytokines, chemokines, and antimicrobial peptides that help to defend against invading microbes and ensure the defense of homeostasis (Bertrand et al., 2009; Hasegawa et al., 2008; Park et al., 2007; Yang et al., 2007).

Although XIAP is involved in regulating NOD signaling, XIAP deficiency manifests in a wide range of clinical immune phenotypes, including EBV-associated hemophagocytic lymphohistiocytosis, Crohn-like bowel disease, severe infectious mononucleosis, splenomegaly, uveitis, periodic fever, fistulating skin abscesses, and severe Giardia enteritis (Speckmann et al., 2013). These clinical phenotypes are not shared by mutations in other components of the NOD-dependent multiprotein complex, and thus suggest other immune-related roles of XIAP.

IAPs other than cIAPs and XIAP also play important roles in regulating apoptosis as well as nonapoptotic signaling events in an Ub-dependent fashion. BRUCE/Apollon is a membrane-associated IAP that carries only one BIR domain. Additionally, it also contains an UBC motif that can function as an Ub-E2, transferring Ub to substrates. In addition to contributing to cytokinesis, BRUCE/Apollon also safeguards cell viability by targeting caspase-9, and the IAP-antagonist protein SMAC/Diablo, for Ub-mediated proteasomal degradation (Bartke, Pohl, Pyrowolakis, & Jentsch, 2004; Hao et al., 2004). In *Drosophila*, the activity of dBRUCE is indispensable for controlled activation of caspases required for spermatide individualization (Arama, Agapite, & Steller, 2003). Further, dBRUCE also targets the IAP-antogonist Reaper and Grim for proteasomal degradation, thereby contributing to the apoptotic threshold (Domingues & Ryoo, 2012; Vernooy et al., 2002).

Although IAPs are the most prominent class of E3 ligases that target caspases for ubiquitylation, IAPs are not the only E3s that control caspases. Components of a testis-specific Cullin-3-based E3 Ub-ligase complex are particularly important for caspase activation during sperm differentiation (Arama, Bader, Rieckhof, & Steller, 2007; Bader, Arama, & Steller, 2010; Kaplan, Gibbs-Bar, Kalifa, Feinstein-Rotkopf, & Arama, 2010). Defects in any of the Cullin-3 components decrease caspase activity and cause male sterility due to individualization defects. Similarly, caspases are activated during mammalian spermatogenesis and disrupting the cell death machinery results in infertile males due to defects during late spermatid maturation. In *Drosophila*, spermatid individualization is a process in which 64 interconnected spermatids separate from one another and eliminate the majority of their cytoplasmic contents (Feinstein-Rotkopf & Arama, 2009). This process requires the activity of Cullin-3, the substrate-binding BTB protein KLHL10, and the pseudosubstrate inhibitor Soti (Arama et al., 2007; Bader et al., 2010; Kaplan et al., 2010). Cullin-3/KLHL10 allows sublethal activation of caspases by targeting the caspase inhibitor dBRUCE for ubiquitylation and proteasomal degradation. Interestingly, it was recently reported that mutations in the human form of KLHL10 are associated with male infertility and low sperm count (Yatsenko et al., 2006), indicating that ubiquitylation may be used in a similar fashion for cell sculpting in mammals.

6. CONCLUDING REMARKS

The maintenance of homeostasis in multicellular organisms depends on a continuous, coordinated response to external and internal insults that challenge cellular and tissue integrity throughout life. Loss of homeostasis is a hallmark of aging, resulting in pathologies often caused by defective or deregulated tissue damage responses. Inflammation and cell death are essential defense responses that are induced by infection or injury. However, inflammation and cell death are also induced by tissue stress and malfunction to maintain tissue homeostasis under a variety of noxious conditions. A number of dedicated sensors have evolved to detect different stressors and induce appropriate adaptive responses. Common to these pathways is the conjugation of nondegradative Ub chains that produce robust signaling networks, which coordinate tissue remodeling and adaptation to tissue stress via transcriptional programs, the induction of apoptosis or necroptosis. Most, if not all, cellular stress responses, in addition to cell-autonomous adaptive changes, produce secreted factors that affect other cells in the tissue.

This coordinates tissue remodeling that replaces malfunctioning or damaged tissues (Grivennikov, Greten, & Karin, 2010).

Whatever the cause of the inflammatory response, its "purpose" is to remove the source of the disturbance, to allow the host to adapt to the abnormal conditions, and, ultimately, to restore functionality and homeostasis to the tissue. Much more needs to be learnt about how the Ub-signaling system impacts on inflammation and adaptation to tissue stress. Further, unraveling how the Ub signal is conjugated, recognized, and disassembled will be critically important to gain a better understanding of the regulatory processes that control cell death, inflammation, and stress responses, and might contribute to disease pathologies such as cancer.

ACKNOWLEDGMENTS

We regret not being able to discuss many studies relevant to this topic. P. M. and O. M. are supported by funds from Breakthrough Breast Cancer and The Institute of Cancer Research, and M. B. is supported by the Deutsche Forschungsgemeinschaft (DFG) (BR3442/2-1). We acknowledge NHS funding to the NIHR Biomedical Research Centre.

REFERENCES

Amoyel, M., & Bach, E. A. (2014). Cell competition: How to eliminate your neighbours. *Development*, *141*(5), 988–1000. http://dx.doi.org/10.1242/dev.079129.

Arama, E., Agapite, J., & Steller, H. (2003). Caspase activity and a specific cytochrome C are required for sperm differentiation in Drosophila. *Developmental Cell*, *4*(5), 687–697.

Arama, E., Bader, M., Rieckhof, G. E., & Steller, H. (2007). A ubiquitin ligase complex regulates caspase activation during sperm differentiation in Drosophila. *PLoS Biology*, *5*(10), e251. http://dx.doi.org/10.1371/journal.pbio.0050251.

Bader, M., Arama, E., & Steller, H. (2010). A novel F-box protein is required for caspase activation during cellular remodeling in Drosophila. *Development*, *137*(10), 1679–1688. http://dx.doi.org/10.1242/dev.050088.

Balkwill, F. (2009). Tumour necrosis factor and cancer. *Nature Reviews Cancer*, *9*(5), 361–371. http://dx.doi.org/10.1038/nrc2628. nrc2628 [pii].

Bartke, T., Pohl, C., Pyrowolakis, G., & Jentsch, S. (2004). Dual role of BRUCE as an anti-apoptotic IAP and a chimeric E2/E3 ubiquitin ligase. *Molecular Cell*, *14*(6), 801–811.

Bauler, L. D., Duckett, C. S., & O'Riordan, M. X. (2008). XIAP regulates cytosol-specific innate immunity to Listeria infection. *PLoS Pathogens*, *4*(8), e1000142. http://dx.doi.org/10.1371/journal.ppat.1000142.

Bertrand, M. J., Doiron, K., Labbe, K., Korneluk, R. G., Barker, P. A., & Saleh, M. (2009). Cellular inhibitors of apoptosis cIAP1 and cIAP2 are required for innate immunity signaling by the pattern recognition receptors NOD1 and NOD2. *Immunity*, *30*(6), 789–801. http://dx.doi.org/10.1016/j.immuni.2009.04.011. S1074-7613(09)00203-9 [pii].

Bhoj, V. G., & Chen, Z. J. (2009). Ubiquitylation in innate and adaptive immunity. *Nature*, *458*(7237), 430–437. http://dx.doi.org/10.1038/nature07959. nature07959 [pii].

Bondar, T., & Medzhitov, R. (2010). p53-mediated hematopoietic stem and progenitor cell competition. *Cell Stem Cell*, *6*(4), 309–322. http://dx.doi.org/10.1016/j.stem.2010.03.002.

Bonnet, M. C., Preukschat, D., Welz, P. S., van Loo, G., Ermolaeva, M. A., Bloch, W., et al. (2011). The adaptor protein FADD protects epidermal keratinocytes from necroptosis in vivo and prevents skin inflammation. *Immunity*, *35*(4), 572–582. http://dx.doi.org/10.1016/j.immuni.2011.08.014. S1074-7613(11)00401-8 [pii].

Broemer, M., Tenev, T., Rigbolt, K. T., Hempel, S., Blagoev, B., Silke, J., et al. (2010). Systematic in vivo RNAi analysis identifies IAPs as NEDD8-E3 ligases. *Molecular Cell*, *40*(5), 810–822. http://dx.doi.org/10.1016/j.molcel.2010.11.011. S1097-2765(10)00849-X [pii].

Carswell, E. A., Old, L. J., Kassel, R. L., Green, S., Fiore, N., & Williamson, B. (1975). An endotoxin-induced serum factor that causes necrosis of tumors. *Proceedings of the National Academy of Sciences of the United States of America*, *72*(9), 3666–3670.

Chai, J., Shiozaki, E., Srinivasula, S. M., Wu, Q., Datta, P., Alnemri, E. S., et al. (2001). Structural basis of caspase-7 inhibition by XIAP. *Cell*, *104*(5), 769–780.

Chai, J., Yan, N., Huh, J. R., Wu, J. W., Li, W., Hay, B. A., et al. (2003). Molecular mechanism of Reaper-Grim-Hid-mediated suppression of DIAP1-dependent Dronc ubiquitination. *Nature Structural Biology*, *10*(11), 892–898.

Chen, G., Shaw, M. H., Kim, Y. G., & Nunez, G. (2009). NOD-like receptors: Role in innate immunity and inflammatory disease. *Annual Review of Pathology*, *4*, 365–398. http://dx.doi.org/10.1146/annurev.pathol.4.110807.092239.

Chovatiya, R., & Medzhitov, R. (2014). Stress, inflammation, and defense of homeostasis. *Molecular Cell*, *54*(2), 281–288. http://dx.doi.org/10.1016/j.molcel.2014.03.030.

Claveria, C., Giovinazzo, G., Sierra, R., & Torres, M. (2013). Myc-driven endogenous cell competition in the early mammalian embryo. *Nature*, *500*(7460), 39–44. http://dx.doi.org/10.1038/nature12389.

Dannappel, M., Vlantis, K., Kumari, S., Polykratis, A., Kim, C., Wachsmuth, L., et al. (2014). RIPK1 maintains epithelial homeostasis by inhibiting apoptosis and necroptosis. *Nature*, *513*, 90–94. http://dx.doi.org/10.1038/nature13608.

Declercq, W., Vanden Berghe, T., & Vandenabeele, P. (2009). RIP kinases at the crossroads of cell death and survival. *Cell*, *138*(2), 229–232. http://dx.doi.org/10.1016/j.cell.2009.07.006. S0092-8674(09)00845-9 [pii].

Dikic, I., Wakatsuki, S., & Walters, K. J. (2009). Ubiquitin-binding domains—From structures to functions. *Nature Reviews Molecular Cell Biology*, *10*(10), 659–671. http://dx.doi.org/10.1038/nrm2767. nrm2767 [pii].

Ditzel, M., Broemer, M., Tenev, T., Bolduc, C., Lee, T. V., Rigbolt, K. T., et al. (2008). Inactivation of effector caspases through nondegradative polyubiquitylation. *Molecular Cell*, *32*(4), 540–553. http://dx.doi.org/10.1016/j.molcel.2008.09.025.

Ditzel, M., Wilson, R., Tenev, T., Zachariou, A., Paul, A., Deas, E., et al. (2003). Degradation of DIAP1 by the N-end rule pathway is essential for regulating apoptosis. *Nature Cell Biology*, *5*(5), 467–473.

Domingues, C., & Ryoo, H. D. (2012). Drosophila BRUCE inhibits apoptosis through nonlysine ubiquitination of the IAP-antagonist REAPER. *Cell Death and Differentiation*, *19*(3), 470–477. http://dx.doi.org/10.1038/cdd.2011.116. cdd2011116 [pii].

Dondelinger, Y., Aguileta, M. A., Goossens, V., Dubuisson, C., Grootjans, S., Dejardin, E., et al. (2013). RIPK3 contributes to TNFR1-mediated RIPK1 kinase-dependent apoptosis in conditions of cIAP1/2 depletion or TAK1 kinase inhibition. *Cell Death and Differentiation*, *20*(10), 1381–1392. http://dx.doi.org/10.1038/cdd.2013.94.

Duprez, L., Takahashi, N., Van Hauwermeiren, F., Vandendriessche, B., Goossens, V., Vanden Berghe, T., et al. (2011). RIP kinase-dependent necrosis drives lethal systemic inflammatory response syndrome. *Immunity*, *35*(6), 908–918. http://dx.doi.org/10.1016/j.immuni.2011.09.020. S1074-7613(11)00508-5 [pii].

Engel, E., Viargues, P., Mortier, M., Taillebourg, E., Coute, Y., Thevenon, D., et al. (2014). Identifying USPs regulating immune signals in Drosophila: USP2 deubiquitinates Imd and promotes its degradation by interacting with the proteasome. *Cell Communication and Signalling*, *12*, 41. http://dx.doi.org/10.1186/PREACCEPT-1588328929121802.

Ermolaeva, M. A., Michallet, M. C., Papadopoulou, N., Utermohlen, O., Kranidioti, K., Kollias, G., et al. (2008). Function of TRADD in tumor necrosis factor receptor 1 signaling and in TRIF-dependent inflammatory responses. *Nature Immunology*, *9*(9), 1037–1046. http://dx.doi.org/10.1038/ni.1638. ni.1638 [pii].

Erturk-Hasdemir, D., Broemer, M., Leulier, F., Lane, W. S., Paquette, N., Hwang, D., et al. (2009). Two roles for the Drosophila IKK complex in the activation of Relish and the induction of antimicrobial peptide genes. *Proceedings of the National Academy of Sciences of the United States of America*, *106*(24), 9779–9784. http://dx.doi.org/10.1073/pnas.0812022106. 0812022106 [pii].

Feinstein-Rotkopf, Y., & Arama, E. (2009). Can't live without them, can live with them: Roles of caspases during vital cellular processes. *Apoptosis*, *14*(8), 980–995. http://dx.doi.org/10.1007/s10495-009-0346-6.

Feng, S., Yang, Y., Mei, Y., Ma, L., Zhu, D. E., Hoti, N., et al. (2007). Cleavage of RIP3 inactivates its caspase-independent apoptosis pathway by removal of kinase domain. *Cellular Signalling*, *19*(10), 2056–2067. http://dx.doi.org/10.1016/j.cellsig.2007.05.016.

Feoktistova, M., Geserick, P., Kellert, B., Dimitrova, D. P., Langlais, C., Hupe, M., et al. (2011). cIAPs block Ripoptosome formation, a RIP1/caspase-8 containing intracellular cell death complex differentially regulated by cFLIP isoforms. *Molecular Cell*, *43*(3), 449–463. http://dx.doi.org/10.1016/j.molcel.2011.06.011.

Ferrandon, D., Imler, J. L., Hetru, C., & Hoffmann, J. A. (2007). The Drosophila systemic immune response: Sensing and signalling during bacterial and fungal infections. *Nature Reviews Immunology*, 7(11), 862–874.

Fuchs, Y., & Steller, H. (2011). Programmed cell death in animal development and disease. *Cell*, *147*(4), 742–758. http://dx.doi.org/10.1016/j.cell.2011.10.033.

Gardam, S., Turner, V. M., Anderton, H., Limaye, S., Basten, A., Koentgen, F., et al. (2011). Deletion of cIAP1 and cIAP2 in murine B lymphocytes constitutively activates cell survival pathways and inactivates the germinal center response. *Blood*, *117*(15), 4041–4051. http://dx.doi.org/10.1182/blood-2010-10-312793. blood-2010-10-312793 [pii].

Geserick, P., Hupe, M., Moulin, M., Wong, W. W., Feoktistova, M., Kellert, B., et al. (2009). Cellular IAPs inhibit a cryptic CD95-induced cell death by limiting RIP1 kinase recruitment. *The Journal of Cell Biology*, *187*(7), 1037–1054. http://dx.doi.org/10.1083/jcb.200904158. jcb.200904158 [pii].

Goyal, L., McCall, K., Agapite, J., Hartwieg, E., & Steller, H. (2000). Induction of apoptosis by Drosophila reaper, hid and grim through inhibition of IAP function. *The EMBO Journal*, *19*(4), 589–597.

Green, D. R., Oberst, A., Dillon, C. P., Weinlich, R., & Salvesen, G. S. (2011). RIPK-dependent necrosis and its regulation by caspases: A mystery in five acts. *Molecular Cell*, *44*(1), 9–16. http://dx.doi.org/10.1016/j.molcel.2011.09.003.

Grivennikov, S. I., Greten, F. R., & Karin, M. (2010). Immunity, inflammation, and cancer. *Cell*, *140*(6), 883–899. http://dx.doi.org/10.1016/j.cell.2010.01.025.

Hao, Y., Sekine, K., Kawabata, A., Nakamura, H., Ishioka, T., Ohata, H., et al. (2004). Apollon ubiquitinates SMAC and caspase-9, and has an essential cytoprotection function. *Nature Cell Biology*, *6*(9), 849–860.

Hasegawa, M., Fujimoto, Y., Lucas, P. C., Nakano, H., Fukase, K., Nunez, G., et al. (2008). A critical role of RICK/RIP2 polyubiquitination in Nod-induced NF-kappaB activation. *The EMBO Journal*, *27*(2), 373–383. http://dx.doi.org/10.1038/sj.emboj.7601962. 7601962 [pii].

Hawkins, C. J., Wang, S. L., & Hay, B. A. (1999). A cloning method to identify caspases and their regulators in yeast: Identification of Drosophila IAP1 as an inhibitor of the

Drosophila caspase DCP-1. *Proceedings of the National Academy of Sciences of the United States of America*, *96*(6), 2885–2890.

Herman-Bachinsky, Y., Ryoo, H. D., Ciechanover, A., & Gonen, H. (2007). Regulation of the Drosophila ubiquitin ligase DIAP1 is mediated via several distinct ubiquitin system pathways. *Cell Death and Differentiation*, *14*(4), 861–871.

Hershko, A., Ciechanover, A., & Varshavsky, A. (2000). Basic Medical Research Award. The ubiquitin system. *Nature Medicine*, *6*(10), 1073–1081.

Hoeller, D., Hecker, C. M., & Dikic, I. (2006). Ubiquitin and ubiquitin-like proteins in cancer pathogenesis. *Nature Reviews Cancer*, *6*(10), 776–788.

Huang, Y., Park, Y. C., Rich, R. L., Segal, D., Myszka, D. G., & Wu, H. (2001). Structural basis of caspase inhibition by XIAP: Differential roles of the linker versus the BIR domain. *Cell*, *104*(5), 781–790.

Huang, Q., Tang, X., Wang, G., Fan, Y., Ray, L., Bergmann, A., et al. (2014). Ubr3 E3 ligase regulates apoptosis by controlling the activity of DIAP1 in Drosophila. *Cell Death and Differentiation*, *21*(12), 1961–1970. http://dx.doi.org/10.1038/cdd.2014.115.

Igaki, T., Kanda, H., Yamamoto-Goto, Y., Kanuka, H., Kuranaga, E., Aigaki, T., et al. (2002). Eiger, a TNF superfamily ligand that triggers the Drosophila JNK pathway. *The EMBO Journal*, *21*(12), 3009–3018. http://dx.doi.org/10.1093/emboj/cdf306.

Igaki, T., & Miura, M. (2014). The Drosophila TNF ortholog Eiger: Emerging physiological roles and evolution of the TNF system. *Seminars in Immunology*, *26*(3), 267–274. http://dx.doi.org/10.1016/j.smim.2014.05.003.

Inohara, N., Koseki, T., Lin, J., del Peso, L., Lucas, P. C., Chen, F. F., et al. (2000). An induced proximity model for NF-kappa B activation in the Nod1/RICK and RIP signaling pathways. *The Journal of Biological Chemistry*, *275*(36), 27823.

Jacobson, M. D., Weil, M., & Raff, M. C. (1997). Programmed cell death in animal development. *Cell*, *88*(3), 347–354.

Kaiser, W. J., Vucic, D., & Miller, L. K. (1998). The Drosophila inhibitor of apoptosis D-IAP1 suppresses cell death induced by the caspase drICE. *FEBS Letters*, *440*(1–2), 243–248.

Kanayama, A., Seth, R. B., Sun, L., Ea, C. K., Hong, M., Shaito, A., et al. (2004). TAB2 and TAB3 activate the NF-kappaB pathway through binding to polyubiquitin chains. *Molecular Cell*, *15*(4), 535–548.

Kaplan, Y., Gibbs-Bar, L., Kalifa, Y., Feinstein-Rotkopf, Y., & Arama, E. (2010). Gradients of a ubiquitin E3 ligase inhibitor and a caspase inhibitor determine differentiation or death in spermatids. *Developmental Cell*, *19*(1), 160–173. http://dx.doi.org/10.1016/j.devcel.2010.06.009.

Kataoka, T., & Tschopp, J. (2004). N-terminal fragment of c-FLIP(L) processed by caspase 8 specifically interacts with TRAF2 and induces activation of the NF-kappaB signaling pathway. *Molecular and Cellular Biology*, *24*(7), 2627–2636.

Kirisako, T., Kamei, K., Murata, S., Kato, M., Fukumoto, H., Kanie, M., et al. (2006). A ubiquitin ligase complex assembles linear polyubiquitin chains. *The EMBO Journal*, *25*(20), 4877–4887.

Kleino, A., Valanne, S., Ulvila, J., Kallio, J., Myllymaki, H., Enwald, H., et al. (2005). Inhibitor of apoptosis 2 and TAK1-binding protein are components of the Drosophila Imd pathway. *The EMBO Journal*, *24*(19), 3423–3434.

Komander, D. (2009). The emerging complexity of protein ubiquitination. *Biochemical Society Transactions*, *37*(Pt. 5), 937–953. http://dx.doi.org/10.1042/BST0370937. BST0370937 [pii].

Komander, D., Clague, M. J., & Urbe, S. (2009). Breaking the chains: Structure and function of the deubiquitinases. *Nature Reviews Molecular Cell Biology*, *10*(8), 550–563. http://dx.doi.org/10.1038/nrm2731. nrm2731 [pii].

Komander, D., Lord, C. J., Scheel, H., Swift, S., Hofmann, K., Ashworth, A., et al. (2008). The structure of the CYLD USP domain explains its specificity for Lys63-linked polyubiquitin and reveals a B box module. *Molecular Cell*, *29*(4), 451–464.

Komander, D., & Rape, M. (2012). The ubiquitin code. *Annual Review of Biochemistry*, *81*, 203–229. http://dx.doi.org/10.1146/annurev-biochem-060310-170328.

Krieg, A., Correa, R. G., Garrison, J. B., Le Negrate, G., Welsh, K., Huang, Z., et al. (2009). XIAP mediates NOD signaling via interaction with RIP2. *Proceedings of the National Academy of Sciences of the United States of America*, *106*(34), 14524–14529. http://dx.doi.org/10.1073/pnas.0907131106. 0907131106 [pii].

Lee, T. V., Fan, Y., Wang, S., Srivastava, M., Broemer, M., Meier, P., et al. (2011). Drosophila IAP1-mediated ubiquitylation controls activation of the initiator caspase DRONC independent of protein degradation. *PLoS Genetics*, 7(9), e1002261. http://dx.doi.org/10.1371/journal.pgen.1002261.

Lin, Y., Devin, A., Rodriguez, Y., & Liu, Z. G. (1999). Cleavage of the death domain kinase RIP by caspase-8 prompts TNF-induced apoptosis. *Genes & Development*, *13*(19), 2514–2526.

Lisi, S., Mazzon, I., & White, K. (2000). Diverse domains of THREAD/DIAP1 are required to inhibit apoptosis induced by REAPER and HID in Drosophila. *Genetics*, *154*(2), 669–678.

Lu, Y., Wu, L. P., & Anderson, K. V. (2001). The antibacterial arm of the drosophila innate immune response requires an IkappaB kinase. *Genes & Development*, *15*(1), 104–110.

Lundgren, J., Masson, P., Realini, C. A., & Young, P. (2003). Use of RNA interference and complementation to study the function of the Drosophila and human 26S proteasome subunit S13. *Molecular and Cellular Biology*, *23*(15), 5320–5330.

Ma, X., Huang, J., Yang, L., Yang, Y., Li, W., & Xue, L. (2012). NOPO modulates Egr-induced JNK-independent cell death in Drosophila. *Cell Research*, *22*(2), 425–431. http://dx.doi.org/10.1038/cr.2011.135.

Ma, X., Li, W., Yu, H., Yang, Y., Li, M., Xue, L., et al. (2014). Bendless modulates JNK-mediated cell death and migration in Drosophila. *Cell Death and Differentiation*, *21*(3), 407–415. http://dx.doi.org/10.1038/cdd.2013.154.

Ma, X., Yang, L., Yang, Y., Li, M., Li, W., & Xue, L. (2013). dUev1a modulates TNF-JNK mediated tumor progression and cell death in Drosophila. *Developmental Biology*, *380*(2), 211–221. http://dx.doi.org/10.1016/j.ydbio.2013.05.013.

Martin, S. J., Henry, C. M., & Cullen, S. P. (2012). A perspective on mammalian caspases as positive and negative regulators of inflammation. *Molecular Cell*, *46*(4), 387–397. http://dx.doi.org/10.1016/j.molcel.2012.04.026. S1097-2765(12)00348-6 [pii].

Martins, V. C., Busch, K., Juraeva, D., Blum, C., Ludwig, C., Rasche, V., et al. (2014). Cell competition is a tumour suppressor mechanism in the thymus. *Nature*, *509*(7501), 465–470. http://dx.doi.org/10.1038/nature13317.

Matsuzawa, A., Tseng, P. H., Vallabhapurapu, S., Luo, J. L., Zhang, W., Wang, H., et al. (2008). Essential cytoplasmic translocation of a cytokine receptor-assembled signaling complex. *Science*, *321*(5889), 663–668. http://dx.doi.org/10.1126/science.1157340. 1157340 [pii].

Medzhitov, R. (2008). Origin and physiological roles of inflammation. *Nature*, *454*(7203), 428–435. http://dx.doi.org/10.1038/nature07201.

Meier, P., Silke, J., Leevers, S. J., & Evan, G. I. (2000). The Drosophila caspase DRONC is regulated by DIAP1. *The EMBO Journal*, *19*(4), 598–611.

Meinander, A., Runchel, C., Tenev, T., Chen, L., Kim, C. H., Ribeiro, P. S., et al. (2012). Ubiquitylation of the initiator caspase DREDD is required for innate immune signalling. *The EMBO Journal*, *31*(12), 2770–2783. http://dx.doi.org/10.1038/emboj.2012.121. emboj2012121 [pii].

Meyer, H. J., & Rape, M. (2014). Enhanced protein degradation by branched ubiquitin chains. *Cell, 157*(4), 910–921. http://dx.doi.org/10.1016/j.cell.2014.03.037.

Micheau, O., Thome, M., Schneider, P., Holler, N., Tschopp, J., Nicholson, D. W., et al. (2002). The long form of FLIP is an activator of caspase-8 at the Fas death-inducing signaling complex. *The Journal of Biological Chemistry, 277*(47), 45162–45171.

Micheau, O., & Tschopp, J. (2003). Induction of TNF receptor I-mediated apoptosis via two sequential signaling complexes. *Cell, 114*(2), 181–190.

Moreno, E., Yan, M., & Basler, K. (2002). Evolution of TNF signaling mechanisms: JNK-dependent apoptosis triggered by Eiger, the Drosophila homolog of the TNF superfamily. *Current Biology, 12*(14), 1263–1268.

Morizane, Y., Honda, R., Fukami, K., & Yasuda, H. (2005). X-linked inhibitor of apoptosis functions as ubiquitin ligase toward mature caspase-9 and cytosolic Smac/DIABLO. *Journal of Biochemistry (Tokyo), 137*(2), 125–132.

Moulin, M., Anderton, H., Voss, A. K., Thomas, T., Wong, W. W., Bankovacki, A., et al. (2012). IAPs limit activation of RIP kinases by TNF receptor 1 during development. *The EMBO Journal, 31*(7), 1679–1691. http://dx.doi.org/10.1038/emboj.2012.18. emboj201218 [pii].

Muro, I., Means, J. C., & Clem, R. J. (2005). Cleavage of the apoptosis inhibitor DIAP1 by the apical caspase DRONC in both normal and apoptotic Drosophila cells. *The Journal of Biological Chemistry, 280*(19), 18683–18688.

Neves, J., Demaria, M., Campisi, J., & Jasper, H. (2015). Of flies, mice, and men: Evolutionarily conserved tissue damage responses and aging. *Developmental Cell, 32*(1), 9–18. http://dx.doi.org/10.1016/j.devcel.2014.11.028.

Oberst, A., Dillon, C. P., Weinlich, R., McCormick, L. L., Fitzgerald, P., Pop, C., et al. (2011). Catalytic activity of the caspase-8-FLIP(L) complex inhibits RIPK3-dependent necrosis. *Nature, 471*(7338), 363–367. http://dx.doi.org/10.1038/nature09852.

O'Donnell, M. A., Perez-Jimenez, E., Oberst, A., Ng, A., Massoumi, R., Xavier, R., et al. (2011). Caspase 8 inhibits programmed necrosis by processing CYLD. *Nature Cell Biology, 13*(12), 1437–1442. http://dx.doi.org/10.1038/ncb2362. ncb2362 [pii].

Oertel, M., Menthena, A., Dabeva, M. D., & Shafritz, D. A. (2006). Cell competition leads to a high level of normal liver reconstitution by transplanted fetal liver stem/progenitor cells. *Gastroenterology, 130*(2), 507–520. http://dx.doi.org/10.1053/j.gastro.2005.10.049. quiz 590.

Oliver, E. R., Saunders, T. L., Tarle, S. A., & Glaser, T. (2004). Ribosomal protein L24 defect in belly spot and tail (Bst), a mouse Minute. *Development, 131*(16), 3907–3920. http://dx.doi.org/10.1242/dev.01268.

Paquette, N., Broemer, M., Aggarwal, K., Chen, L., Husson, M., Erturk-Hasdemir, D., et al. (2010). Caspase-mediated cleavage, IAP binding, and ubiquitination: Linking three mechanisms crucial for Drosophila NF-kappaB signaling. *Molecular Cell, 37*(2), 172–182. http://dx.doi.org/10.1016/j.molcel.2009.12.036. S1097-2765(10)00039-0 [pii].

Park, J. H., Kim, Y. G., McDonald, C., Kanneganti, T. D., Hasegawa, M., Body-Malapel, M., et al. (2007). RICK/RIP2 mediates innate immune responses induced through Nod1 and Nod2 but not TLRs. *Journal of Immunology, 178*(4), 2380–2386. 178/4/2380 [pii].

Pasparakis, M., & Vandenabeele, P. (2015). Necroptosis and its role in inflammation. *Nature, 517*(7534), 311–320. http://dx.doi.org/10.1038/nature14191.

Polykratis, A., Hermance, N., Zelic, M., Roderick, J., Kim, C., Van, T. M., et al. (2014). Cutting edge: RIPK1 kinase inactive mice are viable and protected from TNF-induced necroptosis in vivo. *Journal of Immunology, 193*(4), 1539–1543. http://dx.doi.org/10.4049/jimmunol.1400590.

Rahighi, S., Ikeda, F., Kawasaki, M., Akutsu, M., Suzuki, N., Kato, R., et al. (2009). Specific recognition of linear ubiquitin chains by NEMO is important for NF-kappaB activation.

Cell, 136(6), 1098–1109. http://dx.doi.org/10.1016/j.cell.2009.03.007. S0092-8674 (09)00264-5 [pii].
Ribeiro, P. S., Kuranaga, E., Tenev, T., Leulier, F., Miura, M., & Meier, P. (2007). DIAP2 functions as a mechanism-based regulator of drICE that contributes to the caspase activity threshold in living cells. *The Journal of Cell Biology, 179*(7), 1467–1480. http://dx.doi.org/10.1083/jcb.200706027. jcb.200706027 [pii].
Riedl, S. J., Renatus, M., Schwarzenbacher, R., Zhou, Q., Sun, C., Fesik, S. W., et al. (2001). Structural basis for the inhibition of caspase-3 by XIAP. *Cell, 104*(5), 791–800.
Rodriguez, A., Chen, P., Oliver, H., & Abrams, J. M. (2002). Unrestrained caspase-dependent cell death caused by loss of Diap1 function requires the Drosophila Apaf-1 homolog, Dark. *The EMBO Journal, 21*(9), 2189–2197.
Rutschmann, S., Jung, A. C., Zhou, R., Silverman, N., Hoffmann, J. A., & Ferrandon, D. (2000). Role of Drosophila IKK gamma in a toll-independent antibacterial immune response. *Nature Immunology, 1*(4), 342–347.
Rutschmann, S., Kilinc, A., & Ferrandon, D. (2002). Cutting edge: The toll pathway is required for resistance to gram-positive bacterial infections in Drosophila. *Journal of Immunology, 168*(4), 1542–1546.
Schile, A. J., Garcia-Fernandez, M., & Steller, H. (2008). Regulation of apoptosis by XIAP ubiquitin-ligase activity. *Genes & Development, 22*(16), 2256–2266. http://dx.doi.org/10.1101/gad.1663108. 22/16/2256 [pii].
Shapiro, P. J., Hsu, H. H., Jung, H., Robbins, E. S., & Ryoo, H. D. (2008). Regulation of the Drosophila apoptosome through feedback inhibition. *Nature Cell Biology, 10*(12), 1440–1446. http://dx.doi.org/10.1038/ncb1803. ncb1803 [pii].
Shi, Y. (2002a). A conserved tetrapeptide motif: Potentiating apoptosis through IAP-binding. *Cell Death and Differentiation, 9*(2), 93–95.
Shi, Y. (2002b). Mechanisms of caspase activation and inhibition during apoptosis. *Molecular Cell, 9*(3), 459–470.
Shiozaki, E. N., Chai, J., Rigotti, D. J., Riedl, S. J., Li, P., Srinivasula, S. M., et al. (2003). Mechanism of XIAP-mediated inhibition of caspase-9. *Molecular Cell, 11*(2), 519.
Silke, J. (2011). The regulation of TNF signalling: What a tangled web we weave. *Current Opinion in Immunology, 23*(5), 620–626. http://dx.doi.org/10.1016/j.coi.2011.08.002. S0952-7915(11)00105-1 [pii].
Silke, J., Ekert, P. G., Day, C. L., Hawkins, C. J., Baca, M., Chew, J., et al. (2001). Direct inhibition of caspase 3 is dispensable for the anti-apoptotic activity of XIAP. *The EMBO Journal, 20*(12), 3114–3123.
Silke, J., & Meier, P. (2013). Inhibitor of apoptosis (IAP) proteins-modulators of cell death and inflammation. *Cold Spring Harbor Perspectives in Biology, 5*(2). http://dx.doi.org/10.1101/cshperspect.a008730.
Silverman, N., Zhou, R., Erlich, R. L., Hunter, M., Bernstein, E., Schneider, D., et al. (2003). Immune activation of NF-kappaB and JNK requires Drosophila TAK1. *The Journal of Biological Chemistry, 278*(49), 48928–48934. http://dx.doi.org/10.1074/jbc.M304802200. M304802200 [pii].
Silverman, N., Zhou, R., Stoven, S., Pandey, N., Hultmark, D., & Maniatis, T. (2000). A Drosophila IkappaB kinase complex required for Relish cleavage and antibacterial immunity. *Genes & Development, 14*(19), 2461–2471.
Speckmann, C., Lehmberg, K., Albert, M. H., Damgaard, R. B., Fritsch, M., Gyrd-Hansen, M., et al. (2013). X-linked inhibitor of apoptosis (XIAP) deficiency: The spectrum of presenting manifestations beyond hemophagocytic lymphohistiocytosis. *Clinical Immunology, 149*(1), 133–141. http://dx.doi.org/10.1016/j.clim.2013.07.004.
Stickle, N. H., Chung, J., Klco, J. M., Hill, R. P., Kaelin, W. G., Jr., & Ohh, M. (2004). pVHL modification by NEDD8 is required for fibronectin matrix assembly and suppression of tumor development. *Molecular and Cellular Biology, 24*(8), 3251–3261.

Suzuki, Y., Nakabayashi, Y., Nakata, K., Reed, J. C., & Takahashi, R. (2001). X-linked inhibitor of apoptosis protein (XIAP) inhibits caspase-3 and - 7 in distinct modes. *The Journal of Biological Chemistry*, *276*(29), 27058–27063.

Suzuki, Y., Nakabayashi, Y., & Takahashi, R. (2001). Ubiquitin-protein ligase activity of X-linked inhibitor of apoptosis protein promotes proteasomal degradation of caspase-3 and enhances its anti-apoptotic effect in Fas-induced cell death. *Proceedings of the National Academy of Sciences of the United States of America*, *98*(15), 8662–8667.

Tatham, M. H., Geoffroy, M. C., Shen, L., Plechanovova, A., Hattersley, N., Jaffray, E. G., et al. (2008). RNF4 is a poly-SUMO-specific E3 ubiquitin ligase required for arsenic-induced PML degradation. *Nature Cell Biology*, *10*(5), 538–546. http://dx.doi.org/10.1038/ncb1716.

Tenev, T., Bianchi, K., Darding, M., Broemer, M., Langlais, C., Wallberg, F., et al. (2011). The Ripoptosome, a signaling platform that assembles in response to genotoxic stress and loss of IAPs. *Molecular Cell*, *43*(3), 432–448. http://dx.doi.org/10.1016/j.molcel.2011.06.006. S1097-2765(11)00420-5 [pii].

Tenev, T., Zachariou, A., Wilson, R., Ditzel, M., & Meier, P. (2005). IAPs are functionally non-equivalent and regulate effector caspases through distinct mechanisms. *Nature Cell Biology*, 7(1), 70–77.

Thevenon, D., Engel, E., Avet-Rochex, A., Gottar, M., Bergeret, E., Tricoire, H., et al. (2009). The Drosophila ubiquitin-specific protease dUSP36/Scny targets IMD to prevent constitutive immune signaling. *Cell Host & Microbe*, *6*(4), 309–320. http://dx.doi.org/10.1016/j.chom.2009.09.007. S1931-3128(09)00316-3 [pii].

Thompson, C. B. (1995). Apoptosis in the pathogenesis and treatment of disease. *Science*, *267*(5203), 1456–1462.

Vandenabeele, P., Galluzzi, L., Vanden Berghe, T., & Kroemer, G. (2010). Molecular mechanisms of necroptosis: An ordered cellular explosion. *Nature Reviews Molecular Cell Biology*, *11*(10), 700–714. http://dx.doi.org/10.1038/nrm2970. nrm2970 [pii].

Verhagen, A. M., Kratina, T. K., Hawkins, C. J., Silke, J., Ekert, P. G., & Vaux, D. L. (2007). Identification of mammalian mitochondrial proteins that interact with IAPs via N-terminal IAP binding motifs. *Cell Death and Differentiation*, *14*(2), 348–357. http://dx.doi.org/10.1038/sj.cdd.4402001. 4402001 [pii].

Vernooy, S. Y., Chow, V., Su, J., Verbrugghe, K., Yang, J., Cole, S., et al. (2002). Drosophila Bruce can potently suppress Rpr- and Grim-dependent but not Hid-dependent cell death. *Current Biology*, *12*(13), 1164–1168.

Vidal, S., Khush, R. S., Leulier, F., Tzou, P., Nakamura, M., & Lemaitre, B. (2001). Mutations in the Drosophila dTAK1 gene reveal a conserved function for MAPKKKs in the control of rel/NF-kappaB-dependent innate immune responses. *Genes & Development*, *15*(15), 1900–1912.

Vincent, J. P., Fletcher, A. G., & Baena-Lopez, L. A. (2013). Mechanisms and mechanics of cell competition in epithelia. *Nature Reviews Molecular Cell Biology*, *14*(9), 581–591. http://dx.doi.org/10.1038/nrm3639.

Walczak, H. (2013). Death receptor-ligand systems in cancer, cell death, and inflammation. *Cold Spring Harbor Perspectives in Biology*, *5*(5), a008698. http://dx.doi.org/10.1101/cshperspect.a008698.

Wang, L., Du, F., & Wang, X. (2008). TNF-alpha induces two distinct caspase-8 activation pathways. *Cell*, *133*(4), 693–703. http://dx.doi.org/10.1016/j.cell.2008.03.036. S0092-8674(08)00501-1 [pii].

Wang, S. L., Hawkins, C. J., Yoo, S. J., Muller, H. A., & Hay, B. A. (1999). The Drosophila caspase inhibitor DIAP1 is essential for cell survival and is negatively regulated by HID. *Cell*, *98*(4), 453–463.

Welz, P. S., Wullaert, A., Vlantis, K., Kondylis, V., Fernandez-Majada, V., Ermolaeva, M., et al. (2011). FADD prevents RIP3-mediated epithelial cell necrosis and chronic intestinal inflammation. *Nature*, *477*(7364), 330–334. http://dx.doi.org/10.1038/nature10273. nature10273 [pii].

Wilson, R., Goyal, L., Ditzel, M., Zachariou, A., Baker, D. A., Agapite, J., et al. (2002). The DIAP1 RING finger mediates ubiquitination of Dronc and is indispensable for regulating apoptosis. *Nature Cell Biology*, *4*, 445–450.

Wu, G., Chai, J., Suber, T. L., Wu, J. W., Du, C., Wang, X., et al. (2000). Structural basis of IAP recognition by Smac/DIABLO. *Nature*, *408*(6815), 1008–1012.

Xue, L., Igaki, T., Kuranaga, E., Kanda, H., Miura, M., & Xu, T. (2007). Tumor suppressor CYLD regulates JNK-induced cell death in Drosophila. *Developmental Cell*, *13*(3), 446–454. http://dx.doi.org/10.1016/j.devcel.2007.07.012. S1534-5807(07)00273-0 [pii].

Yan, N., Wu, J. W., Chai, J., Li, W., & Shi, Y. (2004). Molecular mechanisms of DrICE inhibition by DIAP1 and removal of inhibition by Reaper, Hid and Grim. *Nature Structural & Molecular Biology*, *11*(5), 420–428.

Yang, Y., Yin, C., Pandey, A., Abbott, D., Sassetti, C., & Kelliher, M. A. (2007). NOD2 pathway activation by MDP or Mycobacterium tuberculosis infection involves the stable polyubiquitination of Rip2. *The Journal of Biological Chemistry*, *282*(50), 36223–36229. http://dx.doi.org/10.1074/jbc.M703079200. M703079200 [pii].

Yatsenko, A. N., Roy, A., Chen, R., Ma, L., Murthy, L. J., Yan, W., et al. (2006). Non-invasive genetic diagnosis of male infertility using spermatozoal RNA: KLHL10 mutations in oligozoospermic patients impair homodimerization. *Human Molecular Genetics*, *15*(23), 3411–3419. http://dx.doi.org/10.1093/hmg/ddl417.

Zachariou, A., Tenev, T., Goyal, L., Agapite, J., Steller, H., & Meier, P. (2003). IAP-antagonists exhibit non-redundant modes of action through differential DIAP1 binding. *The EMBO Journal*, *22*(24), 6642–6652.

Zhuang, Z. H., Sun, L., Kong, L., Hu, J. H., Yu, M. C., Reinach, P., et al. (2006). Drosophila TAB2 is required for the immune activation of JNK and NF-kappaB. *Cellular Signalling*, *18*(7), 964–970. http://dx.doi.org/10.1016/j.cellsig.2005.08.020. S0898-6568(05)00220-2 [pii].

CHAPTER NINE

The Sound of Silence: Signaling by Apoptotic Cells

Caitlin E. Fogarty, Andreas Bergmann[1]
Department of Molecular, Cell, and Cancer Biology, University of Massachusetts Medical School, Worcester, Massachusetts, USA
[1]Corresponding author: e-mail address: andreas.bergmann@umassmed.edu

Contents

Abstract

Apoptosis is a carefully choreographed process of cellular self-destruction in the absence of inflammation. During the death process, apoptotic cells actively communicate with their environment, signaling to both their immediate neighbors as well as distant sentinels. Some of these signals direct the anti-inflammatory immune response, instructing specific subsets of phagocytes to participate in the limited and careful clearance of dying cellular debris. These immunomodulatory signals can also regulate the activation state of the engulfing phagocytes. Other signals derived from apoptotic cells contribute to tissue growth control with the common goal of maintaining tissue integrity. Derangements in these growth control signals during prolonged apoptosis can lead to excessive cell loss or proliferation. Here, we highlight some of the most intriguing signals produced by apoptotic cells during the course of normal development as well as during physiological disturbances such as atherosclerosis and cancer.

Current Topics in Developmental Biology, Volume 114
ISSN 0070-2153
http://dx.doi.org/10.1016/bs.ctdb.2015.07.013

1. APOPTOSIS: A SILENT DEATH?

Apoptosis is a carefully choreographed process of cellular self-destruction, observed across the spectrum of metazoans from worms to flies to mammals (Abrams, White, Fessler, & Steller, 1993; Ellis & Horvitz, 1986; Kerr, Wyllie, & Currie, 1972). During development, apoptosis shapes developing tissues by removing superfluous cells, sculpting out defined structures, or regulating tissue size (Glucksmann, 1951) (more recently reviewed in Suzanne & Steller, 2013). In adult organisms, apoptosis can trigger death in cells that are no longer functioning properly such as those injured by toxins or transformed by genetic aberrations (reviewed in Fuchs & Steller, 2011). This removal is critical to maintaining tissue integrity and homeostasis, and it is the mechanism of removal that distinguishes apoptosis from other forms of cell death. Cells that are damaged, infected, or otherwise unwanted are capable of initiating a tightly controlled cascade of events, which leads to the cessation of normal cellular activity, the degradation of major macromolecules including DNA, and ultimately the contained fragmentation of the cell so that it may be cleared via phagocytosis (Kerr et al., 1972; Lockshin & Williams, 1965; Schwartz, Smith, Jones, & Osborne, 1993).

Apoptosis was initially distinguished from necrotic cell death based on the quiet nature of its cellular demise. Unlike necrosis where cells spill their contents causing secondary tissue damage and infiltrating immune cells react with such fervor they induce significant inflammation, apoptosis is characterized by an unassuming departure, contained cellular contents, few immune cells, and no detectable inflammation. This contrast earned apoptotic cell death the moniker of "altruistic cell suicide," and so for a time, the characterization of apoptosis as the silent cell death prevailed (Bar, 1996).

To better understand how apoptotic cells can die without causing further damage, we will first review the basics of apoptotic cell death. From worms to humans, there are a variety of ways to initiate the apoptotic cascade—some cascades are triggered by intrinsic developmentally regulated transcriptional programs, others by extrinsic death signals; some are triggered by active induction, others by neglect; some depend on the release of cytochrome C from the mitochondria, others can be driven by accumulation of proapoptotic factors (reviewed in Bergmann, 2010; Conradt, 2009; Czabotar, Lessene, Strasser, & Adams, 2014; Danial & Korsmeyer, 2004; Domingos & Steller, 2007; Steller, 1995; Xu et al., 2009).

What all apoptotic deaths have in common, however, is the activation of caspases. These cysteine-dependent aspartate-directed proteases are the critical effectors of cell death (Miura, Zhu, Rotello, Hartwieg, & Yuan, 1993; Yuan, Shaham, Ledoux, Ellis, & Horvitz, 1993). Caspases are initially produced as zymogens, which are not active until they are proteolytically cleaved. Autocatalytic activation of the initiator Caspase-9 most typically occurs via complex formation with the adaptor protein Apaf-1, along with cytochrome C and dATP (Li et al., 1997). Activated initiator caspases can cleave and activate effector caspases such as Caspase-3 and Caspase-7 (Brustugun, Fladmark, Doskeland, Orrenius, & Zhivotovsky, 1998; Zou, Henzel, Liu, Lutschg, & Wang, 1997). Activated effector caspases carry out the methodical process of executing cell death, directly activating other death enzymes such as nucleases and kinases, inactivating proteins required to sustain normal cellular processes, or indirectly disrupting normal physiological processes by disassembling compartments such as the nucleus and the mitochondria (Coleman et al., 2001; Enari et al., 1998; Gavrieli, Sherman, & Ben-Sasson, 1992; Li, Luo, & Wang, 2001; Liu, Zou, Slaughter, & Wang, 1997; Sebbagh et al., 2001; Susin et al., 1999).

While only ten percent of specific caspase cleavage sites are conserved between worms and humans, there is incredible conservation of the biological pathways which are targeted by effector caspases (Crawford et al., 2012). Among these, there are a number of targets that do not seem to be involved in the actual disassembly of the dying cell, but instead are released into the surrounding microenvironment. Over the past decade, interest in these apoptosis-derived signals has led to the discovery of critical communications between dying cells and their environment. Generally these signals can be broken into two categories: (1) signals that act on immune cells to regulate the clearance of apoptotic debris, prevent inflammation, and limit fibrosis; and (2) signals that act directly on neighboring surviving cells to maintain tissue integrity via growth control. Taken together, it is becoming increasingly evident that apoptosis is not a silent death, but instead apoptotic cells actively instruct the many players involved in executing an efficient, nondisruptive death. Here, we discuss some of the most intriguing signals produced by apoptotic cells that direct these coordinated responses and their importance in development and disease.

2. APOPTOTIC SIGNALS DICTATE IMMUNOLOGICAL RESPONSES TO CELL DEATH

A key component of the apoptotic program is the efficient and controlled clearance of the dying cells before they become dangerous necrotic cells, spilling their contents haphazardly. The phagocytic removal of dying

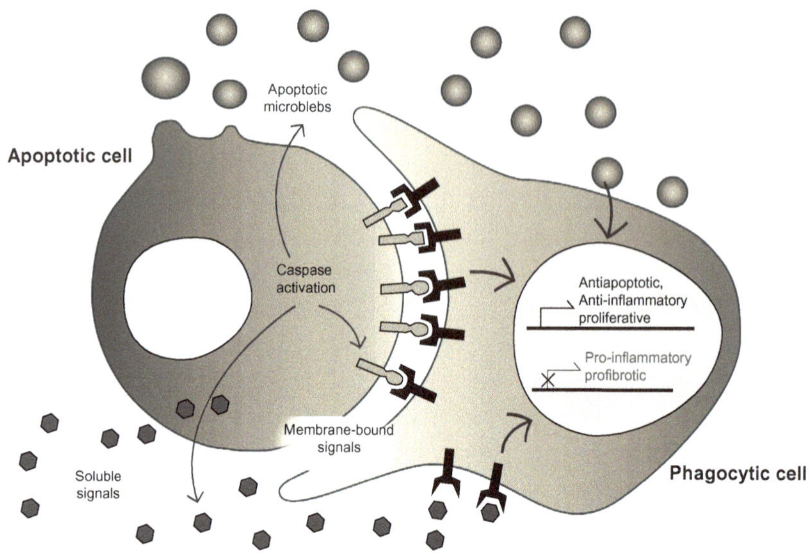

Figure 1 Apoptotic cells communicate with phagocytes to modulate immune responses to cell death. Apoptotic cells communicate with professional phagocytes such as macrophages and nonprofessional phagocytes such as engulfing epithelial cells. The apoptosis-derived signals include (1) soluable chemokines, (2) membrane-bound proteins and phospholipids, and (3) apoptotic microblebs. These signals are capable of directing appropriate anti-inflammatory responses promoting survival of neighboring healthy tissue.

cells is often referred to as efferocytosis meaning "take to the grave" (Thorp, Subramanian, & Tabas, 2011). However, apoptotic cells do not go quietly to their final resting place. Instead, they communicate directly with professional phagocytes, such as macrophages, and nonprofessional phagocytes, such as engulfing epithelial cells. These signals are capable of directing appropriate prophagocytic and anti-inflammatory responses (Fig. 1).

2.1 "Find Me! Eat Me!": Classic Chemoattractants and Phagocytic Signals

There are a myriad of classic signals that can lead phagocytic cells to sites of apoptotic cell death, which have been expertly review elsewhere (Chen, Zhao, & Liu, 2014; Hochreiter-Hufford & Ravichandran, 2013; Peter, Wesselborg, Herrmann, & Lauber, 2010; Poon, Lucas, Rossi, & Ravichandran, 2014). Here, we would like to highlight a few apoptosis-derived chemoattractant signals that are specifically produced via caspase activity.

2.1.1 Lysophosphatidylcholine

Lysophosphatidylcholine (LPC) is a phospholipid that was identified as one of the first truly caspase-derived "find me" signals. Caspase-3-mediated activation of calcium-independent phospholipase A_2 (iPLA_2) leads to the production of a variety of lipid-based signaling molecules. Production of LPC by iPLA_2 specifically leads to the recruitment of phagocytes to the site of apoptosis (Lauber et al., 2003; Mueller, Sheriff, Gaipl, Wesselborg, & Lauber, 2007). Once phagocytes arrive, this phospholipid can also facilitate opsonization of apoptotic cells (Kim, Gershov, Ma, Brot, & Elkon, 2002).

2.1.2 Sphingosine-1-Phosphate

Sphingosine-1-phosphate (S1P) is another lipid-based chemoattractant that is produced directly following caspase activation (Gude et al., 2008). Upon triggering apoptosis, the expression of sphingosine kinase is upregulated in a caspase-dependent manner, leading to excess production of S1P. Gude et al. (2008) found that S1P is a potent attractant for monocytes and macrophages. Interestingly, S1P is also critically involved in regulating epithelial cell extrusion during apoptosis, as described later in this chapter (Gu, Forostyan, Sabbadini, & Rosenblatt, 2011; Gu et al., 2015).

2.1.3 Endothelial Monocyte-Activating Polypeptide II

Endothelial monocyte-activating polypeptide II (EMAP II) is a processed form of pro-EMAP/p43 that is shed by apoptotic cells and has the ability to serve as a monocyte chemoattractant (Knies et al., 1998). Interestingly, work by Behrensdorf, van de Craen, Knies, Vandenabeele, and Clauss (2000) suggested that this chemoattractant is released by caspases, specifically by Caspase-3 and Caspase-7-mediated cleavage (Behrensdorf et al., 2000). However, this study was limited to using mouse recombinant proteins *in vitro*. Later work by Zhang and Schwartz refuted the conclusion that pro-EMAP II is directly cleaved by caspases in humans, finding no evidence of caspase-3- or caspase-7-dependent release in human tumor cell cultures (Zhang & Schwarz, 2002). Adding to the controversy, other published reports have suggested that pro-EMAP II may be either cleaved or released in an unprocessed form, that it may come from either apoptotic cells or necrotic cells, and that it may be cleaved by elastases, metalloproteases, or calpains—all depending on which experimental parameters were tested (Martinet et al., 2010; Matschurat et al., 2003; van Horssen, Eggermont, & ten Hagen, 2006). These seeming conflicts in the literature might simply reflect a versatile "find me" signal and that EMAP II regulation is finely tuned by a variety of factors depending on the context of the dying cell. Future work may elucidate the full spectrum of EMAP II processing and function in recruiting professional phagocytes to sites of cell death, and may provide helpful clues to the apoptotic programs that direct this clearance signal.

2.2 "Listen to Me!": Complex Modulatory Signals from Apoptotic Cells

In addition to the traditional functions of recruiting professional and non-professional phagocytes to the site of cell death, and promoting apoptotic cell removal, several of these apoptosis-derived immunoregulatory signals can also function in more nuanced modulatory pathways. Here, we highlight a few apoptosis-derived signals that were initially characterized in classic clearance pathways but have since been found to profoundly affect their surrounding environment by modulating the activities of engulfing phagocytes and surrounding progenitors.

2.2.1 Phosphatidylserine

Shortly following initiation of the apoptotic program, phosphatidylserine (PS) begins to accumulate on the outer leaflet of the dying cell's membrane (Fadok et al., 1992; Martin et al., 1995). This exposure is critically required for recognition and engulfment of the dying cell by the phagocyte (Fadok, de Cathelineau, Daleke, Henson, & Bratton, 2001). In living cells, plasma membrane phospholipid asymmetry is maintained by flippases. Upon activation of effector caspases, these flippases are cleaved and inactivated, which contributes to PS accumulation on the outer membrane (Bratton et al., 1997; Chen, Mapes, Lee, Skeen-Gaar, & Xue, 2013; Mandal, Mazumder, Das, Kundu, & Basu, 2005; Martin, Finucane, Amarante-Mendes, O'Brien, & Green, 1996; Segawa et al., 2014). In addition, other enzymes such as the Xk-related family of scramblases are activated upon caspase cleavage and actively promote the transfer of PS to the outer leaflet (Suzuki, Imanishi, & Nagata, 2014). Interestingly, PS exposure on the apoptotic cell also dictates what type of immunological response should occur. Upon recognizing the apoptotic "eat me" signal, macrophages actively induce production of anti-inflammatory cytokines including TGFβ and prostaglandins, and suppress production of typical pro-inflammatory cytokines such as TNFα and IL-1β (Fadok et al., 1998). This prevents the further recruitment of other, potentially more damaging, immune cells such as neutrophils. These immunosuppressive programs are dependent of the activation of the PS receptor (Fadok et al., 2000; Huynh, Fadok, & Henson, 2002).

Strikingly, this finding can also be extended to nonprofessional phagocytes. Mammary gland involution involves the massive apoptosis, extrusion, and clearance of most milk-producing mammary epithelial cells within 72 h of ceasing lactation (Walker, Bennett, & Kerr, 1989). In this developmental context of physiological weaning, the dying milk cells are phagocytosed not by macrophages, but primarily by the remaining viable mammary epithelial cells (Monks et al., 2005; Monks, Smith-Steinhart, Kruk, Fadok, & Henson, 2008). Again, the apoptotic cells are capable of inducing this anti-inflammatory state in the tissue via activation of the PS receptor on the nonprofessional phagocytes. The engulfing epithelial cells then produce TGFβ and suppress the production of any pro-inflammatory cytokines (Monks et al., 2005).

2.2.2 Fractalkine

Fractalkine (FKN), also known as CX3CL1, is a membrane-bound chemokine that can facilitate intercellular interactions, or it can be cleaved by the TNFα-converting enzyme ADAM17 (Garton et al., 2001), diffusing away in search of cells bearing its receptor CX3CR1. Specifically, FKN is released by apoptotic cells to recruit professional phagocytes to the site of cell death (Sokolowski, Chabanon-Hicks, Han, Heffron, & Mandell, 2014; Truman et al., 2008; Tsai et al., 2014). Beyond simple recruitment, however, FKN can also enhance the ability of macrophages and microglia to execute their phagocytic functions (Miksa, Amin, Wu, Ravikumar, & Wang, 2007).

Neuronally derived FKN promotes survival of microglia under neurotoxic conditions, inhibiting Fas-mediated death via upregulation of anti-apoptotic Bcl family proteins (Boehme, Lio, Maciejewski-Lenoir, Bacon, & Conlon, 2000). At the same time, FKN also triggers activation of the phagocytic response to clear cellular debris and stress–response pathways to counteract any remaining neurotoxic molecules that caused the initial damage (Noda et al., 2011). In the cardiovascular system, FKN released from apoptotic cells exerts both antiapoptotic and mitogenic effects on neighboring vascular smooth muscle cells (White et al., 2010). These authors demonstrated that FKN released by apoptotic cells binds its receptor CX3CR1 on smooth muscle cells and induces the expression of epiregulin, an epidermal growth factor receptor ligand. FKN may also promote proper wound healing and regeneration by inhibiting fibrotic responses to cell death. Engel and colleagues found CX3CR1 in the kidneys to be important for inhibiting proliferation of profibrotic macrophages (Engel et al., 2015). Taken together, release of FKN from apoptotic cells appears to guide the immune response to minimize further tissue damage and preserve tissue function.

2.2.3 Apoptotic Microblebs

Blebbing of the apoptotic cellular membrane is easily seen on electron micrographs and was one of the early defining features that distinguished apoptotic cells from necrotic cells. Membrane blebbing is driven by caspase-mediated activation of the Rho-associated kinase ROCK I (Coleman et al., 2001; Sebbagh et al., 2001). These approximately five micron particles function as chemoattractants, recruiting monocytes to sites of apoptosis (Segundo et al., 1999) possibly based on the classic attractants they harbor on their membranes (Tsai et al., 2014). Upon calcium-activated release, these microblebs spread out through the intercellular spaces, potentially traveling great distances within the organism (Hoang, Rampon, Freyssinet, Vriz, & Kerbiriou-Nabias, 2011). For example, circulating apoptotic microblebs are found abundantly in patients with cardiovascular disease (Rautou et al., 2011), and directly correlate with the degree of disease-associated vascular endothelial dysfunction and cell death (Werner, Wassmann, Ahlers, Kosiol, & Nickenig, 2006). In an epidemiological study, higher levels of circulating microblebs are associated with greater risk of cardiovascular death, independent of other classic cardiovascular risk factors (Chistiakov, Orekhov, & Bobryshev, 2015; Sinning et al., 2011). Microblebs are therefore an intriguing potential biomarker for those patients in need of more aggressive therapy and risk management.

Beyond serving as a chemoattractant, and potential biomarker for disease, these circulating microblebs are also capable of altering the biological state of other cells, both near and far from the site of apoptosis. Jansen and colleagues reported that microblebs derived from apoptotic endothelial cells can be taken up by target surviving endothelial cells via the PS receptor. They are then capable of inhibiting p38 activity, thus promoting an anti-apoptotic state in the surviving cells (Jansen et al., 2012). The microblebs also carry the instructions to recruit endothelial progenitor cells to sites of excessive cell death, such as atherosclerotic plaques (Hristov, Erl, Linder, & Weber, 2004; Zernecke et al., 2009). These instructions include an enrichment of microRNA-126, which upon transfer to the recipient endothelial cell, promotes the increased production of CXCL12, a progenitor chemoattractant (Zernecke et al., 2009). Recruited progenitors in turn promote regeneration, stabilization of the plaque, and maintenance of vascular function. Therefore, apoptotic microblebs are versatile messengers, carrying whatever final instructions the apoptotic cell wished to communicate, allowing these signals to persist long after the dying cell corpse has been removed.

3. APOPTOTIC CELLS DIRECTLY INFLUENCE TISSUE HOMEOSTASIS AND GROWTH CONTROL

Kerr, Wylie, and Currie initially characterized apoptosis as the kinetic counterpoint to mitosis, a process that actively contributes to tissue homeostasis and maintenance of cell populations (Kerr et al., 1972). Yet, since that time, apoptosis has often been regarded as an altruistic cell death, indirectly affecting growth control through a passive death—a solitary and final action for the greater benefit of the tissue (Czabotar et al., 2014). An emerging field of study focuses on the pathways utilized by apoptotic cells to produce instructive signals influencing the growth state of neighboring survivor cells. Depending on the context, apoptotic cells are capable of producing proapoptotic, antiapoptotic, mitogenic, and morphogenetic signals that act directly on the surrounding tissue, without an immunological intermediate (Fig. 2).

3.1 Apoptosis-Induced Apoptosis: Communal Cell Death

One of the earliest descriptions of programmed cell death was the case of coordinated cell loss during embryogenesis (Glucksmann, 1951).

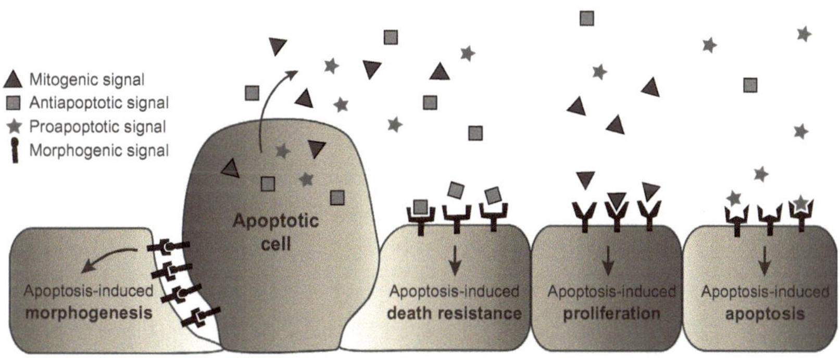

Figure 2 Apoptotic cells direct growth control. Apoptotic cells can produce a variety of signals that can have an effect on the growth state of the surviving tissue. The outcome of this communication can depend on both the amounts of signal produced and the receptivity of the surviving cell to these different stimuli. (See the color plate.)

Throughout development, there are several instances where entire populations of cells initiate apoptosis almost simultaneously. In certain contexts at least, this communal, or cohort, cell death is triggered by proapoptotic signals originating from the initial apoptotic cells (Perez-Garijo, Fuchs, & Steller, 2013). These authors found in developing *Drosophila* epithelial tissues (wing imaginal discs) that induction of cell death in one tissue compartment resulted in additional ectopic apoptosis in distant compartments. This apoptosis-induced apoptosis (AiA) is dependent on the production and release of the *Drosophila* TNFα ortholog, Eiger, from the initial population of dying cells. Interestingly, not all cells in the disc become apoptotic; healthy nonapoptotic cells persist between the two populations of dying cells. These intervening healthy cells are known to be particularly resistant to apoptotic stimuli, whereas the cells subject to AiA are known to be particularly susceptible (Milan, Campuzano, & Garcia-Bellido, 1997; Moon et al., 2005). This susceptibility requirement fits well with the observed phenomenon of rapid, complete, and specific cohort cell death.

Perez-Garijo et al. (2013) also investigated AiA in a mammalian system. The mammalian hair follicle experiences cohort cell death during the normal regressive phase of the hair cycle, and previous studies established that the coordination of this cell death is dependent on TNFα (Botchkareva, Ahluwalia, & Shander, 2006; Lindner et al., 1997; Tong & Coulombe, 2006). TNFα production was limited to apoptotic cells, and inhibition of TNFα signaling reduced the number of apoptotic cells. This apoptosis-derived feed-forward loop provides a mechanism for the observed cohort cell death in the hair follicle. However, the direct connection between caspase activation and TNFα production in both the fly and mammalian hair follicle remains to be determined.

AiA may also contribute to the complex cellular responses observed following radiation treatments or following certain pathological conditions. The radiation bystander effect includes the ability of irradiated cells to induce the death of healthy nonirradiated cells (Hei, Zhou, Chai, Ponnaiya, & Ivanov, 2011; Prise & O'Sullivan, 2009). Additionally, dying cells that are damaged by infection or ischemia are known to propagate death to neighboring cells (Barber, 2001). Most spreading death seems to be derived from necrotic cells, but it is possible in these cases there is a specific apoptotic component that drives AiA. Further research may determine if there are specific apoptosis-derived signals that mediate these pathological forms of cohort death.

3.2 Apoptosis-Induced Death Resistance: The Mahakali Effect

In contrast to the radiation bystander effect of AiA, recent studies in *Drosophila* have also found an apoptosis-induced death resistance (AiDR) program. Jaklevic et al. (2008) first noted that ionizing radiation of the developing wing disc generally increases the levels of *bantam*, a microRNA that stimulates cell proliferation and inhibits apoptosis by repressing the proapoptotic factor *hid* (Brennecke, Hipfner, Stark, Russell, & Cohen, 2003; Jaklevic et al., 2008). Additionally, *bantam* null animals were exceptionally susceptible to ionizing radiation, exhibiting increased apoptosis. However, follow-up studies by the same group revealed the surprising finding that when apoptosis is limited to a specific domain, the increase in *bantam*, and thus the antiapoptotic effect, was nonautonomous (Bilak, Uyetake, & Su, 2014). These authors found that AiDR, which they termed the "Mahakali" effect, was mediated by the receptor tyrosine kinase Tie on the surviving cell, and the apoptosis-dependent production of the Pvf1 ligand. Interestingly, Pvf1 was required, but not sufficient, to induce AiDR, suggesting that other apoptosis-derived signals may also be required to stimulate this prosurvival effect.

Angiopoeitin-1 is the mammalian ligand for Tie-2, and has been described as a nonautonomous apoptosis survival factor in human cell culture (Kwak, So, Lee, Kim, & Koh, 1999). Bilak et al. (2014) suggested that Angiopoeitin-1 may be released by neighboring cells to protect the endothelial cells; however, there is no evidence to date of AiDR in mammals.

3.3 Apoptosis-Induced Proliferation: Caspase-Driven Compensatory Proliferation

Compensatory proliferation can occur during regeneration of lost tissue via additional or accelerated cell divisions. In *Drosophila*, some of the first evidence for compensatory proliferation was uncovered by Haynie and Bryant when they demonstrated that up to sixty percent of cells in developing wing precursor tissue could be eliminated by radiation, and yet extra cell divisions within the surviving tissue resulted in a full-sized and normally functioning adult wing (Haynie & Bryant, 1977).

In 2004, it was reported that in the developing *Drosophila* epithelial tissues, induction of apoptotic cell death could induce nonautonomous proliferation in the surrounding cells (Huh, Guo, & Hay, 2004; Perez-Garijo, Martin, & Morata, 2004; Ryoo, Gorenc, & Steller, 2004). Since that time, there has been a mounting interest in the concept that this proliferation, under certain contexts, is driven explicitly by mitogenic signals produced by the dying cell. There is still controversy today regarding which signals are actually produced by the dying cell, versus which mitogenic factors may be produced elsewhere, but in the past decade numerous studies described below have validated the concept in both flies and mammals. Apoptosis-induced proliferation (AiP) therefore is a form of compensatory proliferation and is defined as the process by which apoptotic cells actively stimulate surviving cells to divide (Mollereau et al., 2013).

3.3.1 AiP in Drosophila Depends on the Environmental Context of the Dying Cell

In vivo studies of apoptotic cells and any apoptosis-derived signals during normal development are challenging due to the fleeting nature of the cell death and signal. However, in *Drosophila*, by capitalizing on the use of the effector caspase inhibitor P35, it is possible to examine the pathways driving AiP (Hay, Wolff, & Rubin, 1994; Huh et al., 2004; Perez-Garijo et al., 2004; Ryoo et al., 2004). Through misexpressing *p35* in one compartment of the developing wing, at the same time as triggering an apoptotic stimulus, these researchers were able to uncouple the initiation of the apoptotic signaling cascade from the actual execution of cell death. This generates an "undead" state in which any apoptosis-derived signals are sustained. In this case, mitogen production was sustained, resulting in excessive proliferation and tissue overgrowth. Importantly, proliferation occurs in both the posterior compartment where the undead apoptotic cells exist, and in the genetically unaffected anterior region (Huh et al., 2004; Ryoo et al., 2004). This strongly suggested the presence of a secreted and diffusible mitogen. Two mitogens identified at the time included Wingless (Wg, an ortholog of Wnt) and Decapentaplegic (Dpp, ortholog of TGFβ) (Perez-Garijo et al., 2004; Ryoo et al., 2004). Follow-up work has also identified a role for Spitz (Spi, the EGF ortholog in *Drosophila*) (Fan et al., 2014). In the undead model, these mitogens are produced in response to c-Jun N-terminal kinase (JNK) activity in the undead cells and are required for tissue overgrowth (Bergantinos, Corominas, & Serras, 2010; Fan et al., 2014; Ryoo et al., 2004). However, in genuine models of AiP that do not rely on *p35* expression, the requirement and source of these mitogens is still under debate (Martin, Perez-Garijo, & Morata, 2009; Perez-Garijo, Martin, Struhl, & Morata, 2005; Perez-Garijo, Shlevkov, & Morata, 2009; Smith-Bolton, Worley, Kanda, & Hariharan, 2009).

A common hypothesis in these studies had been that, following apoptosis induction, there is a bifurcation in the signaling cascade that ultimately results in both nonautonomous proliferation and autonomous execution of cell death. In this specific context, the bifurcation is at the level of the *Drosophila* initiator caspase, Dronc (Caspase-9) (Fan et al., 2014; Huh et al., 2004; Kondo, Senoo-Matsuda, Hiromi, & Miura, 2006; Wells, Yoshida, & Johnston, 2006). In genuine apoptotic cells, Dronc activates the effector caspases Drice and Dcp-1 (Caspase-3 and -7), as well as a currently unidentified target upstream of JNK activation, leading to compensatory proliferation (Fan et al., 2014). In undead cells, effector caspase activity is inhibited, and so Dronc continuously signals for compensatory proliferation, thus driving the tissue to hyperproliferation. Thus, the undead model is a potentially fascinating tool to understand how AiP may play a role in the development of cancers (Bergmann & Steller, 2010; Ryoo & Bergmann, 2012).

Interestingly, this Dronc-dependent signaling cascade only applies to epithelial cells that have not begun terminal differentiation. Undifferentiated, actively proliferating tissue exists in the wing and the anterior of the eye imaginal disc. However, in the posterior eye disc differentiated photoreceptors induce AiP by a completely different mechanism (Fan & Bergmann, 2008a, 2008b). Here, apoptotic photoreceptor cells trigger a Drice- and Dcp-1-dependent cascade that leads to release of Hedgehog (Hh). Hh secreted from these apical cells then stimulates the underlying unspecified progenitor cells to reenter the cell cycle and proliferate.

The idea of context-dependent apoptosis-derived mitogenic signals has opened up this field to a wealth of possibilities. Ongoing work is investigating the spectrum of AiP-derived mitogenic signals in the flies as well as the context-specific determinants of survivor receptivity to proliferative signals.

3.3.2 AiP in Mammals: Phoenix Rising and Tumor Repopulation

Unlike developmental apoptosis which is neatly confined to small cell populations or well-defined compartments, the cell death associated with trauma often cuts across large swaths of cell types at varying developmental stages. Therefore, one of the most visible characteristics of mammalian wound healing and tissue regeneration is the infiltration of immune cells to the wound site (Haertel, Werner, & Schafer, 2014; Muller, Meyer, & Werner, 2012). This inflammatory reaction precedes regeneration, and the infiltrating cells disperse after the wound resolves. Therefore, it was long assumed that this type of regeneration required proliferative signals from the inflammatory cells. Immune cells are certainly capable of producing a wide variety of cytokines and growth factors, which in many cases likely do contribute to healing growth. However, PU.1 knockout mice lack the inflammatory cells typically associated with wound healing, including macrophages and neutrophils (Scott, Simon, Anastasi, & Singh, 1994), yet these animals are capable of fully repairing damage to their skin (Martin et al., 2003). This demonstrated that there is something intrinsic to the wounded tissue that can also stimulate regrowth and healing. Based on this observation, Li et al. hypothesized that the initial proliferative signal could be derived directly from the damaged tissue, specifically apoptotic cells (Li et al., 2010). This group established a method to test in *in vitro* coculture assays the effect of a large number of irradiated dying cells on a smaller population of fluorescently labeled nonirradiated surviving cells. Excitingly, in these assays, dying cells stimulated proliferation of the fluorescently labeled cells via Caspase-3-dependent activation of calcium-independent phospholipase A_2 (iPLA_2). This enzyme had previously been identified as a caspase cleavage target important for the release of phospholipid derived "find me" signals (see Section 2.1.1) (Atsumi et al., 1998; Lauber et al., 2003). In this model of regeneration, however, the important signal produced by iPLA_2 is prostaglandin E_2 (PGE_2) which promotes stem and progenitor cell proliferation (Hagedorn, Durand, Fast, & Zon, 2014). Li et al. (2010) termed this mammalian regenerative AiP pathway "Phoenix Rising."

Following this critical mammalian study, a series of reports found that the hypothetical scenario posited by the invertebrate field does in fact exist in mammals: The Phoenix Rising pathway can be co-opted in cancer and can contribute to tumor repopulation following radiation and chemotherapy (Huang et al., 2011). These findings were validated in breast cancer, melanoma, and pancreatic ductal adenocarcinoma cell lines (Cheng et al., 2015; Donato et al., 2014; Kurtova et al., 2015).

This Caspase-3/iPLA_2/PGE_2 signaling cascade can be found in dying tumor cells promoting growth of surviving tumor cells, but also extends to dying vascular endothelial cells promoting tumor cell growth. One treatment considered for many solid tumors is antiangiogenic therapy to limit oxygen to the developing tumor mass (Sitohy, Nagy, & Dvorak, 2012). However, if dying vascular endothelial cells are also capable of activating the Caspase-3/iPLA_2/PGE_2 cascade, such as in one study of glioma cells, targeting this AiP pathway may be critically important for successful therapy (Mao, Smith, Xie, & Wang, 2013).

In addition to the Caspase-3/iPLA_2/PGE_2 signaling, in certain cancer cell lines such as the Panc1 line derived from ductal carcinoma, there is also a role for 7-PKCδ-Akt/p38 MAPK-stimulated mitogen production (Cheng et al., 2015). Also, there is evidence of Sonic Hedgehog modulating AiP, feeding back on dying cells in a paracrine manner to regulate WNT signaling via SFRP1 (secreted frizzled related protein 1). SFRP1 acts as an antagonist of WNT signaling, which allows for tumor repopulation (Ma, Cheng, Gong, Tian, & Huang, 2015).

These human cell line studies, taken with the *in vivo* work in *Drosophila*, would suggest that AiP is likely a commonly used mechanism to promote regeneration, whether normal or cancerous. Historically, cancer has been called a wound that will not heal (Dvorak, 1986). It is possible that the very pathways designed to promote wound healing are the ones promoting tumor growth. Most notably, given the diverse signals already cataloged, it is very likely that we have only begun to uncover the many paths to AiP, which will vary based on the dying cell, the surviving cell, the developmental context, and the surrounding microenvironment.

3.4 Apoptosis-Induced Morphogenesis: Directed Maintenance of Tissue Integrity

Finally, independent of the growth state of the surrounding environment, apoptotic cells can have a profound effect on the organization of the surviving tissue. Apoptosis-derived signals can direct tissue reorganization and promote maintenance of tissue integrity and function.

3.4.1 Apoptosis-Directed Extrusion from Epithelial Layers

When stromal cells or individual immune cells activate an apoptotic program, their clearance and subsequent absence has relatively little effect on the structure of their surrounding environment. However, if an epithelial cell activates apoptosis, its unexpected loss could result in a gap or discontinuity of the functional layer where it resides. Proper extrusion of the dying cell from this layer maintains epithelial integrity and preserves its barrier function. Work by Rosenblatt, Raff, and Cramer initially determined that proper extrusion was dependent on changes to the cytoskeleton in both the dying and neighboring cells, but that the process originated from the apoptotic cell (Rosenblatt, Raff, & Cramer, 2001). Follow-up studies determined that this coordinated cytoskeletal rearrangement is in fact dependent on a caspase-mediated signal from the apoptotic cell (Andrade & Rosenblatt, 2011). S1P produced by the apoptotic cell (see Section 2.1.2) binds to the G-protein-coupled receptor $S1P_2$ on neighboring live cells (Gu et al., 2011). Interestingly, Gu et al. noted that $S1P_2$

expression is reduced in a number of cancers (Gu et al., 2015). Their investigation centered on evaluating the effects of blocking S1P activity by (1) reducing S1P levels, (2) reducing $S1P_2$ levels, or (3) blocking their interaction by neutralizing antibody. All of these conditions resulted in defects in proper apical extrusion, and commonly resulted in cells being extruded basally. In this scenario, Gu et al. proposed that transformed cells are more likely to persist in this basal compartment and potentially invade and metastasize. Enhancing proper apical extrusion could reduce tumorigenesis.

3.4.2 Apoptosis-Induced Sprouting Morphogenesis

An overlooked domain in cell biology, though not a signal in the usual sense, biophysical properties of dying cells can even serve to communicate with the surrounding environment. Apoptotic cells direct the sprouting morphogenesis of vascular endothelial cells based on the accumulated negative charges on the apoptotic cell membrane (Weihua, Tsan, Schroit, & Fidler, 2005). Specifically, negatively charged apoptotic membranes attract hyperpolarized endothelial cell membrane extensions, which then provide the scaffold for recruited proliferating endothelial cells. Revascularization following injury is important for wound healing, and angiogenesis is critically involved in tumor progression. Better understanding of the biophysical properties of apoptotic cells, and how these properties complement or contradict secreted signals could potentially provide mechanisms to enhance or inhibit these angiogenic events.

4. APOPTOSIS: A LOUD DEATH

Under normal physiological conditions, we should not think of dead apoptotic cells. Instead, there are actively dying apoptotic cells that interact quite extensively with their environment to coordinate a host of physiological processes. These dying cells are eventually cleared and digested by phagocytes, which cease their instructive signals and resolves the death process. Before this clearance, however, apoptotic cells actively produce signals, both secreted and membrane bound, that serve to dictate their own immunologic control and fate, as well as direct growth control to ensure tissue integrity and homeostasis.

A common concern often presented in the literature is the seemingly contradictory effects of dying cells on their neighbors. This has been especially true, when researchers are working in the same tissue such as the *Drosophila* wing discs. How can the same dying cells promote cell death and cell proliferation? The reality of the apoptotic cell is likely much more complex than we have outlined here. Dying cells may produce any combination of these signals depending on their predeath state, the specific mode of apoptotic induction, the presence or absence of immune cells as signals are being released, or feedback from the surrounding immune cells and microenvironment. Additionally, signals from dying apoptotic cells will only have as much effect on neighboring cells as those neighbors are receptive to hearing them. Progrowth and prodeath signals may originate from the same dying cell, with the winning signal dependent on the current state of the neighbor. While apoptotic death is rapid, it is not instantaneous, and in that dying moment, it is anything but silent.

ACKNOWLEDGMENTS

We would like to thank our colleagues for helpful discussions during the course of this work. We apologize to those whose work may not have been cited due to space restrictions. We would like to acknowledge Christopher MacKay for his assistance with the artwork. C.E.F. would like to thank the UMMS MD/PhD program for ongoing support. This work was supported by the NIH (GM068016 and GM107789).

REFERENCES

Abrams, J. M., White, K., Fessler, L. I., & Steller, H. (1993). Programmed cell death during Drosophila embryogenesis. *Development*, *117*(1), 29–43.

Andrade, D., & Rosenblatt, J. (2011). Apoptotic regulation of epithelial cellular extrusion. *Apoptosis*, *16*(5), 491–501. http://dx.doi.org/10.1007/s10495-011-0587-z.

Atsumi, G., Tajima, M., Hadano, A., Nakatani, Y., Murakami, M., & Kudo, I. (1998). Fas-induced arachidonic acid release is mediated by Ca2+ – independent phospholipase A2 but not cytosolic phospholipase A2, which undergoes proteolytic inactivation. *The Journal of Biological Chemistry*, *273*(22), 13870–13877.

Bar, P. R. (1996). Apoptosis—The cell's silent exit. *Life Sciences*, *59*(5–6), 369–378.

Barber, G. N. (2001). Host defense, viruses and apoptosis. *Cell Death and Differentiation*, *8*(2), 113–126. http://dx.doi.org/10.1038/sj.cdd.4400823.

Behrensdorf, H. A., van de Craen, M., Knies, U. E., Vandenabeele, P., & Clauss, M. (2000). The endothelial monocyte-activating polypeptide II (EMAP II) is a substrate for caspase-7. *FEBS Letters*, *466*(1), 143–147.

Bergantinos, C., Corominas, M., & Serras, F. (2010). Cell death-induced regeneration in wing imaginal discs requires JNK signalling. *Development*, *137*(7), 1169–1179. http://dx.doi.org/10.1242/dev.045559. 137/7/1169 [pii].

Bergmann, A. (2010). The role of ubiquitylation for the control of cell death in Drosophila. *Cell Death and Differentiation*, *17*(1), 61–67. http://dx.doi.org/10.1038/cdd.2009.70. cdd200970 [pii].

Bergmann, A., & Steller, H. (2010). Apoptosis, stem cells, and tissue regeneration. *Science Signaling*, *3*(145), re8. http://dx.doi.org/10.1126/scisignal.3145re8. scisignal.3145re8 [pii].

Bilak, A., Uyetake, L., & Su, T. T. (2014). Dying cells protect survivors from radiation-induced cell death in Drosophila. *PLoS Genetics*, *10*(3), e1004220. http://dx.doi.org/10.1371/journal.pgen.1004220.

Boehme, S. A., Lio, F. M., Maciejewski-Lenoir, D., Bacon, K. B., & Conlon, P. J. (2000). The chemokine fractalkine inhibits Fas-mediated cell death of brain microglia. *Journal of Immunology, 165*(1), 397–403.

Botchkareva, N. V., Ahluwalia, G., & Shander, D. (2006). Apoptosis in the hair follicle. *The Journal of Investigative Dermatology, 126*(2), 258–264. http://dx.doi.org/10.1038/sj.jid.5700007.

Bratton, D. L., Fadok, V. A., Richter, D. A., Kailey, J. M., Guthrie, L. A., & Henson, P. M. (1997). Appearance of phosphatidylserine on apoptotic cells requires calcium-mediated nonspecific flip-flop and is enhanced by loss of the aminophospholipid translocase. *The Journal of Biological Chemistry, 272*(42), 26159–26165.

Brennecke, J., Hipfner, D. R., Stark, A., Russell, R. B., & Cohen, S. M. (2003). bantam encodes a developmentally regulated microRNA that controls cell proliferation and regulates the proapoptotic gene hid in Drosophila. *Cell, 113*(1), 25–36.

Brustugun, O. T., Fladmark, K. E., Doskeland, S. O., Orrenius, S., & Zhivotovsky, B. (1998). Apoptosis induced by microinjection of cytochrome c is caspase-dependent and is inhibited by Bcl-2. *Cell Death and Differentiation, 5*(8), 660–668. http://dx.doi.org/10.1038/sj.cdd.4400399.

Chen, Y. Z., Mapes, J., Lee, E. S., Skeen-Gaar, R. R., & Xue, D. (2013). Caspase-mediated activation of Caenorhabditis elegans CED-8 promotes apoptosis and phosphatidylserine externalization. *Nature Communications, 4*, 2726. http://dx.doi.org/10.1038/ncomms3726.

Chen, J., Zhao, Y., & Liu, Y. (2014). The role of nucleotides and purinergic signaling in apoptotic cell clearance—Implications for chronic inflammatory diseases. *Frontiers in Immunology, 5*, 656. http://dx.doi.org/10.3389/fimmu.2014.00656.

Cheng, J., Tian, L., Ma, J., Gong, Y., Zhang, Z., Chen, Z., et al. (2015). Dying tumor cells stimulate proliferation of living tumor cells via caspase-dependent protein kinase Cδ activation in pancreatic ductal adenocarcinoma. *Molecular Oncology, 9*(1), 105–114. http://dx.doi.org/10.1016/j.molonc.2014.07.024.

Chistiakov, D. A., Orekhov, A. N., & Bobryshev, Y. V. (2015). Extracellular vesicles and atherosclerotic disease. *Cellular and Molecular Life Sciences, 72*, 2697–2708. http://dx.doi.org/10.1007/s00018-015-1906-2.

Coleman, M. L., Sahai, E. A., Yeo, M., Bosch, M., Dewar, A., & Olson, M. F. (2001). Membrane blebbing during apoptosis results from caspase-mediated activation of ROCK I. *Nature Cell Biology, 3*(4), 339–345. http://dx.doi.org/10.1038/35070009.

Conradt, B. (2009). Genetic control of programmed cell death during animal development. *Annual Review of Genetics, 43*, 493–523. http://dx.doi.org/10.1146/annurev.genet.42.110807.091533.

Crawford, E. D., Seaman, J. E., Barber, A. E., 2nd, David, D. C., Babbitt, P. C., Burlingame, A. L., et al. (2012). Conservation of caspase substrates across metazoans suggests hierarchical importance of signaling pathways over specific targets and cleavage site motifs in apoptosis. *Cell Death and Differentiation, 19*(12), 2040–2048. http://dx.doi.org/10.1038/cdd.2012.99. cdd201299 [pii].

Czabotar, P. E., Lessene, G., Strasser, A., & Adams, J. M. (2014). Control of apoptosis by the BCL-2 protein family: Implications for physiology and therapy. *Nature Reviews. Molecular Cell Biology, 15*(1), 49–63. http://dx.doi.org/10.1038/nrm3722.

Danial, N. N., & Korsmeyer, S. J. (2004). Cell death: Critical control points. *Cell, 116*(2), 205–219.

Domingos, P. M., & Steller, H. (2007). Pathways regulating apoptosis during patterning and development. *Current Opinion in Genetics & Development, 17*(4), 294–299. http://dx.doi.org/10.1016/j.gde.2007.05.009.

Donato, A. L., Huang, Q., Liu, X., Li, F., Zimmerman, M. A., & Li, C. Y. (2014). Caspase 3 promotes surviving melanoma tumor cell growth after cytotoxic therapy. *The Journal of Investigative Dermatology, 134*(6), 1686–1692. http://dx.doi.org/10.1038/jid.2014.18.

Dvorak, H. F. (1986). Tumors: Wounds that do not heal. Similarities between tumor stroma generation and wound healing. *The New England Journal of Medicine*, *315*(26), 1650–1659. http://dx.doi.org/10.1056/NEJM198612253152606.

Ellis, H. M., & Horvitz, H. R. (1986). Genetic control of programmed cell death in the nematode C. elegans. *Cell*, *44*(6), 817–829.

Enari, M., Sakahira, H., Yokoyama, H., Okawa, K., Iwamatsu, A., & Nagata, S. (1998). A caspase-activated DNase that degrades DNA during apoptosis, and its inhibitor ICAD. *Nature*, *391*(6662), 43–50. http://dx.doi.org/10.1038/34112.

Engel, D. R., Krause, T. A., Snelgrove, S. L., Thiebes, S., Hickey, M. J., Boor, P., et al. (2015). CX3CR1 reduces kidney fibrosis by inhibiting local proliferation of profibrotic macrophages. *Journal of Immunology*, *194*(4), 1628–1638. http://dx.doi.org/10.4049/jimmunol.1402149.

Fadok, V. A., Bratton, D. L., Konowal, A., Freed, P. W., Westcott, J. Y., & Henson, P. M. (1998). Macrophages that have ingested apoptotic cells *in vitro* inhibit proinflammatory cytokine production through autocrine/paracrine mechanisms involving TGF-beta, PGE2, and PAF. *The Journal of Clinical Investigation*, *101*(4), 890–898. http://dx.doi.org/10.1172/JCI1112.

Fadok, V. A., Bratton, D. L., Rose, D. M., Pearson, A., Ezekewitz, R. A., & Henson, P. M. (2000). A receptor for phosphatidylserine-specific clearance of apoptotic cells. *Nature*, *405*(6782), 85–90. http://dx.doi.org/10.1038/35011084.

Fadok, V. A., de Cathelineau, A., Daleke, D. L., Henson, P. M., & Bratton, D. L. (2001). Loss of phospholipid asymmetry and surface exposure of phosphatidylserine is required for phagocytosis of apoptotic cells by macrophages and fibroblasts. *The Journal of Biological Chemistry*, *276*(2), 1071–1077. http://dx.doi.org/10.1074/jbc.M003649200.

Fadok, V. A., Voelker, D. R., Campbell, P. A., Cohen, J. J., Bratton, D. L., & Henson, P. M. (1992). Exposure of phosphatidylserine on the surface of apoptotic lymphocytes triggers specific recognition and removal by macrophages. *Journal of Immunology*, *148*(7), 2207–2216.

Fan, Y., & Bergmann, A. (2008a). Apoptosis-induced compensatory proliferation. The Cell is dead. Long live the Cell! *Trends in Cell Biology*, *18*(10), 467–473. http://dx.doi.org/10.1016/j.tcb.2008.08.001. S0962-8924(08)00212-2 [pii].

Fan, Y., & Bergmann, A. (2008b). Distinct mechanisms of apoptosis-induced compensatory proliferation in proliferating and differentiating tissues in the Drosophila eye. *Developmental Cell*, *14*(3), 399–410. http://dx.doi.org/10.1016/j.devcel.2008.01.003. S1534-5807(08)00031-2 [pii].

Fan, Y., Wang, S., Hernandez, J., Yenigun, V. B., Hertlein, G., Fogarty, C. E., et al. (2014). Genetic models of apoptosis-induced proliferation decipher activation of JNK and identify a requirement of EGFR signaling for tissue regenerative responses in Drosophila. *PLoS Genetics*, *10*(1), e1004131. http://dx.doi.org/10.1371/journal.pgen.1004131. PGENETICS-D-12-01255 [pii].

Fuchs, Y., & Steller, H. (2011). Programmed cell death in animal development and disease (Research Support, Non-U.S. Gov't. Review) *Cell*, *147*(4), 742–758. http://dx.doi.org/10.1016/j.cell.2011.10.033.

Garton, K. J., Gough, P. J., Blobel, C. P., Murphy, G., Greaves, D. R., Dempsey, P. J., et al. (2001). Tumor necrosis factor-alpha-converting enzyme (ADAM17) mediates the cleavage and shedding of fractalkine (CX3CL1). *The Journal of Biological Chemistry*, *276*(41), 37993–38001. http://dx.doi.org/10.1074/jbc.M106434200.

Gavrieli, Y., Sherman, Y., & Ben-Sasson, S. A. (1992). Identification of programmed cell death in situ via specific labeling of nuclear DNA fragmentation. *The Journal of Cell Biology*, *119*(3), 493–501.

Glucksmann, A. (1951). Cell deaths in normal vertebrate ontogeny. *Biological Reviews of the Cambridge Philosophical Society*, *26*(1), 59–86.

Gu, Y., Forostyan, T., Sabbadini, R., & Rosenblatt, J. (2011). Epithelial cell extrusion requires the sphingosine-1-phosphate receptor 2 pathway. *The Journal of Cell Biology, 193*(4), 667–676. http://dx.doi.org/10.1083/jcb.201010075.

Gu, Y., Shea, J., Slattum, G., Firpo, M. A., Alexander, M., Mulvihill, S. J., et al. (2015). Defective apical extrusion signaling contributes to aggressive tumor hallmarks. *eLife, 4*, e04069. http://dx.doi.org/10.7554/eLife.04069.

Gude, D. R., Alvarez, S. E., Paugh, S. W., Mitra, P., Yu, J., Griffiths, R., et al. (2008). Apoptosis induces expression of sphingosine kinase 1 to release sphingosine-1-phosphate as a "come-and-get-me" signal. *The FASEB Journal, 22*(8), 2629–2638. http://dx.doi.org/10.1096/fj.08-107169.

Haertel, E., Werner, S., & Schafer, M. (2014). Transcriptional regulation of wound inflammation. *Seminars in Immunology, 26*(4), 321–328. http://dx.doi.org/10.1016/j.smim.2014.01.005.

Hagedorn, E. J., Durand, E. M., Fast, E. M., & Zon, L. I. (2014). Getting more for your marrow: Boosting hematopoietic stem cell numbers with PGE2. *Experimental Cell Research, 329*(2), 220–226. http://dx.doi.org/10.1016/j.yexcr.2014.07.030.

Hay, B. A., Wolff, T., & Rubin, G. M. (1994). Expression of baculovirus P35 prevents cell death in Drosophila. *Development, 120*(8), 2121–2129.

Haynie, J. L., & Bryant, P. J. (1977). The effects of X-rays on the proliferation dynamics of cells in the imaginal wing disc of Drosophila melanogaster. *Development Genes and Evolution, 183*(2), 85–100.

Hei, T. K., Zhou, H., Chai, Y., Ponnaiya, B., & Ivanov, V. N. (2011). Radiation induced non-targeted response: Mechanism and potential clinical implications. *Current Molecular Pharmacology, 4*(2), 96–105.

Hoang, T. Q., Rampon, C., Freyssinet, J. M., Vriz, S., & Kerbiriou-Nabias, D. (2011). A method to assess the migration properties of cell-derived microparticles within a living tissue. *Biochimica et Biophysica Acta, 1810*(9), 863–866. http://dx.doi.org/10.1016/j.bbagen.2011.05.003.

Hochreiter-Hufford, A., & Ravichandran, K. S. (2013). Clearing the dead: Apoptotic cell sensing, recognition, engulfment, and digestion. *Cold Spring Harbor Perspectives in Biology, 5*(1), a008748. http://dx.doi.org/10.1101/cshperspect.a008748.

Hristov, M., Erl, W., Linder, S., & Weber, P. C. (2004). Apoptotic bodies from endothelial cells enhance the number and initiate the differentiation of human endothelial progenitor cells *in vitro*. *Blood, 104*(9), 2761–2766. http://dx.doi.org/10.1182/blood-2003-10-3614.

Huang, Q., Li, F., Liu, X., Li, W., Shi, W., Liu, F. F., et al. (2011). Caspase 3-mediated stimulation of tumor cell repopulation during cancer radiotherapy. *Nature Medicine, 17*(7), 860–866. http://dx.doi.org/10.1038/nm.2385.

Huh, J. R., Guo, M., & Hay, B. A. (2004). Compensatory proliferation induced by cell death in the Drosophila wing disc requires activity of the apical cell death caspase Dronc in a nonapoptotic role. *Current Biology, 14*(14), 1262–1266.

Huynh, M. L., Fadok, V. A., & Henson, P. M. (2002). Phosphatidylserine-dependent ingestion of apoptotic cells promotes TGF-beta1 secretion and the resolution of inflammation. *The Journal of Clinical Investigation, 109*(1), 41–50. http://dx.doi.org/10.1172/JCI11638.

Jaklevic, B., Uyetake, L., Wichmann, A., Bilak, A., English, C. N., & Su, T. T. (2008). Modulation of ionizing radiation-induced apoptosis by bantam microRNA in Drosophila. *Developmental Biology, 320*(1), 122–130. http://dx.doi.org/10.1016/j.ydbio.2008.04.043.

Jansen, F., Yang, X., Hoyer, F. F., Paul, K., Heiermann, N., Becher, M. U., et al. (2012). Endothelial microparticle uptake in target cells is annexin I/phosphatidylserine receptor dependent and prevents apoptosis. *Arteriosclerosis, Thrombosis, and Vascular Biology, 32*(8), 1925–1935. http://dx.doi.org/10.1161/ATVBAHA.112.253229.

Kerr, J. F., Wyllie, A. H., & Currie, A. R. (1972). Apoptosis: A basic biological phenomenon with wide-ranging implications in tissue kinetics. *British Journal of Cancer, 26*(4), 239–257.

Kim, S. J., Gershov, D., Ma, X., Brot, N., & Elkon, K. B. (2002). I-PLA(2) activation during apoptosis promotes the exposure of membrane lysophosphatidylcholine leading to binding by natural immunoglobulin M antibodies and complement activation. *The Journal of Experimental Medicine, 196*(5), 655–665.

Knies, U. E., Behrensdorf, H. A., Mitchell, C. A., Deutsch, U., Risau, W., Drexler, H. C., et al. (1998). Regulation of endothelial monocyte-activating polypeptide II release by apoptosis. *Proceedings of the National Academy of Sciences of the United States of America, 95*(21), 12322–12327.

Kondo, S., Senoo-Matsuda, N., Hiromi, Y., & Miura, M. (2006). DRONC coordinates cell death and compensatory proliferation. *Molecular and Cellular Biology, 26*(19), 7258–7268.

Kurtova, A. V., Xiao, J., Mo, Q., Pazhanisamy, S., Krasnow, R., Lerner, S. P., et al. (2015). Blocking PGE2-induced tumour repopulation abrogates bladder cancer chemoresistance. *Nature, 517*(7533), 209–213. http://dx.doi.org/10.1038/nature14034.

Kwak, H. J., So, J. N., Lee, S. J., Kim, I., & Koh, G. Y. (1999). Angiopoietin-1 is an apoptosis survival factor for endothelial cells. *FEBS Letters, 448*(2–3), 249–253.

Lauber, K., Bohn, E., Krober, S. M., Xiao, Y. J., Blumenthal, S. G., Lindemann, R. K., et al. (2003). Apoptotic cells induce migration of phagocytes via caspase-3-mediated release of a lipid attraction signal. *Cell, 113*(6), 717–730.

Li, F., Huang, Q., Chen, J., Peng, Y., Roop, D. R., Bedford, J. S., et al. (2010). Apoptotic cells activate the "phoenix rising" pathway to promote wound healing and tissue regeneration. *Science Signaling, 3*(110), ra13. http://dx.doi.org/10.1126/scisignal.2000634.

Li, L. Y., Luo, X., & Wang, X. (2001). Endonuclease G is an apoptotic DNase when released from mitochondria. *Nature, 412*(6842), 95–99. http://dx.doi.org/10.1038/35083620.

Li, P., Nijhawan, D., Budihardjo, I., Srinivasula, S. M., Ahmad, M., Alnemri, E. S., et al. (1997). Cytochrome c and dATP-dependent formation of Apaf-1/caspase-9 complex initiates an apoptotic protease cascade. *Cell, 91*(4), 479–489.

Lindner, G., Botchkarev, V. A., Botchkareva, N. V., Ling, G., van der Veen, C., & Paus, R. (1997). Analysis of apoptosis during hair follicle regression (catagen). *The American Journal of Pathology, 151*(6), 1601–1617.

Liu, X., Zou, H., Slaughter, C., & Wang, X. (1997). DFF, a heterodimeric protein that functions downstream of caspase-3 to trigger DNA fragmentation during apoptosis. *Cell, 89*(2), 175–184.

Lockshin, R. A., & Williams, C. M. (1965). Programmed cell death—I. Cytology of degeneration in the intersegmental muscles of the pernyi silkmoth. *Journal of Insect Physiology, 11*, 123–133.

Ma, J., Cheng, J., Gong, Y., Tian, L., & Huang, Q. (2015). Wnt signaling downregulated after sonic hedgehog activation negatively associates with tumor repopulation. *Disease Models & Mechanisms, 8*(4), 385–391. http://dx.doi.org/10.1242/dmm.018887.

Mandal, D., Mazumder, A., Das, P., Kundu, M., & Basu, J. (2005). Fas-, caspase 8-, and caspase 3-dependent signaling regulates the activity of the aminophospholipid translocase and phosphatidylserine externalization in human erythrocytes. *The Journal of Biological Chemistry, 280*(47), 39460–39467. http://dx.doi.org/10.1074/jbc.M506928200.

Mao, P., Smith, L., Xie, W., & Wang, M. (2013). Dying endothelial cells stimulate proliferation of malignant glioma cells via a caspase 3-mediated pathway. *Oncology Letters, 5*(5), 1615–1620. http://dx.doi.org/10.3892/ol.2013.1223.

Martin, P., D'Souza, D., Martin, J., Grose, R., Cooper, L., Maki, R., et al. (2003). Wound healing in the PU.1 null mouse—Tissue repair is not dependent on inflammatory cells. *Current Biology, 13*(13), 1122–1128.

Martin, S. J., Finucane, D. M., Amarante-Mendes, G. P., O'Brien, G. A., & Green, D. R. (1996). Phosphatidylserine externalization during CD95-induced apoptosis of cells and cytoplasts requires ICE/CED-3 protease activity. *The Journal of Biological Chemistry*, *271*(46), 28753–28756.

Martin, F. A., Perez-Garijo, A., & Morata, G. (2009). Apoptosis in Drosophila: Compensatory proliferation and undead cells. *The International Journal of Developmental Biology*, *53*(8–10), 1341–1347. http://dx.doi.org/10.1387/ijdb.072447fm. 072447fm [pii].

Martin, S. J., Reutelingsperger, C. P., McGahon, A. J., Rader, J. A., van Schie, R. C., LaFace, D. M., et al. (1995). Early redistribution of plasma membrane phosphatidylserine is a general feature of apoptosis regardless of the initiating stimulus: Inhibition by overexpression of Bcl-2 and Abl. *The Journal of Experimental Medicine*, *182*(5), 1545–1556.

Martinet, W., De Meyer, I., Cools, N., Timmerman, V., Bult, H., Bosmans, J., et al. (2010). Cell death-mediated cleavage of the attraction signal p43 in human atherosclerosis: Implications for plaque destabilization. *Arteriosclerosis, Thrombosis, and Vascular Biology*, *30*(7), 1415–1422. http://dx.doi.org/10.1161/ATVBAHA.110.206029.

Matschurat, S., Knies, U. E., Person, V., Fink, L., Stoelcker, B., Ebenebe, C., et al. (2003). Regulation of EMAP II by hypoxia. *The American Journal of Pathology*, *162*(1), 93–103. http://dx.doi.org/10.1016/S0002-9440(10)63801-1.

Miksa, M., Amin, D., Wu, R., Ravikumar, T. S., & Wang, P. (2007). Fractalkine-induced MFG-E8 leads to enhanced apoptotic cell clearance by macrophages. *Molecular Medicine*, *13*(11–12), 553–560. http://dx.doi.org/10.2119/2007-00019.Miksa.

Milan, M., Campuzano, S., & Garcia-Bellido, A. (1997). Developmental parameters of cell death in the wing disc of Drosophila. *Proceedings of the National Academy of Sciences of the United States of America*, *94*(11), 5691–5696.

Miura, M., Zhu, H., Rotello, R., Hartwieg, E. A., & Yuan, J. (1993). Induction of apoptosis in fibroblasts by IL-1 beta-converting enzyme, a mammalian homolog of the C. elegans cell death gene ced-3. *Cell*, *75*(4), 653–660.

Mollereau, B., Perez-Garijo, A., Bergmann, A., Miura, M., Gerlitz, O., Ryoo, H. D., et al. (2013). Compensatory proliferation and apoptosis-induced proliferation: A need for clarification. *Cell Death and Differentiation*, *20*(1), 181. http://dx.doi.org/10.1038/cdd.2012.82. cdd201282 [pii].

Monks, J., Rosner, D., Geske, F. J., Lehman, L., Hanson, L., Neville, M. C., et al. (2005). Epithelial cells as phagocytes: Apoptotic epithelial cells are engulfed by mammary alveolar epithelial cells and repress inflammatory mediator release. *Cell Death and Differentiation*, *12*(2), 107–114. http://dx.doi.org/10.1038/sj.cdd.4401517.

Monks, J., Smith-Steinhart, C., Kruk, E. R., Fadok, V. A., & Henson, P. M. (2008). Epithelial cells remove apoptotic epithelial cells during post-lactation involution of the mouse mammary gland. *Biology of Reproduction*, *78*(4), 586–594. http://dx.doi.org/10.1095/biolreprod.107.065045.

Moon, N. S., Frolov, M. V., Kwon, E. J., Di Stefano, L., Dimova, D. K., Morris, E. J., et al. (2005). Drosophila E2F1 has context-specific pro- and antiapoptotic properties during development. *Developmental Cell*, *9*(4), 463–475. http://dx.doi.org/10.1016/j.devcel.2005.08.015.

Mueller, R. B., Sheriff, A., Gaipl, U. S., Wesselborg, S., & Lauber, K. (2007). Attraction of phagocytes by apoptotic cells is mediated by lysophosphatidylcholine. *Autoimmunity*, *40*(4), 342–344. http://dx.doi.org/10.1080/08916930701356911.

Muller, A. K., Meyer, M., & Werner, S. (2012). The roles of receptor tyrosine kinases and their ligands in the wound repair process. *Seminars in Cell & Developmental Biology*, *23*(9), 963–970. http://dx.doi.org/10.1016/j.semcdb.2012.09.015.

Noda, M., Doi, Y., Liang, J., Kawanokuchi, J., Sonobe, Y., Takeuchi, H., et al. (2011). Fractalkine attenuates excito-neurotoxicity via microglial clearance of damaged neurons

and antioxidant enzyme heme oxygenase-1 expression. *The Journal of Biological Chemistry, 286*(3), 2308–2319. http://dx.doi.org/10.1074/jbc.M110.169839.
Perez-Garijo, A., Fuchs, Y., & Steller, H. (2013). Apoptotic cells can induce non-autonomous apoptosis through the TNF pathway. *eLife, 2*, e01004. http://dx.doi.org/10.7554/eLife.01004. 01004 [pii].
Perez-Garijo, A., Martin, F. A., & Morata, G. (2004). Caspase inhibition during apoptosis causes abnormal signalling and developmental aberrations in Drosophila. *Development, 131*(22), 5591–5598.
Perez-Garijo, A., Martin, F. A., Struhl, G., & Morata, G. (2005). Dpp signaling and the induction of neoplastic tumors by caspase-inhibited apoptotic cells in Drosophila. *Proceedings of the National Academy of Sciences of the United States of America, 102*(49), 17664–17669. http://dx.doi.org/10.1073/pnas.0508966102.
Perez-Garijo, A., Shlevkov, E., & Morata, G. (2009). The role of Dpp and Wg in compensatory proliferation and in the formation of hyperplastic overgrowths caused by apoptotic cells in the Drosophila wing disc. *Development, 136*(7), 1169–1177. http://dx.doi.org/10.1242/dev.034017. dev.034017 [pii].
Peter, C., Wesselborg, S., Herrmann, M., & Lauber, K. (2010). Dangerous attraction: Phagocyte recruitment and danger signals of apoptotic and necrotic cells. *Apoptosis, 15*(9), 1007–1028. http://dx.doi.org/10.1007/s10495-010-0472-1.
Poon, I. K., Lucas, C. D., Rossi, A. G., & Ravichandran, K. S. (2014). Apoptotic cell clearance: Basic biology and therapeutic potential. *Nature Reviews. Immunology, 14*(3), 166–180. http://dx.doi.org/10.1038/nri3607.
Prise, K. M., & O'Sullivan, J. M. (2009). Radiation-induced bystander signalling in cancer therapy. *Nature Reviews. Cancer, 9*(5), 351–360. http://dx.doi.org/10.1038/nrc2603.
Rautou, P. E., Vion, A. C., Amabile, N., Chironi, G., Simon, A., Tedgui, A., et al. (2011). Microparticles, vascular function, and atherothrombosis. *Circulation Research, 109*(5), 593–606. http://dx.doi.org/10.1161/CIRCRESAHA.110.233163.
Rosenblatt, J., Raff, M. C., & Cramer, L. P. (2001). An epithelial cell destined for apoptosis signals its neighbors to extrude it by an actin- and myosin-dependent mechanism. *Current Biology, 11*(23), 1847–1857.
Ryoo, H. D., & Bergmann, A. (2012). The role of apoptosis-induced proliferation for regeneration and cancer. *Cold Spring Harbor Perspectives in Biology, 4*(8), a008797. http://dx.doi.org/10.1101/cshperspect.a008797.
Ryoo, H. D., Gorenc, T., & Steller, H. (2004). Apoptotic cells can induce compensatory cell proliferation through the JNK and the Wingless signaling pathways. *Developmental Cell, 7*(4), 491–501.
Schwartz, L. M., Smith, S. W., Jones, M. E., & Osborne, B. A. (1993). Do all programmed cell deaths occur via apoptosis? *Proceedings of the National Academy of Sciences of the United States of America, 90*(3), 980–984.
Scott, E. W., Simon, M. C., Anastasi, J., & Singh, H. (1994). Requirement of transcription factor PU.1 in the development of multiple hematopoietic lineages. *Science, 265*(5178), 1573–1577.
Sebbagh, M., Renvoize, C., Hamelin, J., Riche, N., Bertoglio, J., & Breard, J. (2001). Caspase-3-mediated cleavage of ROCK I induces MLC phosphorylation and apoptotic membrane blebbing. *Nature Cell Biology, 3*(4), 346–352. http://dx.doi.org/10.1038/35070019.
Segawa, K., Kurata, S., Yanagihashi, Y., Brummelkamp, T. R., Matsuda, F., & Nagata, S. (2014). Caspase-mediated cleavage of phospholipid flippase for apoptotic phosphatidylserine exposure. *Science, 344*(6188), 1164–1168. http://dx.doi.org/10.1126/science.1252809.
Segundo, C., Medina, F., Rodriguez, C., Martinez-Palencia, R., Leyva-Cobian, F., & Brieva, J. A. (1999). Surface molecule loss and bleb formation by human germinal center

B cells undergoing apoptosis: Role of apoptotic blebs in monocyte chemotaxis. *Blood*, *94*(3), 1012–1020.

Sinning, J. M., Losch, J., Walenta, K., Bohm, M., Nickenig, G., & Werner, N. (2011). Circulating CD31+/Annexin V+ microparticles correlate with cardiovascular outcomes. *European Heart Journal*, *32*(16), 2034–2041. http://dx.doi.org/10.1093/eurheartj/ehq478.

Sitohy, B., Nagy, J. A., & Dvorak, H. F. (2012). Anti-VEGF/VEGFR therapy for cancer: Reassessing the target. *Cancer Research*, *72*(8), 1909–1914. http://dx.doi.org/10.1158/0008-5472.CAN-11-3406.

Smith-Bolton, R. K., Worley, M. I., Kanda, H., & Hariharan, I. K. (2009). Regenerative growth in Drosophila imaginal discs is regulated by Wingless and Myc. *Developmental Cell*, *16*(6), 797–809. http://dx.doi.org/10.1016/j.devcel.2009.04.015.

Sokolowski, J. D., Chabanon-Hicks, C. N., Han, C. Z., Heffron, D. S., & Mandell, J. W. (2014). Fractalkine is a "find-me" signal released by neurons undergoing ethanol-induced apoptosis. *Frontiers in Cellular Neuroscience*, *8*, 360. http://dx.doi.org/10.3389/fncel.2014.00360.

Steller, H. (1995). Mechanisms and genes of cellular suicide. *Science*, *267*(5203), 1445–1449.

Susin, S. A., Lorenzo, H. K., Zamzami, N., Marzo, I., Snow, B. E., Brothers, G. M., et al. (1999). Molecular characterization of mitochondrial apoptosis-inducing factor. *Nature*, *397*(6718), 441–446. http://dx.doi.org/10.1038/17135.

Suzanne, M., & Steller, H. (2013). Shaping organisms with apoptosis. *Cell Death and Differentiation*, *20*(5), 669–675. http://dx.doi.org/10.1038/cdd.2013.11.

Suzuki, J., Imanishi, E., & Nagata, S. (2014). Exposure of phosphatidylserine by Xk-related protein family members during apoptosis. *The Journal of Biological Chemistry*, *289*(44), 30257–30267. http://dx.doi.org/10.1074/jbc.M114.583419.

Thorp, E., Subramanian, M., & Tabas, I. (2011). The role of macrophages and dendritic cells in the clearance of apoptotic cells in advanced atherosclerosis. *European Journal of Immunology*, *41*(9), 2515–2518. http://dx.doi.org/10.1002/eji.201141719.

Tong, X., & Coulombe, P. A. (2006). Keratin 17 modulates hair follicle cycling in a TNFalpha-dependent fashion. *Genes & Development*, *20*(10), 1353–1364. http://dx.doi.org/10.1101/gad.1387406.

Truman, L. A., Ford, C. A., Pasikowska, M., Pound, J. D., Wilkinson, S. J., Dumitriu, I. E., et al. (2008). CX3CL1/fractalkine is released from apoptotic lymphocytes to stimulate macrophage chemotaxis. *Blood*, *112*(13), 5026–5036. http://dx.doi.org/10.1182/blood-2008-06-162404.

Tsai, W. H., Shih, C. H., Feng, S. Y., Li, I. T., Chang, S. C., Lin, Y. C., et al. (2014). CX3CL1(+) microparticles mediate the chemoattraction of alveolar macrophages toward apoptotic acute promyelocytic leukemic cells. *Cellular Physiology and Biochemistry*, *33*(3), 594–604. http://dx.doi.org/10.1159/000358637.

van Horssen, R., Eggermont, A. M., & ten Hagen, T. L. (2006). Endothelial monocyte-activating polypeptide-II and its functions in (patho)physiological processes. *Cytokine & Growth Factor Reviews*, *17*(5), 339–348. http://dx.doi.org/10.1016/j.cytogfr.2006.08.001.

Walker, N. I., Bennett, R. E., & Kerr, J. F. (1989). Cell death by apoptosis during involution of the lactating breast in mice and rats. *The American Journal of Anatomy*, *185*(1), 19–32. http://dx.doi.org/10.1002/aja.1001850104.

Weihua, Z., Tsan, R., Schroit, A. J., & Fidler, I. J. (2005). Apoptotic cells initiate endothelial cell sprouting via electrostatic signaling. *Cancer Research*, *65*(24), 11529–11535. http://dx.doi.org/10.1158/0008-5472.CAN-05-2718.

Wells, B. S., Yoshida, E., & Johnston, L. A. (2006). Compensatory proliferation in Drosophila imaginal discs requires Dronc-dependent p53 activity. *Current Biology*, *16*(16), 1606–1615.

Werner, N., Wassmann, S., Ahlers, P., Kosiol, S., & Nickenig, G. (2006). Circulating CD31 +/annexin V+ apoptotic microparticles correlate with coronary endothelial function in patients with coronary artery disease. *Arteriosclerosis, Thrombosis, and Vascular Biology, 26*(1), 112–116. http://dx.doi.org/10.1161/01.ATV.0000191634.13057.15.

White, G. E., Tan, T. C., John, A. E., Whatling, C., McPheat, W. L., & Greaves, D. R. (2010). Fractalkine has anti-apoptotic and proliferative effects on human vascular smooth muscle cells via epidermal growth factor receptor signalling. *Cardiovascular Research, 85*(4), 825–835. http://dx.doi.org/10.1093/cvr/cvp341.

Xu, D., Woodfield, S. E., Lee, T. V., Fan, Y., Antonio, C., & Bergmann, A. (2009). Genetic control of programmed cell death (apoptosis) in Drosophila. *Fly (Austin), 3*(1), 78–90. http://dx.doi.org/10.4161/fly.3.1.7800.

Yuan, J., Shaham, S., Ledoux, S., Ellis, H. M., & Horvitz, H. R. (1993). The C. elegans cell death gene ced-3 encodes a protein similar to mammalian interleukin-1 beta-converting enzyme. *Cell, 75*(4), 641–652.

Zernecke, A., Bidzhekov, K., Noels, H., Shagdarsuren, E., Gan, L., Denecke, B., et al. (2009). Delivery of microRNA-126 by apoptotic bodies induces CXCL12-dependent vascular protection. *Science Signaling, 2*(100), ra81. http://dx.doi.org/10.1126/scisignal.2000610.

Zhang, F. R., & Schwarz, M. A. (2002). Pro-EMAP II is not primarily cleaved by caspase-3 and −7. *American Journal of Physiology. Lung Cellular and Molecular Physiology, 282*(6), L1239–L1244. http://dx.doi.org/10.1152/ajplung.00141.2001.

Zou, H., Henzel, W. J., Liu, X., Lutschg, A., & Wang, X. (1997). Apaf-1, a human protein homologous to C. elegans CED-4, participates in cytochrome c-dependent activation of caspase-3. *Cell, 90*(3), 405–413.

CHAPTER TEN

Clearance of Apoptotic Cells and Pyrenocytes

Satoshi Toda, Chihiro Nishi, Yuichi Yanagihashi, Katsumori Segawa, Shigekazu Nagata[1]
Laboratory of Biochemistry and Immunology, Immunology Frontier Research Center, Osaka University, Osaka, Japan
[1]Corresponding author: e-mail address: snagata@ifrec.osaka-u.ac.jp

Contents

Abstract

Apoptotic cells are engulfed and digested by macrophages to maintain homeostasis in animals. If dead cells are not engulfed swiftly, they undergo secondary necrosis and release intracellular components that activate the immune system. Apoptotic cells are efficiently cleared due to phosphatidylserine (PtdSer) exposed on the cell surface that acts as an "eat me" signal. PtdSer is exposed through the activation of phospholipid scramblase and the inactivation of phospholipid flippase, which are both caspase-mediated events. Macrophages express a variety of molecules to recognize PtdSer,

Current Topics in Developmental Biology, Volume 114
ISSN 0070-2153
http://dx.doi.org/10.1016/bs.ctdb.2015.07.017

and use a sophisticated mechanism to engulf apoptotic cells. In red blood cells, the nucleus is lost when it is extruded as a pyrenocyte during definitive erythropoiesis. These pyrenocytes (nuclei surrounded by plasma membrane) also expose PtdSer on their surface and are efficiently engulfed by macrophages in a PtdSer-dependent manner. Macrophages transfer the engulfed apoptotic cell or pyrenocyte into lysosomes, where the components of the dead cell or pyrenocyte are degraded. If lysosomes cannot digest the DNA from apoptotic cells or pyrenocytes, the undigested DNA accumulates in the lysosome and activates macrophages to produce type I interferon (IFN) via a STING-dependent pathway; in embryos, this causes severe anemia. Here, we discuss how macrophages clear apoptotic cells and pyrenocytes.

1. INTRODUCTION

Many cells are overproduced during animal development. Extra cells undergo a type of cell death that is programmed through physiological settings. In development, this programmed cell death is a crucial part of digit formation, sexual differentiation, metamorphosis, neural network formation, and immune system development (Jacobson, Weil, & Raff, 1997). Several types of programmed cell death have been proposed, including apoptosis, necrosis, autophagic cell death, and cornification (Kroemer et al., 2009). Apoptosis, a major type of programmed cell death, was originally defined by the distinctive morphology of the dying cells: the cell body shrinks, chromatin condenses, and cell membrane blebs (Kerr, Wyllie, & Currie, 1972). In contrast, the morphological features of necrosis include cell swelling and rupture of the plasma membrane.

Every day, billions of toxic, useless, or senescent cells undergo apoptosis in the human adult. Apoptotic cells are engulfed and degraded by professional phagocytes such as macrophages and immature dendritic cells. Fibroblasts and endothelial cells also engulf apoptotic cells, although much less efficiently. Apoptotic cells are cleared so swiftly that it is difficult to find free apoptotic cells *in vivo*, even in tissues where large numbers of cells undergo apoptosis (Surh & Sprent, 1994). The efficiency with which apoptotic cells are engulfed is due to an "eat me" signal exposed on the dying cell's surface. Although many molecules have been proposed as an "eat me" signal (Savill & Fadok, 2000), phosphatidylserine (PtdSer) is the best suited to this function. Macrophages rely on sophisticated machineries to recognize PtdSer and engulf PtdSer-bearing apoptotic cells. Once engulfed, a dead cell is embedded in phagosomes and transferred to lysosomes, where the cellular components are degraded to nucleotides, amino acids, and lipids that can be reused. Pyrenocytes, which are membrane-wrapped nuclei expelled from mammalian red blood cells during the final stages of definitive erythropoiesis (Chasis & Mohandas, 2008), also expose surface PtdSer and are recognized and engulfed by macrophages (Yoshida, Kawane, et al., 2005).

If apoptotic cells and pyrenocytes are not engulfed quickly, the plasma membrane ruptures and releases intracellular materials that activate the immune system, leading to systemic lupus erythematosus (SLE)-type autoimmune diseases (Munoz, Lauber, Schiller, Manfredi, & Herrmann, 2010; Nagata, Hanayama, & Kawane, 2010). If apoptotic cells and pyrenocytes are engulfed by macrophages but are not properly digested in the lysosomes, innate immunity is activated in a manner that appears to lead to autoinflammation and a type of lysosomal storage disease (Nagata et al., 2010).

2. APOPTOSIS AND SIGNAL TRANSDUCTION

2.1 Apoptosis

Apoptosis is mediated by two independent pathways: intrinsic and extrinsic (Ow, Green, Hao, & Mak, 2008). Both pathways activate caspases, which are members of a cysteine protease family. The intrinsic pathway is activated during animal development or by genotoxic agents such as anticancer drugs and γ-rays. These cues upregulate BH3-only proteins in the Bcl-2 family, which activate Bax or Bak oligomerization. The oligomerized Bax/Bak then releases cytochrome *c* from the mitochondria. Cytochrome *c* associates with Apaf-1 to form apoptosomes, which are wheel-like particles with sevenfold symmetry (Acehan et al., 2002). The apoptosomes activate an initiator caspase, caspase 9, which subsequently activates the downstream effector caspases 3 and 7.

The extrinsic pathway is activated by death factors such as Fas ligand (FasL), tumor necrosis factor (TNF), and TNF-related apoptosis-inducing ligand (TRAIL) (Nagata, 1997; Strasser, Jost, & Nagata, 2009). FasL binds to its receptor, Fas, to form a death-inducing signaling complex (DISC) consisting of Fas, Fas-associated protein with death domain, and procaspase 8 (Krammer, 2000). In the DISC, procaspase 8 is autocatalytically processed to active caspase 8. Type I apoptotic cells produce high levels of caspase 8, and caspase 8 activates caspase 3 to execute apoptosis. In type II cells, the caspase 8 activation is insufficient to directly activate caspase 3; instead, caspase 8 cleaves the BH3-only protein Bid, and the truncated Bid activates

Bax/Bak to release cytochrome *c* from mitochondria, as in the intrinsic apoptotic pathway.

Caspases 3, 6, and 7 cleave more than 400 cellular protein substrates at the most downstream point of the apoptotic pathway (Luthi & Martin, 2007; Timmer & Salvesen, 2007). Although this caspase-mediated substrate cleavage is considered to be responsible for the apoptotic characteristics and for killing apoptotic cells, the significance of the cleavage is clear in only a few cases, as next described.

2.2 Apoptotic DNA Degradation

A distinctive hallmark of apoptosis is the fragmentation of chromosomal DNA into 180-bp nucleosomal units (Wyllie, 1980); this fragmentation is the work of caspase-activated DNase (CAD), also called DNA fragmentation factor-40 (DFF-40) (Fig. 1) (Enari et al., 1998; Liu, Zou, Slaughter, & Wang, 1997; Sakahira, Enari, & Nagata, 1998). In healthy cells, CAD resides as a complex with its inhibitor, known as inhibitor of CAD (ICAD) or DFF-45. ICAD binds to nascent CAD polypeptides and chaperones their folding on the ribosome. Thus, CAD is released from the ribosomes only in complex with ICAD, not as a functional molecule. In dying cells, active caspase 3 or 7 cleaves ICAD at two positions (Sakahira et al., 1998) to release CAD as a homodimer. The CAD dimer, which acts in the nucleus when the chromatin structure is still intact, has a scissor-like structure with the enzymatic active site located at the cleft of the scissors (Woo et al., 2004). Thus, the active site cannot access DNA on nucleosomes, but can access DNA at the spacer regions between nucleosomes, explaining the fragmentation of DNA into nucleosomal units during apoptosis. Fragmented DNA from apoptotic cells can be identified with TUNEL (terminal transferase-mediated dUTP nick end labeling) staining (Loo, 2010), in which terminal transferase adds a chemically labeled nucleotide to the 3′ end of the DNA fragments. The TUNEL staining is widely used to detect apoptotic cells *in vitro* and *in vivo*, along with staining of active caspase (Nicholls & Hyman, 2014).

2.3 PtdSer Exposure

PtdSer exposure (Fadok et al., 1992), another hallmark of apoptosis, links apoptosis with the clearance of apoptotic cells. Macrophages specifically recognize “eat me” signal(s) on the surface of apoptotic cells and respond by

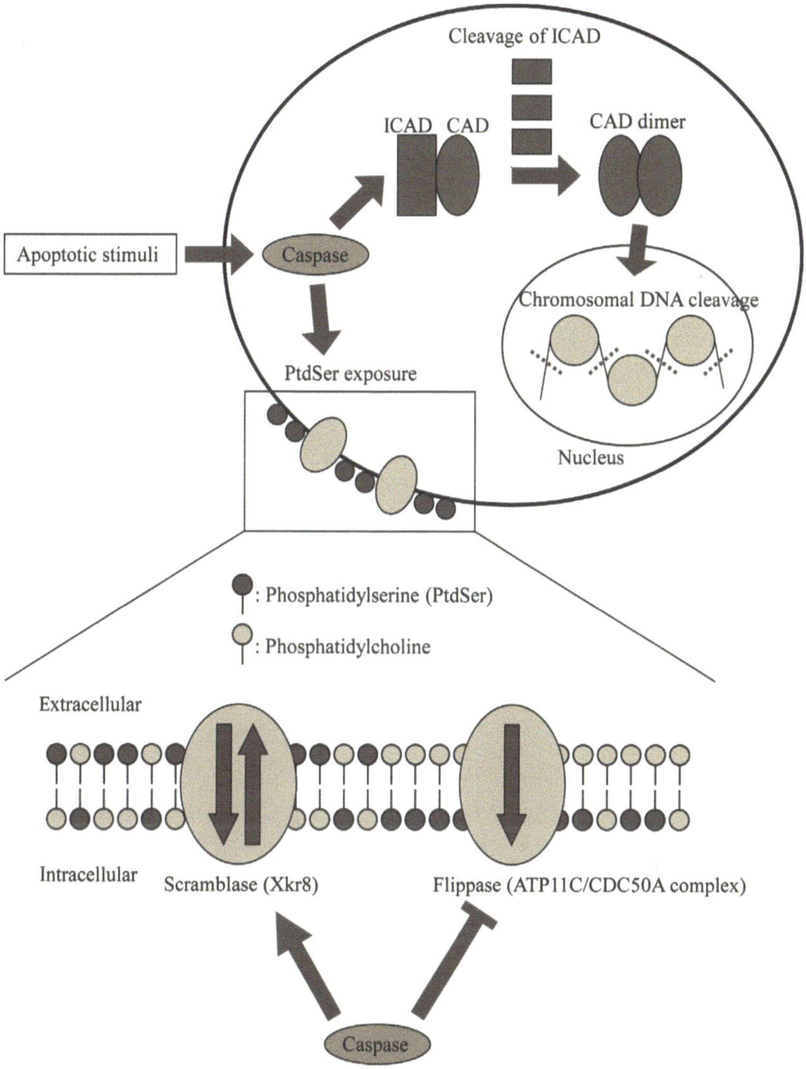

Figure 1 Apoptotic DNA degradation and PtdSer exposure. In healthy cells, CAD forms a heterodimer with ICAD. Apoptotic stimuli activate effector caspases that cleave ICAD at two positions, releasing CAD. The released CAD forms a homodimer that cleaves chromosomal DNA in the nucleus into nucleosomal units. In healthy cells, PtdSer is localized to the inner leaflet of the plasma membrane by the action of ATP-dependent flippase. When cells receive apoptotic stimuli, effector caspases cleave and activate Xkr8 scramblase, while cleaving and inactivating ATP11C flippase, quickly exposing PtdSer to the surface of the apoptotic cell. (See the color plate.)

engulfing them. The strongest candidate for the "eat me" signal is PtdSer, since masking PtdSer inhibits the engulfment of apoptotic cells (Asano et al., 2004; Fadok et al., 1992). In healthy cells, PtdSer is localized to the inner leaflet of the plasma membrane (Leventis & Grinstein, 2010). This asymmetrical PtdSer distribution is disrupted by scramblases in response to apoptotic stimuli. Scramblases are proposed to nonspecifically and bidirectionally transport phospholipids between the inner and outer leaflets in an ATP-independent manner. In 1996, Basse, Stout, Sims, and Wiedmer (1996) reported finding a 37-kDa protein with scramblase activity in a cell-free system and called it phospholipid scramblase (PLSCR1). However, PLSCR1 is a cytoplasmic protein, and cells lacking PLSCR1 retain normal phospholipid scrambling activity (Zhou, Zhao, Wiedmer, & Sims, 2002), arguing against PLSCR1's identification as a phospholipid scramblase. Since PLSCR1 belongs to a large family, it was postulated that other family members might function as redundant scramblases. However, Acharya (2006) mutated both PLSCR homologs in the *Drosophila* genome and found that the mutant retained full scramblase activity, indicating that PLSCR is not a scramblase (Bevers & Williamson, 2010).

We recently identified two families of scramblases: TMEM16F and its homologs, and Xk-related protein 8 (Xkr8) and its homologs. TMEM16F, a membrane protein that carries eight transmembrane regions and localizes to plasma membranes, is ubiquitously expressed in various tissues and cells (Suzuki, Umeda, Sims, & Nagata, 2010) and supports phospholipid scrambling in response to Ca^{2+}. TMEM16F also scrambles phospholipids in activated platelets to promote blood clotting. In the same family, TMEM16C, 16D, 16G, and 16J support Ca^{2+}-dependent phospholipid scrambling and are expressed in specific tissues, such as the brain and intestines, but their physiological functions remain unknown (Suzuki, Fujii, et al., 2013).

Xkr8, which has six transmembrane regions, localizes to the plasma membrane (Suzuki, Denning, Imanishi, Horvitz, & Nagata, 2013) and carries a recognition sequence for caspases 3 and 7 in its C-terminal tail. Xkr8 must be cleaved by caspase to promote phospholipid scrambling and expose PtdSer. In some human lymphoma cell lines, *XKR8* gene expression is epigenetically repressed by heavy methylation at CpG islands in the promoter region, and PtdSer is not exposed during apoptosis; Xkr8-deficient mouse fetal thymocyte cell lines also fail to expose PtdSer during apoptosis. Consequently, these cells are not engulfed by macrophages during apoptosis. CED-8, an Xkr8 homolog in *Caenorhabditis elegans*, also mediates apoptotic PtdSer exposure and is involved in engulfing cell corpses (Suzuki, Denning, et al., 2013). Of the eight members of the mouse Xkr family, Xkr4, Xkr8, and Xkr9 carry a C-terminal caspase-recognition site and can support apoptotic PtdSer exposure (Suzuki, Imanishi, & Nagata, 2014). Unlike Xkr8, which is ubiquitous in various tissues, Xkr 4 and Xkr9 are specifically expressed in the brain and intestines, respectively. Whether these molecules have specific physiological roles in these tissues remains to be determined.

In healthy cells, asymmetrical phospholipid distribution between the inner and outer leaflets of the plasma membranes is maintained by an ATP-dependent aminophospholipid translocase or flippase (Tanaka, Fujimura-Kamada, & Yamamoto, 2011). By genetic screening with the human near-haploid KBM7 cell line, we recently identified ATP11C (a P4-type ATPase) and CDC50A (the ATP11C β-subunit) as molecules responsible for flippase activity (Segawa, Kurata, et al., 2014). Human ATP11C has three caspase-recognition sites for caspases 3, 6, and 7, and cleavage by caspase during apoptosis inactivates its flippase activity. A caspase-resistant ATP11C mutant retains flippase activity, and expressing this mutant in apoptotic cells prevents PtdSer exposure (Segawa, Kurata, et al., 2014). Thus, PtdSer exposure on apoptotic cells is mediated by scramblase activation and flippase inactivation, which are both caspase-dependent events (Fig. 1).

ATP11C is a member of a large (14 members in human and 15 members in mouse) P4-type ATPase family (Tanaka et al., 2011). More than 80% of the flippase activity is lost in ATP11C-deficient human KBM7 or mouse WR19L cells. However, cells lacking ATP11C still maintain asymmetrical phospholipid distribution, suggesting that other family members also have flippase activity. CDC50A functions as a chaperon for ATP11C. There are three CDC50 members, but only CDC50A is expressed in KBM7 and WR19L cells. CDC50A appears to be required for P4-ATPases, including ATP11C, to localize to plasma membranes. Thus, cells lacking CDC50A lose flippase activity almost entirely and expose PtdSer on their surface. Notably, these CDC50A-null, PtdSer-exposing living cells are engulfed by macrophages (Segawa, Kurata, et al., 2014), indicating that PtdSer exposure is sufficient as an "eat me" signal, or that an "eat me" signal can overcome a "don't eat me" signal (Chao, Majeti, & Weissman, 2011).

3. THE ENGULFMENT OF APOPTOTIC CELLS

3.1 Assay for Engulfed Apoptotic Cells

To identify molecules or signal transduction pathways involved in the engulfment of apoptotic cells, a reliable system for assaying the engulfment is needed. In early assays, macrophages were incubated with apoptotic cells, and the number of engulfed cells was counted under a microscope, which is an accurate, but time-consuming procedure (Fadok et al., 1992). Later, in a simpler method, macrophages were incubated with fluorescently labeled apoptotic cells, treated with trypsin, and analyzed by FACS; however, there was no guarantee that the unengulfed apoptotic cells were completely removed by trypsin. We designed an assay based on the two steps by which an apoptotic cell's chromosomal DNA is digested (Hanayama et al., 2002): cell-autonomously in apoptotic cells by CAD and then noncell-autonomously in macrophages by lysosomal DNase II. When CAD-deficient apoptotic cells are used as prey, DNA degradation is observed only in apoptotic cells that have been engulfed by macrophages, and is easily and quantitatively assayed by TUNEL staining. This method was successfully used to identify milk-fat globule (MFG) epidermal growth factor (EGF) 8 (MFG-E8) and Tim4 as molecules involved in apoptotic cell engulfment. Miksa et al. developed an even simpler, more convenient method (Miksa, Komura, Wu, Shah, & Wang, 2009) using pHrodo® Red probes, which fluoresce only in acidic conditions, not at neutral pH. Thus, pHrodo-conjugated apoptotic cells fluoresce only when they are engulfed and transported into lysosomes. We are currently using this reliable method, with slight modification (Toda, Hanayama, & Nagata, 2012).

3.2 Molecules That Form a Bridge Between Phagocytes and Apoptotic Cells

Several secreted proteins have been identified that bind apoptotic cells and phagocytes, creating a bridge between the two cells (Fig. 2A). MFG-E8 is produced by thioglycolate-elicited peritoneal macrophages (Hanayama et al., 2002). MFG-E8 binds PtdSer on apoptotic cells via two C-terminal factor-VIII-homologous domains (C1 and C2), and binds integrin-$\alpha_v\beta_3$ or integrin-$\alpha_v\beta_5$ on macrophages via an RGD (Arg-Gly-Asp) motif in its N-terminal EGF domain (Hanayama et al., 2002). MFG-E8 is present on tingible body macrophages in the germinal center of the spleen (Hanayama, Tanaka, Miyasaka, et al., 2004; Kranich et al.,

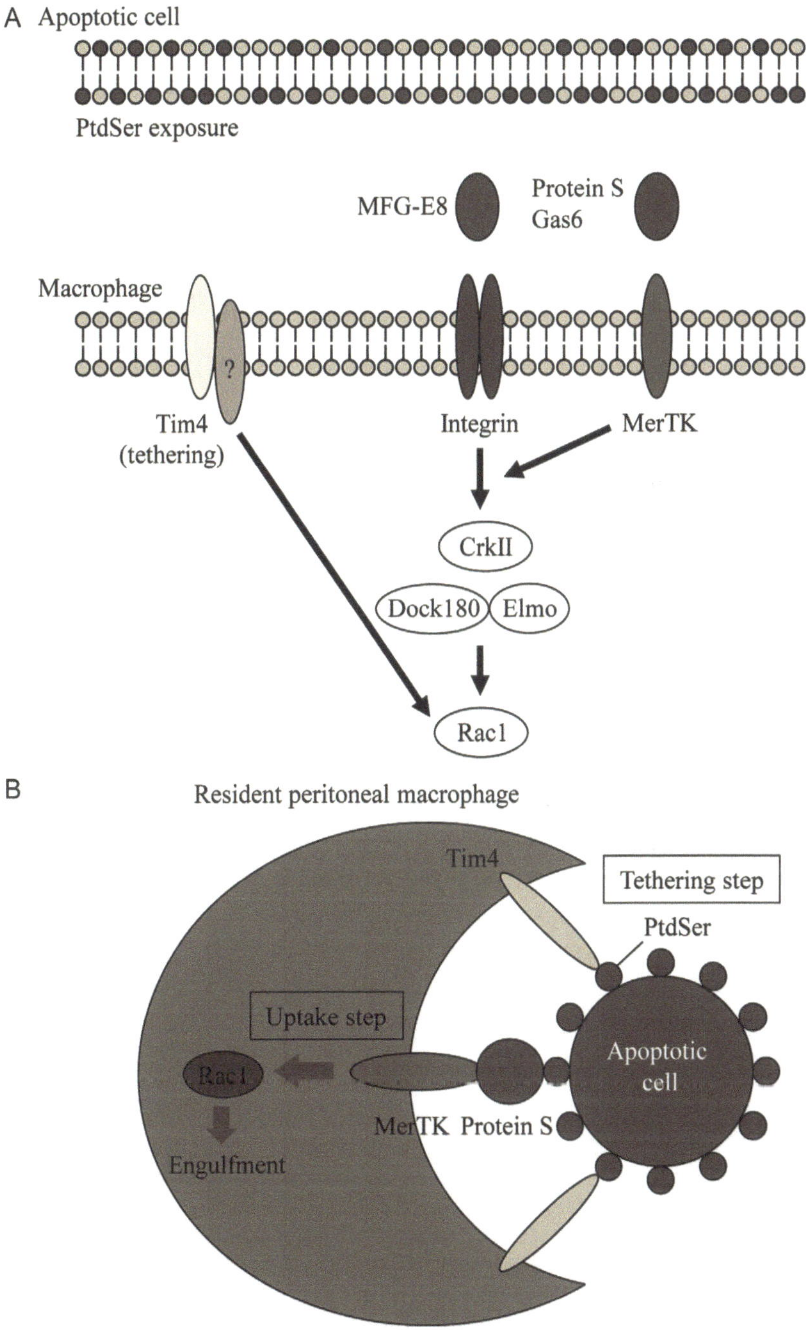

Figure 2 Engulfment of apoptotic cells. (A) The secreted proteins MFG-E8, Protein S, and Gas6 form bridges between apoptotic cells and phagocytes by recognizing and binding

(Continued)

2008; Miyasaka, Hanayama, Tanaka, & Nagata, 2004), which engulf activated and apoptotic B lymphocytes. MFG-E8-deficient female mice develop age-dependent SLE-type autoimmune disease, especially in the B6/129-mixed background (Hanayama, Tanaka, Miyasaka, et al., 2004). They also spontaneously develop age-dependent dermatitis associated with the activation of $CD8^+$ T cells (Peng & Elkon, 2011). Similarly, the myeloid-specific deletion of integrin-α_v inactivates the ability of gut-associated macrophages to engulf apoptotic cells, causing colitis (Lacy-Hulbert et al., 2007). Thus, unengulfed apoptotic cells can activate the immune system, leading to autoimmune diseases.

MFG-E8, which associates with the MFGs secreted from mammary glands (Patton & Keenan, 1975), increases at pregnancy and peaks when the lactating mammary glands undergo involution to return to a prepregnant state (Aoki et al., 1997). When MFGs expose PtdSer (probably because they do not generate the ATP required by flippase), MFG-E8 binds the MFG for clearance by mammary epithelial cells (Hanayama & Nagata, 2005). MFG-E8-null mice have defective mammary gland involution and accumulate a large number of MFGs in the mammary duct, often developing mastitis (Hanayama & Nagata, 2005).

Two related plasma proteins, growth arrest-specific 6 (Gas6) and Protein S, bind to PtdSer (Anderson et al., 2003; Nakano et al., 1997). Both proteins have an N-terminal Gla domain, which contains multiple glutamic acid residues that undergo vitamin K-dependent γ-carboxylation. The γ-carboxylated Gla domain binds PtdSer in the presence of calcium. Protein S was originally identified as an anticoagulation factor that works with Protein C to degrade coagulation factors (Dahlbäck & Villoutreix, 2005). Gas6 is also involved in clotting, and its deficiency causes platelet dysfunction in thrombosis (Angelillo-Scherrer et al., 2001). Protein S and Gas6 function in the engulfment of apoptotic cells (Anderson et al., 2003; Ishimoto, Ohashi, Mizuno, & Nakano, 2000; Scott et al., 2001; Seitz, Camenisch, Lemke,

Figure 2—Cont'd PtdSer on apoptotic cells, and binding integrins or TAM kinase receptors on phagocytes. Tim1 and Tim4 are type I membrane proteins that directly recognize PtdSer. During the engulfment of apoptotic cells, Rac1 is activated and promotes actin polymerization to form a phagocytic cup. Signaling molecules such as CrkII, Dock180, Elmo1, and GULP may act upstream of Rac1. (B) In a two-step process, Tim4 tethers apoptotic cells to the surface of mouse resident peritoneal macrophages, and Protein S binds to PtdSer on the tethered apoptotic cells to stimulate MerTK, which activates intracellular phagocytic signaling to engulf the apoptotic cell. Thus, two PtdSer-binding systems, the Tim4 and Protein S/MerTK systems, cooperate to engulf apoptotic cells. (See the color plate.)

Earp, & Matsushima, 2007) by binding TAM tyrosine kinase receptors (Axl and MerTK), which are expressed in various tissue macrophages and in microglia (Gautier et al., 2012). Protein S binds MerTK via a C-terminal SHBG (sex hormone-binding globulin)-like domain (Lemke & Rothlin, 2008). Thus, like MFG-E8, Protein S and Gas6 create a bridge between apoptotic cells and phagocytes.

MerTK is indispensable for thioglycolate-elicited or resident peritoneal macrophages to efficiently engulf apoptotic cells (Nishi, Toda, Segawa, & Nagata, 2014; Scott et al., 2001). Mice with a complete knockout of the TAM receptor genes display a wide spectrum of autoimmune diseases characterized by hyperactivated antigen-presenting cells (Lu & Lemke, 2001). Every day, retinal pigment epithelial cells phagocytose large quantities of photoreceptor cell outer segments (Strauss, 2005); this process also requires MerTK, and a MerTK deficiency or a dual Protein S/Gas6 deficiency causes retinal degeneration (Burstyn-Cohen et al., 2012). As with exosomes and MFGs, it is likely that these outer photoreceptor cell segments expose PtdSer and are cleared by epithelial cells in a PtdSer-dependent manner.

3.3 PtdSer Receptors

In addition to the soluble proteins that bind PtdSer, some type I membrane proteins directly bind PtdSer and function as PtdSer receptors (PSRs). In 2000, Fadok et al. (2000) reported a protein designated as PSR, and found that a monoclonal antibody against the receptor (mAb 217) inhibited the uptake of apoptotic cells by activated human macrophages. Using this antibody to screen a phage display library, Fadok et al. identified a peptide that is a component of human PSR. PSR was thought to be a type II membrane protein, and was reported to enhance the uptake of apoptotic cells. Subsequently, Li, Sarkisian, Mehal, Rakic, and Flavell (2003) and Wang et al. (2003) established PSR-knockout mice and nematodes, respectively, which have defects in engulfing apoptotic cells; these defects are lethal in PSR-deficient mice. Meanwhile, PSR was found to carry a Jumonji domain (Cikala et al., 2004), which is involved in hydroxylating the asparagine residue of some transcription factors, and to be present in the nucleus (Cikala et al., 2004; Cui, Qin, Liu, Pan, & Pei, 2004). Bose et al. (2004) independently established a strain of PSR-deficient mice, and showed that PSR-null cells express the mAb 217 antigen and can engulf apoptotic cells, thus ruling out PSR as a PtdSer receptor involved in apoptosis. In 2007, PSR, now also called JMJD6, was shown to be a histone arginine methylase (Chang, Chen,

Zhao, & Bruick, 2007). Recently, Yang et al. (2014) reported that in *C. elegans*, PSR-1 binds PtdSer and enhances the engulfment of apoptotic cells. However, PSR's affinity for PtdSer is only 1% or less than that of MFG-E8 and Tim4. Thus, it is unlikely that PSR is involved in the engulfment of apoptotic cells, at least in mammals.

We established a hamster monoclonal antibody (Kat-5) against mouse resident peritoneal macrophages that inhibits apoptotic cell engulfment. By screening a mouse resident peritoneal macrophage cDNA library by expression cloning, we identified Tim4 (T-cell immunoglobulin- and mucin-domain-containing molecule 4) as the antigen recognized by this antibody (Miyanishi et al., 2007). Tim4 was first reported as a Tim1 ligand expressed in Th2 cells (Meyers et al., 2005). Tim4 contains a signal sequence, IgV (immunoglobulin V), mucin-like, and transmembrane domains, and a cytoplasmic region. The Tim family has 8 (Tim1–8) and 3 (Tim1, Tim3, and Tim4) members in mice and humans, respectively (Freeman, Casasnovas, Umetsu, & Dekruyff, 2010). Tim1, like Tim4, tightly binds PtdSer with Kd of around 2 nM. Activated T cells and macrophages produce exosomes or membrane vesicles that expose PtdSer (Théry, Ostrowski, & Segura, 2009). Both Tim4-expressing macrophages and Tim1-expressing T cells bind exosomes via PtdSer. Exosomes thus mediate an interaction between Tim1 and Tim4, explaining the apparent ligand-like function of Tim4 for Tim1 (Miyanishi et al., 2007). Transforming mouse NIH3T3 fibroblasts with Tim1 or Tim4 renders them able to engulf apoptotic cells. Ichimura et al. (2008) independently found that Tim1, which is also called kidney injury molecule 1 (Kim1) and is expressed in kidney epithelial cells after ischemic injury, mediates the engulfment of apoptotic bodies. Immunohistochemical and RNA analyses indicated that Tim4 is expressed in resident peritoneal macrophages, thymic macrophages, marginal zone macrophages, tingible body macrophages in the spleen, and Kupffer cells in the liver (Miyanishi et al., 2007; Miyanishi, Segawa, & Nagata, 2012; Wong et al., 2010). Mouse resident peritoneal macrophages cultured *in vitro* quickly lose Tim4. As suggested by Okabe and Medzhitov (2014), a peritoneal-specific signal such as retinoic acid may regulate Tim4's expression. We and others (Miyanishi et al., 2012; Rodriguez-Manzanet et al., 2010; Wong et al., 2010) independently established Tim4-deficient mice, and showed that Tim4$^{-/-}$ peritoneal macrophages cannot engulf apoptotic cells.

Tim1 and Tim4 bind PtdSer via an IgV domain, and the analysis of their structures revealed a pocket for PtdSer (Santiago et al., 2007). Tim3 also carries a PtdSer-binding site. Nakayama et al. (2009) and DeKruyff et al.

(2010) showed that Tim3 also binds PtdSer. Tim3 is expressed in thioglycolate-elicited peritoneal exudate macrophages and splenic $CD8^+$ dendritic cells, and is reported to be involved in these phagocytes' engulfment of apoptotic cells (Nakayama et al., 2009). On the other hand, our results indicate that Tim3's ability to bind PtdSer is low compared to that of Tim1 or Tim4 (Miyanishi et al., 2007). Tim3 inhibits Th1- and Tc1-mediated immune responses (Sánchez-Fueyo et al., 2003). Whether Tim3's ability to bind PtdSer contributes to this function is not clear.

Other proposed PSRs include brain-specific angiogenesis inhibitor 1 (BAI1) (Park et al., 2007) and stabilin-1 and -2 (Park et al., 2008; Park, Jung, et al., 2009). BAI1, an adhesion GPCR (G-protein coupled receptor) with a long N-terminal region, is highly expressed in the brain and is involved in synaptogenesis (Stephenson, Purcell, & Hall, 2014). Unlike Tim4, MFG-E8, Protein S, or Gas6 which specifically bind PtdSer, BAI1 binds not only PtdSer but also phosphatidic acids, cardiolipin, phosphatidylinositol 4-phosphate, and sulfatide (Park et al., 2007). BAI1's involvement in apoptotic cell engulfment still needs to be demonstrated using BAI1-deficient macrophages. Stabilin-1 and -2 are fasciclin-like scavenger receptors. Stabilin-2 is a receptor for hyaluronan, which is expressed in sinusoidal endothelial cells in the liver (Harris, Weigel, & Weigel, 2008). Lee, Park, Jung, Bae, and Kim (2011) reported that stabilin-1 and -2 expressed in hepatic sinusoidal cells assist the PtdSer-dependent clearance of aged erythrocytes. This interesting possibility would be investigated using stabilin-1 and -2 double knockout mice (Schledzewski et al., 2011).

3.4 Two-Step Engulfment of Apoptotic Cells

In 2001, Hoffmann et al. (2001) proposed a two-step model of engulfment, in which phagocytes first tether the apoptotic cell, followed by tickling (PtdSer-mediated macropinocytosis). In this model, ligation of a PSR by PtdSer is essential for uptake of the apoptotic cell. When Tim4, a PSR, is expressed in adherent cells such as NIH3T3 or mouse embryonic fibroblasts (MEFs), the transformants efficiently engulf apoptotic cells (Miyanishi et al., 2007; Park, Hochreiter-Hufford, & Ravichandran, 2009). Tim4's very short cytoplasmic region (43 amino acids) is dispensable for its ability to support the engulfment of apoptotic cells (Park, Hochreiter-Hufford, et al., 2009), suggesting that Tim4 does not mediate the uptake signal, and that an endogenous molecule expressed in the NIH3T3 or MEF cell transduces the signal. In contrast, integrin-$\alpha_v\beta_3$ (which is bound by MFG-E8) and TAM family receptors (which are bound by Protein S and/or Gas6) are known to transduce various signals (Kim, Ye, & Ginsberg, 2011; Lemke & Rothlin, 2008). Tingible body macrophages in the spleen express both Tim4 and MFG-E8 (Hanayama, Tanaka, Miyasaka, et al., 2004; Miyanishi et al., 2012), while resident peritoneal macrophages express both Tim4 and MerTK (Nishi et al., 2014).

To examine the relationship between tethering and tickling in apoptotic cell engulfment, we used a pro-B cell line (Ba/F3) that grows in suspension and lacks the ability to engulf apoptotic cells as a host to reconstitute the engulfment system (Toda et al., 2012). We found that Ba/F3 transformants expressing Tim4 strongly bind apoptotic cells but cannot engulf them, whereas transformants expressing integrin-$\alpha_v\beta_3$ do not bind apoptotic cells in the presence of MFG-E8 and thus do not engulf them. Simultaneously expressing Tim4 and integrin-$\alpha_v\beta_3$ allows Ba/F3 cells to efficiently engulf apoptotic cells in the presence of MFG-E8. Similarly, MerTK-expressing cells cannot bind or recruit apoptotic cells in the presence of Protein S unless MerTK is coexpressed with Tim4 (Nishi et al., 2014). Accordingly, MerTK$^{-/-}$ mouse resident peritoneal macrophages, which express Tim4, can recruit or tether apoptotic cells, but do not engulf them. On the other hand, Tim4$^{-/-}$ resident peritoneal macrophages do not bind apoptotic cells (Nishi et al., 2014). These results strongly support the two-step theory of engulfment at the molecular level (Fig. 2B).

In addition to Tim4, Tim1 functions as a PSR. The TAM family consists of three members, Tyro3, Axl, and MerTK (Lemke & Rothlin, 2008), and there is a homolog of MFG-E8, Del-1 (Hanayama, Tanaka, Miwa, & Nagata, 2004). Macrophages and immature dendritic cells are heterogeneous (Gordon & Plűddemann, 2013), and it is important to determine whether different phagocytes use different pairs of PSRs and signaling molecules in this two-step engulfment process.

3.5 Signaling Pathways Involved in the Engulfment of Apoptotic Cells

As in mammals (Asano et al., 2004; Fadok et al., 1992), masking PtdSer inhibits the engulfment of apoptotic cells in *C. elegans* (Venegas & Zhou, 2007) and *Drosophila* (Tung et al., 2013), indicating that PtdSer is a conserved "eat me" signal in metazoans. In fact, CED-8, a homolog of the caspase-dependent scramblase Xkr8, is indispensable in *C. elegans* for PtdSer exposure (Chen, Mapes, Lee, Skeen-Gaar, & Xue, 2013; Suzuki, Denning, et al., 2013), and for efficiently engulfing cell corpses (Stanfield & Horvitz,

2000). Analyses of *C. elegans* mutants that accumulate cell corpses revealed two parallel, partially redundant signaling pathways that mediate the engulfment of apoptotic cells: the CED-1/-6/-7 and CED-2/-5/-10/-12 pathways (Reddien & Horvitz, 2004). The mammalian homologs of CED-1, -6, and -7 are MEGF10 (multiple EGF-like-domains 10), GULP (phosphotyrosine-binding domain-containing engulfment adaptor protein), and the ABC (ATP-binding cassette) transporter, respectively, while CrkII, Dock180, Rac1, and Elmo are mammalian homologs of CED-2, -5, -10, and -12, respectively.

In the first pathway, CED-1 and MEGF10, which are transmembrane proteins carrying multiple EGF-like domains in their extracellular regions, are likely to be involved in recognizing apoptotic cells (Venegas & Zhou, 2007). While MEGF10 can interact with GULP and ABC to promote the engulfment of apoptotic cells, the direct binding of CED1 or MEGF10 to PtdSer has not been demonstrated (Hamon et al., 2006). Although the molecules that recognize dying cells in the second pathway have not been identified in *C. elegans*, the CrkII/Dock180/Rac1/Elmo pathway is involved in the integrin-$\alpha_v\beta_5$- and MerTK-mediated engulfment of apoptotic cells in mammalian cells (Akakura et al., 2004; Wu, Singh, Georgescu, & Birge, 2005) (Fig. 2A).

Monitoring of the Rac1 activation using a fluorescence resonance energy transfer probe in MFG-E8/integrin-$\alpha_v\beta_3$-mediated apoptotic cell engulfment showed that Rac1 is activated when actins polymerize to form a phagocytic cup (Nakaya, Kitano, Matsuda, & Nagata, 2008). The closure of the phagocytic cup after internalization of the apoptotic cell is associated with Rac1 downregulation. These results support the idea that a signal from integrin activates Dock180, a guanine-nucleotide exchange factor for Rac1, and Rac1 regulates the actin polymerization (Cote & Vuori, 2007). The internalized apoptotic cells are transferred into lysosomes in a process involving RhoA, Rab5, dynamin, and various other molecules (Kitano, Nakaya, Nakamura, Nagata, & Matsuda, 2008; Nakaya, Tanaka, Okabe, Hanayama, & Nagata, 2006). Whether this process corresponds to phagosome maturation, in which phagosomes are trafficked through acidic membrane structures (Kinchen & Ravichandran, 2008), remains to be studied. Furthermore, living cells are often engulfed by phagocytes in various situations, and some engulfed cells are released from the phagocyte before reaching to lysosomes (Overholtzer et al., 2007; Segawa, Kurata, et al., 2014). It might be interesting to study physiological and pathological roles of the engulfment of living cells, and at which stage an internalized cell can return to the outside.

4. THE ENGULFMENT OF PYRENOCYTES

4.1 Definitive Erythropoiesis in Erythroblastic Islands

Early in mammalian embryogenesis, red blood cells are produced in the yolk sac as primitive erythrocytes that have a nucleus and embryonic hemoglobin. Erythropoiesis switches to the fetal liver in the later stages of embryogenesis, and to the bone marrow after birth. In the fetal liver and bone marrow, definitive erythropoiesis occurs, during which the nuclei are removed and fetal and adult hemoglobins are produced (Palis, 2008). This definitive erythropoiesis occurs at anatomical units called erythroblastic islands, each of which has a macrophage at the center (Fig. 3). Erythroblasts proliferate and

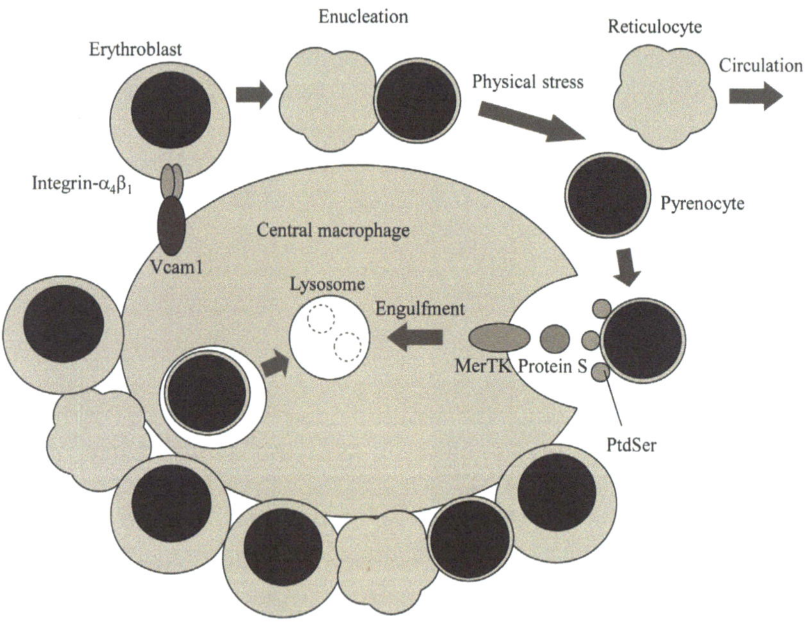

Figure 3 Enucleation and engulfment of pyrenocytes in erythroblastic islands. Erythroblasts bind to a central macrophage via adhesion molecules, such as Vcam1 and integrin-$\alpha_4\beta_1$, to form an erythroblastic island. Erythroblasts proliferate and differentiate on the central macrophage and finally undergo enucleation into a reticulocyte and pyrenocyte, which are separated by physical stress. The newly separated pyrenocyte exposes PtdSer on its surface and is engulfed by the central macrophage in a process similar to the engulfment of apoptotic cells (Protein S/MerTK), although the tethering step (Tim4) may not be required due to the close interaction between pyrenocytes and macrophages.

differentiate on the macrophage. Adhesion molecules such as integrin-$\alpha_4\beta_1$ (on erythroblasts) and Vcam1 (on central macrophages) (Sadahira, Yoshino, & Monobe, 1995) mediate the association between erythroblasts and macrophages. Central macrophages not only anchor the erythroblasts to themselves but also support erythroblast proliferation and differentiation (Chasis & Mohandas, 2008). At the final stage of definitive erythropoiesis, the nucleus of the erythroblast moves to one side of the cell and is extruded as a pyrenocyte, a nucleus enclosed in a plasma membrane (Wang et al., 2012). The enucleation process is cell-autonomous and requires cytoskeletal rearrangement (Ji, Jayapal, & Lodish, 2008; Konstantinidis et al., 2012). It is not clear how enucleation is triggered, or how the cytoskeletal machinery moves the nucleus within the cell or extrudes it to form a pyrenocyte.

4.2 The Engulfment of Pyrenocytes in Erythroblastic Islands

In erythroblastic islands, pyrenocytes but not reticulocytes are engulfed by the central macrophage. As with apoptotic cells, pyrenocytes expose PtdSer on their surface as an "eat me" signal (Yoshida, Kawane, et al., 2005). Enucleation consists of two steps: the cell-autonomous extrusion of the nucleus and the separation of the reticulocyte and pyrenocyte by shear stress. The pyrenocyte exposes PtdSer on its surface immediately after being separated from the reticulocyte. As mentioned above, the asymmetrical distribution of phospholipids at plasma membranes is maintained by an ATP-dependent flippase. Reticulocytes carry the glycolysis system, which generates ATP; pyrenocytes do not. Thus, pyrenocytes lose ATP as soon as they separate from the reticulocyte, which would inactivate the flippase. The loss of ATP also increases the intracellular Ca^{2+} concentration, since the pump that excludes intracellular Ca^{2+} requires ATP. In apoptotic cells, caspase inactivates flippase and activates scramblase, while in pyrenocytes, flippase seems to be inactivated by the loss of ATP, and Ca^{2+}-dependent scramblase is activated by the increase in intracellular Ca^{2+}. To prove this hypothesis, it will be necessary to identify the flippase and scramblase molecules in pyrenocytes.

The question arises of how the PtdSer exposed on pyrenocytes is recognized by macrophages in the erythroblastic islands. As reported for the culture of human erythroblasts (Migliaccio, Masselli, Varricchio, & Whitsett, 2012), mouse erythroblastic islands cultured in normal medium in the presence of erythropoietin stimulated erythroblast proliferation and differentiation, but the pyrenocytes were not separated from the reticulocytes

(Toda, Segawa, & Nagata, 2014). However, when the erythroblastic islands were cultured in a highly viscous medium containing 1% methylcellulose, the pyrenocytes were efficiently separated and engulfed by the macrophages. The viscosity may provide sufficient shear stress to separate pyrenocytes from reticulocytes. In this system, the engulfment of pyrenocytes was mediated by MerTK and Protein S, while Tim4 was dispensable. These results suggest that macrophages use a similar mechanism when engulfing apoptotic cells or pyrenocytes, but that tethering may not be necessary for engulfing pyrenocytes in erythroblastic islands, probably due to the proximity of pyrenocytes to the central macrophage. In any case, the system developed for mouse erythroblastic islands, in which erythroblasts are cultured on Vcam1- and MerTK-expressing fibroblasts in 1% methylcellulose, can be used to effectively produce large quantities of enucleated red blood cells *in vitro*.

5. DNA DEGRADATION IN MACROPHAGES

5.1 DNA Degradation in Macrophages

Chromosomal DNA in apoptotic cells is cleaved cell-autonomously into 180-bp nucleosomal units by CAD. Although CAD-null cells do not undergo DNA fragmentation during apoptosis, their DNA is digested by DNase II in macrophage lysosomes after the dead cells are engulfed (Kawane et al., 2003; McIlroy et al., 2000). Similarly, although pyrenocyte DNA is not degraded cell-autonomously, it is digested by DNase II in macrophage lysosomes. DNase II, which is ubiquitously expressed in various tissues and cells, is solely responsible for digesting the DNA of engulfed apoptotic cells and pyrenocytes within the macrophage. The degradation of the dead cell's DNA by DNase II is well conserved in *Drosophila* (Mukae, Yokoyama, Yokokura, Sakoyama, & Nagata, 2002) and *C. elegans* (Wu, Stanfield, & Horvitz, 2000).

DNase II-deficient mice are embryonic-lethal or die at the last stage of embryogenesis (Kawane et al., 2001). Macrophages in the thymus and interdigits, and microglia in the brain of DNase II-deficient embryos accumulate large quantities of fragmented DNA (Fig. 4), indicating that programmed cell death occurs in these regions, and the cells are engulfed by macrophages (Kawane et al., 2003). In addition, central macrophages in erythroblastic islands in the fetal liver of DNase II-null embryos carry naked, intact DNA in their lysosomes, confirming that pyrenocytes are engulfed by the

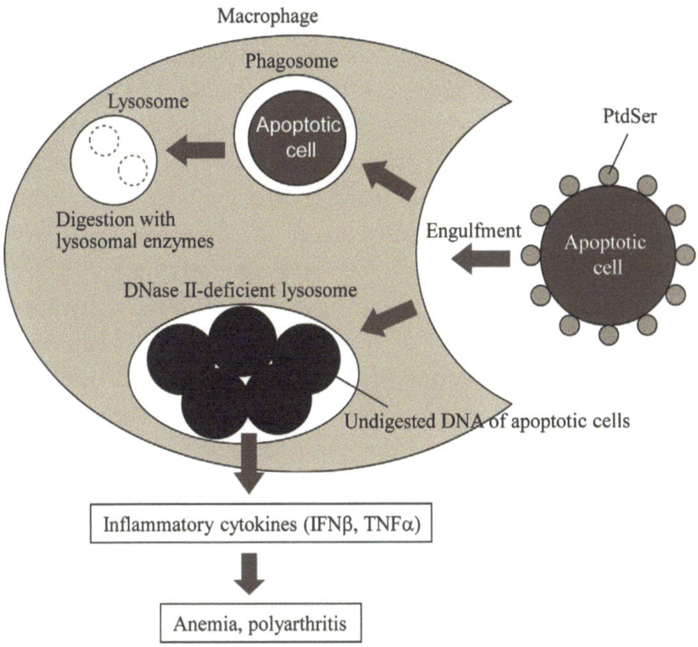

Figure 4 Degradation of the DNA of the engulfed apoptotic cells and pyrenocytes by DNase II in lysosomes. Apoptotic cells are engulfed by macrophages, enclosed in phagosomes, and transferred to lysosomes, where they are degraded into components such as amino acids, nucleotides, and lipids. The DNA of engulfed apoptotic cells and pyrenocytes is digested by DNase II. In DNase II-deficient macrophages, lysosomes accumulate undigested DNA from apoptotic cells and pyrenocytes, which activates the macrophages to produce cytokines such as IFNβ and TNFα. IFNβ causes lethal anemia in embryos, while TNFα appears to cause polyarthritis in adults.

macrophages in the erythroblastic islands. In DNase II-null macrophages, all components except DNA are digested in the lysosomes.

5.2 Anemia and Polyarthritis Caused by Deficient Lysosomal DNA Degradation

Lethality in DNase II-null mice occurs because many erythroid precursor cells undergo apoptosis in the late stages of embryogenesis, causing severe anemia (Kawane et al., 2001). A set of type I interferon (IFN)-inducible genes is strongly upregulated in DNase II-deficient embryos, and a deficiency in the IFN type I receptor (IFNIR) rescues the lethality of DNase II-null mice, indicating that erythroblasts are killed by type I IFNs (Yoshida, Okabe, Kawane, Fukuyama, & Nagata, 2005). In fact, macrophages containing undigested DNA produce IFNβ. Type I IFNs, such as

IFNβ, are cytokines that confer resistance to viral infection, but are also cytotoxic (Borden et al., 2007). Since erythroid precursor cells are sensitive to the cytotoxic activity of type I IFNs (Binder, Fehr, Hengartner, & Zinkernagel, 1997), it is likely that the IFNβ produced by macrophages carrying undigested DNA kills erythroid precursor cells.

Mice deficient in both DNase II and IFNIR develop normally but suffer from polyarthritis as they age (Kawane et al., 2006). Mice in which the DNase II gene is deleted inducibly after birth also develop polyarthritis with high levels of inflammatory cytokines and chemokines in the joints, and of rheumatoid factors in the serum. The arthritis in DNase II-deficient mice is not mediated by the acquired immune system, but by inflammatory cytokines (TNFα, IL-6, and IL-1β) (Kawane, Tanaka, Kitahara, Shimaoka, & Nagata, 2010). Blocking one of these inflammatory cytokines inhibits the expression of the other cytokine genes and the development of arthritis, suggesting that these cytokines activate each other's genes in macrophages and synovial cells. It is likely that cytokines accumulated in the joints stimulate synovial cell growth and macrophage differentiation into osteoclasts, destroying the bones. Macrophages carrying undigested DNA produce TNFα (Kawane et al., 2006), and TNFα transgenic mice develop polyarthritis (Keffer et al., 1991), suggesting that the polyarthritis is triggered by TNFα produced by the macrophages. The etiology of human rheumatoid arthritis is not known. The properties of arthritis in DNase II-null mice are similar to those of human arthritis, particularly of systemic-onset juvenile idiopathic arthritis and Still's disease (Vastert, Kuis, & Grom, 2009). It would be interesting to examine whether macrophage lysosomes function properly in human patients with these diseases.

5.3 Innate Immunity and DNA Degradation in Lysosomes

Analyses of DNase II-deficient mice indicate that DNA accumulation in the lysosomes activates macrophages to express IFNβ and TNFα. The signal that induces IFNβ gene expression is independent of Toll-like receptor (Okabe, Kawane, Akira, Taniguchi, & Nagata, 2005) but requires interferon regulatory factor (IRF)3 or IRF7 (Okabe, Kawane, & Nagata, 2008). A null mutation of STING fully rescues the lethality of DNase II-null embryos (Ahn, Gutman, Saijo, & Barber, 2012), indicating STING's involvement in this pathway. STING, a transmembrane protein localized to the endoplasmic reticulum, is essential for the innate immune response to DNA and RNA viruses (Ishikawa & Barber, 2008; Ishikawa, Ma, & Barber, 2009). In response to cytosolic DNA, STING forms a dimer, activates TBK1 (TANK-binding kinase 1), and stimulates IRF3 to produce IFNβ (Tanaka & Chen, 2012). Recently, cyclic GMP-AMP (cGAMP) synthase (cGAS) was identified as a cytosolic DNA sensor that induces the production of cGAMP, which functions as a second messenger to activate STING (Sun, Wu, Du, Chen, & Chen, 2013; Wu et al., 2013). We have evidence that the cGAMP-cGAS system functions as a DNA sensor for type I IFN production in DNase II-null embryos (Motani, Ito, & Nagata, 2015), and it would be interesting to determine how DNA accumulation in lysosomes activates cGAS.

6. PERSPECTIVES

The failure of macrophages to swiftly engulf apoptotic cells or pyrenocytes appears to result in SLE-type autoimmune diseases in humans and mice (Gaipl et al., 2005; Munoz et al., 2010; Nagata et al., 2010; Rothlin & Lemke, 2010). If apoptotic cells—or their DNA, at least—are not properly degraded, it causes autoinflammation, leading to severe anemia and polyarthritis (Kawane, Motani, & Nagata, 2014). Here, we have discussed the molecular mechanisms that operate to clear apoptotic cells and pyrenocytes. PtdSer is probably the only "eat me" signal that prompts macrophages to engulf apoptotic cells and pyrenocytes. We now understand the molecular machinery that exposes PtdSer on the surface of the dying cell (Segawa, Suzuki, & Nagata, 2014). Although a Ca^{2+}-dependent phospholipid scramblase is probably involved in the PtdSer exposure on pyrenocytes, this enzyme has yet to be identified.

Several molecules have been shown to specifically bind PtdSer. Their contribution to the engulfment of apoptotic cells has not been fully confirmed using macrophages lacking them, nor has it been determined whether these molecules are simply redundant. If they are not redundant, how different PtdSer-binding proteins and PSRs cooperate in different macrophages to engulf apoptotic cells and pyrenocytes should be clarified. PtdSer is known to work not only as an "eat me" signal but also as a scaffold for blood-clotting factors. In fact, some PtdSer-binding proteins involved in the engulfment of apoptotic cells are also involved in blood clotting, although how these molecules serve both functions remains to be clarified.

In the phagocytosis of apoptotic cells or pyrenocytes, the macrophage forms a phagocytic cup of polymerized actin, the dead cell sinks through the cup, and the cup closes (Nakaya et al., 2008). The dead cell is then transferred to a lysosome for degradation. The processes that form and close the phagocytic cup, and transfer the dead cell into the lysosome, are elusive. Finally, we have discussed the engulfment of the apoptotic cell, but this event is preceded by the macrophage approaching the dead cell. It has been proposed that macrophages are recruited by ATP or UTP released from apoptotic cells (Elliott et al., 2009). However, the ATP level in the supernatant of FasL-treated apoptotic cells is very low (Yamaguchi, Maruyama, Urade, & Nagata, 2014). Whether this low ATP level is sufficient to recruit macrophages (or any other molecules involved in this process) remains to be clarified.

Compared with the year 2000, this field concerned with the engulfment of apoptotic cells and pyrenocytes has certainly advanced. However, there are still many points that require clarification, and further investigation will contribute to our understanding of autoimmune diseases such as SLE and arthritis.

REFERENCES

Acehan, D., Jiang, X., Morgan, D. G., Heuser, J. E., Wang, X., & Akey, C. W. (2002). Three-dimensional structure of the apoptosome: Implications for assembly, procaspase-9 binding, and activation. *Molecular Cell*, *9*, 423–432.

Acharya, U. (2006). Drosophila melanogaster Scramblases modulate synaptic transmission. *The Journal of Cell Biology*, *173*, 69–82.

Ahn, J., Gutman, D., Saijo, S., & Barber, G. N. (2012). STING manifests self DNA-dependent inflammatory disease. *Proceedings of the National Academy of Sciences of the United States of America*, *109*, 19386–19391.

Akakura, S., Singh, S., Spataro, M., Akakura, R., Kim, J. I., Albert, M. L., et al. (2004). The opsonin MFG-E8 is a ligand for the alphavbeta5 integrin and triggers DOCK180-dependent Rac1 activation for the phagocytosis of apoptotic cells. *Experimental Cell Research*, *292*, 403–416.

Anderson, H. A., Maylock, C. A., Williams, J. A., Paweletz, C. P., Shu, H., & Shacter, E. (2003). Serum-derived protein S binds to phosphatidylserine and stimulates the phagocytosis of apoptotic cells. *Nature Immunology*, *4*, 87–91.

Angelillo-Scherrer, A., de Frutos, P., Aparicio, C., Melis, E., Savi, P., Lupu, F., et al. (2001). Deficiency or inhibition of Gas6 causes platelet dysfunction and protects mice against thrombosis. *Nature Medicine*, *7*, 215–221.

Aoki, N., Ishii, T., Ohira, S., Yamaguchi, Y., Negi, M., Adachi, T., et al. (1997). Stage specific expression of milk fat globule membrane glycoproteins in mouse mammary gland: Comparison of MFG-E8, butyrophilin, and CD36 with a major milk protein, beta-casein. *Biochimica et Biophysica Acta*, *1334*, 182–190.

Asano, K., Miwa, M., Miwa, K., Hanayama, R., Nagase, H., Nagata, S., et al. (2004). Masking of phosphatidylserine inhibits apoptotic cell engulfment and induces autoantibody production in mice. *The Journal of Experimental Medicine*, *200*, 459–467.

Basse, F., Stout, J. G., Sims, P. J., & Wiedmer, T. (1996). Isolation of an erythrocyte membrane protein that mediates Ca^{2+}-dependent transbilayer movement of phospholipid. *The Journal of Biological Chemistry*, *271*, 17205–17210.

Bevers, E. M., & Williamson, P. L. (2010). Phospholipid scramblase: An update. *FEBS Letters, 584*, 2724–2730.
Binder, D., Fehr, J., Hengartner, H., & Zinkernagel, R. M. (1997). Virus-induced transient bone marrow aplasia: Major role of interferon-alpha/beta during acute infection with the noncytopathic lymphocytic choriomeningitis virus. *The Journal of Experimental Medicine, 185*, 517–530.
Borden, E. C., Sen, G. C., Uze, G., Silverman, R. H., Ransohoff, R. M., Foster, G. R., et al. (2007). Interferons at age 50: Past, current and future impact on biomedicine. *Nature Reviews. Drug Discovery, 6*, 975–990.
Bose, J., Gruber, A. D., Helming, L., Schiebe, S., Wegener, I., Hafner, M., et al. (2004). The phosphatidylserine receptor has essential functions during embryogenesis but not in apoptotic cell removal. *Journal of Biology, 3*, 15.
Burstyn-Cohen, T., Lew, E. D., Través, P. G., Burrola, P. G., Hash, J. C., & Lemke, G. (2012). Genetic dissection of TAM receptor–ligand interaction in retinal pigment epithelial cell phagocytosis. *Neuron, 76*, 1123–1132.
Chang, B., Chen, Y., Zhao, Y., & Bruick, R. K. (2007). JMJD6 is a histone arginine demethylase. *Science, 318*, 444–447.
Chao, M. P., Majeti, R., & Weissman, I. L. (2011). Programmed cell removal: A new obstacle in the road to developing cancer. *Nature Reviews. Cancer, 12*, 58–67.
Chasis, J. A., & Mohandas, N. (2008). Erythroblastic islands: Niches for erythropoiesis. *Blood, 112*, 470–478.
Chen, Y.-Z., Mapes, J., Lee, E.-S., Skeen-Gaar, R. R., & Xue, D. (2013). Caspase-mediated activation of Caenorhabditis elegans CED-8 promotes apoptosis and phosphatidylserine externalization. *Nature Communications, 4*, 1–9.
Cikala, M., Alexandrova, O., David, C. N., Pröschel, M., Stiening, B., Cramer, P., et al. (2004). The phosphatidylserine receptor from *Hydra* is a nuclear protein with potential Fe(II) dependent oxygenase activity. *BMC Cell Biology, 5*, 26–35.
Cote, J. F., & Vuori, K. (2007). GEF what? Dock180 and related proteins help Rac to polarize cells in new ways. *Trends in Cell Biology, 17*, 383–393.
Cui, P., Qin, B., Liu, N., Pan, G., & Pei, D. (2004). Nuclear localization of the phosphatidylserine receptor protein via multiple nuclear localization signals. *Experimental Cell Research, 293*, 154–163.
Dahlbäck, B., & Villoutreix, B. O. (2005). The anticoagulant protein C pathway. *FEBS Letters, 579*, 3310–3316.
DeKruyff, R. H., Bu, X., Ballesteros, A., Santiago, C., Chim, Y. L. E., Lee, H. H., et al. (2010). T cell/transmembrane, Ig, and mucin-3 allelic variants differentially recognize phosphatidylserine and mediate phagocytosis of apoptotic cells. *Journal of Immunology, 184*, 1918–1930.
Elliott, M. R., Chekeni, F. B., Trampont, P. C., Lazarowski, E. R., Kadl, A., Walk, S. F., et al. (2009). Nucleotides released by apoptotic cells act as a find-me signal to promote phagocytic clearance. *Nature, 461*, 282–286.
Enari, M., Sakahira, H., Yokoyama, H., Okawa, K., Iwamatsu, A., & Nagata, S. (1998). A caspase-activated DNase that degrades DNA during apoptosis, and its inhibitor ICAD. *Nature, 391*, 43–50.
Fadok, V., Bratton, D., Rose, D., Pearson, A., Ezekewitz, R., & Henson, P. (2000). A receptor for phosphatidylserine-specific clearance of apoptotic cells. *Nature, 405*, 85–90.
Fadok, V. A., Voelker, D. R., Campbell, P. A., Cohen, J. J., Bratton, D. L., & Henson, P. M. (1992). Exposure of phosphatidylserine on the surface of apoptotic lymphocytes triggers specific recognition and removal by macrophages. *Journal of Immunology, 148*, 2207–2216.
Freeman, G. J., Casasnovas, J. M., Umetsu, D. T., & Dekruyff, R. H. (2010). TIM genes: A family of cell surface phosphatidylserine receptors that regulate innate and adaptive immunity. *Immunological Reviews, 235*, 172–189.

Gaipl, U. S., Voll, R. E., Sheriff, A., Franz, S., Kalden, J. R., & Herrmann, M. (2005). Impaired clearance of dying cells in systemic lupus erythematosus. *Autoimmunity Reviews*, *4*, 189–194.

Gautier, E. L., Shay, T., Miller, J., Greter, M., Jakubzick, C., Ivanov, S., et al. (2012). Gene-expression profiles and transcriptional regulatory pathways that underlie the identity and diversity of mouse tissue macrophages. *Nature Immunology*, *13*, 1118–1128.

Gordon, S., & Plűddemann, A. (2013). Tissue macrophage heterogeneity: Issues and prospects. *Seminars in Immunopathology*, *35*, 533–540.

Hamon, Y., Trompier, D., Ma, Z., Venegas, V., Pophillat, M., Mignotte, V., et al. (2006). Cooperation between engulfment receptors: The case of ABCA1 and MEGF10. *PLoS One*, *1*. e120.

Hanayama, R., & Nagata, S. (2005). Impaired involution of mammary glands in the absence of milk fat globule EGF factor 8. *Proceedings of the National Academy of Sciences of the United States of America*, *102*, 16886–16891.

Hanayama, R., Tanaka, M., Miwa, K., & Nagata, S. (2004). Expression of developmental endothelial locus-1 in a subset of macrophages for engulfment of apoptotic cells. *Journal of Immunology*, *172*, 3876–3882.

Hanayama, R., Tanaka, M., Miwa, K., Shinohara, A., Iwamatsu, A., & Nagata, S. (2002). Identification of a factor that links apoptotic cells to phagocytes. *Nature*, *417*, 182–187.

Hanayama, R., Tanaka, M., Miyasaka, K., Aozasa, K., Koike, M., Uchiyama, Y., et al. (2004). Autoimmune disease and impaired uptake of apoptotic cells in MFG-E8-deficient mice. *Science*, *304*, 1147–1150.

Harris, E. N., Weigel, J. A., & Weigel, P. H. (2008). The human hyaluronan receptor for endocytosis (HARE/Stabilin-2) is a systemic clearance receptor for heparin. *The Journal of Biological Chemistry*, *283*, 17341–17350.

Hoffmann, P. R., deCathelineau, A. M., Ogden, C. A., Leverrier, Y., Bratton, D. L., Daleke, D. L., et al. (2001). Phosphatidylserine (PS) induces PS receptor-mediated macropinocytosis and promotes clearance of apoptotic cells. *The Journal of Cell Biology*, *155*, 649–659.

Ichimura, T., Asseldonk, E. J., Humphreys, B. D., Gunaratnam, L., Duffield, J. S., & Bonventre, J. V. (2008). Kidney injury molecule-1 is a phosphatidylserine receptor that confers a phagocytic phenotype on epithelial cells. *The Journal of Clinical Investigation*, *118*, 1657–1668.

Ishikawa, H., & Barber, G. N. (2008). STING is an endoplasmic reticulum adaptor that facilitates innate immune signalling. *Nature*, *455*, 674–678.

Ishikawa, H., Ma, Z., & Barber, G. N. (2009). STING regulates intracellular DNA-mediated, type I interferon-dependent innate immunity. *Nature*, *461*, 788–792.

Ishimoto, Y., Ohashi, K., Mizuno, K., & Nakano, T. (2000). Promotion of the uptake of PS liposomes and apoptotic cells by a product of growth arrest-specific gene, gas6. *Journal of Biochemistry*, *127*, 411–417.

Jacobson, M. D., Weil, M., & Raff, M. C. (1997). Programmed cell death in animal development. *Cell*, *88*, 347–354.

Ji, P., Jayapal, S. R., & Lodish, H. F. (2008). Enucleation of cultured mouse fetal erythroblasts requires Rac GTPases and mDia2. *Nature*, *10*, 314–321.

Kawane, K., Fukuyama, H., Kondoh, G., Takeda, J., Ohsawa, Y., Uchiyama, Y., et al. (2001). Requirement of DNase II for definitive erythropoiesis in the mouse fetal liver. *Science*, *292*, 1546–1549.

Kawane, K., Fukuyama, H., Yoshida, H., Nagase, H., Ohsawa, Y., Uchiyama, Y., et al. (2003). Impaired thymic development in mouse embryos deficient in apoptotic DNA degradation. *Nature Immunology*, *4*, 138–144.

Kawane, K., Motani, K., & Nagata, S. (2014). DNA degradation and its defects. *Cold Spring Harbor Perspectives in Biology*, *6*, a016394.

Kawane, K., Ohtani, M., Miwa, K., Kizawa, T., Kanbara, Y., Yoshioka, Y., et al. (2006). Chronic polyarthritis caused by mammalian DNA that escapes from degradation in macrophages. *Nature, 443*, 998–1002.

Kawane, K., Tanaka, H., Kitahara, Y., Shimaoka, S., & Nagata, S. (2010). Cytokine-dependent but acquired immunity-independent arthritis caused by DNA escaped from degradation. *Proceedings of the National Academy of Sciences of the United States of America, 107*, 19432–19437.

Keffer, J., Probert, L., Cazlaris, H., Georgopoulos, S., Kaslaris, E., Kioussis, D., et al. (1991). Transgenic mice expressing human tumour necrosis factor: A predictive genetic model of arthritis. *The EMBO Journal, 10*, 4025–4031.

Kerr, J. F., Wyllie, A. H., & Currie, A. R. (1972). Apoptosis: A basic biological phenomenon with wide-ranging implications in tissue kinetics. *British Journal of Cancer, 26*, 239–257.

Kim, C., Ye, F., & Ginsberg, M. H. (2011). Regulation of integrin activation. *Annual Review of Cell and Developmental Biology, 27*, 321–345.

Kinchen, J. M., & Ravichandran, K. S. (2008). Phagosome maturation: Going through the acid test. *Nature Reviews. Molecular Cell Biology, 9*, 781–795.

Kitano, M., Nakaya, M., Nakamura, T., Nagata, S., & Matsuda, M. (2008). Imaging of Rab5 activity identifies essential regulators for phagosome maturation. *Nature, 453*, 241–245.

Konstantinidis, D. G., Pushkaran, S., Johnson, J. F., Cancelas, J. A., Manganaris, S., Harris, C. E., et al. (2012). Signaling and cytoskeletal requirements in erythroblast enucleation. *Blood, 119*, 6118–6127.

Krammer, P. H. (2000). CD95's deadly mission in the immune system. *Nature, 407*, 789–795.

Kranich, J., Krautler, N. J., Heinen, E., Polymenidou, M., Bridel, C., Schildknecht, A., et al. (2008). Follicular dendritic cells control engulfment of apoptotic bodies by secreting Mfge8. *The Journal of Experimental Medicine, 205*, 1293–1302.

Kroemer, G., Galluzzi, L., Vandenabeele, P., Abrams, J., Alnemri, E. S., Baehrecke, E. H., et al. (2009). Classification of cell death: Recommendations of the nomenclature committee on cell death 2009. *Cell Death and Differentiation, 16*, 3–11.

Lacy-Hulbert, A., Smith, A. M., Tissire, H., Barry, M., Crowley, D., Bronson, R. T., et al. (2007). Ulcerative colitis and autoimmunity induced by loss of myeloid alphav integrins. *Proceedings of the National Academy of Sciences of the United States of America, 104*, 15823–15828.

Lee, S.-J., Park, S.-Y., Jung, M.-Y., Bae, S. M., & Kim, I.-S. (2011). Mechanism for phosphatidylserine-dependent erythrophagocytosis in mouse liver. *Blood, 117*, 5215–5223.

Lemke, G., & Rothlin, C. V. (2008). Immunobiology of the TAM receptors. *Nature Reviews Immunology, 8*, 327–336.

Leventis, P. A., & Grinstein, S. (2010). The distribution and function of phosphatidylserine in cellular membranes. *Annual Review of Biophysics, 39*, 407–427.

Li, M. O., Sarkisian, M. R., Mehal, W. Z., Rakic, P., & Flavell, R. A. (2003). Phosphatidylserine receptor is required for clearance of apoptotic cells. *Science, 302*, 1560–1563.

Liu, X., Zou, H., Slaughter, C., & Wang, X. (1997). DFF, a heterodimeric protein that functions downstream of caspase-3 to trigger DNA fragmentation during apoptosis. *Cell, 89*, 175–184.

Loo, D. T. (2010). *In situ detection of apoptosis by the TUNEL assay: An overview of techniques.* Totowa, NJ: Humana Press. pp. 3–13.

Lu, Q., & Lemke, G. (2001). Homeostatic regulation of the immune system by receptor tyrosine kinases of the Tyro 3 family. *Science, 293*, 306–311.

Luthi, A. U., & Martin, S. J. (2007). The CASBAH: A searchable database of caspase substrates. *Cell Death and Differentiation, 14*, 641–650.

McIlroy, D., Tanaka, M., Sakahira, H., Fukuyama, H., Suzuki, M., Yamamura, K., et al. (2000). An auxiliary mode of apoptotic DNA fragmentation provided by phagocytes. *Genes & Development, 14*, 549–558.

Meyers, J. H., Chakravarti, S., Schlesinger, D., Illes, Z., Waldner, H., Umetsu, S. E., et al. (2005). TIM-4 is the ligand for TIM-1, and the TIM-1–TIM-4 interaction regulates T cell proliferation. *Nature Immunology, 6*, 455–464.

Migliaccio, A. R., Masselli, E., Varricchio, L., & Whitsett, C. (2012). Ex-vivo expansion of red blood cells: How real for transfusion in humans? *Blood Reviews, 26*, 81–95.

Miksa, M., Komura, H., Wu, R., Shah, K. G., & Wang, P. (2009). A novel method to determine the engulfment of apoptotic cells by macrophages using pHrodo succinimidyl ester. *Journal of Immunological Methods, 342*, 71–77.

Miyanishi, M., Segawa, K., & Nagata, S. (2012). Synergistic effect of Tim4 and MFG-E8 null mutations on the development of autoimmunity. *International Immunology, 24*, 551–559.

Miyanishi, M., Tada, K., Koike, M., Uchiyama, Y., Kitamura, T., & Nagata, S. (2007). Identification of Tim4 as a phosphatidylserine receptor. *Nature, 450*, 435–439.

Miyasaka, K., Hanayama, R., Tanaka, M., & Nagata, S. (2004). Expression of milk fat globule epidermal growth factor 8 in immature dendritic cells for engulfment of apoptotic cells. *European Journal of Immunology, 34*, 1414–1422.

Motani, K., Ito, S., & Nagata, S. (2015). DNA-mediated cyclic GMP-AMP synthase-dependent and -independent regulation of innate immune responses. *The Journal of Immunology, 194*, 4914–4923.

Mukae, N., Yokoyama, H., Yokokura, T., Sakoyama, Y., & Nagata, S. (2002). Activation of the innate immunity in Drosophila by endogenous chromosomal DNA that escaped apoptotic degradation. *Genes & Development, 16*, 2662–2671.

Munoz, L. E., Lauber, K., Schiller, M., Manfredi, A. A., & Herrmann, M. (2010). The role of defective clearance of apoptotic cells in systemic autoimmunity. *Nature Reviews. Rheumatology, 6*, 280–289.

Nagata, S. (1997). Apoptosis by death factor. *Cell, 88*, 355–365.

Nagata, S., Hanayama, R., & Kawane, K. (2010). Autoimmunity and the clearance of dead cells. *Cell, 140*, 619–630.

Nakano, T., Ishimoto, Y., Kishino, J., Umeda, M., Inoue, K., Nagata, K., et al. (1997). Cell adhesion to phosphatidylserine mediated by a product of growth arrest-specific gene 6. *The Journal of Biological Chemistry, 272*, 29411–29414.

Nakaya, M., Kitano, M., Matsuda, M., & Nagata, S. (2008). Spatiotemporal activation of Rac1 for engulfment of apoptotic cells. *Proceedings of the National Academy of Sciences of the United States of America, 105*, 9198–9203.

Nakaya, M., Tanaka, M., Okabe, Y., Hanayama, R., & Nagata, S. (2006). Opposite effects of Rho family GTPases on engulfment of apoptotic cells by macrophages. *The Journal of Biological Chemistry, 281*, 8836–8842.

Nakayama, M., Akiba, H., Takeda, K., Kojima, Y., Hashiguchi, M., Azuma, M., et al. (2009). Tim-3 mediates phagocytosis of apoptotic cells and cross-presentation. *Blood, 113*, 3821–3830.

Nicholls, S. B., & Hyman, B. T. (2014). Measuring caspase activity in vivo. *Methods in Enzymology, 544*, 251–269. http://dx.doi.org/10.1016/B978-0-12-417158-9.00010-8.

Nishi, C., Toda, S., Segawa, K., & Nagata, S. (2014). Tim4- and MerTK-mediated engulfment of apoptotic cells by mouse resident peritoneal macrophages. *Molecular and Cellular Biology, 34*, 1512–1520.

Okabe, Y., Kawane, K., Akira, S., Taniguchi, T., & Nagata, S. (2005). Toll-like receptor-independent gene induction program activated by mammalian DNA escaped from apoptotic DNA degradation. *The Journal of Experimental Medicine, 202*, 1333–1339.

Okabe, Y., Kawane, K., & Nagata, S. (2008). IFN regulatory factor (IRF) 3/7-dependent and -independent gene induction by mammalian DNA that escapes degradation. *European Journal of Immunology*, *38*, 3150–3158.

Okabe, Y., & Medzhitov, R. (2014). Tissue-specific signals control reversible program of localization and functional polarization of macrophages. *Cell*, *157*, 832–844.

Overholtzer, M., Mailleux, A. A., Mouneimne, G., Normand, G., Schnitt, S. J., King, R. W., et al. (2007). A nonapoptotic cell death process, entosis, that occurs by cell-in-cell invasion. *Cell*, *131*, 966–979.

Ow, Y. P., Green, D. R., Hao, Z., & Mak, T. W. (2008). Cytochrome c: Functions beyond respiration. *Nature Reviews. Molecular Cell Biology*, *9*, 532–542.

Palis, J. (2008). Ontogeny of erythropoiesis. *Current Opinion in Hematology*, *15*, 155–161.

Park, D., Hochreiter-Hufford, A., & Ravichandran, K. (2009). The phosphatidylserine receptor TIM-4 does not mediate direct signaling. *Current Biology*, *19*, 346–351.

Park, S. Y., Jung, M. Y., Kim, H. J., Lee, S. J., Kim, S. Y., Lee, B. H., et al. (2008). Rapid cell corpse clearance by stabilin-2, a membrane phosphatidylserine receptor. *Cell Death and Differentiation*, *15*, 192–201.

Park, S.-Y., Jung, M.-Y., Lee, S.-J., Kang, K. B., Gratchev, A., Riabov, V., et al. (2009). Stabilin-1 mediates phosphatidylserine-dependent clearance of cell corpses in alternatively activated macrophages. *Journal of Cell Science*, *122*, 3365–3373.

Park, D., Tosello-Trampont, A. C., Elliott, M. R., Lu, M., Haney, L. B., Ma, Z., et al. (2007). BAI1 is an engulfment receptor for apoptotic cells upstream of the ELMO/Dock180/Rac module. *Nature*, *450*, 430–434.

Patton, S., & Keenan, T. W. (1975). The milk fat globule membrane. *Biochimica et Biophysica Acta*, *415*, 273–309.

Peng, Y., & Elkon, K. B. (2011). Autoimmunity in MFG-E8-deficient mice is associated with altered trafficking and enhanced cross-presentation of apoptotic cell antigens. *The Journal of Clinical Investigation*, *121*, 2221–2241.

Reddien, P. W., & Horvitz, H. R. (2004). The engulfment process of programmed cell death in Caenorhabditis elegans. *Annual Review of Cell and Developmental Biology*, *20*, 193–221.

Rodriguez-Manzanet, R., Sanjuan, M., Wu, H., Quintana, F., Xiao, S., Anderson, A., et al. (2010). T and B cell hyperactivity and autoimmunity associated with niche-specific defects in apoptotic body clearance in TIM-4-deficient mice. *Proceedings of the National Academy of Sciences of the United States of America*, *107*, 8706–8711.

Rothlin, C. V., & Lemke, G. (2010). TAM receptor signaling and autoimmune disease. *Current Opinion in Immunology*, *22*, 740–746.

Sadahira, Y., Yoshino, T., & Monobe, Y. (1995). Very late activation antigen 4-vascular cell adhesion molecule 1 interaction is involved in the formation of erythroblastic islands. *The Journal of Experimental Medicine*, *181*, 411–415.

Sakahira, H., Enari, M., & Nagata, S. (1998). Cleavage of CAD inhibitor in CAD activation and DNA degradation during apoptosis. *Nature*, *391*, 96–99.

Sánchez-Fueyo, A., Tian, J., Picarella, D., Domenig, C., Zheng, X. X., Sabatos, C. A., et al. (2003). Tim-3 inhibits T helper type 1-mediated auto- and alloimmune responses and promotes immunological tolerance. *Nature Immunology*, *4*, 1093–1101.

Santiago, C., Ballesteros, A., Martínez-Muñoz, L., Mellado, M., Kaplan, G., Freeman, G., et al. (2007). Structures of T cell immunoglobulin mucin protein 4 show a metal-Ion-dependent ligand binding site where phosphatidylserine binds. *Immunity*, *27*, 941–951.

Savill, J., & Fadok, V. (2000). Corpse clearance defines the meaning of cell death. *Nature*, *407*, 784–788.

Schledzewski, K., Géraud, C., Arnold, B., Wang, S., Gröne, H.-J., Kempf, T., et al. (2011). Deficiency of liver sinusoidal scavenger receptors stabilin-1 and -2 in mice causes glomerulofibrotic nephropathy via impaired hepatic clearance of noxious blood factors. *The Journal of Clinical Investigation*, *121*, 703–714.

Scott, R. S., McMahon, E. J., Pop, S. M., Reap, E. A., Caricchio, R., Cohen, P. L., et al. (2001). Phagocytosis and clearance of apoptotic cells is mediated by MER. *Nature*, *411*, 207–211.

Segawa, K., Kurata, S., Yanagihashi, Y., Brummelkamp, T., Matsuda, F., & Nagata, S. (2014). Caspase-mediated cleavage of phospholipid flippase for apoptotic phosphatidylserine exposure. *Science*, *344*, 1164–1168.

Segawa, K., Suzuki, J., & Nagata, S. (2014). Flippases and scramblases in the plasma membrane. *Cell Cycle*, *13*, 2990–2991.

Seitz, H. M., Camenisch, T. D., Lemke, G., Earp, H. S., & Matsushima, G. K. (2007). Macrophages and dendritic cells use different Axl/Mertk/Tyro3 receptors in clearance of apoptotic cells. *Journal of Immunology*, *178*, 5635–5642.

Stanfield, G. M., & Horvitz, H. R. (2000). The ced-8 gene controls the timing of programmed cell deaths in C. elegans. *Molecular Cell*, *5*, 423–433.

Stephenson, J. R., Purcell, R. H., & Hall, R. A. (2014). The BAI subfamily of adhesion GPCRs: Synaptic regulation and beyond. *Trends in Pharmacological Sciences*, *35*, 208–215.

Strasser, A., Jost, P. J., & Nagata, S. (2009). The many roles of FAS receptor signaling in the immune system. *Immunity*, *30*, 180–192.

Strauss, O. (2005). The retinal pigment epithelium in visual function. *Physiological Reviews*, *85*, 845–881.

Sun, L., Wu, J., Du, F., Chen, X., & Chen, Z. J. (2013). Cyclic GMP-AMP synthase is a cytosolic DNA sensor that activates the type I interferon pathway. *Science*, *339*, 786–791.

Surh, C. D., & Sprent, J. (1994). T-cell apoptosis detected *in situ* during positive and negative selection in the thymus. *Nature*, *372*, 100–103.

Suzuki, J., Denning, D. P., Imanishi, E., Horvitz, H. R., & Nagata, S. (2013). Xk-related protein 8 and CED-8 promote phosphatidylserine exposure in apoptotic cells. *Science*, *341*, 403–406.

Suzuki, J., Fujii, T., Imao, T., Ishihara, K., Kuba, H., & Nagata, S. (2013). Calcium-dependent phospholipid scramblase activity of TMEM16 protein family members. *The Journal of Biological Chemistry*, *288*, 13305–13316.

Suzuki, J., Imanishi, E., & Nagata, S. (2014). Exposure of phosphatidylserine by Xk-related protein family members during apoptosis. *The Journal of Biological Chemistry*, *289*, 30257–30267.

Suzuki, J., Umeda, M., Sims, P. J., & Nagata, S. (2010). Calcium-dependent phospholipid scrambling by TMEM16F. *Nature*, *468*, 834–838.

Tanaka, Y., & Chen, Z. J. (2012). STING specifies IRF3 phosphorylation by TBK1 in the cytosolic DNA signaling pathway. *Science Signaling*, *5*, ra20.

Tanaka, K., Fujimura-Kamada, K., & Yamamoto, T. (2011). Functions of phospholipid flippases. *Journal of Biochemistry*, *149*, 131–143.

Théry, C., Ostrowski, M., & Segura, E. (2009). Membrane vesicles as conveyors of immune responses. *Nature Reviews. Immunology*, *9*, 581–593.

Timmer, J. C., & Salvesen, G. S. (2007). Caspase substrates. *Cell Death and Differentiation*, *14*, 66–72.

Toda, S., Hanayama, R., & Nagata, S. (2012). Two-step engulfment of apoptotic cells. *Molecular and Cellular Biology*, *32*, 118–125.

Toda, S., Segawa, K., & Nagata, S. (2014). MerTK-mediated engulfment of pyrenocytes by central macrophages in erythroblastic islands. *Blood*, *123*, 3963–3971.

Tung, T. T., Nagaosa, K., Fujita, Y., Kita, A., Mori, H., Okada, R., et al. (2013). Phosphatidylserine recognition and induction of apoptotic cell clearance by drosophila engulfment receptor draper. *Journal of Biochemistry*, *153*, 483–491.

Vastert, S. J., Kuis, W., & Grom, A. A. (2009). Systemic JIA: New developments in the understanding of the pathophysiology and therapy. *Best Practice & Research. Clinical Rheumatology, 23*, 655–664.
Venegas, V., & Zhou, Z. (2007). Two alternative mechanisms that regulate the presentation of apoptotic cell engulfment signal in Caenorhabditis elegans. *Molecular Biology of the Cell, 18*, 3180–3192.
Wang, J., Ramirez, T., Ji, P., Jayapal, S. R., Lodish, H. F., & Murata-Hori, M. (2012). Mammalian erythroblast enucleation requires PI3K-dependent cell polarization. *Journal of Cell Science, 125*, 340–349.
Wang, X., Wu, Y. C., Fadok, V. A., Lee, M. C., Gengyo-Ando, K., Cheng, L. C., et al. (2003). Cell corpse engulfment mediated by C. elegans phosphatidylserine receptor through CED-5 and CED-12. *Science, 302*, 1563–1566.
Wong, K., Valdez, P., Tan, C., Yeh, S., Hongo, J., & Ouyang, W. (2010). Phosphatidylserine receptor Tim-4 is essential for the maintenance of the homeostatic state of resident peritoneal macrophages. *Proceedings of the National Academy of Sciences of the United States of America, 107*, 8712–8717.
Woo, E.-J., Kim, Y.-G., Kim, M.-S., Han, W.-D., Shin, S., Robinson, H., et al. (2004). Structural mechanism for inactivation and activation of CAD/DFF40 in the apoptotic pathway. *Molecular Cell, 14*, 531–539.
Wu, Y., Singh, S., Georgescu, M.-M., & Birge, R. B. (2005). A role for Mer tyrosine kinase in alphavbeta5 integrin-mediated phagocytosis of apoptotic cells. *Journal of Cell Science, 118*, 539–553.
Wu, Y. C., Stanfield, G. M., & Horvitz, H. R. (2000). NUC-1, a Caenorhabditis elegans DNase II homolog, functions in an intermediate step of DNA degradation during apoptosis. *Genes & Development, 14*, 536–548.
Wu, J., Sun, L., Chen, X., Du, F., Shi, H., Chen, C., et al. (2013). Cyclic GMP-AMP is an endogenous second messenger in innate immune signaling by cytosolic DNA. *Science, 339*, 826–830.
Wyllie, A. H. (1980). Glucocorticoid-induced thymocyte apoptosis is associated with endogenous endonuclease activation. *Nature, 284*, 555–556.
Yamaguchi, H., Maruyama, T., Urade, Y., & Nagata, S. (2014). Immunosuppression via adenosine receptor activation by adenosine monophosphate released from apoptotic cells. *eLife, 3*. e02172.
Yang, H., Chen, Y.-Z., Zhang, Y., Wang, X., Zhao, X., Godfroy, J. I., et al. (2014). A lysine-rich motif in the phosphatidylserine receptor PSR-1 mediates recognition and removal of apoptotic cells. *Nature Communications, 6*, 1–12.
Yoshida, H., Kawane, K., Koike, M., Mori, Y., Uchiyama, Y., & Nagata, S. (2005). Phosphatidylserine-dependent engulfment by macrophages of nuclei from erythroid precursor cells. *Nature, 437*, 754–758.
Yoshida, H., Okabe, Y., Kawane, K., Fukuyama, H., & Nagata, S. (2005). Lethal anemia caused by interferon-beta produced in mouse embryos carrying undigested DNA. *Nature Immunology, 6*, 49–56.
Zhou, Q., Zhao, J., Wiedmer, T., & Sims, P. J. (2002). Normal hemostasis but defective hematopoietic response to growth factors in mice deficient in phospholipid scramblase 1. *Blood, 99*, 4030–4038.

CHAPTER ELEVEN

Apoptotic Cell Clearance in Development

Jeny Shklover, Flonia Levy-Adam, Estee Kurant[1]
Department of Genetics and Developmental Biology, The Rappaport Family Institute for Research in the Medical Sciences, Faculty of Medicine, Technion—Israel Institute of Technology, Haifa, Israel
[1]Corresponding author: e-mail address: kurante@tx.technion.ac.il

Contents

Abstract

Programmed cell death and its specific form apoptosis play an important role during development of multicellular organisms. They are crucial for morphogenesis and organ sculpting as well as for adjusting cell number in different systems. Removal of apoptotic

Current Topics in Developmental Biology, Volume 114
ISSN 0070-2153
http://dx.doi.org/10.1016/bs.ctdb.2015.07.024

cells is the last critical step of apoptosis. Apoptotic cells are properly and efficiently recognized and eliminated through phagocytosis, which is performed by professional and nonprofessional phagocytes. Phagocytosis of apoptotic cells or apoptotic cell clearance is a dynamic multistep process, involving interactions between phagocytic receptors and ligands on apoptotic cells, which are highly conserved in evolution. However, this process is extremely redundant in mammals, containing multiple factors playing similar roles in the process. Using model organisms such as *Caenorhabditis elegans*, *Drosophila melanogaster*, zebrafish, and mouse permits addressing fundamental questions in developmental cell clearance by a comprehensive approach including powerful genetics and cell biological tools enriched by live imaging. Recent studies in model organisms have enhanced significantly our understanding of the molecular and cellular basis of apoptotic cell clearance during development. Here, we review the current knowledge and illuminate the great potential of the research performed in genetic models, which opens new directions in developmental biology.

1. INTRODUCTION

Apoptosis during development, or developmental apoptosis, has been studied extensively and much progress has been made in understanding the molecular and cellular basis of the process (for review, see Adachi-Yamada & O'Connor, 2004; Arya & White, 2015; Conradt, 2009; Domingos & Steller, 2007; Meier, Finch, & Evan, 2000; Ryoo & Baehrecke, 2010; Steller, 2008; Suzanne & Steller, 2013; Tittel & Steller, 2000). However, the final step of apoptosis, clearance of apoptotic cells by phagocytes during development, has obtained much less attention. Apoptotic cells are barely seen unengulfed by phagocytes in normal tissues, indicating that their removal occurs concurrently with progression of apoptosis (Franc, 2002; Gregory & Devitt, 2004; Henson & Hume, 2006; Lauber, Blumenthal, Waibel, & Wesselborg, 2004). Defects in the clearance of apoptotic cells may lead to inflammation, autoimmune responses, and developmental abnormalities (Elliott & Ravichandran, 2010; Hanayama & Nagata, 2005; Hanayama et al., 2004; Juncadella et al., 2013; Lu et al., 2011; Mahoney & Rosen, 2005; Mevorach, 2010; Munoz, Peter, Herrmann, Wesselborg, & Lauber, 2010; Nagata, Hanayama, & Kawane, 2010; Nathan & Ding, 2010; Poon, Lucas, Rossi, & Ravichandran, 2014). Clearance of apoptotic cells is performed by two types of phagocytes, "professional phagocytes," such as macrophages and immature dendritic cells, whose main function is phagocytosis, and "nonprofessional" neighboring cells that in addition to their defined functions in tissues are also able to perform phagocytosis (for review, see Fullard, Kale, & Baker, 2009; Henson & Hume, 2006; Kinchen, 2010; Kurant, 2011; Mangahas & Zhou, 2005; Nakanishi & Shiratsuchi, 2004).

Removal of apoptotic cells is a highly dynamic process, which proceeds in a few distinct steps (Fig. 1), starting with recruitment of phagocytes to apoptotic cells through secreted "find-me" signals (Lauber et al., 2003; Munoz et al., 2010; Peter et al., 2008; Peter, Wesselborg, Herrmann, & Lauber, 2010; Ravichandran, 2011; Weigert et al., 2010), followed by recognition and binding of "eat-me" signals exposed on apoptotic cell surface by transmembrane phagocytic receptors on the phagocytic site or secreted bridging molecules. This binding leads to receptor clustering and cytoskeleton rearrangement in the phagocyte, which results in engulfment of apoptotic cells by phagocytes, phagosome formation, maturation, and finally degradation of apoptotic cells inside the phagosome (for review, see Elliott & Ravichandran, 2010; Erwig & Henson, 2007, 2008; Henson & Hume, 2006; Hochreiter-Hufford & Ravichandran, 2013; Kinchen et al., 2008; Kinchen & Ravichandran, 2007, 2008b; Krysko, D'Herde, & Vandenabeele, 2006; Lauber et al., 2004; Ravichandran, 2011; Zhou & Yu, 2008).

Initial insights into the molecular mechanisms of apoptotic cell clearance came from genetic screens in *Caenorhabditis elegans*, where two partially

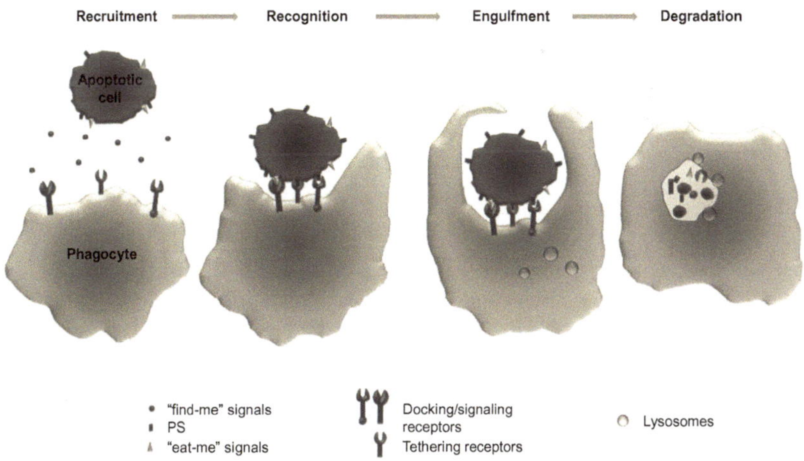

Figure 1 Schematic representation of different stages of apoptotic cell clearance. "Find-me" signals are secreted from apoptotic cells. "Eat-me" signals are exposed on the surface of apoptotic cells. Interactions between phagocytic receptors on phagocytic membrane and "eat-me" ligands on apoptotic cells lead to clustering of receptors, engulfment of apoptotic cell, phagosome formation, maturation, and finally degradation of the apoptotic cell. (See the color plate.)

redundant and highly evolutionarily conserved genetic pathways were identified. A phagocytic receptor CED-1 (Zhou, Hartwieg, & Horvitz, 2001; Draper in flies, LRP/MEGF10 in mammals), its adapter CED-6 (Liu & Hengartner, 1998; dCed-6 in flies, hCED6/GULP in mammals), dynamin (Yu, Odera, Chuang, Lu, & Zhou, 2006; *shibire* in flies, Dyn-1 in mammals), and an ABC transporter CED-7 (Wu & Horvitz, 1998a; ABCA1 or ABCA7 in mammals) were initially identified to comprise the first pathway. Recently, Clathrin and the AP2 complex were shown to play a critical role in cytoskeleton rearrangement and initiation of phagosome maturation downstream of CED-1 and CED-6 (Chen, Jian, et al., 2013; Pinto & Hengartner, 2012; Shen et al., 2013). The second pathway involves the actin-regulating protein complex CED-2 (Reddien & Horvitz, 2000) (DCrk in flies, CrkII in mammals)/CED-5 (Wu & Horvitz, 1998b) (Myoblast City in flies, Dock180 in mammals)/CED-12 (Gumienny et al., 2001) (dCed-12 in flies, ELMO in mammals), which can be activated by the phosphatidylserine (PS) receptor (PSR-1) (Wang et al., 2003) or two signaling modules, INA-1/integrin-SRC-1/Src (Hsu & Wu, 2010) or UNC-73/TRIO-MIG-2/RhoG (deBakker et al., 2004), to promote CED-10-mediated engulfment. Both pathways converge in activation of CED-10 (Kinchen et al., 2005; Reddien & Horvitz, 2000; dRac in flies, Rac1 in mammals; for review, see Kinchen; Kinchen & Ravichandran, 2007; Mangahas & Zhou, 2005; Reddien & Horvitz, 2004; Stuart & Ezekowitz, 2005; Yu et al., 2006; Zhou, Mangahas, & Yu, 2004). The recently discovered single worm homolog of the mammalian nm23 family, NDK-1, functions at the meeting point of the CED-10 and CED-1 pathways in cytoskeleton rearrangement (Fancsalszky et al., 2014). There is evidence for synergy between the two pathways during early embryogenesis where a noncanonical Wnt pathway acts through CED-2 as well as through an engulfment coreceptor of CED-1 (Cabello et al., 2010).

Apoptotic cell clearance is a tightly regulated process. Several negative regulators of apoptotic cell clearance have been recently exposed in *C. elegans* genetic studies, which again have led the way in addressing central questions about developmental cell clearance. Some of these, such as the orthologue of mammalian tyrosine kinase Abl (ABL-1), the *C. elegans* E3 ubiquitin ligase, and adaptor protein SLI-1 (Cbl in mammals), act in parallel to the classical phagocytic pathways, inhibiting engulfment of apoptotic cells through previously unidentified mechanisms (Anderson, Zhou, Sawin, Horvitz, & Hurwitz, 2012; Hurwitz et al., 2009). Negative regulators acting within the known pathways include PDR-1 (Parkin in mammals), which has been reported to ubiquitinate CED-10 for degradation in the CED-2/CED-5/CED-10/CED-12 classical pathway (Cabello et al., 2014), and a member of the Myotubularin family (MTM), which negatively regulates engulfment through CED-5/CED-12/CED-10 pathway as well by dephosphorylation of PtdIns(3)P on the plasma membrane (Zou et al., 2009). However, the exact interactions between distinct inhibitory mechanisms are not yet clear.

In order to study the role of apoptotic cell clearance during development, *in vivo* studies, which enable following this dynamic process at different time points in the context of the whole organism, are highly desirable. In addition, in contrast to apoptosis itself, where a cell undergoes autonomous morphological and molecular changes, apoptotic cell clearance involves at least two types of cells, apoptotic cells and phagocytes, and can be best studied only when both cell types are monitored concomitantly and their relationship is taken into account. Therefore, genetic model organisms, such as *C. elegans*, *Drosophila*, zebrafish, and mouse, provide a great advantage in research aiming to uncover the molecular basis of developmental apoptotic cell clearance.

Many fundamental questions regarding apoptotic cell clearance during development still remain unclear. How do phagocytes know which cells to engulf? How do they distinguish between living and dying cells? How do specific cell populations acquire their ability to phagocytose apoptotic cells? Is apoptosis itself involved in establishment of phagocytes as potent removers of apoptotic cells? Do professional and nonprofessional phagocytes develop similarly or do they employ different developmental programs? Do the same pathways act in different phagocytic cell populations during development or do different programs work in different types of phagocytes? How do phagocytes behave when stress-induced excess apoptosis occurs during development? Are they competent to remove the extra apoptotic cells? Can phagocytosis influence cell death during development? Is it required for execution of apoptosis? In this review, we bring together the latest data to try to shed light on these important questions.

2. PROFESSIONAL AND NONPROFESSIONAL PHAGOCYTES DURING DEVELOPMENT

When cells undergo developmental apoptosis, they are properly and efficiently cleared by professional or nonprofessional phagocytes. From an evolutionary point of view, the demand for apoptotic cell clearance increases from worms to vertebrates both in quantitative and qualitative terms. During *C. elegans* development, only about 10% of cells die through apoptosis and they are efficiently removed by neighboring cells. In worms, there are only nonprofessional phagocytes, which remove apoptotic cells (Lettre & Hengartner, 2006; Mangahas & Zhou, 2005; Zhou et al., 2004). During *Drosophila* development, a much higher percent of cells (more than 30%) undergo apoptosis at different stages of the life cycle: embryogenesis, metamorphosis, and the emerging adult (Abrams, White, Fessler, & Steller, 1993; Bangs & White, 2000; Peterson, Carney, Taylor, & White, 2002; Rogulja-Ortmann, Luer, Seibert, Rickert, & Technau, 2007; Steller, 2008; Steller & Grether, 1994; Togane et al., 2012; Truman, Thorn, & Robinow, 1992). This increase in apoptosis goes along with evolving of professional phagocytes, macrophages, and nonprofessional glia in the central nervous system (CNS) and ectodermal cells, which are engaged in apoptotic cell clearance during distinct stages of development. In vertebrates, even more developmental apoptosis takes place, which is accompanied by the increased complexity of the innate immune system responsible for removal of apoptotic cells (Arya & White, 2015; Henson & Hume, 2006; Stuart & Ezekowitz, 2005). This complexity results in a substantial redundancy of factors involved in phagocytosis and as a consequence, difficulties in studying its molecular basis by addressing the roles of specific players. To overcome these difficulties, genetic model organisms are of great interest and value.

Professional phagocytes are not always established and abundant at the time and location of developmental apoptosis. Therefore, nonprofessional neighbors often clear apoptotic cells during development. For example, in human developing retina, it was found that most apoptotic cells were not taken up by macrophages but by neighboring retinal cells (Francisco-Morcillo, Bejarano-Escobar, Rodriguez-Leon, Navascues, & Martin-Partido, 2014; Penfold & Provis, 1986), and in the *Drosophila* embryonic CNS glia are the main phagocytes of apoptotic neurons (Kurant, Axelrod, Leaman, & Gaul, 2008). An interesting example of nonprofessional phagocytosis during development is removal of apoptotic neurons by phagocytic satellite glial precursors in developing mouse dorsal root ganglia. In this process, the Jedi phagocytic receptor, a mammalian homolog of Draper, has been shown to promote phagocytosis via a clathrin-dependent mechanism (Sullivan et al., 2014; Wu et al., 2009). A striking example of nonprofessional phagocytosis of apoptotic neurons has been reported by Lu and colleagues during adult mouse neurogenesis where a specific type of neural progenitor cell performs phagocytosis in neurogenic zones. Moreover, abnormal phagocytosis resulted in accumulation of apoptotic cells in the neurogenic zones and inhibited adult neurogenesis (Lu et al., 2011).

An important question of what are the differences in phagocytic performance between professional macrophages and nonprofessional phagocytic neighbors has been addressed in a few studies. For example, *in vitro* studies testing rat microglia and astrocytes have shown that astrocytes are much less efficient in phagocytosis of apoptotic lymphocytes compared to microglia (Magnus, Chan, Linker, Toyka, & Gold, 2002). Another example was demonstrated during clearance of apoptotic cells in interdigital spaces, where mesenchymal cells can replace professional macrophages when the latter are not properly developed (Wood et al., 2000). However, the neighboring mesenchymal cells were much less efficient in removal of apoptotic cells during digit sculpting (Wood et al., 2000). Moreover, Parnaik and colleagues demonstrated *in vitro* and *in vivo* that nonprofessional lens epithelial cells, which clear apoptotic cells during normal lens development, as well as additional types of nonprofessional phagocytes such as baby hamster kidney cells, rat brain astrocytes, and COS cells perform much more slowly in clearance of apoptotic cells (Parnaik, Raff, & Scholes, 2000). The authors suggested that apoptotic cells change with time, which is reflected by different signals for professional and nonprofessional phagocytes resulting in delayed phagocytosis by nonprofessional phagocytes (Parnaik et al., 2000).

Professional phagocytes are usually highly motile and can quickly approach a specific site. However, developmental tissues and organs go through tissue rearrangements and other changes, which may affect motility and accessibility of phagocytes to specific sites of apoptosis. Often apoptotic cells are removed by phagocytic neighbors due to inaccessibility of the tissue or organ to macrophages. For example, the *Drosophila* embryonic CNS goes through massive neuronal apoptosis starting at mid-embryogenesis (Abrams et al., 1993; Rogulja-Ortmann et al., 2007). At this stage, professional macrophages play an important role in apoptotic cell clearance required for CNS morphogenesis (Sears, Kennedy, & Garrity, 2003). However, at late embryogenesis, specific glial cells (surface glia) cover the developing nerve cord and prevent entrance of macrophages into the CNS. Starting this time point, glia become the main phagocytes in the embryonic CNS. It has been demonstrated that differences in phagocytic ability between professional circulating macrophages and CNS resident glia are largely a reflection of their differential motility (Kurant et al., 2008).

3. HOW DO PHAGOCYTES "SENSE" APOPTOTIC CELLS?

During development apoptotic cells can be removed by neighboring tissue-resident phagocytes, which are situated in their close vicinity, as in *C. elegans* embryo or *Drosophila* embryonic CNS or by professional motile phagocytes such as dendritic cells or macrophages, which have to be attracted to apoptotic cells. This attraction occurs through secreted "find-me" signals, which have been discovered in mammals, mostly by *in vitro* analysis of supernatants of apoptotic cell cultures. Specifically, several soluble molecules which are secreted by apoptotic cells have been implicated in attraction of professional phagocytes to sites of apoptosis: the lipid lysophosphatidylcholine (LPC; Lauber et al., 2003), sphingosine 1-phosphate (S1P; Gude et al., 2008; Weigert et al., 2010), the fractalkine CX3CL1 (Truman et al., 2008), and the nucleotides ATP and UTP (Elliott et al., 2009). Most of them are secreted in a caspase-dependent manner at very early stages of apoptosis. There are very detailed reviews describing these different signals and their mode of action (Munoz et al., 2010; Peter et al., 2010; Ravichandran, 2011). Additional signals, which attract professional phagocytes to the specific sites of wounding, damage, or inflammation, may interfere with signals from apoptotic cells. This very important topic is not the focus of this review.

4. HOW DO PHAGOCYTES RECOGNIZE APOPTOTIC CELLS?

Phagocytes must recognize their targets with superb level of specificity in order to remove only apoptotic cells and not living cells. How is this specificity achieved?

4.1 Ligands on Apoptotic Cells

Apoptotic cells expose specific ligands on their membrane, called "eat-me" signals, which distinguish them from live cells. The most well-known and evolutionary conserved "eat-me" signal is PS (Fadok, Bratton, Frasch, Warner, & Henson, 1998; Fadok, Bratton, & Henson, 2001; Wu, Tibrewal, & Birge, 2006), which is exposed on the outer leaflet of the plasma membrane in apoptotic cells, while in viable cells it is normally maintained in the inner leaflet of the membrane bilayer (Balasubramanian & Schroit,

1998; Gardai, Bratton, Ogden, & Henson, 2006; Hochreiter-Hufford & Ravichandran, 2013; Ravichandran, 2010, 2011). The PS exposure on the outer leaflet of the apoptotic cell membrane is due to increased flip-flop of PS in and out plasma membrane and inactivation of the aminophospholipid translocase responsible for flipping PS back to the inner leaflet (Bratton et al., 1999, 1997; Segawa et al., 2014). PS-mediated phagocytosis has been described for mammals (Fadok et al., 1992), worms (Venegas & Zhou, 2007; Wang et al., 2007), and recently, for *Drosophila* (Shklyar, Levy-Adam, Mishnaevski, & Kurant, 2013; Tung et al., 2013), demonstrating its universality in different organisms. Other "eat-me" signals identified on apoptotic cell surface include changes in charge, in glycosylation patterns of cell surface proteins (Kinchen & Ravichandran, 2007), binding of serum proteins such as thrombospondin and complement C1q to the apoptotic cell (Mevorach, Mascarenhas, Gershov, & Elkon, 1998; Ravichandran & Lorenz, 2007), alterations in intercellular proteins such as ICAM3 (Devitt et al., 1998; Gregory, Devitt, & Moffatt, 1998; Moffatt, Devitt, Bell, Simmons, & Gregory, 1999) and oxidized LDL-like moieties (Chang, Binder, Torzewski, & Witztum, 2002; Gardai et al., 2006; Hochreiter-Hufford & Ravichandran, 2013). In addition, intracellular proteins such as Calreticulin (Gardai et al., 2005; Martins et al., 2010; Obeid et al., 2007), Annexin I (Arur et al., 2003), and Pretaporter in *Drosophila* (Kuraishi et al., 2009) have been shown to serve as "eat-me" signals on apoptotic cells when translocated from the ER to the plasma membrane.

4.2 Is PS Exposure Sufficient for Engulfment?

PS is exposed at very early stages of apoptosis and is required for engulfment (Hochreiter-Hufford & Ravichandran, 2013; Kay & Grinstein, 2013; Ravichandran, 2011). However, living cells may transiently expose PS on their surfaces, though, in much lower concentration compared to apoptotic cells (Borisenko et al., 2003). A very important question, which has attracted much attention in the field, is whether PS alone is sufficient for engulfment. Previous experiments with PS-coated liposomes argued for sufficiency of PS for engulfment (Ishimoto, Ohashi, Mizuno, & Nakano, 2000). Moreover, recent work in human cell culture reported that abnormal flippase activity may lead to PS exposure on living cells, which was shown to be sufficient for engulfment (Segawa et al., 2014). On the other hand, studies where PS was artificially incorporated into the outer leaflet of live cell plasma membranes

(Borisenko et al., 2003) or was exposed on live BaF3 cells by gain of function of a calcium-dependent phospholipid scramblase (TMEM16F; Segawa, 2011; Suzuki, Umeda, Sims, & Nagata, 2010) demonstrated that PS exposure alone is not sufficient for engulfment of these cells. Moreover, PS can be detected on living cell surfaces, such as on macrophages and activated lymphocytes as part of other receptor-mediated signaling events, without induction of phagocytosis (Dillon, Mancini, Rosen, & Schlissel, 2000; Hamon et al., 2000). In addition, recently, it has been demonstrated that live neurons in the *Drosophila* embryonic CNS expose PS on their membranes but they are never engulfed in the absence of caspase activation in these cells (Shklyar et al., 2013). It seems that PS exposure on the plasma membrane is not sufficient for engulfment of living cells during normal development. However, given the discrepancy between the data obtained from different experimental systems, more studies are required to fully address this question.

4.3 Caspase Activation and PS Exposure

An intriguing question in the field regards the requirement of caspase activity for PS exposure on apoptotic cells. PS exposure on apoptotic cells is considered caspase dependent in worms and mammals (Chen, Mapes, Lee, Skeen-Gaar, & Xue, 2013; Segawa et al., 2014; Suzuki, Denning, Imanishi, Horvitz, & Nagata, 2013). However, current data suggest that this requirement is context dependent. An elegant live imaging study in zebrafish, which used different markers for apoptotic neurons during their microglial phagocytosis *in vivo*, demonstrated that AnnexinV, which specifically binds PS, appeared on apoptotic cells only following activation of effector caspases labeled with an anti-activated caspase 3 antibody (Mazaheri et al., 2014). However, in *Drosophila* embryos, lacking caspase activity, AnnexinV binds PS on some live neurons, suggesting that caspase activation is not required for PS exposure on these cells (Shklyar et al., 2013). These differences may result from differential affinity of the AnnexinV used in different species; in *Drosophila*, it was injected into live embryos whereas in zebrafish it was genetically expressed and secreted to the brain extracellular space (Mazaheri et al., 2014). Moreover, the study of van Delft and colleagues in *Casp9−/−* mice demonstrated that intact slowly dying *Casp9−/−* mutant thymocytes exposed PS, which led to their noninflammatory phagocytosis (van Delft, Smith, Lahoud, Huang, & Adams, 2010). It seems that PS can be exposed on the cell membrane through caspase-independent mechanisms;

however, the much higher amounts of PS on apoptotic cell surfaces require caspase activation, which in addition, leads to appearance of other caspase-dependent ligands (Shklyar et al., 2013), such as Calreticulin (Gardai et al., 2005) and Pretaporter (Kuraishi et al., 2009) recognized by specific phagocytic receptors.

4.4 Phagocytic Receptors and Bridging Molecules

Specific recognition of apoptotic cells is achieved through transmembrane phagocytic receptors or secreted bridging molecules, which recognize "eat-me" signals on apoptotic surfaces (Kinchen & Ravichandran, 2007, 2008a, 2008b; Lauber et al., 2004; Ravichandran, 2011; Ravichandran & Lorenz, 2007; Stuart & Ezekowitz, 2005). In mammals, a large number of transmembrane receptors and soluble bridging molecules have been shown to play a role in recognition and engulfment of apoptotic particles (Hanayama et al., 2002; Miyanishi et al., 2007; Park et al., 2008, 2009, 2007). Different types of the transmembrane receptors recognize PS, such as a family of T cell immunoglobulin and mucin receptors TIM-4, TIM-1, and TIM-3 (DeKruyff et al., 2010; Freeman, Casasnovas, Umetsu, & DeKruyff, 2010; Ichimura et al., 2008; Kobayashi et al., 2007; Miyanishi et al., 2007; Rodriguez-Manzanet et al., 2010; Wong et al., 2010); a member of the adhesion type II GPCR family brain angiogenesis inhibitor 1 (BAI1; Park et al., 2007); and proteins containing atypical epidermal growth factor (EGF) motifs called Nimrod (NIM) repeats, including Stabilin-2, CED-1 in *C. elegans* and Draper and Six-Microns-Under (SIMU) in *Drosophila* (Kurant et al., 2008; Park et al., 2008; Shklyar et al., 2013; Tung et al., 2013; Zhou et al., 2001). Scavenger receptors such as CD36, CD68, and (hSR-B1)/CLA-1 have been shown to recognize PS as well (Imachi et al., 2000; Savill, Hogg, Ren, & Haslett, 1992).

Secreted bridging molecules recognize PS on apoptotic cells and interact with transmembrane receptors on phagocytes, thereby tethering apoptotic cells to phagocytes and participating in engulfment. The most prominent PS-recognizing bridging molecule is human milk fat globule factor-E8 (MFG-E8), which is secreted by activated macrophages and binds to integrin $\alpha_v\beta_3$ on phagocytes (Akakura et al., 2004; Hanayama & Nagata, 2005; Hanayama et al., 2002). Interestingly, in *C. elegans* integrin receptor likely recognizes PS on apoptotic cells with no bridging molecules (Hsu & Wu, 2010). Additional examples of PS-binding bridging molecules are Gas6 and Protein S, which are recognized by the Tyro-3-Axl-Mer family of receptors (denoted as TAM receptors) on phagocytes (Lemke & Rothlin, 2008; Scott et al., 2001) and β2-glycoprotein I (β2GPI) and thrombospondins (Balasubramanian & Schroit, 1998; Mevorach, 2000; Savill et al., 1992), which interact with scavenger transmembrane receptors on phagocytes. To date, no bridging molecules have been identified in worms and flies.

In vertebrates, the redundancy of the recognizing phagocytic receptors is remarkably high and in many cases lack of a single receptor function does not result in abnormal apoptotic cell clearance (Grimsley & Ravichandran, 2003; Platt & Gordon, 2001; Scott et al., 2001; Stuart & Ezekowitz, 2005). In *Drosophila* and worms, the redundancy of receptors is much lower, which enables addressing the functions of specific receptors in phagocytosis.

4.5 Tethering and Signaling/Docking Receptors

Phagocytic receptors may be classified as tethering receptors, which are transmembrane proteins mostly without any signaling ability, and signaling/docking receptors. As such, phagocytic receptors may function at different steps of phagocytosis. By using loss-of-function mutants of specific receptors, it is possible to understand at which step of the process they are participating. Tethering receptors are mostly involved in recognition and engulfment steps since their intracellular portions are dispensable for their function in phagocytosis that is mainly to tie apoptotic cells to phagocytes. In *Drosophila*, a good example of such receptor is SIMU, which is a transmembrane tethering receptor that is also able to act as a secreted bridging molecule (Kurant et al., 2008). In *simu* mutants, an increased amount of apoptotic particles appears unengulfed by macrophages or by phagocytic glia in the CNS, demonstrating that SIMU functions at the recognition and engulfment steps of phagocytosis (Kurant et al., 2008). Examples of tethering receptors in vertebrates are Stabilin-2 and TIM-4 (Kobayashi et al., 2007; Park et al., 2008).

Docking/signaling receptors such as Draper in *Drosophila* and BAI1 and integrins $\alpha_V\beta_3$ and $\alpha_V\beta_5$ in vertebrates are able to signal and recruit the intracellular machinery involved in engulfment as well as in degradation steps of phagocytosis. Live imaging studies in zebrafish revealed that BAI1 and TIM-4 phagocytic receptors act in different modes during engulfment; BAI1 is required for phagosome formation whereas TIM-4 stabilizes the phagosome (Mazaheri et al., 2014). These distinct functions of tethering and docking receptors during apoptotic cell clearance reveal an evolutionary expansion of this process at the molecular level in *Drosophila* and vertebrates compared to *C. elegans* where the engulfment receptor CED-1 is required for both phagosome formation and maturation (Yu, Lu, & Zhou, 2008).

5. INTRACELLULAR PHAGOCYTIC MACHINERY

5.1 Cytoskeleton Rearrangement During Engulfment

The initial recognition of dying cells by phagocytes is followed by intracellular signaling controlling cytoskeleton rearrangement. The signaling pathways described above for *C. elegans* are highly conserved in other species (Kinchen, 2010; Kinchen & Ravichandran, 2010; Mangahas & Zhou, 2005). However, additional molecular mechanisms involved in cytoskeleton rearrangement have been elucidated in *Drosophila* embryonic macrophages, where two parallel signaling cascades were described. One cascade involves Pallbearer (Pall), which is an F-box protein that acts as an E3 ubiquitin ligase in complex with SkpA, dCul1 (Lin19), and UbcD1 (Effete). Mutation in each of the components of this Skp Cullin F-box complex results in defective phagocytosis of apoptotic corpses by embryonic macrophages. Moreover, phagocytic defects in *pall* mutant were rescued by mutation in phosphorylated ribosomal protein S6 (RpS6), which negatively regulates phagocytosis by ubiquitylation and proteosomal degradation of Pall independently of RpS6 ribosomal function (Silva, Au-Yeung, Van Goethem, Burden, & Franc, 2007). Experiments with S2 cells and embryonic macrophages *in vivo* demonstrated that Pall and RpS6 regulate phagocytosis through F-actin cytoskeleton rearrangement and RAC2 GTPase upregulation (Xiao et al., 2015).

A second event of internalization is a Ca^{2+} signaling pathway, which is initiated by direct binding of dCed-6 and Shark to a cytoplasmic tail of Draper (Van Goethem, Silva, Xiao, & Franc, 2012). This leads to coupling of ER and plasma membranes, facilitated by Undertaker (Retinophilin), and followed by Ca^{2+} release from ER through the Ryanodine receptor-encoding gene Rya-r44F, Ca^{2+} sensor dSTIM, and extracellular Ca^{2+} influx via dOrai channels (Cuttell et al., 2008; Fullard et al., 2009). Interestingly, genetic interactions between *rya-r44F*, *undertaker*, *drpr*, and the *Drosophila* TRPP cation channel *almost there* (*amo/pkd2*) have been reported during removal of apoptotic particles by macrophages. However, dCed-12, which is also involved in phagocytosis by macrophages likely functions in a parallel genetic pathway similarly to its *C. elegans* homolog (Van Goethem et al., 2012).

5.2 Phagosome Formation and Maturation

Internalization of apoptotic particles leads to formation of a membrane organelle phagosome, which undergoes multiple changes termed phagosome maturation resulting in degradation of apoptotic particles inside the phagosome. These changes mostly involve acidification of phagosomes that requires the function of Rab-5, Rab-7, and Arf6 GTPases as well as LAMP-1/2 lysosomal markers, which are heavily regulated by guanine nucleotide exchange proteins, GTPase-activating proteins, and multiple effectors (for detailed review, see Kinchen & Ravichandran, 2008b; Zhou & Yu, 2008). In addition, many factors are brought to the phagosome, including V-ATPases and acidic proteases such as cathepsins. Studies in *C. elegans* and zebrafish revealed that V-ATPases may act at different stages of phagosome maturation. In worms, vha-16 (the D subunit of the V0ATPase) is required for early stages of maturation (Kinchen & Ravichandran, 2008b), whereas in zebrafish, the A1 subunit of the V-ATPase is required for fusion of phagosomes with lysosomes (Peri & Nusslein-Volhard, 2008). However, the exact mechanism by which V-ATPases regulate phagosome maturation remains unclear (Kinchen & Ravichandran, 2008b).

The role of Rab-5 and Rab-7 in phagosome maturation has been elucidated in *C. elegans* and mammalian cells, where they act at distinct partially overlapping stages of maturation. Recruitment of Rab-5 to the phagosome is considered an earliest event of maturation, which is regulated by Dyn-1 and the PtdIns(3)-kinase VPS-34 (human PI3KC3/hVPS34; Kinchen et al., 2008). Interestingly, recent studies in *C. elegans*, mammalian cells, and *Drosophila* demonstrated that Dyn-1, which was previously reported as a factor involved in internalization of apoptotic particles (Yu et al., 2006), plays an important role in phagosome maturation. Abnormal function of Dyn-1 leads to arrest of phagosome maturation and accumulation of apoptotic particles inside phagosomes, which can be monitored *in vivo* in worms and *Drosophila* CNS (Kinchen et al., 2008; Shklover, Mishnaevski, Levy-Adam, & Kurant, 2015; Yu et al., 2008). Recently, studies in *C. elegans* highlighted the role of Arf-like small GTPase Arl8 (ARL-8) during fusion of phagosomes with lysosomes (Sasaki et al., 2013) and the role of lysosomal cathepsin CPL-1 in apoptotic cells removal (Yu & Schuldiner, 2014). Additional multiple regulators of Rab-5 and Rab-7 function and their recruitment to phagosomes are reviewed in detail in Kinchen and Ravichandran (2008b) and Zhou and Yu (2008).

The important question of whether phagosome maturation during developmental apoptotic cell clearance is regulated by phagocytic receptors has been addressed in *C. elegans* and *Drosophila*. Using live imaging of worm embryos, it has been demonstrated that the CED-1 receptor and CED-5/Dock180, which functions in a parallel engulfment pathway downstream of PSR-1 and integrin receptor, are required for Rab-7 recruitment to the phagosomes, indicating their role in phagosome maturation (Yu et al., 2008, 2006). During *Drosophila* embryogenesis, *draper* (CED-1 homolog) mutants exhibit large numbers of apoptotic particles accumulating inside phagosomes, suggesting abnormal phagosome maturation (Kurant et al., 2008). Moreover, quantification of phagosome acidification in *draper* embryonic CNS by LysoTracker labeling showed much lower staining as compared to wild-type embryos, revealing *draper*'s role in acidification of phagosomes (Shklyar B. & Kurant E. unpublished data). It seems that docking/signaling receptors, which interact with intracellular proteins, may play a role not only in internalization of apoptotic particles but in their degradation as well. To gain deeper insight into the effect of phagocytic receptors on phagosome maturation, further *in vivo* studies are desirable, which are able to monitor localization of apoptotic particles inside or outside the phagosomes and measure different markers of the multistep process of phagosome acidification.

6. HOW DO PHAGOCYTES BECOME PHAGOCYTIC?

What defines phagocytic cells as phagocytes? The phagocytic ability depends on expression of specific receptors and factors of the intracellular phagocytic machinery, which have been discussed above. Most of the phagocytic receptors are exclusively expressed in phagocytic cell populations, whereas intracellular molecules are much more abundant. How this specific expression of receptors is controlled during development remains elusive. Is it triggered by appearance of apoptotic cells or signals coming from them? Or is it part of the developmental programs underlying establishment of phagocytic cell populations?

C. elegans transmembrane phagocytic receptor CED-1, which is required for recognition, engulfment, and degradation of apoptotic cells, is broadly expressed in cell types that can function as engulfing cells (Zhou et al., 2001). This is not surprising since almost every cell in the worm is capable of phagocytosing. The situation is different in flies and vertebrates where only specific cell populations, professional, or nonprofessional phagocytes are dedicated to engulfing and degrading apoptotic cells.

A recently published study in *Drosophila* focused on how embryonic glia and macrophages gain their ability to phagocytose during embryogenesis (Shklyar et al., 2014). Several phagocytic receptors for apoptotic cells are known in *Drosophila*, which exhibit highly specific expression in phagocytic cell populations during embryogenesis. The CD36 homolog Croquemort (Crq) is expressed mostly in professional phagocytes, the macrophages (Franc, Heitzler, Ezekowitz, & White, 1999). Two other receptors, Draper and SIMU, are expressed in professional macrophages and nonprofessional ectoderm and glia (Freeman, Delrow, Kim, Johnson, & Doe, 2003; Kurant et al., 2008). Shklyar and colleagues reported that the phagocytic ability of embryonic glia is highly dependent on specific expression of the phagocytic receptors SIMU and Draper (Shklyar et al., 2014). SIMU has only one isoform, whose expression is tightly regulated, with high levels exclusively during stages of developmental apoptosis, suggesting that it functions solely in apoptotic cell clearance (Kurant et al., 2008). Draper has three isoforms, differentially expressed during all stages of *Drosophila* development, which participate in different clearance mechanisms including neuronal pruning, synapse removal, and axonal degeneration (Awasaki et al., 2006; Logan et al., 2012; Ziegenfuss et al., 2008). The specific expression of SIMU during particular stages of developmental apoptosis raises the question of the role of apoptosis in the regulation of phagocytic receptor expression. Evaluation of transcript and protein levels of SIMU and Draper in mutants lacking caspase activation or containing elevated apoptosis demonstrated that apoptosis *per se* does not affect receptor expression (Shklyar et al., 2014). Furthermore, this study revealed that specific developmental programs that are responsible for differentiation of phagocytic cell populations, glia and macrophages, control their capacity to phagocytose (Shklyar et al., 2014). This suggests that the ability of phagocytes to remove apoptotic cells is established during development independently and ahead of apoptotic cell appearance, which might explain the rapid elimination of apoptotic cells and why they are barely detected in normal tissues.

7. WHY ARE LIVING CELLS NOT REMOVED BY PHAGOCYTES?

The importance of "eat-me" signals for recognition and removal of apoptotic cells is very high; however, some of these signals such as PS appear on living cells, which could potentially lead to elimination of needed live cells. One way to avoid elimination of live cells is "don't-eat-me" signals, exposed on their surfaces. *In vitro* studies in vertebrate cells reported that exposure of transmembrane molecules CD47 or CD31 on the cell surface prevents engulfment of these cells by phagocytes (Brown et al., 2002; Gardai et al., 2005). Moreover, loss or inactivation of CD47 resulted in phagocytosis of nonapoptotic cells (Gardai et al., 2005), suggesting the intriguing possibility that cell removal is a default process, which is actively prevented by inhibitory signals on live cells. However, the role of "don't-eat-me" signals *in vivo* has not been addressed yet. Up to date, there is no evidence for "don't-eat-me" signals in worms and *Drosophila*. Yet, a different mechanism has been recently suggested in the fly embryo, which might prevent phagocytosis of living cells. This mechanism involves an initial binding with relatively low affinity of phagocytic receptors to their ligand unless clustering with additional receptors occurs (Fig. 2; Shklyar et al., 2013). In the *Drosophila* embryo lacking caspase activation, some living cells expose PS on their surface. Even though the tethering receptor SIMU is present on the membranes of phagocytic glia, its affinity to PS is low and engulfment does not take place in these embryos. When cells undergo apoptosis they expose

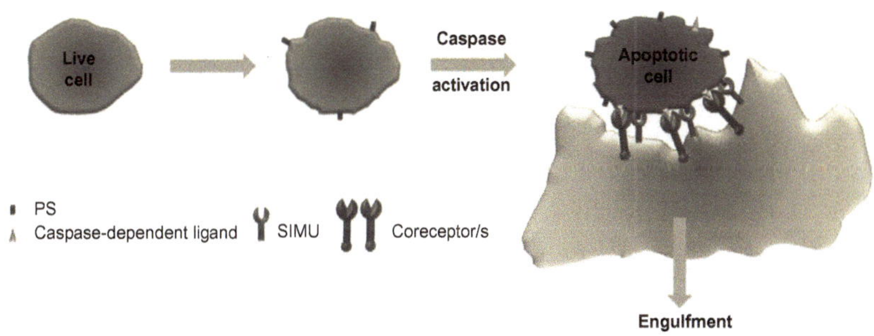

Figure 2 PS and caspase activation are required for engulfment. Schematic representation of PS and additional ligands on an apoptotic cell exposed following caspase activation that are recognized by different receptors on the phagocytic surface. Clustering of receptors leads to engulfment of the apoptotic cell by the phagocyte. (See the color plate.)

caspase-dependent signals, including higher amounts of PS, on their surfaces, which are recognized by signaling phagocytic coreceptor/s of SIMU. This leads to SIMU clustering and substantial increase of its affinity to PS resulting in engulfment of the apoptotic cell (Fig. 2; Shklyar et al., 2013). This mechanism might serve in *Drosophila* to prevent engulfment of living cells exposing PS where caspases are not active. Identification of the signaling coreceptor/s of SIMU will further elucidate the molecular basis of this proposed mechanism.

8. GLIAL PHAGOCYTOSIS OF APOPTOTIC NEURONS IN THE DEVELOPING CNS

Apoptotic cell clearance plays a critical role during CNS development. In vertebrates, more than 50% of neurons die through apoptosis (Boya & de la Rosa, 2005; Diaz, Serna, De Pablo, & de la Rosa, 2000; Oppenheim et al., 1999) and are engulfed and degraded mostly by professional phagocytes of the brain, microglia (Ashwell, 1990; Bilimoria & Stevens, 2015; Caldero, Brunet, Ciutat, Hereu, & Esquerda, 2009; Egensperger, Maslim, Bisti, Hollander, & Stone, 1996; Marin-Teva et al., 2004; Peri & Nusslein-Volhard, 2008; Sierra et al., 2010). The role of microglia in response to CNS damage, neuronal pruning, and different pathological situations has been thoroughly described in several recently published reviews (Aguzzi, Barres, & Bennett, 2013; Bilimoria & Stevens, 2015; Sierra, Abiega, Shahraz, & Neumann, 2013). However, removal of apoptotic cells by microglia during normal development has garnered much less attention.

Early studies have described the phagocytic role of microglia during development in different species, including mouse retina (Hume, Perry, & Gordon, 1983), mouse cerebellum (Ashwell, 1990), quail retina (Marin-Teva, Cuadros, Calvente, Almendros, & Navascues, 1999), human telencephalon (Rakic & Zecevic, 2000), and the cerebral cortex of neonatal rat pups (Upender & Naegele, 1999). In these studies, dying cells appeared in correlation with microglial markers, suggesting their engulfment and degradation by microglial cells. A study in the chick embryo revealed that apoptosis of developing motoneurons stimulates a microglial response resulting in engulfment of about 40% of apoptotic neurons by microglia (Caldero et al., 2009). Importantly, the phagocytic ability of microglia is not constant in different areas of the developing brain and may vary from 100% in the hippocampus at postnatal day 0 (Dalmau, Vela, Gonzalez, Finsen, & Castellano, 2003; Sierra et al., 2010) to 50% in the developing cerebellum (Marin-Teva et al., 2004; Wakselman et al., 2008). A still open question is whether the remaining apoptotic cells are cleared by microglia at a later stage, or whether they are removed by nonprofessional phagocytes (Sierra et al., 2013).

8.1 Glial Phagocytosis of Apoptotic Neurons During *Drosophila* Development

Similarly to vertebrates, *Drosophila* CNS goes through massive developmental apoptosis, which adjusts the cell number during three main stages of development (embryogenesis, metamorphosis, and emerging adult; Abrams et al., 1993; Rogulja-Ortmann et al., 2007; Tissot & Stocker, 2000; Togane et al., 2012). There are no microglia in flies and most apoptotic neurons are cleared by glia of ectodermal origin (Kurant et al., 2008), which in terms of phagocytosis are highly functionally homologous to microglia (Kurant, 2011; Kurant et al., 2008). Embryonic glia are established as potent phagocytes by a developmental program involving specific expression of the SIMU and Draper phagocytic receptors (Shklyar et al., 2014). *simu* and *draper* mutant phenotypes during embryogenesis strongly suggest that at least at this developmental stage SIMU is required for engulfment of apoptotic cells and Draper is mostly involved in the degradation step of phagocytosis. This specific function of SIMU and Draper at distinct steps of phagocytosis correlates with *simu* and *draper* expression during different developmental stages, when SIMU is expressed exclusively at stages of developmental apoptosis and apoptotic cell clearance and Draper is expressed throughout the whole fly life cycle (Shklyar et al., 2014). It is possible that in processes where *simu* is not involved, for example, during larval synaptic remodeling or axonal degeneration in the adult brain, *draper* is required for engulfment as well. Close proximity of glia and the engulfed material in the CNS makes it difficult to determine whether *draper* is required for engulfment or degradation steps in the mentioned above processes. Unengulfed material and material accumulated inside phagocytic glia may appear similar in mutants, using specific phagosome markers may help to address this interesting point.

During metamorphosis, a large number of developing neurons die through apoptosis (Tissot & Stocker, 2000; Togane et al., 2012). In addition, neuronal pruning, which includes axonal and dendritic pruning, takes place at this stage of development (Fischbach & Technau, 1984; Kimura & Truman, 1990; Marin, Watts, Tanaka, Ito, & Luo, 2005; Truman, 1990; Truman et al., 1992; Watts, Hoopfer, & Luo, 2003; Williams & Truman,

2005a, 2005b). Glia play a prominent role in both these processes (Awasaki & Ito, 2004; Awasaki et al., 2006; Cantera & Technau, 1996; Freeman, 2006; Hoopfer et al., 2006; Watts, Schuldiner, Perrino, Larsen, & Luo, 2004). Since the topic of our review is apoptotic cell clearance, we will not discuss developmental pruning, which is explained in great detail elsewhere.

Neuronal cell death occurs in the first quarter of pupal development (Togane et al., 2012; Truman, 1990; Truman et al., 1992). Electron microscopy analysis of the *Drosophila* CNS during early pupal stages revealed that all glial subtypes appear to be engaged in engulfment and degradation of neuronal cell corpses, except for midline glial cells (Cantera & Technau, 1996). It is difficult to distinguish between glia enveloping healthy neurons and the same glia phagocytosing dying neurons. No professional macrophages were reported to function in pupal brain during early metamorphosis (Cantera & Technau, 1996). A recent study highlights the novel role of *Drosophila* astrocytes in clearance of apoptotic vCrz+ neurons during metamorphosis (Hakim, Yaniv, & Schuldiner, 2014; Tasdemir-Yilmaz & Freeman, 2014) where two phagocytic pathways, Draper signaling and the Crk/Mbc/dCed-12 complex, differentially mediate clearance of apoptotic cell bodies (mostly by the Draper pathway) or neurites (largely through the Crk/Mbc/dCed-12 complex). This study reveals that fly astrocytes engage different genetic programs for apoptotic cell clearance and neuronal pruning (Tasdemir-Yilmaz & Freeman, 2014).

In emerging adult flies, there are neurons that accomplish their developmental function immediately after eclosion and die through apoptosis (Peterson et al., 2002). How these apoptotic cells are removed remains mostly unknown.

9. ADDITIONAL EXAMPLES OF APOPTOTIC CELL CLEARANCE DURING DEVELOPMENT

9.1 Clearance by Macrophages

Despite the fact that the role of macrophages in tissue homeostasis and repair is widely appreciated and described (Epelman, Lavine, & Randolph, 2014; Lavin & Merad, 2013; Pollard, 2009; Wynn, Chawla, & Pollard, 2013), direct evidence regarding the role of apoptotic cell clearance by macrophages during development is limited. Early observations demonstrated that distribution of macrophages correlates precisely with the pattern of cell death during formation of a foot with separated digits from a hindlimb footplate in a mouse embryo between stages E12.5 and E14.5 (Hopkinson-Woolley, Hughes, Gordon, & Martin, 1994). Phagocytic function of these macrophages was demonstrated using specific markers (Wood et al., 2000). An additional role of macrophages in apoptotic cell clearance has been shown during early postnatal development of mammary glands in mice where an outgrowth of terminal end buds is accompanied by epithelial apoptosis and phagocytosis of apoptotic epithelial cells by macrophages. A null mutation in colony-stimulating factor 1 (CSF-1), which is the major growth factor of macrophages, caused defects in postnatal mammary glands development (Gouon-Evans, Rothenberg, & Pollard, 2000; Jones et al., 2013; Pollard, 2009).

9.2 Cell Competition

The role of apoptotic cell clearance during the process of cell competition in *Drosophila*, which controls organ sizes (Diaz & Moreno, 2005), remains controversial. It has been shown previously that expression of the phagocytosis genes *draper*, *wasp*, *psr*, *mbc/dock180*, and *rac1* is required in wild-type cells for elimination of more slowly growing neighboring cells, and that these genes are sufficient for removal of wild-type cells when they are overexpressed (Li & Baker, 2007). However, a more recent study demonstrated that these phagocytosis genes are dispensable for cell competition and that clearance of apoptotic cells during this process is performed by recruited macrophages (Lolo, Casas-Tinto, & Moreno, 2013).

9.3 Autophagy-Related Cell Clearance

An intriguing study regarding apoptotic cell clearance during the earliest event of developmental apoptosis in mammals, embryonic cavitation, revealed the requirement of genes normally acting in autophagy, which is an additional type of programmed cell death (He & Klionsky, 2009; Marino, Niso-Santano, Baehrecke, & Kroemer, 2014). By examining apoptosis and apoptotic cell clearance in simple embryoid bodies, the authors demonstrated that lack of autophagy genes did not affect cell death but resulted in abnormal clearance of apoptotic cells due to their abnormal exposure of PS and lower secretion of "find-me" signals. Moreover, knockout mice of the *atg5* autophagy gene displayed defects in apoptotic cell clearance during embryogenesis (Qu et al., 2007). The molecular mechanisms of this interesting interaction between autophagy and apoptotic cell clearance during development are still not clear.

A reverse example of interaction between autophagy and apoptotic cell clearance has been demonstrated during salivary gland degradation in *Drosophila*. In this study, the phagocytic receptor Draper has been shown to be cell autonomously required downstream or in parallel to caspases for the induction of autophagy in dying salivary gland cells. Importantly, it was dispensable for autophagy associated with cell survival (McPhee & Baehrecke, 2010; McPhee, Logan, Freeman, & Baehrecke, 2010). The Draper-mediated signaling mechanism leading to autophagy induction in salivary glands remains to be elucidated.

10. STRESS-INDUCED UPREGULATION OF PHAGOCYTOSIS DURING DEVELOPMENT

During development, different defects caused by mutations, starvation, or environmental stresses such as radiation may occur, leading to excess apoptosis (Arya & White, 2015). The increased number of apoptotic cells compared to the normal physiological situation must be efficiently removed through phagocytosis in order to prevent broader damage to the surrounding tissue. Whether phagocytes are able to remove extra apoptotic cells and what pathways are responsible for promotion of phagocytosis during development have been recently addressed in several studies. Shklover et al. recently described a novel mechanism in the *Drosophila* embryonic CNS, which may take place when increased numbers of neurons die through apoptosis and phagocytic glia are required to clear them (Shklover, Mishnaevski, et al., 2015). This mechanism is able to synchronize between higher levels of neuronal death in the embryonic CNS and upregulation of phagocytic ability in neighboring glia (Fig. 3). It was demonstrated that *Drosophila* c-Jun N-terminal kinase (dJNK) pathway overactivation in embryonic neurons leads to activation of dJNK signaling in glial cells, where normally it is

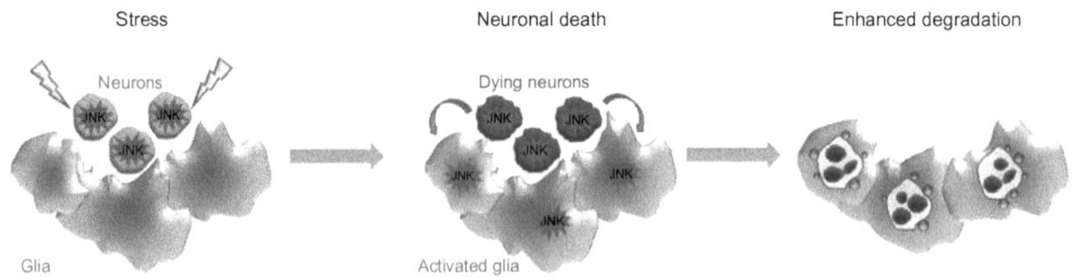

Figure 3 A model for synchronized dJNK pathway activation in the *Drosophila* CNS. dJNK signaling in neurons leads to dJNK pathway activation in glia, leading to upregulation of glial phagocytosis.

not needed for phagocytosis (Shklover, Mishnaevski, et al., 2015). JNK signaling activation in glia promotes the degradation step of phagocytosis with no effect on phagocytic receptors SIMU and Draper (Shklover, Mishnaevski, et al., 2015). The authors suggested that upregulation of glial phagocytosis is achieved by promoting degradation of apoptotic particles since the engulfment ability of embryonic glia is already very high during normal embryogenesis (Shklover, Mishnaevski, et al., 2015).

An additional example of stress-induced upregulation of phagocytosis is clearance of apoptotic germ line cells by follicle cells in the *Drosophila* ovary following starvation. In healthy egg chambers where normally phagocytosis does not take place, almost no Draper expression is detected and no activity of the dJNK pathway has been identified (Etchegaray, 2012 #89). However, following starvation, Draper-mediated upregulation of dJNK signaling in follicle cells promotes phagocytosis of apoptotic germ line cells by follicle cells (Etchegaray et al., 2012).

Importantly, in both examples, dJNK signaling is able to rescue the *draper* mutant phenotype, placing the dJNK pathway downstream of Draper function in phagocytosis likely in the degradation step. In addition, the role of the JNK pathway in *Drosophila* phagocytosis has been demonstrated during cell competition (Ohsawa et al., 2011) and clearance of degenerating axons in the adult CNS (Macdonald, Doherty, Hackett, & Freeman, 2013). *In vitro* studies in vertebrate cells demonstrated JNK pathway activity in professional and nonprofessional phagocytosis (Patel et al., 2006, 2010).

11. ANTI-INFLAMMATORY RESPONSE FOLLOWING APOPTOTIC CELL CLEARANCE

One of the most important features of apoptotic cell clearance is its noninflammatory nature. This is reasonable as the elimination of apoptotic cells takes place during normal development and tissue homeostasis. Current understanding of immunosuppression by apoptotic cells focuses on determinants exposed on their surface more than factors secreted from them (Birge & Ucker, 2008). However, the immunosuppressive determinants may be the same or different from "eat-me" signals exposed on apoptotic cells. The molecular mechanisms of highly complex immunosuppressive signaling in phagocytes are not fully understood. Worms and flies lack inflammation and an adaptive immune response; therefore, they can hardly serve as a basis for addressing these questions. In contrast, the suppression of inflammatory response during phagocytosis of apoptotic cells in mammals has drawn much attention (for review, see Birge & Ucker, 2008; Hochreiter-Hufford & Ravichandran, 2013; Savill & Fadok, 2000). The interaction between PS and its receptor (PSR) on macrophages has been reported as responsible for the release of anti-inflammatory cytokines (Fadok et al., 2000) and does not require the engulfment of the apoptotic cell *per se*. This phenomenon has been described also for nonprofessional phagocytes such as mammary epithelial cells (MECs) following lactation (Monks et al., 2005). MECs, which phagocytose their neighbor counterparts that undergo apoptosis, produce TGFβ1 in response to stimulation of the PSR by apoptotic cells or an anti-PSR antibody. In addition, *in vitro* studies demonstrate that apoptotic cells may repress secretion of inflammatory cytokines (Kim, Elkon, & Ma, 2004; Voll et al., 1997). More work is needed to elucidate the mechanisms underlying the immunosuppressive nature of apoptotic cells *in vivo* during development.

12. PHAGOCYTOSIS-PROMOTED PCD

An intriguing question regarding the role of phagocytosis in developmental PCD can be separated into two more specific questions. One is whether phagocytes are required for PCD by providing secreted molecules causing cell death and the second is whether engulfment *per se* leads to cell loss. The first evidence for macrophage-mediated cell death has been reported in the developing mouse eye where ablation of macrophages significantly delayed apoptosis of two normally transient structures (Lang, Lustig, Francois, Sellinger, & Plesken, 1994; Lang & Bishop, 1993). More recent studies revealed a molecular mechanism wherein the WNT pathway ligand WNT7 expressed in macrophages activates WNT signaling in adjacent vascular endothelial cells, which leads to vascular regression of the temporary vessels of the developing eye (Lobov et al., 2005). It is possible that macrophages influence life–death decision in different cells using this mechanism.

Several examples of microglia-mediated neuronal cell death by releasing various killing factors have been demonstrated during development; in rat spinal cord through secretion of tumor necrosis factor-α (TNF-α) (Sedel, Bechade, Vyas, & Triller, 2004), in chick retina by nerve growth factor (Frade & Barde, 1998; Kettenmann, Hanisch, Noda, & Verkhratsky, 2011), and in the developing mouse cerebellum where microglial release of superoxide ions executes the Purkinje cells death program (Claude, Linnartz-Gerlach, Kudin, Kunz, & Neumann, 2013; Marin-Teva et al., 2004). Moreover, microglial expression of integrin CD11b and immunoreceptor DAP12 is required for developmental neuronal apoptosis in the developing mouse hippocampus by regulating the production of microglial superoxide ions (Wakselman et al., 2008). On the other hand, a recent study in the *Drosophila* embryonic CNS has reported that abnormal glial phagocytosis in *simu* mutants was unable to prevent induced neuronal death (Shklover, Levy-Adam, & Kurant, 2015).

Phagocytosis-mediated cell death in mammals is similar to the previously demonstrated engulfment-promoted cell death in *C. elegans* (Hoeppner, Hengartner, & Schnabel, 2001; Reddien, Cameron, & Horvitz, 2001), suggesting that these pathways are evolutionarily conserved. In both cases, an initial apoptotic program involving caspase activation begins autonomously and phagocytic cells are needed to accomplish the death program.

However, it has been demonstrated that phagocytosis-promoted cell elimination may also occur without activation of apoptosis. During development of the cerebral cortex in rhesus monkey and rat, microglia seem to control neuronal number by phagocytosis of live neural precursor cells, which do not exhibit any markers of apoptosis, including TUNEL (TdT-mediated dUTP nick end-labeling) staining, caspase activation, or PS exposure on their surfaces. Moreover, activation or deactivation/elimination of glial cells affects the number of precursor cells, leading to abnormalities in neural development and behavioral outcome, which demonstrates the important role of microglial phagocytosis in neural development (Cunningham, Martinez-Cerdeno, & Noctor, 2013). Additional examples of glial phagocytosis of live neurons, called "phagoptosis," which may contribute to neuronal loss during development, were recently described during inflammation (Brown & Neher, 2014). Viable but stressed cells transiently exposing PS on their surfaces are engulfed by activated microglia through the MGF-E8 bridging molecule (Fricker et al., 2012), demonstrating death by phagocytosis.

Phagocytosis-induced cell death has been recently demonstrated in the *Drosophila* ovary, where overexpression of Draper or activation of the dJNK pathway in follicle cells induced nonautonomous death of egg chambers even without starvation (Etchegaray et al., 2012). While existence of this phenomenon during development of different species points at the evolutionary conserved nature of phagocytosis-mediated cell death, its molecular mechanisms await to be elucidated.

13. CONCLUDING REMARKS

During normal development of multicellular organisms, a great number of apoptotic cells are precisely and efficiently removed through immunologically silent phagocytosis by professional phagocytes and nonprofessional neighboring cells. The importance of precise clearance of apoptotic cells during development is very high since unremoved cells may prevent morphogenesis of tissues or organs (Sears et al., 2003) and induce immune reaction, which may damage developing tissue (Nagata et al., 2010). On the other hand, removal of live nonapoptotic functional cells is dangerous and may lead to defects in development (Brown & Neher, 2012).

As described above, apoptotic cell clearance is a dynamic multistep process, which is highly redundant in mammals mostly at the level of phagocytic receptors and their ligands on apoptotic cells. The distinct steps of apoptotic cell clearance discussed here involve multiple factors highly conserved in evolution, which makes research in model organisms extremely valuable for understanding the molecular and cellular basis of developmental apoptotic cell clearance. Moreover, to appreciate the role of this process during development, the research must be done *in vivo* in the developing organism, and the genetic models perfectly suit this goal.

The studies discussed in this review shed light on fundamental questions in developmental biology dealing with how unneeded cells are recognized and eliminated during development. Based on the recent research in this relatively young field, we may summarize that substantial progress has been achieved mostly in identification of the molecules involved in the process. Future studies are needed to gain a more mechanistic understanding of the interactions between these molecules and their developmental regulation. This basic scientific approach will help advance the current knowledge of developmental defects resulting from abnormal clearance of unneeded cells, with potential applications for intervention.

ACKNOWLEDGMENTS

We would like to thank T. Schultheiss for comments on the manuscript and the Kurant laboratory members for discussions and constructive criticism. We gratefully acknowledge financial support for the work in our lab from the Israel Science Foundation (Grant No. 427/11), the Rappaport Institute and from the Allen and Jewell Prince Center for Neurodegenerative Disorders of the Brain.

REFERENCES

Abrams, J. M., White, K., Fessler, L. I., & Steller, H. (1993). Programmed cell death during Drosophila embryogenesis. *Development*, *117*(1), 29–43.

Adachi-Yamada, T., & O'Connor, M. B. (2004). Mechanisms for removal of developmentally abnormal cells: Cell competition and morphogenetic apoptosis. *Journal of Biochemistry*, *136*(1), 13–17.

Aguzzi, A., Barres, B. A., & Bennett, M. L. (2013). Microglia: Scapegoat, saboteur, or something else? *Science*, *339*(6116), 156–161.

Akakura, S., Singh, S., Spataro, M., Akakura, R., Kim, J. I., Albert, M. L., et al. (2004). The opsonin MFG-E8 is a ligand for the alphavbeta5 integrin and triggers DOCK180-dependent Rac1 activation for the phagocytosis of apoptotic cells. *Experimental Cell Research*, *292*(2), 403–416.

Anderson, C., Zhou, S., Sawin, E., Horvitz, H. R., & Hurwitz, M. E. (2012). SLI-1 Cbl inhibits the engulfment of apoptotic cells in C. elegans through a ligase-independent function. *PLoS Genetics*, *8*(12), e1003115.

Arur, S., Uche, U. E., Rezaul, K., Fong, M., Scranton, V., Cowan, A. E., et al. (2003). Annexin I is an endogenous ligand that mediates apoptotic cell engulfment. *Developmental Cell*, *4*(4), 587–598.

Arya, R., & White, K. (2015). Cell death in development: Signaling pathways and core mechanisms. *Seminars in Cell & Developmental Biology*, *39*, 12–19.

Ashwell, K. (1990). Microglia and cell death in the developing mouse cerebellum. *Brain Research. Developmental Brain Research*, *55*(2), 219–230.

Awasaki, T., & Ito, K. (2004). Engulfing action of glial cells is required for programmed axon pruning during Drosophila metamorphosis. *Current Biology*, *14*(8), 668–677.

Awasaki, T., Tatsumi, R., Takahashi, K., Arai, K., Nakanishi, Y., Ueda, R., et al. (2006). Essential role of the apoptotic cell engulfment genes draper and ced-6 in programmed axon pruning during Drosophila metamorphosis. *Neuron*, *50*(6), 855–867.

Balasubramanian, K., & Schroit, A. J. (1998). Characterization of phosphatidylserine-dependent beta2-glycoprotein I macrophage interactions. Implications for apoptotic cell clearance by phagocytes. *The Journal of Biological Chemistry*, *273*(44), 29272–29277.

Bangs, P., & White, K. (2000). Regulation and execution of apoptosis during Drosophila development. *Developmental Dynamics*, *218*(1), 68–79.

Bilimoria, P. M., & Stevens, B. (2015). Microglia function during brain development: New insights from animal models. *Brain Research*, *1617*, 7–17.

Birge, R. B., & Ucker, D. S. (2008). Innate apoptotic immunity: The calming touch of death. *Cell Death and Differentiation*, *15*(7), 1096–1102.

Borisenko, G. G., Matsura, T., Liu, S. X., Tyurin, V. A., Jianfei, J., Serinkan, F. B., et al. (2003). Macrophage recognition of externalized phosphatidylserine and phagocytosis of apoptotic Jurkat cells—Existence of a threshold. *Archives of Biochemistry and Biophysics*, *413*(1), 41–52.

Boya, P., & de la Rosa, E. J. (2005). Cell death in early neural life. *Birth Defects Research. Part C, Embryo Today*, *75*(4), 281–293.

Bratton, D. L., Fadok, V. A., Richter, D. A., Kailey, J. M., Frasch, S. C., Nakamura, T., et al. (1999). Polyamine regulation of plasma membrane phospholipid flip-flop during apoptosis. *The Journal of Biological Chemistry*, *274*(40), 28113–28120.

Bratton, D. L., Fadok, V. A., Richter, D. A., Kailey, J. M., Guthrie, L. A., & Henson, P. M. (1997). Appearance of phosphatidylserine on apoptotic cells requires calcium-mediated nonspecific flip-flop and is enhanced by loss of the aminophospholipid translocase. *The Journal of Biological Chemistry*, *272*(42), 26159–26165.

Brown, S., Heinisch, I., Ross, E., Shaw, K., Buckley, C. D., & Savill, J. (2002). Apoptosis disables CD31-mediated cell detachment from phagocytes promoting binding and engulfment. *Nature*, *418*(6894), 200–203.

Brown, G. C., & Neher, J. J. (2012). Eaten alive! Cell death by primary phagocytosis: 'Phagoptosis'. *Trends in Biochemical Sciences*, *37*(8), 325–332.

Brown, G. C., & Neher, J. J. (2014). Microglial phagocytosis of live neurons. *Nature Reviews. Neuroscience*, *15*(4), 209–216.

Cabello, J., Neukomm, L. J., Gunesdogan, U., Burkart, K., Charette, S. J., Lochnit, G., et al. (2010). The Wnt pathway controls cell death engulfment, spindle orientation, and migration through CED-10/Rac. *PLoS Biology*, *8*(2), e1000297.

Cabello, J., Samann, J., Gomez-Orte, E., Erazo, T., Coppa, A., Pujol, A., et al. (2014). PDR-1/hParkin negatively regulates the phagocytosis of apoptotic cell corpses in Caenorhabditis elegans. *Cell Death & Disease*, *5*, e1120.

Caldero, J., Brunet, N., Ciutat, D., Hereu, M., & Esquerda, J. E. (2009). Development of microglia in the chick embryo spinal cord: Implications in the regulation of motoneuronal survival and death. *Journal of Neuroscience Research*, *87*(11), 2447–2466.

Cantera, R., & Technau, G. M. (1996). Glial cells phagocytose neuronal debris during the metamorphosis of the central nervous system in Drosophila melanogaster. *Development Genes and Evolution*, *206*(4), 277–280.

Chang, M. K., Binder, C. J., Torzewski, M., & Witztum, J. L. (2002). C-reactive protein binds to both oxidized LDL and apoptotic cells through recognition of a common ligand: Phosphorylcholine of oxidized phospholipids. *Proceedings of the National Academy of Sciences of the United States of America*, *99*(20), 13043–13048.

Chen, D., Jian, Y., Liu, X., Zhang, Y., Liang, J., Qi, X., et al. (2013). Clathrin and AP2 are required for phagocytic receptor-mediated apoptotic cell clearance in Caenorhabditis elegans. *PLoS Genetics*, *9*(5), e1003517.

Chen, Y. Z., Mapes, J., Lee, E. S., Skeen-Gaar, R. R., & Xue, D. (2013). Caspase-mediated activation of Caenorhabditis elegans CED-8 promotes apoptosis and phosphatidylserine externalization. *Nature Communications*, *4*, 2726.

Claude, J., Linnartz-Gerlach, B., Kudin, A. P., Kunz, W. S., & Neumann, H. (2013). Microglial CD33-related Siglec-E inhibits neurotoxicity by preventing the phagocytosis-associated oxidative burst. *The Journal of Neuroscience*, *33*(46), 18270–18276.

Conradt, B. (2009). Genetic control of programmed cell death during animal development. *Annual Review of Genetics*, *43*, 493–523.

Cunningham, C. L., Martinez-Cerdeno, V., & Noctor, S. C. (2013). Microglia regulate the number of neural precursor cells in the developing cerebral cortex. *The Journal of Neuroscience*, *33*(10), 4216–4233.

Cuttell, L., Vaughan, A., Silva, E., Escaron, C. J., Lavine, M., Van Goethem, E., et al. (2008). Undertaker, a Drosophila Junctophilin, links Draper-mediated phagocytosis and calcium homeostasis. *Cell*, *135*(3), 524–534.

Dalmau, I., Vela, J. M., Gonzalez, B., Finsen, B., & Castellano, B. (2003). Dynamics of microglia in the developing rat brain. *The Journal of Comparative Neurology*, *458*(2), 144–157.

deBakker, C. D., Haney, L. B., Kinchen, J. M., Grimsley, C., Lu, M., Klingele, D., et al. (2004). Phagocytosis of apoptotic cells is regulated by a UNC-73/TRIO-MIG-2/RhoG signaling module and armadillo repeats of CED-12/ELMO. *Current Biology*, *14*(24), 2208–2216.

DeKruyff, R. H., Bu, X., Ballesteros, A., Santiago, C., Chim, Y. L., Lee, H. H., et al. (2010). T cell/transmembrane, Ig, and mucin-3 allelic variants differentially recognize phosphatidylserine and mediate phagocytosis of apoptotic cells. *Journal of Immunology*, *184*(4), 1918–1930.

Devitt, A., Moffatt, O. D., Raykundalia, C., Capra, J. D., Simmons, D. L., & Gregory, C. D. (1998). Human CD14 mediates recognition and phagocytosis of apoptotic cells. *Nature*, *392*(6675), 505–509.

Diaz, B., Serna, J., De Pablo, F., & de la Rosa, E. J. (2000). In vivo regulation of cell death by embryonic (pro)insulin and the insulin receptor during early retinal neurogenesis. *Development, 127*(8), 1641–1649.
Diaz, B., & Moreno, E. (2005). The competitive nature of cells. *Experimental Cell Research, 306*(2), 317–322.
Dillon, S. R., Mancini, M., Rosen, A., & Schlissel, M. S. (2000). Annexin V binds to viable B cells and colocalizes with a marker of lipid rafts upon B cell receptor activation. *Journal of Immunology, 164*(3), 1322–1332.
Domingos, P. M., & Steller, H. (2007). Pathways regulating apoptosis during patterning and development. *Current Opinion in Genetics & Development, 17*(4), 294–299.
Egensperger, R., Maslim, J., Bisti, S., Hollander, H., & Stone, J. (1996). Fate of DNA from retinal cells dying during development: Uptake by microglia and macroglia (Muller cells). *Brain Research. Developmental Brain Research, 97*(1), 1–8.
Elliott, M. R., Chekeni, F. B., Trampont, P. C., Lazarowski, E. R., Kadl, A., Walk, S. F., et al. (2009). Nucleotides released by apoptotic cells act as a find-me signal to promote phagocytic clearance. *Nature, 461*(7261), 282–286.
Elliott, M. R., & Ravichandran, K. S. (2010). Clearance of apoptotic cells: Implications in health and disease. *The Journal of Cell Biology, 189*(7), 1059–1070.
Epelman, S., Lavine, K. J., & Randolph, G. J. (2014). Origin and functions of tissue macrophages. *Immunity, 41*(1), 21–35.
Erwig, L. P., & Henson, P. M. (2007). Immunological consequences of apoptotic cell phagocytosis. *The American Journal of Pathology, 171*(1), 2–8.
Erwig, L. P., & Henson, P. M. (2008). Clearance of apoptotic cells by phagocytes. *Cell Death and Differentiation, 15*(2), 243–250.
Etchegaray, J. I., Timmons, A. K., Klein, A. P., Pritchett, T. L., Welch, E., Meehan, T. L., et al. (2012). Draper acts through the JNK pathway to control synchronous engulfment of dying germline cells by follicular epithelial cells. *Development, 139*(21), 4029–4039.
Fadok, V. A., Bratton, D. L., Frasch, S. C., Warner, M. L., & Henson, P. M. (1998). The role of phosphatidylserine in recognition of apoptotic cells by phagocytes. *Cell Death and Differentiation, 5*(7), 551–562.
Fadok, V. A., Bratton, D. L., & Henson, P. M. (2001). Phagocyte receptors for apoptotic cells: Recognition, uptake, and consequences. *The Journal of Clinical Investigation, 108*(7), 957–962.
Fadok, V. A., Bratton, D. L., Rose, D. M., Pearson, A., Ezekewitz, R. A., & Henson, P. M. (2000). A receptor for phosphatidylserine-specific clearance of apoptotic cells. *Nature, 405*(6782), 85–90.
Fadok, V. A., Voelker, D. R., Campbell, P. A., Cohen, J. J., Bratton, D. L., & Henson, P. M. (1992). Exposure of phosphatidylserine on the surface of apoptotic lymphocytes triggers specific recognition and removal by macrophages. *Journal of Immunology, 148*(7), 2207–2216.
Fancsalszky, L., Monostori, E., Farkas, Z., Pourkarimi, E., Masoudi, N., Hargitai, B., et al. (2014). NDK-1, the homolog of NM23-H1/H2 regulates cell migration and apoptotic engulfment in C. elegans. *PLoS One, 9*(3), e92687.
Fischbach, K. F., & Technau, G. (1984). Cell degeneration in the developing optic lobes of the sine oculis and small-optic-lobes mutants of Drosophila melanogaster. *Developmental Biology, 104*(1), 219–239.
Frade, J. M., & Barde, Y. A. (1998). Microglia-derived nerve growth factor causes cell death in the developing retina. *Neuron, 20*(1), 35–41.
Franc, N. C. (2002). Phagocytosis of apoptotic cells in mammals, Caenorhabditis elegans and Drosophila melanogaster: Molecular mechanisms and physiological consequences. *Frontiers in Bioscience, 7*, d1298–d1313.

Franc, N. C., Heitzler, P., Ezekowitz, R. A., & White, K. (1999). Requirement for croquemort in phagocytosis of apoptotic cells in Drosophila. *Science*, *284*(5422), 1991–1994.

Francisco-Morcillo, J., Bejarano-Escobar, R., Rodriguez-Leon, J., Navascues, J., & Martin-Partido, G. (2014). Ontogenetic cell death and phagocytosis in the visual system of vertebrates. *Developmental Dynamics*, *243*(10), 1203–1225.

Freeman, M. R. (2006). Sculpting the nervous system: glial control of neuronal development. *Current Opinion in Neurobiology*, *16*(1), 119–125.

Freeman, G. J., Casasnovas, J. M., Umetsu, D. T., & DeKruyff, R. H. (2010). TIM genes: A family of cell surface phosphatidylserine receptors that regulate innate and adaptive immunity. *Immunological Reviews*, *235*(1), 172–189.

Freeman, M. R., Delrow, J., Kim, J., Johnson, E., & Doe, C. Q. (2003). Unwrapping glial biology: Gcm target genes regulating glial development, diversification, and function. *Neuron*, *38*(4), 567–580.

Fricker, M., Neher, J. J., Zhao, J. W., Thery, C., Tolkovsky, A. M., & Brown, G. C. (2012). MFG-E8 mediates primary phagocytosis of viable neurons during neuroinflammation. *The Journal of Neuroscience*, *32*(8), 2657–2666.

Fullard, J., Kale, A., & Baker, N. (2009). Clearance of apoptotic corpses. *Apoptosis*, *14*(8), 1029–1037.

Gardai, S. J., Bratton, D. L., Ogden, C. A., & Henson, P. M. (2006). Recognition ligands on apoptotic cells: A perspective. *Journal of Leukocyte Biology*, *79*(5), 896–903.

Gardai, S. J., McPhillips, K. A., Frasch, S. C., Janssen, W. J., Starefeldt, A., Murphy-Ullrich, J. E., et al. (2005). Cell-surface calreticulin initiates clearance of viable or apoptotic cells through trans-activation of LRP on the phagocyte. *Cell*, *123*(2), 321–334.

Gouon-Evans, V., Rothenberg, M. E., & Pollard, J. W. (2000). Postnatal mammary gland development requires macrophages and eosinophils. *Development*, *127*(11), 2269–2282.

Gregory, C. D., & Devitt, A. (2004). The macrophage and the apoptotic cell: An innate immune interaction viewed simplistically? *Immunology*, *113*(1), 1–14.

Gregory, C. D., Devitt, A., & Moffatt, O. (1998). Roles of ICAM-3 and CD14 in the recognition and phagocytosis of apoptotic cells by macrophages. *Biochemical Society Transactions*, *26*(4), 644–649.

Grimsley, C., & Ravichandran, K. S. (2003). Cues for apoptotic cell engulfment: Eat-me, don't eat-me and come-get-me signals. *Trends in Cell Biology*, *13*(12), 648–656.

Gude, D. R., Alvarez, S. E., Paugh, S. W., Mitra, P., Yu, J., Griffiths, R., et al. (2008). Apoptosis induces expression of sphingosine kinase 1 to release sphingosine-1-phosphate as a "come-and-get-me" signal. *FASEB Journal*, *22*(8), 2629–2638.

Gumienny, T. L., Brugnera, E., Tosello-Trampont, A. C., Kinchen, J. M., Haney, L. B., Nishiwaki, K., et al. (2001). CED-12/ELMO, a novel member of the CrkII/Dock180/Rac pathway, is required for phagocytosis and cell migration. *Cell*, *107*(1), 27–41.

Hakim, Y., Yaniv, S. P., & Schuldiner, O. (2014). Astrocytes play a key role in Drosophila mushroom body axon pruning. *PLoS One*, *9*(1), 21.

Hamon, Y., Broccardo, C., Chambenoit, O., Luciani, M. F., Toti, F., Chaslin, S., et al. (2000). ABC1 promotes engulfment of apoptotic cells and transbilayer redistribution of phosphatidylserine. *Nature Cell Biology*, *2*(7), 399–406.

Hanayama, R., & Nagata, S. (2005). Impaired involution of mammary glands in the absence of milk fat globule EGF factor 8. *Proceedings of the National Academy of Sciences of the United States of America*, *102*(46), 16886–16891.

Hanayama, R., Tanaka, M., Miwa, K., Shinohara, A., Iwamatsu, A., & Nagata, S. (2002). Identification of a factor that links apoptotic cells to phagocytes. *Nature*, *417*(6885), 182–187.

Hanayama, R., Tanaka, M., Miyasaka, K., Aozasa, K., Koike, M., Uchiyama, Y., et al. (2004). Autoimmune disease and impaired uptake of apoptotic cells in MFG-E8-deficient mice. *Science*, *304*(5674), 1147–1150.
He, C., & Klionsky, D. J. (2009). Regulation mechanisms and signaling pathways of autophagy. *Annual Review of Genetics*, *43*, 67–93.
Henson, P. M., & Hume, D. A. (2006). Apoptotic cell removal in development and tissue homeostasis. *Trends in Immunology*, *27*(5), 244–250.
Hochreiter-Hufford, A., & Ravichandran, K. S. (2013). Clearing the dead: apoptotic cell sensing, recognition, engulfment, and digestion. *Cold Spring Harbor Perspectives in Biology*, *5*(1), a008748.
Hoeppner, D. J., Hengartner, M. O., & Schnabel, R. (2001). Engulfment genes cooperate with ced-3 to promote cell death in Caenorhabditis elegans. *Nature*, *412*(6843), 202–206.
Hoopfer, E. D., McLaughlin, T., Watts, R. J., Schuldiner, O., O'Leary, D. D., & Luo, L. (2006). Wld(s) protection distinguishes axon degeneration following injury from naturally occurring developmental pruning. *Neuron*, *50*(6), 883–895.
Hopkinson-Woolley, J., Hughes, D., Gordon, S., & Martin, P. (1994). Macrophage recruitment during limb development and wound healing in the embryonic and foetal mouse. *Journal of Cell Science*, *107*(Pt. 5), 1159–1167.
Hsu, T. Y., & Wu, Y. C. (2010). Engulfment of apoptotic cells in C. elegans is mediated by integrin alpha/SRC signaling. *Current Biology*, *20*(6), 477–486.
Hume, D. A., Perry, V. H., & Gordon, S. (1983). Immunohistochemical localization of a macrophage-specific antigen in developing mouse retina: Phagocytosis of dying neurons and differentiation of microglial cells to form a regular array in the plexiform layers. *The Journal of Cell Biology*, *97*(1), 253–257.
Hurwitz, M. E., Vanderzalm, P. J., Bloom, L., Goldman, J., Garriga, G., & Horvitz, H. R. (2009). Abl kinase inhibits the engulfment of apoptotic [corrected] cells in Caenorhabditis elegans. *PLoS Biology*, 7(4), e99.
Ichimura, T., Asseldonk, E. J., Humphreys, B. D., Gunaratnam, L., Duffield, J. S., & Bonventre, J. V. (2008). Kidney injury molecule-1 is a phosphatidylserine receptor that confers a phagocytic phenotype on epithelial cells. *The Journal of Clinical Investigation*, *118*(5), 1657–1668.
Imachi, H., Murao, K., Hiramine, C., Sayo, Y., Sato, M., Hosokawa, H., et al. (2000). Human scavenger receptor B1 is involved in recognition of apoptotic thymocytes by thymic nurse cells. *Laboratory Investigation*, *80*(2), 263–270.
Ishimoto, Y., Ohashi, K., Mizuno, K., & Nakano, T. (2000). Promotion of the uptake of PS liposomes and apoptotic cells by a product of growth arrest-specific gene, gas6. *Journal of Biochemistry*, *127*(3), 411–417.
Jones, C. V., Williams, T. M., Walker, K. A., Dickinson, H., Sakkal, S., Rumballe, B. A., et al. (2013). M2 macrophage polarisation is associated with alveolar formation during postnatal lung development. *Respiratory Research*, *14*, 41.
Juncadella, I. J., Kadl, A., Sharma, A. K., Shim, Y. M., Hochreiter-Hufford, A., Borish, L., et al. (2013). Apoptotic cell clearance by bronchial epithelial cells critically influences airway inflammation. *Nature*, *493*(7433), 547–551.
Kay, J. G., & Grinstein, S. (2013). Phosphatidylserine-mediated cellular signaling. *Advances in Experimental Medicine and Biology*, *991*, 177–193.
Kettenmann, H., Hanisch, U. K., Noda, M., & Verkhratsky, A. (2011). Physiology of microglia. *Physiological Reviews*, *91*(2), 461–553.
Kim, S., Elkon, K. B., & Ma, X. (2004). Transcriptional suppression of interleukin-12 gene expression following phagocytosis of apoptotic cells. *Immunity*, *21*(5), 643–653.
Kimura, K. I., & Truman, J. W. (1990). Postmetamorphic cell death in the nervous and muscular systems of Drosophila melanogaster. *The Journal of Neuroscience*, *10*(2), 403–411.

Kinchen, J. M. (2010). A model to die for: Signaling to apoptotic cell removal in worm, fly and mouse. *Apoptosis*, *15*(9), 998–1006.
Kinchen, J. M., Cabello, J., Klingele, D., Wong, K., Feichtinger, R., Schnabel, H., et al. (2005). Two pathways converge at CED-10 to mediate actin rearrangement and corpse removal in C. elegans. *Nature*, *434*(7029), 93–99.
Kinchen, J. M., Doukoumetzidis, K., Almendinger, J., Stergiou, L., Tosello-Trampont, A., Sifri, C. D., et al. (2008). A pathway for phagosome maturation during engulfment of apoptotic cells. *Nature Cell Biology*, *10*(5), 556–566.
Kinchen, J. M., & Ravichandran, K. S. (2007). Journey to the grave: Signaling events regulating removal of apoptotic cells. *Journal of Cell Science*, *120*(Pt. 13), 2143–2149.
Kinchen, J. M., & Ravichandran, K. S. (2008a). Phagocytic signaling: You can touch, but you can't eat. *Current Biology*, *18*(12), R521–R524.
Kinchen, J. M., & Ravichandran, K. S. (2008b). Phagosome maturation: Going through the acid test. *Nature Reviews. Molecular Cell Biology*, *9*(10), 781–795.
Kinchen, J. M., & Ravichandran, K. S. (2010). Identification of two evolutionarily conserved genes regulating processing of engulfed apoptotic cells. *Nature*, *464*(7289), 778–782.
Kobayashi, N., Karisola, P., Pena-Cruz, V., Dorfman, D. M., Jinushi, M., Umetsu, S. E., et al. (2007). TIM-1 and TIM-4 glycoproteins bind phosphatidylserine and mediate uptake of apoptotic cells. *Immunity*, *27*(6), 927–940.
Krysko, D. V., D'Herde, K., & Vandenabeele, P. (2006). Clearance of apoptotic and necrotic cells and its immunological consequences. *Apoptosis*, *11*(10), 1709–1726.
Kuraishi, T., Nakagawa, Y., Nagaosa, K., Hashimoto, Y., Ishimoto, T., Moki, T., et al. (2009). Pretaporter, a Drosophila protein serving as a ligand for Draper in the phagocytosis of apoptotic cells. *The EMBO Journal*, *28*(24), 3868–3878.
Kurant, E. (2011). Keeping the CNS clear: Glial phagocytic functions in Drosophila. *Glia*, *59*(9), 1304–1311.
Kurant, E., Axelrod, S., Leaman, D., & Gaul, U. (2008). Six-microns-under acts upstream of Draper in the glial phagocytosis of apoptotic neurons. *Cell*, *133*(3), 498–509.
Lang, R. A., & Bishop, J. M. (1993). Macrophages are required for cell death and tissue remodeling in the developing mouse eye. *Cell*, *74*(3), 453–462.
Lang, R., Lustig, M., Francois, F., Sellinger, M., & Plesken, H. (1994). Apoptosis during macrophage-dependent ocular tissue remodelling. *Development*, *120*(12), 3395–3403.
Lauber, K., Blumenthal, S. G., Waibel, M., & Wesselborg, S. (2004). Clearance of apoptotic cells: Getting rid of the corpses. *Molecular Cell*, *14*(3), 277–287.
Lauber, K., Bohn, E., Krober, S. M., Xiao, Y. J., Blumenthal, S. G., Lindemann, R. K., et al. (2003). Apoptotic cells induce migration of phagocytes via caspase-3-mediated release of a lipid attraction signal. *Cell*, *113*(6), 717–730.
Lavin, Y., & Merad, M. (2013). Macrophages: Gatekeepers of tissue integrity. *Cancer Immunology Research*, *1*(4), 201–209.
Lemke, G., & Rothlin, C. V. (2008). Immunobiology of the TAM receptors. *Nature Reviews. Immunology*, *8*(5), 327–336.
Lettre, G., & Hengartner, M. O. (2006). Developmental apoptosis in C. elegans: A complex CEDnario. *Nature Reviews. Molecular Cell Biology*, 7(2), 97–108.
Li, W., & Baker, N. E. (2007). Engulfment is required for cell competition. *Cell*, *129*(6), 1215–1225.
Liu, Q. A., & Hengartner, M. O. (1998). Candidate adaptor protein CED-6 promotes the engulfment of apoptotic cells in C. elegans. *Cell*, *93*(6), 961–972.
Lobov, I. B., Rao, S., Carroll, T. J., Vallance, J. E., Ito, M., Ondr, J. K., et al. (2005). WNT7b mediates macrophage-induced programmed cell death in patterning of the vasculature. *Nature*, *437*(7057), 417–421.

Logan, M. A., Hackett, R., Doherty, J., Sheehan, A., Speese, S. D., & Freeman, M. R. (2012). Negative regulation of glial engulfment activity by Draper terminates glial responses to axon injury. *Nature Neuroscience, 15*(5), 722–730.
Lolo, F. N., Casas-Tinto, S., & Moreno, E. (2013). Cell competition time line: Winners kill losers, which are extruded and engulfed by hemocytes. *Cell Reports, 2*(3), 526–539.
Lu, Z., Elliott, M. R., Chen, Y., Walsh, J. T., Klibanov, A. L., Ravichandran, K. S., et al. (2011). Phagocytic activity of neuronal progenitors regulates adult neurogenesis. *Nature Cell Biology, 13*(9), 1076–1083.
Macdonald, J. M., Doherty, J., Hackett, R., & Freeman, M. R. (2013). The c-Jun kinase signaling cascade promotes glial engulfment activity through activation of draper and phagocytic function. *Cell Death and Differentiation, 20*(9), 1140–1148.
Magnus, T., Chan, A., Linker, R. A., Toyka, K. V., & Gold, R. (2002). Astrocytes are less efficient in the removal of apoptotic lymphocytes than microglia cells: Implications for the role of glial cells in the inflamed central nervous system. *Journal of Neuropathology and Experimental Neurology, 61*(9), 760–766.
Mahoney, J. A., & Rosen, A. (2005). Apoptosis and autoimmunity. *Current Opinion in Immunology, 17*(6), 583–588.
Mangahas, P. M., & Zhou, Z. (2005). Clearance of apoptotic cells in Caenorhabditis elegans. *Seminars in Cell & Developmental Biology, 16*(2), 295–306.
Marin, E. C., Watts, R. J., Tanaka, N. K., Ito, K., & Luo, L. (2005). Developmentally programmed remodeling of the Drosophila olfactory circuit. *Development, 132*(4), 725–737.
Marino, G., Niso-Santano, M., Baehrecke, E. H., & Kroemer, G. (2014). Self-consumption: The interplay of autophagy and apoptosis. *Nature Reviews. Molecular Cell Biology, 15*(2), 81–94.
Marin-Teva, J. L., Cuadros, M. A., Calvente, R., Almendros, A., & Navascues, J. (1999). Naturally occurring cell death and migration of microglial precursors in the quail retina during normal development. *The Journal of Comparative Neurology, 412*(2), 255–275.
Marin-Teva, J. L., Dusart, I., Colin, C., Gervais, A., van Rooijen, N., & Mallat, M. (2004). Microglia promote the death of developing Purkinje cells. *Neuron, 41*(4), 535–547.
Martins, I., Kepp, O., Galluzzi, L., Senovilla, L., Schlemmer, F., Adjemian, S., et al. (2010). Surface-exposed calreticulin in the interaction between dying cells and phagocytes. *Annals of the New York Academy of Sciences, 1209*(1), 77–82.
Mazaheri, F., Breus, O., Durdu, S., Haas, P., Wittbrodt, J., Gilmour, D., et al. (2014). Distinct roles for BAI1 and TIM-4 in the engulfment of dying neurons by microglia. *Nature Communications, 5*, 4046.
McPhee, C. K., & Baehrecke, E. H. (2010). The engulfment receptor Draper is required for autophagy during cell death. *Autophagy, 6*(8), 1192–1193.
McPhee, C. K., Logan, M. A., Freeman, M. R., & Baehrecke, E. H. (2010). Activation of autophagy during cell death requires the engulfment receptor Draper. *Nature, 465*(7301), 1093–1096.
Meier, P., Finch, A., & Evan, G. (2000). Apoptosis in development. *Nature, 407*(6805), 796–801.
Mevorach, D. (2000). Opsonization of apoptotic cells. Implications for uptake and autoimmunity. *Annals of the New York Academy of Sciences, 926*, 226–235.
Mevorach, D. (2010). Clearance of dying cells and systemic lupus erythematosus: The role of C1q and the complement system. *Apoptosis, 15*(9), 1114–1123.
Mevorach, D., Mascarenhas, J. O., Gershov, D., & Elkon, K. B. (1998). Complement-dependent clearance of apoptotic cells by human macrophages. *The Journal of Experimental Medicine, 188*(12), 2313–2320.
Miyanishi, M., Tada, K., Koike, M., Uchiyama, Y., Kitamura, T., & Nagata, S. (2007). Identification of Tim4 as a phosphatidylserine receptor. *Nature, 450*(7168), 435–439.

Moffatt, O. D., Devitt, A., Bell, E. D., Simmons, D. L., & Gregory, C. D. (1999). Macrophage recognition of ICAM-3 on apoptotic leukocytes. *Journal of Immunology, 162*(11), 6800–6810.

Monks, J., Rosner, D., Jon Geske, F., Lehman, L., Hanson, L., Neville, M. C., et al. (2005). Epithelial cells as phagocytes: Apoptotic epithelial cells are engulfed by mammary alveolar epithelial cells and repress inflammatory mediator release. *Cell Death and Differentiation, 12*(2), 107–114.

Munoz, L. E., Peter, C., Herrmann, M., Wesselborg, S., & Lauber, K. (2010). Scent of dying cells: The role of attraction signals in the clearance of apoptotic cells and its immunological consequences. *Autoimmunity Reviews, 9*(6), 425–430.

Nagata, S., Hanayama, R., & Kawane, K. (2010). Autoimmunity and the clearance of dead cells. *Cell, 140*(5), 619–630.

Nakanishi, Y., & Shiratsuchi, A. (2004). Phagocytic removal of apoptotic spermatogenic cells by Sertoli cells: Mechanisms and consequences. *Biological & Pharmaceutical Bulletin, 27*(1), 13–16.

Nathan, C., & Ding, A. (2010). Nonresolving inflammation. *Cell, 140*(6), 871–882.

Obeid, M., Tesniere, A., Ghiringhelli, F., Fimia, G. M., Apetoh, L., Perfettini, J. L., et al. (2007). Calreticulin exposure dictates the immunogenicity of cancer cell death. *Nature Medicine, 13*(1), 54–61.

Ohsawa, S., Sugimura, K., Takino, K., Xu, T., Miyawaki, A., & Igaki, T. (2011). Elimination of oncogenic neighbors by JNK-mediated engulfment in Drosophila. *Developmental Cell, 20*(3), 315–328.

Oppenheim, R. W., Homma, S., Marti, E., Prevette, D., Wang, S., Yaginuma, H., et al. (1999). Modulation of early but not later stages of programmed cell death in embryonic avian spinal cord by sonic hedgehog. *Molecular and Cellular Neurosciences, 13*(5), 348–361.

Patel, V. A., Lee, D. J., Feng, L., Antoni, A., Lieberthal, W., Schwartz, J. H., et al. (2010). Recognition of apoptotic cells by epithelial cells: conserved versus tissue-specific signaling responses. *The Journal of Biological Chemistry, 285*(3), 1829–1840.

Park, S. Y., Jung, M. Y., Kim, H. J., Lee, S. J., Kim, S. Y., Lee, B. H., et al. (2008). Rapid cell corpse clearance by stabilin-2, a membrane phosphatidylserine receptor. *Cell Death and Differentiation, 15*(1), 192–201.

Park, S. Y., Jung, M. Y., Lee, S. J., Kang, K. B., Gratchev, A., Riabov, V., et al. (2009). Stabilin-1 mediates phosphatidylserine-dependent clearance of cell corpses in alternatively activated macrophages. *Journal of Cell Science, 122*(Pt. 18), 3365–3373.

Park, D., Tosello-Trampont, A. C., Elliott, M. R., Lu, M., Haney, L. B., Ma, Z., et al. (2007). BAI1 is an engulfment receptor for apoptotic cells upstream of the ELMO/Dock180/Rac module. *Nature, 450*(7168), 430–434.

Parnaik, R., Raff, M. C., & Scholes, J. (2000). Differences between the clearance of apoptotic cells by professional and non-professional phagocytes. *Current Biology, 10*(14), 857–860.

Penfold, P. L., & Provis, J. M. (1986). Cell death in the development of the human retina: Phagocytosis of pyknotic and apoptotic bodies by retinal cells. *Graefe's Archive for Clinical and Experimental Ophthalmology, 224*(6), 549–553.

Peri, F., & Nusslein-Volhard, C. (2008). Live imaging of neuronal degradation by microglia reveals a role for v0-ATPase a1 in phagosomal fusion in vivo. *Cell, 133*(5), 916–927.

Peter, C., Waibel, M., Radu, C. G., Yang, L. V., Witte, O. N., Schulze-Osthoff, K., et al. (2008). Migration to apoptotic "find-me" signals is mediated via the phagocyte receptor G2A. *The Journal of Biological Chemistry, 283*(9), 5296–5305.

Peter, C., Wesselborg, S., Herrmann, M., & Lauber, K. (2010). Dangerous attraction: Phagocyte recruitment and danger signals of apoptotic and necrotic cells. *Apoptosis, 15*(9), 1007–1028.

Peterson, C., Carney, G. E., Taylor, B. J., & White, K. (2002). reaper is required for neuroblast apoptosis during Drosophila development. *Development*, *129*(6), 1467–1476.

Pinto, S. M., & Hengartner, M. O. (2012). Cleaning up the mess: Cell corpse clearance in Caenorhabditis elegans. *Current Opinion in Cell Biology*, *24*(6), 881–888.

Platt, N., & Gordon, S. (2001). Is the class A macrophage scavenger receptor (SR-A) multifunctional?—The mouse's tale. *The Journal of Clinical Investigation*, *108*(5), 649–654.

Pollard, J. W. (2009). Trophic macrophages in development and disease. *Nature Reviews. Immunology*, *9*(4), 259–270.

Poon, I. K., Lucas, C. D., Rossi, A. G., & Ravichandran, K. S. (2014). Apoptotic cell clearance: Basic biology and therapeutic potential. *Nature Reviews. Immunology*, *14*(3), 166–180.

Qu, X., Zou, Z., Sun, Q., Luby-Phelps, K., Cheng, P., Hogan, R. N., et al. (2007). Autophagy gene-dependent clearance of apoptotic cells during embryonic development. *Cell*, *128*(5), 931–946.

Rakic, S., & Zecevic, N. (2000). Programmed cell death in the developing human telencephalon. *The European Journal of Neuroscience*, *12*(8), 2721–2734.

Ravichandran, K. S. (2010). Find-me and eat-me signals in apoptotic cell clearance: Progress and conundrums. *The Journal of Experimental Medicine*, *207*(9), 1807–1817.

Ravichandran, K. S. (2011). Beginnings of a good apoptotic meal: The find-me and eat-me signaling pathways. *Immunity*, *35*(4), 445–455.

Ravichandran, K. S., & Lorenz, U. (2007). Engulfment of apoptotic cells: Signals for a good meal. *Nature Reviews. Immunology*, 7(12), 964–974.

Reddien, P. W., Cameron, S., & Horvitz, H. R. (2001). Phagocytosis promotes programmed cell death in C. elegans. *Nature*, *412*(6843), 198–202.

Reddien, P. W., & Horvitz, H. R. (2000). CED-2/CrkII and CED-10/Rac control phagocytosis and cell migration in Caenorhabditis elegans. *Nature Cell Biology*, *2*(3), 131–136.

Reddien, P. W., & Horvitz, H. R. (2004). The engulfment process of programmed cell death in Caenorhabditis elegans. *Annual Review of Cell and Developmental Biology*, *20*, 193–221.

Rodriguez-Manzanet, R., Sanjuan, M. A., Wu, H. Y., Quintana, F. J., Xiao, S., Anderson, A. C., et al. (2010). T and B cell hyperactivity and autoimmunity associated with niche-specific defects in apoptotic body clearance in TIM-4-deficient mice. *Proceedings of the National Academy of Sciences of the United States of America*, *107*(19), 8706–8711.

Rogulja-Ortmann, A., Luer, K., Seibert, J., Rickert, C., & Technau, G. M. (2007). Programmed cell death in the embryonic central nervous system of Drosophila melanogaster. *Development*, *134*(1), 105–116.

Ryoo, H. D., & Baehrecke, E. H. (2010). Distinct death mechanisms in Drosophila development. *Current Opinion in Cell Biology*, *22*(6), 889–895.

Sasaki, A., Nakae, I., Nagasawa, M., Hashimoto, K., Abe, F., Saito, K., et al. (2013). Arl8/ARL-8 functions in apoptotic cell removal by mediating phagolysosome formation in Caenorhabditis elegans. *Molecular Biology of the Cell*, *24*(10), 1584–1592.

Savill, J., & Fadok, V. (2000). Corpse clearance defines the meaning of cell death. *Nature*, *407*(6805), 784–788.

Savill, J., Hogg, N., Ren, Y., & Haslett, C. (1992). Thrombospondin cooperates with CD36 and the vitronectin receptor in macrophage recognition of neutrophils undergoing apoptosis. *The Journal of Clinical Investigation*, *90*(4), 1513–1522.

Scott, R. S., McMahon, E. J., Pop, S. M., Reap, E. A., Caricchio, R., Cohen, P. L., et al. (2001). Phagocytosis and clearance of apoptotic cells is mediated by MER. *Nature*, *411*(6834), 207–211.

Sears, H. C., Kennedy, C. J., & Garrity, P. A. (2003). Macrophage-mediated corpse engulfment is required for normal Drosophila CNS morphogenesis. *Development*, *130*(15), 3557–3565.

Sedel, F., Bechade, C., Vyas, S., & Triller, A. (2004). Macrophage-derived tumor necrosis factor alpha, an early developmental signal for motoneuron death. *The Journal of Neuroscience, 24*(9), 2236–2246.

Segawa, K., Kurata, S., Yanagihashi, Y., Brummelkamp, T. R., Matsuda, F., & Nagata, S. (2014). Caspase-mediated cleavage of phospholipid flippase for apoptotic phosphatidylserine exposure. *Science, 344*(6188), 1164–1168.

Segawa, K., Suzuki, J., & Nagata, S. (2011). Constitutive exposure of phosphatidylserine on viable cells. *Proceedings of the National Academy of Sciences of the United States of America, 108*(48), 19246–19251.

Shen, Q., He, B., Lu, N., Conradt, B., Grant, B. D., & Zhou, Z. (2013). Phagocytic receptor signaling regulates clathrin and epsin-mediated cytoskeletal remodeling during apoptotic cell engulfment in C. elegans. *Development, 140*(15), 3230–3243.

Shklover, J., Levy-Adam, F., & Kurant, E. (2015). The role of Drosophila TNF Eiger in developmental and damage-induced neuronal apoptosis. *FEBS Letters, 589*(8), 871–879.

Shklover, J., Mishnaevski, K., Levy-Adam, F., & Kurant, E. (2015). JNK pathway activation is able to synchronize neuronal death and glial phagocytosis in Drosophila. *Cell Death & Disease, 6*, e1649.

Shklyar, B., Levy-Adam, F., Mishnaevski, K., & Kurant, E. (2013). Caspase activity is required for engulfment of apoptotic cells. *Molecular and Cellular Biology, 33*(16), 3191–3201.

Shklyar, B., Sellman, Y., Shklover, J., Mishnaevski, K., Levy-Adam, F., & Kurant, E. (2014). Developmental regulation of glial cell phagocytic function during Drosophila embryogenesis. *Developmental Biology, 393*(2), 255–269.

Sierra, A., Abiega, O., Shahraz, A., & Neumann, H. (2013). Janus-faced microglia: Beneficial and detrimental consequences of microglial phagocytosis. *Frontiers in Cellular Neuroscience, 7*, 6.

Sierra, A., Encinas, J. M., Deudero, J. J., Chancey, J. H., Enikolopov, G., Overstreet-Wadiche, L. S., et al. (2010). Microglia shape adult hippocampal neurogenesis through apoptosis-coupled phagocytosis. *Cell Stem Cell, 7*(4), 483–495.

Silva, E., Au-Yeung, H. W., Van Goethem, E., Burden, J., & Franc, N. C. (2007). Requirement for a Drosophila E3-ubiquitin ligase in phagocytosis of apoptotic cells. *Immunity, 27*(4), 585–596.

Steller, H. (2008). Regulation of apoptosis in Drosophila. *Cell Death and Differentiation, 15*(7), 1132–1138.

Steller, H., & Grether, M. E. (1994). Programmed cell death in Drosophila. *Neuron, 13*(6), 1269–1274.

Stuart, L. M., & Ezekowitz, R. A. (2005). Phagocytosis: Elegant complexity. *Immunity, 22*(5), 539–550.

Sullivan, C. S., Scheib, J. L., Ma, Z., Dang, R. P., Schafer, J. M., Hickman, F. E., et al. (2014). The adaptor protein GULP promotes Jedi-1-mediated phagocytosis through a clathrin-dependent mechanism. *Molecular Biology of the Cell, 25*(12), 1925–1936.

Suzanne, M., & Steller, H. (2013). Shaping organisms with apoptosis. *Cell Death and Differentiation, 20*(5), 669–675.

Suzuki, J., Denning, D. P., Imanishi, E., Horvitz, H. R., & Nagata, S. (2013). Xk-related protein 8 and CED-8 promote phosphatidylserine exposure in apoptotic cells. *Science, 341*(6144), 403–406.

Suzuki, J., Umeda, M., Sims, P. J., & Nagata, S. (2010). Calcium-dependent phospholipid scrambling by TMEM16F. *Nature, 468*(7325), 834–838.

Tasdemir-Yilmaz, O. E., & Freeman, M. R. (2014). Astrocytes engage unique molecular programs to engulf pruned neuronal debris from distinct subsets of neurons. *Genes & Development, 28*(1), 20–33.

Tissot, M., & Stocker, R. F. (2000). Metamorphosis in drosophila and other insects: The fate of neurons throughout the stages. *Progress in Neurobiology, 62*(1), 89–111.

Tittel, J. N., & Steller, H. (2000). A comparison of programmed cell death between species. *Genome Biology*, *1*(3). REVIEWS0003.
Togane, Y., Ayukawa, R., Hara, Y., Akagawa, H., Iwabuchi, K., & Tsujimura, H. (2012). Spatio-temporal pattern of programmed cell death in the developing Drosophila optic lobe. *Development, Growth & Differentiation*, *54*(4), 503–518.
Truman, J. W. (1990). Metamorphosis of the central nervous system of Drosophila. *Journal of Neurobiology*, *21*(7), 1072–1084.
Truman, L. A., Ford, C. A., Pasikowska, M., Pound, J. D., Wilkinson, S. J., Dumitriu, I. E., et al. (2008). CX3CL1/fractalkine is released from apoptotic lymphocytes to stimulate macrophage chemotaxis. *Blood*, *112*(13), 5026–5036.
Truman, J. W., Thorn, R. S., & Robinow, S. (1992). Programmed neuronal death in insect development. *Journal of Neurobiology*, *23*(9), 1295–1311.
Tung, T. T., Nagaosa, K., Fujita, Y., Kita, A., Mori, H., Okada, R., et al. (2013). Phosphatidylserine recognition and induction of apoptotic cell clearance by Drosophila engulfment receptor Draper. *Journal of Biochemistry*, *153*(5), 483–491.
Upender, M. B., & Naegele, J. R. (1999). Activation of microglia during developmentally regulated cell death in the cerebral cortex. *Developmental Neuroscience*, *21*(6), 491–505.
van Delft, M. F., Smith, D. P., Lahoud, M. H., Huang, D. C., & Adams, J. M. (2010). Apoptosis and non-inflammatory phagocytosis can be induced by mitochondrial damage without caspases. *Cell Death and Differentiation*, *17*(5), 821–832.
Van Goethem, E., Silva, E. A., Xiao, H., & Franc, N. C. (2012). The Drosophila TRPP cation channel, PKD2 and Dmel/Ced-12 act in genetically distinct pathways during apoptotic cell clearance. *PLoS One*, 7(2), e31488.
Venegas, V., & Zhou, Z. (2007). Two alternative mechanisms that regulate the presentation of apoptotic cell engulfment signal in Caenorhabditis elegans. *Molecular Biology of the Cell*, *18*(8), 3180–3192.
Voll, R. E., Herrmann, M., Roth, E. A., Stach, C., Kalden, J. R., & Girkontaite, I. (1997). Immunosuppressive effects of apoptotic cells. *Nature*, *390*(6658), 350–351.
Wakselman, S., Bechade, C., Roumier, A., Bernard, D., Triller, A., & Bessis, A. (2008). Developmental neuronal death in hippocampus requires the microglial CD11b integrin and DAP12 immunoreceptor. *The Journal of Neuroscience*, *28*(32), 8138–8143.
Wang, X., Wang, J., Gengyo-Ando, K., Gu, L., Sun, C. L., Yang, C., et al. (2007). C. elegans mitochondrial factor WAH-1 promotes phosphatidylserine externalization in apoptotic cells through phospholipid scramblase SCRM-1. *Nature Cell Biology*, *9*(5), 541–549.
Wang, X., Wu, Y. C., Fadok, V. A., Lee, M. C., Gengyo-Ando, K., Cheng, L. C., et al. (2003). Cell corpse engulfment mediated by C. elegans phosphatidylserine receptor through CED-5 and CED-12. *Science*, *302*(5650), 1563–1566.
Watts, R. J., Hoopfer, E. D., & Luo, L. (2003). Axon pruning during Drosophila metamorphosis: Evidence for local degeneration and requirement of the ubiquitin-proteasome system. *Neuron*, *38*(6), 871–885.
Watts, R. J., Schuldiner, O., Perrino, J., Larsen, C., & Luo, L. (2004). Glia engulf degenerating axons during developmental axon pruning. *Current Biology*, *14*(8), 678–684.
Weigert, A., Cremer, S., Schmidt, M. V., von Knethen, A., Angioni, C., Geisslinger, G., et al. (2010). Cleavage of sphingosine kinase 2 by caspase-1 provokes its release from apoptotic cells. *Blood*, *115*(17), 3531–3540.
Williams, D. W., & Truman, J. W. (2005a). Cellular mechanisms of dendrite pruning in Drosophila: Insights from in vivo time-lapse of remodeling dendritic arborizing sensory neurons. *Development*, *132*(16), 3631–3642.
Williams, D. W., & Truman, J. W. (2005b). Remodeling dendrites during insect metamorphosis. *Journal of Neurobiology*, *64*(1), 24–33.

Wong, K., Valdez, P. A., Tan, C., Yeh, S., Hongo, J. A., & Ouyang, W. (2010). Phosphatidylserine receptor Tim-4 is essential for the maintenance of the homeostatic state of resident peritoneal macrophages. *Proceedings of the National Academy of Sciences of the United States of America*, *107*(19), 8712–8717.

Wood, W., Turmaine, M., Weber, R., Camp, V., Maki, R. A., McKercher, S. R., et al. (2000). Mesenchymal cells engulf and clear apoptotic footplate cells in macrophageless PU.1 null mouse embryos. *Development*, *127*(24), 5245–5252.

Wu, H. H., Bellmunt, E., Scheib, J. L., Venegas, V., Burkert, C., Reichardt, L. F., et al. (2009). Glial precursors clear sensory neuron corpses during development via Jedi-1, an engulfment receptor. *Nature Neuroscience*, *12*(12), 1534–1541.

Wu, Y. C., & Horvitz, H. R. (1998a). The C. elegans cell corpse engulfment gene ced-7 encodes a protein similar to ABC transporters. *Cell*, *93*(6), 951–960.

Wu, Y. C., & Horvitz, H. R. (1998b). C. elegans phagocytosis and cell-migration protein CED-5 is similar to human DOCK180. *Nature*, *392*(6675), 501–504.

Wu, Y., Tibrewal, N., & Birge, R. B. (2006). Phosphatidylserine recognition by phagocytes: A view to a kill. *Trends in Cell Biology*, *16*(4), 189–197.

Wynn, T. A., Chawla, A., & Pollard, J. W. (2013). Macrophage biology in development, homeostasis and disease. *Nature*, *496*(7446), 445–455.

Xiao, H., Wang, H., Silva, E. A., Thompson, J., Guillou, A., Yates, J. R., Jr., et al. (2015). The Pallbearer E3 ligase promotes actin remodeling via RAC in efferocytosis by degrading the ribosomal protein S6. *Developmental Cell*, *32*(1), 19–30.

Yu, X., Lu, N., & Zhou, Z. (2008). Phagocytic receptor CED-1 initiates a signaling pathway for degrading engulfed apoptotic cells. *PLoS Biology*, *6*(3), e61.

Yu, X., Odera, S., Chuang, C. H., Lu, N., & Zhou, Z. (2006). C. elegans Dynamin mediates the signaling of phagocytic receptor CED-1 for the engulfment and degradation of apoptotic cells. *Developmental Cell*, *10*(6), 743–757.

Yu, F., & Schuldiner, O. (2014). Axon and dendrite pruning in Drosophila. *Current Opinion in Neurobiology*, *27*, 192–198.

Zhou, Z., Hartwieg, E., & Horvitz, H. R. (2001). CED-1 is a transmembrane receptor that mediates cell corpse engulfment in C. elegans. *Cell*, *104*(1), 43–56.

Zhou, Z., Mangahas, P. M., & Yu, X. (2004). The genetics of hiding the corpse: Engulfment and degradation of apoptotic cells in C. elegans and D. melanogaster. *Current Topics in Developmental Biology*, *63*, 91–143.

Zhou, Z., & Yu, X. (2008). Phagosome maturation during the removal of apoptotic cells: Receptors lead the way. *Trends in Cell Biology*, *18*(10), 474–485.

Ziegenfuss, J. S., Biswas, R., Avery, M. A., Hong, K., Sheehan, A. E., Yeung, Y. G., et al. (2008). Draper-dependent glial phagocytic activity is mediated by Src and Syk family kinase signalling. *Nature*, *453*(7197), 935–939.

Zou, W., Lu, Q., Zhao, D., Li, W., Mapes, J., Xie, Y., et al. (2009). Caenorhabditis elegans myotubularin MTM-1 negatively regulates the engulfment of apoptotic cells. *PLoS Genetics*, *5*(10), e1000679.

CHAPTER TWELVE

The Morphogenetic Role of Apoptosis

Bruno Monier*,†, Magali Suzanne*,†,1
*Université de Toulouse, UPS, LBCMCP, Toulouse, France
†CNRS, LBCMCP, Toulouse, France
[1]Corresponding author: e-mail address: magali.suzanne@univ-tlse3.fr

Contents

Abstract

Beyond safeguarding the organism from cell misbehavior and controlling cell number, apoptosis (or programmed cell death) plays key roles during animal development. In particular, it has long been acknowledged that apoptosis participates in tissue remodeling. Yet, until recently, this contribution to morphogenesis was considered as "passive," consisting simply in the local removal of unnecessary cells leading to a new shape. In recent years, applying live imaging methods to study the dynamics of apoptosis in various contexts has considerably modified our vision, revealing that in fact, dying cells remodel their neighborhood actively. Here, we first focus on the intrinsic cellular properties of apoptotic cells during their dismantling, in particular the role of the cytoskeleton during their characteristic morphological changes. Second, we review the various roles of apoptosis during developmental morphogenetic processes and pinpoint the crucial role of live imaging in revealing new concepts, in particular apoptosis as a generator of mechanical forces to control tissue dynamics.

Current Topics in Developmental Biology, Volume 114
ISSN 0070-2153
http://dx.doi.org/10.1016/bs.ctdb.2015.07.027

1. INTRODUCTION

From its commitment to die to its clearance by phagocytes or neighboring cells, an apoptotic epithelial cell experiences two main phases that are roughly concomitant: execution and extrusion. Execution corresponds to the phase during which caspases coordinate organelle fragmentation, material destruction, and packing in preparation for efficient disposal of cell remnants. During the execution phase, dying cells experience very dramatic changes in shape (Kerr, Wyllie, & Currie, 1972). Execution requires that a cell loses adhesion—both from its neighbors and the substratum—and shrinks. At that stage, the cytoskeleton is particularly active and leads in many cell types to the dynamic blebbing of the plasma membrane. Then, depending on the cell type, the apoptotic cell may, or may not, fragment (Häcker, 2000; Mills, Stone, & Pittman, 1999). Epithelial cells are eventually squeezed out and removed from the epithelium, a process known as extrusion (Rosenblatt, Raff, & Cramer, 2001).

Despite the wealth of knowledge regarding the molecular machinery activated to trigger the dismantling of the cell, very little is known about the cell biology of apoptosis during morphogenetic processes (Suzanne & Steller, 2013). Indeed, although apoptosis has been extensively studied at the cellular level principally in cultured cells, apoptotic cell dynamics and the influence that apoptotic cells have on their surroundings within a tissue are far from being understood. Pioneering studies in whole organisms came from *C. elegans* and showed that apoptosis is under genetic control (Ellis & Horvitz, 1986; Metzstein, Stanfield, & Horvitz, 1998). It is now well established that this molecular program is conserved throughout evolution (Metzstein et al., 1998) and a variety of morphogenetic events have been shown to depend on apoptosis. However, a vast majority of studies focused exclusively on the regulation of apoptosis and on determining the different components involved in the apoptotic pathway. It is only very recently that it has become more and more evident that the function of apoptosis has been underestimated and cannot just be restricted to cell elimination (Suzanne & Steller, 2013; Teng & Toyama, 2011).

Live imaging has been essential in revealing this new perspective. Following the dynamics of different cellular events has been a breakthrough in recent years, and this is particularly true in the case of apoptosis. Indeed, apoptosis is a very dynamic process leading to cell disappearance, hence, particularly difficult to track in fixed tissues. Whether clearance in itself participates to tissue morphogenesis is currently unknown. However, recent reports have linked morphogenetic processes to the extrusion phase (Toyama, Peralta, Wells, Kiehart, & Edwards, 2008) and to a newly defined step during execution (Monier et al., 2015).

In this review, we first summarize the morphological changes occurring at the cellular level in apoptotic cells. Then, in the second part, we review in detail what we know about the link between morphogenesis, extrusion, and execution.

2. PART I: CYTOSKELETAL CONTROL OF APOPTOTIC CELL DYNAMICS

The known or suspected morphogenetic properties of apoptotic cells are a consequence of the physical modifications that apoptotic cells experience during their remodeling. Therefore, we review in this first part what is currently known about the cytoskeletal control of apoptotic cell shape changes, first from the studies of cells in culture, and second in more integrated contexts.

2.1 Cytoskeletal Control of Cell Shape Remodeling in Individual Apoptotic Cells

Cell shape remodeling during execution relies on an active and dynamic cytoskeleton. For simplicity, the various phases of execution have been essentially studied on individual cells in culture. The actomyosin cytoskeleton has clearly been established as a main player during apoptosis (Coleman & Olson, 2002; Mills et al., 1999) while intermediate filaments and microtubules were long-thought to be dismantled early in the execution phase (Byun et al., 2001; Caulín, Salvesen, & Oshima, 1997; Mills, Lee, & Pittman, 1998), possibly to help prevent resistance during apoptotic cell shape changes. Indeed, cytoskeletal components as well as some of their key regulators are targets of caspases (Fischer, Jänicke, & Schulze-Osthoff, 2003). However, we now know that caspase cleavage does not always lead to protein inactivation; it can produce peptide fragments with new activity (see below). More recent evidence shows that, at least for microtubules, a network reforms late during execution (Moss & Lane, 2006).

The first obvious role of the actomyosin cytoskeleton during the apoptotic execution phase concerns blebbing, the dynamic bulging, and retraction of the plasma membrane. Blebbing, which is also observed for example during migration or division (Fackler & Grosse, 2008; Sedzinski et al., 2011), is caused by locally increased actomyosin cortical tension or by cell internal hydrostatic pressure at local sites where interactions between the actin cytoskeleton and the plasma membrane are weakened (Charras, Yarrow, Horton, Mahadevan, & Mitchison, 2005; Paluch, Sykes, Prost, & Bornens, 2006). Apoptotic blebbing can be separated into two sequential steps (Moss & Lane, 2006): the first one is highly dynamic and looks similar to what has been described in healthy cells (although it is not found in all apoptotic cell types) and in the second one, nuclear fragments and ER accumulate in late-forming blebs that likely prefigure apoptotic bodies (Lane, Allan, & Woodman, 2005).

Early blebbing relies upon F-actin cortical structures and activation of nonmuscle myosin II through phosphorylation (Mills et al., 1998). Upstream of myosin II and necessary for apoptotic blebbing (Coleman & Olson, 2002; Coleman et al., 2001; Sebbagh et al., 2001) is the kinase ROCK, usually acting downstream of Rho-GTPases in order to enhance myosin II contractility (Amano et al., 1996; Kawano et al., 1999). Importantly, in mammals, ROCK I (but not ROCK II) is cleaved by Caspase 3 during apoptosis, leading to the removal of the ROCK C-terminal, autoinhibitory region. This releases a constitutively active form of the kinase that is sufficient to induce blebbing when expressed in nonapoptotic cells (Coleman & Olson, 2002; Coleman et al., 2001; Sebbagh et al., 2001). Subtleties have been observed concerning the cytoskeletal machinery involved in the late-blebbing phase, possibly due to differences in cell type (Moss & Lane, 2006). Both actomyosin and microtubules have been reported to be necessary for the presence of late, chromatin-rich blebs (Bonanno et al., 2000; Croft et al., 2005), with microtubules rebuilding a "late" cortical array under the plasma membrane. However, depending on the cell line, drugs targeting either the F-actin/ROCK/myosin II or microtubule networks impinge on late bleb formation (Lane et al., 2005; Moss, Betin, Malesinski, & Lane, 2006; Moss & Lane, 2006; Sánchez-Alcázar et al., 2007). Because brief depolymerization of microtubules with nocodazole is sufficient to cause a relocalization of condensed chromatin fragments to the cell center, it was hypothesized that microtubules might in some way resist to retractile pressures exerted on late blebs (Moss et al., 2006). Although some clarifications are required, a framework is now established to fully decipher the respective contribution of actin and microtubule cytoskeletons during the whole apoptotic blebbing process.

The dismantlement of a cell during apoptosis is tightly regulated and follows a stereotypical sequence of events that culminates with fragmentation of the cell into apoptotic bodies. Apoptotic cell fragmentation has long been known to require actin (Cotter, Lennon, Glynn, & Green, 1992) while the late apoptotic microtubule network is necessary to prevent the too-early cleavage of various caspase targets such as E-Cadherin or FAK, possibly by creating a "caspase-free" zone at the cortex in order to avoid the premature disruption of the cell (Oropesa-Ávila et al., 2013). Interestingly, the Pannexin 1 (PANX1) plasma membrane channel is also required to prevent a too-early destruction of apoptotic cells. PANX1, upon activation by caspase-dependent cleavage, selectively transports biomolecules across the membrane, especially ATP as a "find-me" signal to attract phagocytes (Chekeni et al., 2010). Surprisingly, PANX1 inhibition leads to enhanced formation of apoptotic bodies, while overexpression of the wild-type molecule decreases their formation (Poon et al., 2014). PANX1 inhibition also allowed the discovery of "apoptopodia," a string-like structure linking forming apoptotic bodies to the main cell body. Blocking apoptotic body formation by inhibiting ROCK or depolymerizing F-actin in a PANX1 negative context does not affect apoptopodia. Altogether, these results show that PANX1 regulates fragmentation by preventing the formation of too many apoptotic bodies as well as apoptopodia structures (Poon et al., 2014). Possibly, restraining cellular disassembly by PANX1 activity is necessary to ensure that everything is correctly packed before the very end.

This work opens several questions. In particular, what is the nature of these apoptopodia and how does PANX1 inhibit their formation at the molecular level? Since actin is not involved, and intermediate filaments are supposedly dismantled, it is tempting to speculate that apoptopodia are microtubule-dependent structures. If so, PANX1 could restrain the growth or activity of those microtubules proposed to push organelle fragments to the cell periphery during the late-blebbing phase (Moss et al., 2006). This hypothesis is consistent with the fact that the late apoptotic microtubule network correlates with high-ATP cellular levels (Oropesa et al., 2011). This point needs further investigation, but it suggests that PANX1 activity counteracts an inward pushing force during the late stages of the execution phase. Interestingly, during cell division, cells round up and the cell cortex is remodeled, becoming stiffer (Kunda, Pelling, Liu, & Baum, 2008). Such modifications of the physical properties of mitotic cells, independently of their ability to divide, are critical for morphogenesis: in *Drosophila* tracheal pits (i.e., the epithelial placodes from which the fruit fly trachea derive), the rounding of mitotic cells accelerates local invagination of the epithelium (Kondo & Hayashi, 2013). Given the cytoskeletal dynamics in the late apoptotic cell, it would be interesting to investigate the physical properties of dying cells, as these could be important during morphogenetic processes, as exemplified by the role of mitotic cell rounding during tracheal pit invagination.

2.2 Apoptotic Cell Dynamics Within Epithelial Monolayers

The different execution steps described above are focused on individual apoptotic cells from essentially culture cell experiments. Thus, the cells are considered independently of their surroundings. However, in an organism, dying cells have to deal with the surrounding tissue, and vice versa. A number of studies have characterized the interaction of apoptotic cells with their direct neighbors during the extrusion process. Extrusion (i.e. the removal of dying cells from an epithelium) implicates the squeezing out of the dying cell by the coordinated contraction of the surrounding epithelium and the complete loss of adhesion to release the dead cell remnants (Gu & Rosenblatt, 2012; Rosenblatt et al., 2001). Dying cell expulsion is generally done apically in vertebrate epithelial monolayers, although it can be directed toward the basal side, as is the case in *Drosophila* (Gu & Rosenblatt, 2012; Marshall, Lloyd, Delalande, Näthke, & Rosenblatt, 2011; Slattum, McGee, & Rosenblatt, 2009).

In their pioneering study, Rosenblatt et al. (2001) reported that the apoptotic cell surface shrinks, with the apical surface of neighboring live cells filling the space, leading to a transient "rosette" formation. Eventually, the apoptotic cell is found out of the plane of the epithelium. Although the dynamics of the process were not precisely characterized, they found that an actomyosin ring assembles at the interface between MDCK UV-induced apoptotic cells and their live neighbors. Analysis of mosaic cell lines showed that both dying and live neighboring cells contribute to the formation of the apoptotic actin ring. However, expulsion *per se* is mediated only by the neighboring living cells, the cytoskeletal activity of the apoptotic cell being unnecessary for its own removal from the monolayer (Rosenblatt et al., 2001).

Interestingly, it has been proposed that apoptotic cells send a secreted signal to their neighbors to induce apoptotic ring formation, and subsequent expulsion. Sphingosine-1-phosphate has been reported as such an apoptotic signal (Gu, Forostyan, Sabbadini, & Rosenblatt, 2011). Yet, cell wounding using laser ablation (i.e., physical injury of the cell) also leads to the formation of an actomyosin ring in neighboring cells (Tamada, Perez, Nelson, &

Sheetz, 2007), supporting the hypothesis that additional mechanisms, such as mechanical tension, may also trigger actomyosin ring formation.

Two recent studies have reinvestigated epithelial cell extrusion by combining induction of apoptosis at the single-cell level and time-lapse microscopy to follow apoptotic and nonapoptotic cell dynamics, thus refining the overall picture of the extrusion process. Detailed analysis of F-actin dynamics in the dying cell or its neighbors, thanks to mosaic expression of actin reporters, revealed that apical surface shrinkage of the dying cell is accompanied by the formation of an actin ring inside the dying cell itself, but not in its neighbors. The formation of an actin ring within living cells is only seen later, during the expulsion stage. This suggests that, in a first step, living cells do not compress the apical surface of the dying cell but rather that the dying cell autonomously contracts its apex, pulling on its neighbors (Kuipers et al., 2014). Consistently, blocking Rho signaling in living neighbors does not impinge upon apoptotic cell apical shrinkage, while Rho inhibition within the dying cell does. While it looks like a discrepancy at first glance, studies by Rosenblatt and colleagues and Kuipers and colleagues do not focus on the same stage during extrusion: the first report focuses on whether the dying cell is eventually expelled from the monolayer (Rosenblatt et al., 2001), while the second report specifically considers the dynamics of the apoptotic cell before expulsion (Kuipers et al., 2014).

Consistent with these results, loss of mechanical coupling between the dying cell and its neighbors prevents the typical rosette formation with living cells all stretching toward the site of dying cell extrusion (Lubkov & Bar-Sagi, 2014). Introducing an E-Cadherin/Alpha-Catenin fusion within an E-Cadherin devoid cell line is sufficient to trigger rosette formation at the vicinity of apoptotic cells (Lubkov & Bar-Sagi, 2014). Moreover, inducing cell shape changes associated with apoptosis but without inducing cell death, a trick that can be achieved through expression of a deleted version of ROCK II (Sebbagh, Hamelin, Bertoglio, Solary, & Bréard, 2005), is sufficient to trigger rosette formation and eventually cell extrusion (Lubkov & Bar-Sagi, 2014).

Those results are consistent with a two-step model whereby the dying cell first actively constricts its apical surface (Kuipers et al., 2014) and then triggers the formation of an actomyosin ring in its living neighbors that is believed to move basally to expel the dead cell apically (Fig. 1). Apoptotic cell apical constriction appears inessential for neighbors to proceed to final expulsion (exemplified by normal extrusion of apoptotic cells in which myosin II is specifically inhibited during *Drosophila* wing disc development

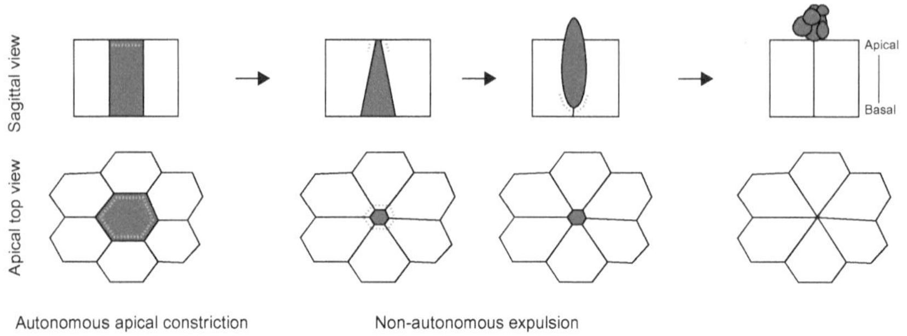

Figure 1 *A two-step mechanism driving epithelial cell extrusion.* Apoptotic cell (red (dark gray in the print version)) extrusion from an epithelium is first caused by the formation of a Rho-dependent apical actomyosin ring (green (gray in the print version) dashed line) within the apoptotic cell which leads to the constriction of its apical surface. Second, a Rho-dependent actomyosin ring forms within living adjacent neighbors. This second actomyosin ring moves basally to expel the dying cell apically (see Gu & Rosenblatt, 2012 for an alternative mechanism driving basal cell expulsion). The apoptotic cell is eventually dismantled and cleared.

(Monier et al., 2015)). Importantly though, during extrusion, the dying cell both produces a force to shrink its apical surface (Kuipers et al., 2014) and is mechanically coupled to its living neighbors (Lubkov & Bar-Sagi, 2014). Thus, extrusion is a critical period during which an apoptotic cell can actively influence its environment through the generation of a mechanical force in the apical plane of the epithelium, which can play a key role during morphogenesis (see the role of apoptotic cell extrusion in dorsal closure discussed below).

2.3 A New Step During the Early Execution Phase Is Responsible of Apoptotic Force Generation

New tools now allow a detailed study of cell dynamics during apoptosis, as well as the behavior of surrounding living cells in physiological conditions. Such a strategy has recently been applied to study apoptosis during the development of the *Drosophila* leg (Monier et al., 2015). To begin, all the main stages of apoptosis were reported during this process, including cell shrinkage, loss of adhesion, blebbing, and fragmentation (Fig. 2). Consistent with previous work (Lubkov & Bar-Sagi, 2014), the first morphological event observed was apical surface shrinkage, without loss of adhesion. Interestingly, a peak-like structure made of adhesion molecules E-Cadherin, alpha-, and beta-catenin then forms within the dying cell. This structure is still linked to the remaining network of adherens junctions of living cells

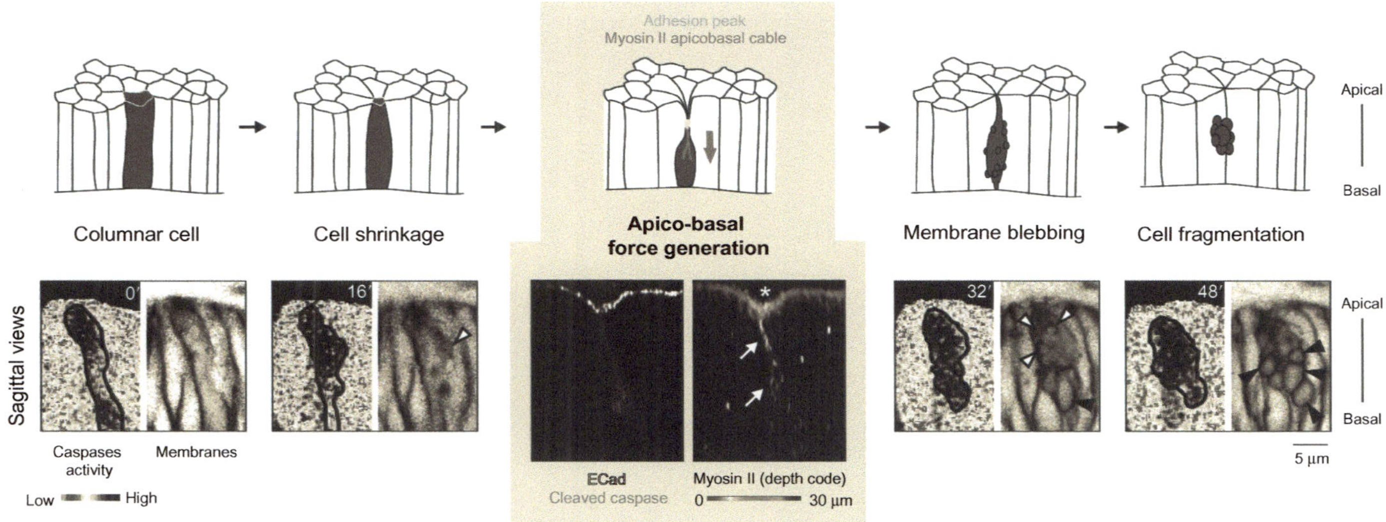

Figure 2 *A new step during the early execution phase generates an apico-basal force within Drosophila epithelia*. Schematization (top) and corresponding confocal sagittal views (bottom) illustrating the successive phases of the execution phase in *Drosophila* epithelia. The new step generating an apico-basal force during apoptosis is boxed in gray. Bottom images either represent successive steps of a movie of an apoptotic cell in the pupal leg where caspase activity is monitored using a FRET probe and membranes using a lipid-binding dye or an apoptotic cell in the pupal wing costained for E-Cadherin (white), active caspases (magenta), and myosin II (red to blue color code indicating depth). White and black arrowheads point at blebs and apoptotic bodies, respectively. White arrows point at the myosin II cable (note the deformation of the apical surface of surrounding cells at that stage, asterisk). *Figure adapted from Monier et al. (2015).* (See the color plate.)

and is concomitant with an apical deformation of the surrounding epithelium. As extrusion occurs basally in the fruit fly, these observations could either have fitted with living cells pushing the dying cell to extrude it basally or, alternatively, with the presence of an active apico-basal pulling mechanism inside the dying cell. To discriminate between these possibilities, the dynamics of myosin II were characterized. Surprisingly, a transient apico-basal myosin II cable-like structure forms at the time the adhesion peak is observed. This actomyosin cable emanates from this adherens junction structure. Upon contraction, the myosin II apico-basal cable pulls the apical surface of the surrounding cells basally (Fig. 2). A few minutes later, the cell and the myosin II structure fragment, the adhesion peak detaches and the apical surface of the epithelium is released. Ectopic induction of cell death in the wing disc recapitulates apical deformation of the surrounding epithelium, but only when myosin II is active. Indeed, expressing a dominant negative form of myosin II specifically in ectopic apoptotic cells is sufficient to abrogate ectopic deformation of the wing disc surface.

The apico-basal myosin II cable appears to be a general characteristic of apoptotic cells from columnar epithelia in *Drosophila*: they are also observed in other tissues such as the embryonic ectoderm as well as antenna and wing imaginal discs. At least in the wing, cell death is sparse and does not play a morphogenetic role. Importantly, these myosin II apico-basal structures are no longer observed in the developing leg when caspase activity is blocked, while ectopic expression of the proapoptotic gene *reaper* in the wing leads to ectopic apico-basal myosin II cable formation. Thus, in the fly, the apico-basal actomyosin cable-like structure appears as a general component of the apoptotic machinery that forms early during the execution phase, producing a mechanical force (Monier et al., 2015).

Numerous questions still remain unanswered. For example, it is currently unknown whether myosin II apico-basal enrichment corresponds to a cable forming within the cytoplasm or to an increase at the level of the lateral cortex. Interestingly, Sherrard and colleagues reported that in ascidians, the apico-basal shortening of invaginating endodermal cells correlates with a myosin II enrichment at the lateral cortex (Sherrard, Robin, Lemaire, & Munro, 2010). Because the apoptotic cell adopts a "pear-like" shape, with the lateral sides of the cell becoming closer on its apical domain, this could lead to a strong cortical accumulation of myosin II looking like an intracellular cable. Furthermore, a complete picture of this apico-basal structure awaits identification of the basal anchoring point. Indeed, for the contracting myosin II cable to produce a force, a basal or lateral anchoring structure is necessary to provide either resistance or even traction. Further studies are therefore necessary to clarify the exact nature of the apoptotic myosin II cable and gain better understanding of the apoptotic force-producing molecular machinery.

Another intriguing question is about the role of this apico-basal actomyosin structure. Because it is found in both morphogenetic and non-morphogenetic apoptotic events, it suggests a fundamental role in the apoptotic process. It is not likely to be required for cell extrusion, as final expulsion from the epithelium is caused by neighboring, living cells (Rosenblatt et al., 2001). In good agreement, inactivation of myosin II specifically in dying cells does not impair their extrusion in imaginal fly tissues (Monier et al., 2015). However, the structure could participate in organelle repositioning during early apoptotic phases, or in organelle and cell fragmentation. Organelle dismantlement during execution involves caspase-dependent cleavage of key components but also a contractile actomyosin cytoskeleton as observed during nucleus or Golgi apparatus fragmentation (Croft et al., 2005; Orlando & Pittman, 2006). For instance, the Lamins, which form the skeleton underlying the nuclear envelope, are caspases targets (Buendia, Santa-Maria, & Courvalin, 1999; Kihlmark, Imreh, & Hallberg, 2001; Rao, Perez, & White, 1996) that have to be cleaved to allow active ROCK-induced nuclear fragmentation (Croft et al., 2005). Drugs inhibiting ROCK or myosin II contractility block nuclear fragmentation while overexpressing an active version of ROCK is sufficient to destroy the nucleus in some circumstances (Croft et al., 2005). Yet, how the actomyosin cytoskeleton is organized for nucleus fragmentation is unknown. Closer investigation of the apico-basal myosin II cable dynamics in fly epithelia might provide valuable information regarding organelle fragmentation and the potential implication of tension in orchestrating this fragmentation.

3. PART II: INFLUENCE OF APOPTOTIC CELLS ON THEIR SURROUNDINGS

In this part of the review, we retrace how our point of view concerning the role of apoptosis has evolved in the morphogenetic field, from a very passive role of apoptosis as a cell remover mechanism to a much more active role as a nonautonomous actor impacting tissue remodeling.

1. The "carving" model or apoptosis viewed as a passive mechanism

One classical example of apoptosis-dependent morphogenesis is the formation of digits in vertebrates (Fig. 3). The embryonic limb is composed of

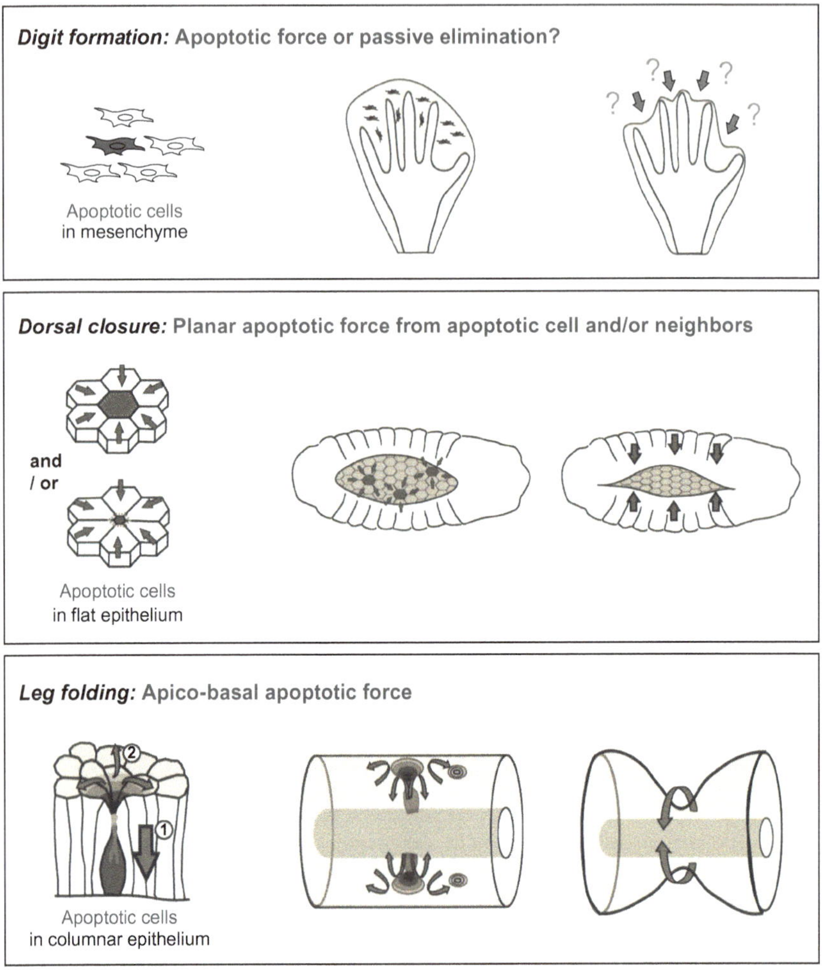

Figure 3 *Consequence of the apoptotic force on tissue remodeling.* Schematization of the consequence of apoptosis on three different morphogenetic processes: digit formation (left), dorsal closure (center), and leg folding (right). The type of cells undergoing apoptosis together with the force exerted by apoptotic cells are schematized in the upper part, the whole tissue is schematized in the lower part. In digit formation, the participation of an apoptotic force to interdigital tissue regression (green arrows) has been proposed but not demonstrated. During dorsal closure, an apoptosis extrusion force accelerates the migration of the lateral epithelium (green arrows); however, the participation of apoptotic cells versus neighbors has not been tested yet. During leg folding, apoptotic cells generate an apico-basal force (1) transmitted to the rest of the epithelium through an increase in tissue tension (2) which leads to tissue constriction (the different forces are represented by green arrows). (See the color plate.)

mesodermal cells covered by an ectodermal layer forming a bud on the lateral surface of the embryonic body. The mesodermal cells of the bud either differentiate in developing cartilage or undergo apoptosis freeing the digits from the limb plate. Cell death occurs in well-defined regions: between the digits, in the most anterior and posterior parts of the bud and in the zone of joint formation. Interestingly, the extent of cell death varies among species, correlating with the final morphology of the digits (webbed or free) (Montero & Hurlé, 2010).

A well-known regulator of interdigital cell death in the developing limb is retinoic acid (RA) which is dependent on upregulation of BMPs. However, digit individualization relies on cell death and regression of the interdigital tissue, as well as differential growth of digits and interdigital domains (Hernández-Martínez, Castro-Obregón, & Covarrubias, 2009; Salas-Vidal, Valencia, & Covarrubias, 2001). Several members of the Fgf family play a role in limb development as survival factors, yet Fgf8 appears to be the most important factor controlling limb growth; downregulation of Fgf8 in the interdigital ectoderm is essential to trigger interdigital cell death (Hernández-Martínez et al., 2009). Thus, RA and Fgf8 are antagonistic regulators of interdigital cell death. Furthermore, the protection level provided by Fgf8 is RA dose dependent suggesting that a balance between Fgf8 in the ectoderm promoting growth and RA in the underlying mesenchyme promoting cell death is required to allow the correct differentiation of digits (Hernández-Martínez et al., 2009).

Differential growth of digital and interdigital regions could suggest that both cell death and growth could be interconnected. Interestingly, it has been proposed that FGFs could have a dual role in the overlying ectoderm as survival signal on the one hand and as "apoptotic signal sensitizers" on the other (Montero & Hurlé, 2010). Indeed, applying specific FGF-inhibitor is sufficient to inhibit interdigital cell death, indicating that in the absence of FGF, BMPs are not sufficient to trigger apoptosis (Montero et al., 2001). Hence, growth signal could control apoptosis in the developing limb. However, interdigital cell death is not likely to influence interdigital growth since decreasing cell death in the interdigital regions leads to normal digit growth and partial digit individualization in adults (equivalent to the developmental stage S11) (Salas-Vidal et al., 2001).

Digit shaping also involves extracellular matrix (ECM) remodeling which may involve noncaspase proteases such as matrix metalloproteinases (MMPs), the principal proteases involved in matrix remodeling (Hernández-Martínez & Covarrubias, 2011). Moreover, it has been shown that regulation of matrix remodeling by FGF and BMP signaling precedes and induces cell death by anoikis (Díaz-Mendoza, Lorda-Diez, Montero, García-Porrero, & Hurlé, 2013).

In these studies, the impact of cell death on the surrounding tissue and the active role of apoptosis in tissue remodeling were not yet considered. The model predicts that tissue remodeling by apoptosis relies exclusively on the elimination of unwanted cells with no impact on the surroundings, hence the establishment of "the carving model" in which apoptosis acts like a stone sculptor, chipping off small fragments to give rise to a new shape. Although the theory of an apoptotic force involved in interdigital regression was first proposed in 2001 (Stenn & Paus, 2001), it has not been supported by any data. Further studies will be needed to determine the potential influence of cell death on tissue remodeling in this system, including matrix reorganization and cell shape changes (Fig. 3).

Another example of apoptosis-dependent morphogenesis is the formation of grooves in the *Drosophila* embryo. In *Drosophila*, the embryo is segmented along its anteroposterior axis; segmental grooves formed by apical constriction physically separate the different segments (Bate & Martinez-Arias, 1993; Larsen, Hirst, Alexandre, & Vincent, 2003). The apoptotic pattern is dynamic in the embryo (Abrams, White, Fessler, & Steller, 1993) but extensive cell death, coinciding with a strong expression of the proapoptotic gene *reaper* (*rpr*), has been observed at the maxillary segment boundaries (Lohmann, McGinnis, Bodmer, & McGinnis, 2002). In *Deformed* (*Dfd*) mutants, a strong reduction of cell death and the disappearance of *rpr* expression at the maxillary/mandibular boundary is observed showing that *Dfd* regulates cell death through the direct regulation of *rpr* expression in the embryo. The essential role of cell death in this system was elegantly shown by rescue of the *Dfd* mutant phenotype (fusion of maxillary and mandibulary) by the induction of *rpr* expression at the border. Finally, a similar role to the one of *Dfd* was shown for another homeotic gene, *Abdominal-B* that appears to regulate the boundaries separating A6, A7, and A8 segments through the regulation of *rpr* (Lohmann et al., 2002).

Based on anterior studies on digit formation, it was proposed that apoptosis could be carving grooves separating embryonic segments the same way it carves digits, by simple elimination of cells in excess (Alonso, 2002; Lohmann et al., 2002). However, the embryonic epidermis is composed of a monolayer epithelium. As described above, the elimination of a subset of cells from a monolayer epithelium is insufficient to explain tissue shape modification and fold formation. Indeed, to maintain the integrity of the epithelium during apoptotic cell elimination, apoptotic cell neighbors extend to fill the space left by apoptotic cell disappearance. This alone cannot explain the mechanism of groove formation. For a fold to be formed, the cells that remain to form the groove have to change their shape not only in the plane of the epithelium but also in depth. Thus, further analysis will be required in this system to characterize the cellular rearrangements occurring in response to local apoptosis.

2. The first clues about the influence of apoptosis on the surrounding tissue

The first clues about the influence of apoptotic cells on their direct surroundings came from studies on palate formation. The formation of the secondary palate in mammals consists in the fusion of two lateral shelves, composed of two epithelial layers, the basal layer and the periderm, that surround mesenchymal tissue. These epithelial layers first grow vertically on each side of the tongue, then elevate in the horizontal position above the tongue and establish contact. After contact, adhesion between shelves progresses to form the midline epithelial seam (MES). The MES is then progressively eliminated, forming transient epithelial islands which end up disappearing; this results in shelf fusion and separation of the nasal and oral cavities. Finally, when fusion of the two shelves is completed, the secondary palate fuses with the primary palate. Failure of these processes results in cleft palate, one of the most common congenital anomalies in humans.

Different cellular processes have been proposed to be involved in palate fusion in addition to apoptosis, including migration and epithelio-mesenchymal transition (although the role of epithelio-mesenchymal transition is controversial (Iseki, 2011)). The role of apoptosis is critical in shelf fusion, as reducing apoptosis results in unfused palate shelves and palate cleft. Initial studies principally focused on the regulation of apoptosis during this process, showing that apoptosis is regulated by RA and occurs specifically in epithelial cells only after shelf contact (Cuervo, Valencia, Chandraratna, & Covarrubias, 2002).

The consequence of apoptosis on tissue remodeling was considered for the first time with the interesting observation that basal lamina degradation occurs concomitantly to cell death in the medial edge epithelia (which can be viewed as the leading edge of the two lateral shelves). It was first hypothesized that basal lamina degradation induces apoptosis due to loss of cell adhesion, as observed in other tissues such as the amphibian intestine (Ishizuya-Oka & Shi, 2011), a process known as anoikis. However, blocking basal lamina degradation by inhibiting MMPs has no effect on cell death, although it inhibits shelf fusion. Thus, basal lamina degradation appears to be a secondary event activated by the dying cells through the production of MMPs. Consistently, no basal lamina degradation was observed when cell death was inhibited. This new process corresponding to a degradation of basal lamina in response to apoptosis has been called "cataptosis" (Cuervo & Covarrubias, 2004).

It would be interesting now to test if this process of cataptosis is involved in other apoptosis-dependent morphogenetic processes such as the formation of the cardiac outflow tract in vertebrates or genitalia rotation in *Drosophila.* During cardiac morphogenesis, apoptosis is required for the shortening and remodeling of the cardiac outflow tract and the inhibition of apoptosis leads to vascular defects and possibly to arrhythmia in the adult. Interestingly, it has been observed that apoptosis coincides with ECM remodeling but nothing is known about the relationship between ECM and cell death in this system (Rothenberg, Fisher, & Watanabe, 2003). During genitalia rotation, apoptosis is required for the initiation of tissue movement, allowing a huge reorganization of the tissue (Kuranaga et al., 2011; Suzanne et al., 2010). Genitalia rotation occurs during male development and consists of a 360° rotation of the genital plate, divided as two 180° movements of two concentric rings. For both movements, apoptosis has been shown to occur at the boundary of the rotating tissue and is necessary to release the rotating part from the rest of the tissue. These works reveal a new role for apoptosis as a way to give more fluidity to the tissue, necessary to undergo considerable remodeling. Although adhesion dynamics have not been characterized in this model, it is tempting to speculate that apoptosis weakens cell adhesion to free the rotating part, perhaps by locally influencing the ECM.

3. The revelation of an active participation of apoptosis during morphogenesis using live imaging

Although the pioneering studies described above had clearly established the essential role of apoptosis in developmental processes, the precise cellular mechanisms underlying the dynamic rearrangement of tissues in response to cell death were far from being understood, in particular, the consequences of apoptosis on neighboring cells. The use of live imaging to follow the development of different structures and organs has been an important step to identify the outcome of cell death, revealing an active impact on cell rearrangement and morphogenesis.

3.1 Dorsal Closure

The first evidence supporting the idea that apoptotic cell dynamics influence neighboring cells came from the study of dorsal closure in *Drosophila* (Toyama et al., 2008), which consists in the migration and fusion of two lateral epithelial sheets that progressively cover the entire dorsal opening of the embryo. Before migration, the dorsal most part of the embryo is transiently covered by the amnioserosa, an extraembryonic tissue destined to degenerate at the end of the process. The most dorsal row of migrating cells, the leading edge, forms an actomyosin cable that has been proposed to participate in dorsal closure by generating a purse-string force on the more lateral cells (Rodriguez-Diaz et al., 2008).

The presence of apoptosis in the amnioserosa has been long known (Abrams et al., 1993); however, it is only recently that the proportion of apoptotic cells (Kiehart, Galbraith, Edwards, Rickoll, & Montague, 2000) and their participation (Toyama et al., 2008) in dorsal closure has been clearly established. Around 10% of amnioserosa cells die by apoptosis, showing the characteristic hallmarks of apoptotic cells including shrinkage, blebbing, and fragmentation (Kiehart et al., 2000). These cells have the particularity to form a very thin epithelial layer or "squamous epithelium." Thus, although apoptotic cells do not represent a huge proportion of the whole tissue, the disappearance of a few apoptotic cells is sufficient to induce a very strong reorganization of their direct and indirect neighbors. Indeed, five to seven cells that neighbor each apoptotic cell are distorted and elongate toward the disappearing apoptotic cell to maintain a continuous epithelial sheet, adopting a rosette geometry once the apoptotic cell delaminates, similarly to what has been described in vertebrate epithelial monolayer (Lubkov & Bar-Sagi, 2014; Rosenblatt et al., 2001). Distortion propagates to the following row of neighbors (Toyama et al., 2008). Thus, a vast majority of amnioserosa cells are influenced by the apoptotic process.

Using genetic means, either by reducing apoptosis or increasing apoptosis specifically in the amnioserosa, the authors further show that apoptosis affects the speed of closure, thus indicating that cell death in the amnioserosa influences the migrating properties of the neighboring lateral epithelial tissue (Toyama et al., 2008). To compare the different mutant backgrounds, both the rate of closure and the force produced by the amnioserosa on the leading edge were quantified. The results indicate that about one third of the force produced by the amnioserosa on the leading edge is attributable to apoptosis. The authors thus concluded that the delamination of apoptotic cells from the amnioserosa generates a force that both facilitates cell extrusion and promotes dorsal closure. In parallel to this work, cytoskeleton dynamics in the amnioserosa were studied in detail and it was shown that cells from this tissue undergo apical constriction during dorsal closure based on the assembly/disassembly of an actomyosin apical meshwork that drives the gradual reduction of the cell apical surface, as the leading edge progresses toward the midline (Gorfinkiel & Blanchard, 2011). Thus, in addition to the force generated by the leading edge actomyosin cable, the amnioserosa is thought to generate two kinds of forces, one due to apical constriction and the other due to apoptosis.

This represents the first example of a developmental process in which apoptosis has been identified as generating a force in a tissue (Fig. 3). It has been proposed that the force generated by apoptotic cell extrusion is due to the actomyosin ring that forms around apoptotic cells. The actomyosin ring is thought to induce a pushing force favoring apoptotic cell extrusion, as well as a pulling force on the lateral epithelium contributing to dorsal closure. However, the contribution of ring contraction in neighbors versus apoptotic cell constriction in generating the extrusion force has not been evaluated yet. Thus, in the light of more recent work on cell extrusion (Kuipers et al., 2014), it would be interesting to ask whether apoptotic cells themselves participate actively to dorsal closure through cell autonomous apical constriction.

These data constitute the first demonstration that the apoptotic process can represent the initiator of a force propagated to the surrounding tissue. Taking this into consideration, it would now be interesting to revisit the work on interdigital regression and hair follicle morphogenesis in which this mechanism was initially proposed (Stenn & Paus, 2001), although never demonstrated (Teng & Toyama, 2011).

3.2 Neural Closure

The first ever live imaging of apoptosis in mammalian tissue has recently been performed during mouse neural closure. Although no clues on the cellular mechanisms have been identified, we summarize here what is known on this model system and discuss it in the view of other recent studies focusing on cell rearrangement dynamics during neural closure and epithelial apoptotic cell dynamics in leg folding.

The first step of neural tube closure in mammals is the bending of the neural plate toward the midline and of two lateral regions that form the dorsal ridges. The dorsal ridges then meet dorsally and fuse to form the neural tube. In humans, defects in neural tube closure lead to lethal defects such as anencephaly and *spina bifida* (Yamaguchi & Miura, 2013; Yamaguchi et al., 2011). Although pharmacological inhibition of apoptosis was found to prevent neural tube closure in chick and mouse, the role of apoptosis has been controversial (Massa et al., 2009; Yamaguchi et al., 2011). Indeed, in Apaf1 and casp3 mutants, neural closure is not totally blocked. Still, clear defects of closure are associated with strong reduction of apoptosis by genetic means (Massa et al., 2009; Yamaguchi et al., 2011). Live imaging analysis revealed that apoptosis occurs principally in the dorsal ridge and in the most dorsal region of the neural tube, at the boundary domain between the neural plate and the ectoderm (Yamaguchi et al., 2011). Interestingly, this pattern is lost in *casp3* and *Apaf1* homozygote mutants and coincides with a strong reduction of neural tube bending at the dorsal ridges and a delay in neural closure in the cranial region. In spite of these pioneering studies following the dynamics of apoptosis on living mouse embryos, no cellular mechanism explaining the bending and closure speed defects has been uncovered yet. Interestingly, two parallel studies focusing on cytoskeleton dynamics have shown that apical constriction is essential for neural tube bending (Haigo, Hildebrand, Harland, & Wallingford, 2003; Nishimura, Honda, & Takeichi, 2012). In *Xenopus*, apical constriction occurs at the midline and at the dorsal ridges and relies on the actin-binding protein Shroom (Haigo et al., 2003); in chicken, the PCP-regulating cadherin Celsr1 organizes the polarized distribution of RhoGEF and ROCK I leading to the polarized distribution and contraction of apical actomyosin, cell rearrangement, and midline bending (Nishimura et al., 2012). Further studies are needed to determine how modeling of the closing neural tube is achieved through apoptosis and to test the connection between apoptosis and actomyosin dynamics in this system. Recent data demonstrates that apoptosis actively participates in tissue bending in *Drosophila*. Thus, we propose a similar mechanism (see below), whereby apoptotic cells from the neural tube would generate an intracellular force that increases tissue tension, eventually leading to tube bending.

3.3 Leg Folding

Altogether, the previous observations suggest that apoptosis can generate a major sculpting force during morphogenesis. However, they tell us very little about how apoptosis might contribute to tissue morphogenesis cell autonomously. Following apoptotic cell dynamics by live imaging in the model of leg joint formation in *Drosophila*, we have recently shown that apoptotic cells directly participate to tissue remodeling, independently of their extrusion phase (Monier et al., 2015). Leg joint formation relies on epithelium folding, a key step in morphogenesis that was known so far to be either driven by contraction of a medioapical actomyosin meshwork, or adherens junction repositioning, or accelerated by cell rounding during mitosis (Kondo & Hayashi, 2013; Martin, Kaschube, & Wieschaus, 2009; Wang, Khan, Kaschube, & Wieschaus, 2012). However, our work revealed a novel cellular mechanism of epithelial folding that relies on a nonautonomous, apoptosis-dependent mechanism (Monier et al., 2015).

The *Drosophila* leg develops from a group of cells initially forming a flat monolayer epithelium called the leg imaginal disc. This group of cells progressively differentiates and a series of concentric folds are formed in a stereotyped order (Rauskolb & Irvine, 1999). These folds correspond to the future joint regions separating the different leg segments in the adult. Apoptosis plays an important role in fold formation in the most distal part of the leg. Indeed, if apoptosis is reduced or blocked, these folds are not properly formed, leading in the most extreme cases to absence of fold formation and absence of joints in the adult. The proapoptotic genes *reaper* (*rpr*) and *hid* are expressed in a ring of cells in each future fold in which apoptosis occurs specifically (Guarner et al., 2014; Manjón, Sánchez-Herrero, & Suzanne, 2007). *rpr* expression pattern has been shown to depend on Notch, JNK, and Dpp signaling in the distal leg (Manjón et al., 2007); however, the regulation of apoptosis in this model is far from being fully understood. Indeed, while *rpr* is expressed in around 150 cells, only approximately 30 of these cells undergo apoptosis; to date, nothing is known about which cells are chosen to die within this predetermined *rpr*-expressing domain. Nonetheless, the apoptotic pattern has been spatially and temporally characterized, focusing on the fold separating the two last tarsal segments (T4–T5). A wave of apoptosis progresses from the most ventral region of the leg disc to the lateral then dorsal most regions, preceding fold formation that follows in a similar pattern, from the most ventral to the most dorsal region (Monier et al., 2015).

As mentioned above, at the tissue level, apoptosis inhibition leads to the absence of fold formation. At a smaller scale, the absence of apoptosis prevents the cellular reorganization that takes place normally during fold formation which includes cell stretching, apical surface reduction, and the preferential orientation adopted by cells from the fold domain. Furthermore, apoptosis inhibition leads to a reduction of cellular tension at the surface of the epithelium and to an absence of actomyosin apical accumulation (Monier et al., 2015). Altogether, these results show that apoptotic cells influence their surroundings in a way that resembles what has been observed during dorsal closure in the embryo. Importantly though, the specific contribution of apoptotic cells versus direct neighbors has been established in this tissue. We showed that, in addition to the apoptotic ring of actomyosin responsible for cell extrusion, an apico-basal force, generated specifically by apoptotic cells early during the execution phase (see Section 2), is crucial for fold formation (Fig. 3). To differentiate between these two forces, cell death was ectopically induced in a naïve flat tissue, the wing. In this tissue, the induction of apoptosis mimics fold formation as observed in the leg disc. However, if the activity of myosin II is specifically blocked in apoptotic cells and not in their surroundings, this is sufficient to entirely block fold formation, whereas cell extrusion is not impaired. In this experiment, the different pools of actomyosin were identified and characterized. When ectopic apoptosis is induced, the entire actomyosin dynamics observed in the leg is reproduced including: (i) the formation of an apico-basal actomyosin cable-like structure in apoptotic cells (most probably responsible for apoptotic force generation), (ii) the accumulation of an actomyosin ring in and around apoptotic cells responsible for dying cell extrusion, and (iii) the apical accumulation of actomyosin in between apoptotic cells responsible for tissue remodeling. In the parallel experiment in which apoptosis is induced while myosin II is blocked specifically in dying cells, neither the apico-basal cable nor the nonautonomous apical accumulation of actomyosin is visible, although the extrusion ring is still formed with extrusion taking place normally.

Thus, in addition to the apical distortion of direct neighboring cells as previously described (Toyama et al., 2008), the elimination of an apoptotic cell creates an autonomous myosin II-dependent apico-basal intracellular force which transiently deforms the apical surface of adjacent cells (Monier et al., 2015). This force leads to myosin II apical accumulation at junctions and to an increase in tissue tension. Based on previous publications showing that transient mechanical forces are sufficient to induce myosin II recruitment at the apical surface of an epithelium (Fernandez-Gonzalez, Simoes, Röper, Eaton, & Zallen, 2009; Pouille, Ahmadi, Brunet, & Farge, 2009), we propose that the transient apico-basal apoptotic force acts as a biomechanical signal to trigger nonautonomous myosin II recruitment in an orthogonal plane, that is, at the level of adherens junctions of neighboring cells.

To test if the apoptotic apico-basal force could indeed constitute the initial signal inducing tissue shape modifications, a physical model was elaborated based on the 2D vertex model (Farhadifar, Röper, Aigouy, Eaton, & Jülicher, 2007). This model further highlights the crucial role of the apico-basal force generated autonomously in apoptotic cells. Indeed, without the local transient apico-basal deformations induced by apoptotic cells, the model does not predict any invagination at the tissue level, indicating that the simple disappearance of 30 cells from a continuous ring-like domain is not sufficient to create an invagination. However, when each dying cell is programmed to exert a transient apico-basal force coupled with an increase of contractility in two rows of cells, this leads to a clear invagination response all around the cylindrical tissue, showing that the transient pulling forces generated at a cellular scale and translated at the tissue scale are essential to modify tissue shape.

Thus, this work both reveals that a force is generated inside apoptotic cells and strongly suggests that this force constitutes the initial signal giving rise to the whole remodeling of the tissue, through redistribution of acto-myosin and tension in the tissue (Fig. 3). The extrusion ring appears to have a secondary role in this particular process. It will be interesting now to reexamine other models of apoptosis-dependent morphogenesis such as dorsal closure and digit formation to establish the respective contribution of apoptotic cells and their neighbors in tissue remodeling.

4. CONCLUSION

In opposition to necrosis, apoptosis was initially considered as a clean, discrete way of eliminating undesirable cells, without intracellular content release, thus limiting the influence of cell elimination on the surrounding tissue. However, the integration of biomechanical studies in the cell biology field together with the possibility of following tissue remodeling in living tissues has recently led us to revisit this passive view of apoptosis.

Indeed, studies in integrated systems have revealed that apoptosis is in fact a highly dynamic process characterized by important morphological changes at the cellular level. Focusing on the role of apoptotic cell neighbors, dying cell extrusion was shown to influence cell rearrangement. More recently, it has been shown that, during execution, apoptotic cells themselves generate a force, either in the plane of the apical surface to induce apical constriction or in the apico-basal axis deforming the apical surface and finally inducing tissue folding. Following on from these new results, it will be interesting in the near future to reassess the different models of apoptosis-dependent morphogenesis bearing in mind the potential influence of each apoptotic step, thus establishing the respective contribution of apoptotic cells and their neighbors to tissue remodeling.

Furthermore, it will be interesting to determine whether the biomechanic role of apoptotic has been conserved throughout evolution. Indeed, apoptotis-dependent epithelium folding has been described in vertebrates during neural tube bending and in *Drosophila* during leg joint formation. Since the generation of an apico-basal apoptotic force has been shown to play a crucial role in *Drosophila*, it is now tempting to propose that neural tube bending also depends on the same kind of cellular mechanisms. Thus, it will be interesting to test if the cellular mechanisms described during leg joint formation are conserved in vertebrates, thus generalizing the mechanism of apoptosis-dependent epithelium folding.

REFERENCES

Abrams, J. M., White, K., Fessler, L. I., & Steller, H. (1993). Programmed cell death during Drosophila embryogenesis. *Development, 117*, 29–43.

Alonso, C. R. (2002). Hox proteins: Sculpting body parts by activating localized cell death. *Current Biology, 12*, R776–R778.

Amano, M., Ito, M., Kimura, K., Fukata, Y., Chihara, K., Nakano, T., et al. (1996). Phosphorylation and activation of myosin by Rho-associated kinase (Rho-kinase). *The Journal of Biological Chemistry, 271*, 20246–20249.

Bate, M., & Martinez-Arias, A. (1993). The development of drosophila melanogaster. *Development and Patterning of the Larval Epidermis of Drosophila*, (Vol. 1), New York, NY: Cold Spring Harbor Laboratory Press.

Bonanno, E., Ruzittu, M., Carlà, E. C., Montinari, M. R., Pagliara, P., & Dini, L. (2000). Cell shape and organelle modification in apoptotic U937 cells. *European Journal of Histochemistry, 44*, 237–246.

Buendia, B., Santa-Maria, A., & Courvalin, J. C. (1999). Caspase-dependent proteolysis of integral and peripheral proteins of nuclear membranes and nuclear pore complex proteins during apoptosis. *Journal of Cell Science, 112*(Pt. 11), 1743–1753.

Byun, Y., Chen, F., Chang, R., Trivedi, M., Green, K. J., & Cryns, V. L. (2001). Caspase cleavage of vimentin disrupts intermediate filaments and promotes apoptosis. *Cell Death and Differentiation, 8*, 443–450. http://dx.doi.org/10.1038/sj.cdd.4400840.

Caulín, C., Salvesen, G. S., & Oshima, R. G. (1997). Caspase cleavage of keratin 18 and reorganization of intermediate filaments during epithelial cell apoptosis. *The Journal of Cell Biology, 138*, 1379–1394.

Charras, G. T., Yarrow, J. C., Horton, M. A., Mahadevan, L., & Mitchison, T. J. (2005). Non-equilibration of hydrostatic pressure in blebbing cells. *Nature, 435*, 365–369. http://dx.doi.org/10.1038/nature03550.

Chekeni, F. B., Elliott, M. R., Sandilos, J. K., Walk, S. F., Kinchen, J. M., Lazarowski, E. R., et al. (2010). Pannexin 1 channels mediate 'find-me' signal release and membrane permeability during apoptosis. *Nature, 467*, 863–867. http://dx.doi.org/10.1038/nature09413.

Coleman, M. L., & Olson, M. F. (2002). Rho GTPase signalling pathways in the morphological changes associated with apoptosis. *Cell Death and Differentiation*, *9*, 493–504. http://dx.doi.org/10.1038/sj/cdd/4400987.

Coleman, M. L., Sahai, E. A., Yeo, M., Bosch, M., Dewar, A., & Olson, M. F. (2001). Membrane blebbing during apoptosis results from caspase-mediated activation of ROCK I. *Nature Cell Biology*, *3*, 339–345. http://dx.doi.org/10.1038/35070009.

Cotter, T. G., Lennon, S. V., Glynn, J. M., & Green, D. R. (1992). Microfilament-disrupting agents prevent the formation of apoptotic bodies in tumor cells undergoing apoptosis. *Cancer Research*, *52*, 997–1005.

Croft, D. R., Coleman, M. L., Li, S., Robertson, D., Sullivan, T., Stewart, C. L., et al. (2005). Actin-myosin-based contraction is responsible for apoptotic nuclear disintegration. *The Journal of Cell Biology*, *168*, 245–255. http://dx.doi.org/10.1083/jcb.200409049.

Cuervo, R., & Covarrubias, L. (2004). Death is the major fate of medial edge epithelial cells and the cause of basal lamina degradation during palatogenesis. *Development*, *131*, 15–24. http://dx.doi.org/10.1242/dev.00907.

Cuervo, R., Valencia, C., Chandraratna, R. A., & Covarrubias, L. (2002). Programmed cell death is required for palate shelf fusion and is regulated by retinoic acid. *Developmental Biology*, *245*, 145–156. http://dx.doi.org/10.1006/dbio.2002.0620.

Díaz-Mendoza, M. J., Lorda-Diez, C. I., Montero, J. A., García-Porrero, J. A., & Hurlé, J. M. (2013). Interdigital cell death in the embryonic limb is associated with depletion of Reelin in the extracellular matrix. *Cell Death & Disease*, *4*, e800. http://dx.doi.org/10.1038/cddis.2013.322.

Ellis, H. M., & Horvitz, H. R. (1986). Genetic control of programmed cell death in the nematode *C. elegans*. *Cell*, *44*, 817–829.

Fackler, O. T., & Grosse, R. (2008). Cell motility through plasma membrane blebbing. *The Journal of Cell Biology*, *181*, 879–884. http://dx.doi.org/10.1083/jcb.200802081.

Farhadifar, R., Röper, J. C., Aigouy, B., Eaton, S., & Jülicher, F. (2007). The influence of cell mechanics, cell-cell interactions, and proliferation on epithelial packing. *Current Biology*, *17*, 2095–2104. http://dx.doi.org/10.1016/j.cub.2007.11.049.

Fernandez-Gonzalez, R., Simoes, S. E. M., Röper, J. C., Eaton, S., & Zallen, J. A. (2009). Myosin II dynamics are regulated by tension in intercalating cells. *Developmental Cell*, *17*, 736–743. http://dx.doi.org/10.1016/j.devcel.2009.09.003.

Fischer, U., Jänicke, R. U., & Schulze-Osthoff, K. (2003). Many cuts to ruin: A comprehensive update of caspase substrates. *Cell Death and Differentiation*, *10*, 76–100. http://dx.doi.org/10.1038/sj.cdd.4401160.

Gorfinkiel, N., & Blanchard, G. B. (2011). Dynamics of actomyosin contractile activity during epithelial morphogenesis. *Current Opinion in Cell Biology*, *23*, 531–539. http://dx.doi.org/10.1016/j.ceb.2011.06.002.

Gu, Y., Forostyan, T., Sabbadini, R., & Rosenblatt, J. (2011). Epithelial cell extrusion requires the sphingosine-1-phosphate receptor 2 pathway. *The Journal of Cell Biology*, *193*, 667–676. http://dx.doi.org/10.1083/jcb.201010075.

Gu, Y., & Rosenblatt, J. (2012). New emerging roles for epithelial cell extrusion. *Current Opinion in Cell Biology*, *24*, 865–870. http://dx.doi.org/10.1016/j.ceb.2012.09.003.

Guarner, A., Manjón, C., Edwards, K., Steller, H., Suzanne, M., & Sánchez-Herrero, E. (2014). The zinc finger homeodomain-2 gene of Drosophila controls Notch targets and regulates apoptosis in the tarsal segments. *Developmental Biology*, *385*, 350–365. http://dx.doi.org/10.1016/j.ydbio.2013.10.011.

Häcker, G. (2000). The morphology of apoptosis. *Cell and Tissue Research*, *301*, 5–17.

Haigo, S. L., Hildebrand, J. D., Harland, R. M., & Wallingford, J. B. (2003). Shroom induces apical constriction and is required for hingepoint formation during neural tube closure. *Current Biology*, *13*, 2125–2137.

Hernández-Martínez, R., Castro-Obregón, S., & Covarrubias, L. (2009). Progressive interdigital cell death: Regulation by the antagonistic interaction between fibroblast growth factor 8 and retinoic acid. *Development*, *136*, 3669–3678. http://dx.doi.org/10.1242/dev.041954.

Hernández-Martínez, R., & Covarrubias, L. (2011). Interdigital cell death function and regulation: New insights on an old programmed cell death model. *Development, Growth & Differentiation*, *53*, 245–258. http://dx.doi.org/10.1111/j.1440-169X.2010.01246.x.

Iseki, S. (2011). Disintegration of the medial epithelial seam: Is cell death important in palatogenesis? *Development, Growth & Differentiation*, *53*, 259–268. http://dx.doi.org/10.1111/j.1440-169X.2010.01245.x.

Ishizuya-Oka, A., & Shi, Y. B. (2011). Evolutionary insights into postembryonic development of adult intestinal stem cells. *Cell & Bioscience*, *1*, 37. http://dx.doi.org/10.1186/2045-3701-1-37.

Kawano, Y., Fukata, Y., Oshiro, N., Amano, M., Nakamura, T., Ito, M., et al. (1999). Phosphorylation of myosin-binding subunit (MBS) of myosin phosphatase by Rho-kinase in vivo. *The Journal of Cell Biology*, *147*, 1023–1038.

Kerr, J. F., Wyllie, A. H., & Currie, A. R. (1972). Apoptosis: A basic biological phenomenon with wide-ranging implications in tissue kinetics. *British Journal of Cancer*, *26*, 239–257.

Kiehart, D. P., Galbraith, C. G., Edwards, K. A., Rickoll, W. L., & Montague, R. A. (2000). Multiple forces contribute to cell sheet morphogenesis for dorsal closure in Drosophila. *The Journal of Cell Biology*, *149*, 471–490.

Kihlmark, M., Imreh, G., & Hallberg, E. (2001). Sequential degradation of proteins from the nuclear envelope during apoptosis. *Journal of Cell Science*, *114*, 3643–3653.

Kondo, T., & Hayashi, S. (2013). Mitotic cell rounding accelerates epithelial invagination. *Nature*, *494*, 125–129. http://dx.doi.org/10.1038/nature11792.

Kuipers, D., Mehonic, A., Kajita, M., Peter, L., Fujita, Y., Duke, T., et al. (2014). Epithelial repair is a two-stage process driven first by dying cells and then by their neighbours. *Journal of Cell Science*, *127*, 1229–1241. http://dx.doi.org/10.1242/jcs.138289.

Kunda, P., Pelling, A. E., Liu, T., & Baum, B. (2008). Moesin controls cortical rigidity, cell rounding, and spindle morphogenesis during mitosis. *Current Biology*, *18*, 91–101. http://dx.doi.org/10.1016/j.cub.2007.12.051.

Kuranaga, E., Matsunuma, T., Kanuka, H., Takemoto, K., Koto, A., Kimura, K., et al. (2011). Apoptosis controls the speed of looping morphogenesis in Drosophila male terminalia. *Development*, *138*, 1493–1499. http://dx.doi.org/10.1242/dev.058958.

Lane, J. D., Allan, V. J., & Woodman, P. G. (2005). Active relocation of chromatin and endoplasmic reticulum into blebs in late apoptotic cells. *Journal of Cell Science*, *118*, 4059–4071. http://dx.doi.org/10.1242/jcs.02529.

Larsen, C. W., Hirst, E., Alexandre, C., & Vincent, J. P. (2003). Segment boundary formation in Drosophila embryos. *Development*, *130*, 5625–5635. http://dx.doi.org/10.1242/dev.00867.

Lohmann, I., McGinnis, N., Bodmer, M., & McGinnis, W. (2002). The Drosophila Hox gene deformed sculpts head morphology via direct regulation of the apoptosis activator reaper. *Cell*, *110*, 457–466.

Lubkov, V., & Bar-Sagi, D. (2014). E-cadherin-mediated cell coupling is required for apoptotic cell extrusion. *Current Biology*, *24*, 868–874. http://dx.doi.org/10.1016/j.cub.2014.02.057.

Manjón, C., Sánchez-Herrero, E., & Suzanne, M. (2007). Sharp boundaries of Dpp signalling trigger local cell death required for Drosophila leg morphogenesis. *Nature Cell Biology*, *9*, 57–63. http://dx.doi.org/10.1038/ncb1518.

Marshall, T. W., Lloyd, I. E., Delalande, J. M., Näthke, I., & Rosenblatt, J. (2011). The tumor suppressor adenomatous polyposis coli controls the direction in which a cell extrudes from an epithelium. *Molecular Biology of the Cell, 22*, 3962–3970. http://dx.doi.org/10.1091/mbc.E11-05-0469.

Martin, A. C., Kaschube, M., & Wieschaus, E. F. (2009). Pulsed contractions of an actin-myosin network drive apical constriction. *Nature, 457*, 495–499. http://dx.doi.org/10.1038/nature07522.

Massa, V., Toyama, Y., Abravanel, D. L., Wiemann, J. M., Wells, A. R., Tulu, U. S., et al. (2009). Apoptosis is not required for mammalian neural tube closure. *Proceedings of the National Academy of Sciences of the United States of America, 106*, 8233–8238. http://dx.doi.org/10.1073/pnas.0900333106.

Metzstein, M. M., Stanfield, G. M., & Horvitz, H. R. (1998). Genetics of programmed cell death in *C. elegans*: Past, present and future. *Trends in Genetics, 14*, 410–416.

Mills, J. C., Lee, V. M., & Pittman, R. N. (1998). Activation of a PP2A-like phosphatase and dephosphorylation of tau protein characterize onset of the execution phase of apoptosis. *Journal of Cell Science, 111*(Pt. 5), 625–636.

Mills, J. C., Stone, N. L., & Pittman, R. N. (1999). Extranuclear apoptosis. The role of the cytoplasm in the execution phase. *The Journal of Cell Biology, 146*, 703–708.

Monier, B., Gettings, M., Gay, G., Mangeat, T., Schott, S., Guarner, A., et al. (2015). Apico-basal forces exerted by apoptotic cells drive epithelium folding. *Nature, 518*, 245–248. http://dx.doi.org/10.1038/nature14152.

Montero, J. A., Gañan, Y., Macias, D., Rodriguez-Leon, J., Sanz-Ezquerro, J. J., Merino, R., et al. (2001). Role of FGFs in the control of programmed cell death during limb development. *Development, 128*, 2075–2084.

Montero, J. A., & Hurlé, J. M. (2010). Sculpturing digit shape by cell death. *Apoptosis, 15*, 365–375. http://dx.doi.org/10.1007/s10495-009-0444-5.

Moss, D. K., Betin, V. M., Malesinski, S. D., & Lane, J. D. (2006). A novel role for microtubules in apoptotic chromatin dynamics and cellular fragmentation. *Journal of Cell Science, 119*, 2362–2374. http://dx.doi.org/10.1242/jcs.02959.

Moss, D. K., & Lane, J. D. (2006). Microtubules: Forgotten players in the apoptotic execution phase. *Trends in Cell Biology, 16*, 330–338. http://dx.doi.org/10.1016/j.tcb.2006.05.005.

Nishimura, T., Honda, H., & Takeichi, M. (2012). Planar cell polarity links axes of spatial dynamics in neural-tube closure. *Cell, 149*, 1084–1097. http://dx.doi.org/10.1016/j.cell.2012.04.021.

Orlando, K. A., & Pittman, R. N. (2006). Rho kinase regulates phagocytosis, surface expression of GlcNAc, and Golgi fragmentation of apoptotic PC12 cells. *Experimental Cell Research, 312*, 3298–3311. http://dx.doi.org/10.1016/j.yexcr.2006.06.033.

Oropesa, M., de la Mata, M., Maraver, J. G., Cordero, M. D., Cotán, D., Rodríguez-Hernández, A., et al. (2011). Apoptotic microtubule network organization and maintenance depend on high cellular ATP levels and energized mitochondria. *Apoptosis, 16*, 404–424. http://dx.doi.org/10.1007/s10495-011-0577-1.

Oropesa-Ávila, M., Fernández-Vega, A., de la Mata, M., Maraver, J. G., Cordero, M. D., Cotán, D., et al. (2013). Apoptotic microtubules delimit an active caspase free area in the cellular cortex during the execution phase of apoptosis. *Cell Death & Disease, 4*, e527. http://dx.doi.org/10.1038/cddis.2013.58.

Paluch, E., Sykes, C., Prost, J., & Bornens, M. (2006). Dynamic modes of the cortical actomyosin gel during cell locomotion and division. *Trends in Cell Biology, 16*, 5–10. http://dx.doi.org/10.1016/j.tcb.2005.11.003.

Poon, I. K., Chiu, Y. H., Armstrong, A. J., Kinchen, J. M., Juncadella, I. J., Bayliss, D. A., et al. (2014). Unexpected link between an antibiotic, pannexin channels and apoptosis. *Nature, 507*, 329–334. http://dx.doi.org/10.1038/nature13147.

Pouille, P. A., Ahmadi, P., Brunet, A. C., & Farge, E. (2009). Mechanical signals trigger Myosin II redistribution and mesoderm invagination in Drosophila embryos. *Science Signaling, 2*, ra16. http://dx.doi.org/10.1126/scisignal.2000098.

Rao, L., Perez, D., & White, E. (1996). Lamin proteolysis facilitates nuclear events during apoptosis. *The Journal of Cell Biology*, *135*, 1441–1455.
Rauskolb, C., & Irvine, K. D. (1999). Notch-mediated segmentation and growth control of the Drosophila leg. *Developmental Biology*, *210*, 339–350. http://dx.doi.org/10.1006/dbio.1999.9273.
Rodriguez-Diaz, A., Toyama, Y., Abravanel, D. L., Wiemann, J. M., Wells, A. R., Tulu, U. S., et al. (2008). Actomyosin purse strings: Renewable resources that make morphogenesis robust and resilient. *HFSP Journal*, *2*, 220–237. http://dx.doi.org/10.2976/1.2955565.
Rosenblatt, J., Raff, M. C., & Cramer, L. P. (2001). An epithelial cell destined for apoptosis signals its neighbors to extrude it by an actin- and myosin-dependent mechanism. *Current Biology*, *11*, 1847–1857.
Rothenberg, F., Fisher, S. A., & Watanabe, M. (2003). Sculpting the cardiac outflow tract. *Birth Defects Research Part C, Embryo Today*, *69*, 38–45. http://dx.doi.org/10.1002/bdrc.10007.
Salas-Vidal, E., Valencia, C., & Covarrubias, L. (2001). Differential tissue growth and patterns of cell death in mouse limb autopod morphogenesis. *Developmental Dynamics*, *220*, 295–306. http://dx.doi.org/10.1002/dvdy.1108.
Sánchez-Alcázar, J. A., Rodríguez-Hernández, A., Cordero, M. D., Fernández-Ayala, D. J., Brea-Calvo, G., Garcia, K., et al. (2007). The apoptotic microtubule network preserves plasma membrane integrity during the execution phase of apoptosis. *Apoptosis*, *12*, 1195–1208. http://dx.doi.org/10.1007/s10495-006-0044-6.
Sebbagh, M., Hamelin, J., Bertoglio, J., Solary, E., & Bréard, J. (2005). Direct cleavage of ROCK II by granzyme B induces target cell membrane blebbing in a caspase-independent manner. *The Journal of Experimental Medicine*, *201*, 465–471. http://dx.doi.org/10.1084/jem.20031877.
Sebbagh, M., Renvoizé, C., Hamelin, J., Riché, N., Bertoglio, J., & Bréard, J. (2001). Caspase-3-mediated cleavage of ROCK I induces MLC phosphorylation and apoptotic membrane blebbing. *Nature Cell Biology*, *3*, 346–352. http://dx.doi.org/10.1038/35070019.
Sedzinski, J., Biro, M., Oswald, A., Tinevez, J. Y., Salbreux, G., & Paluch, E. (2011). Polar actomyosin contractility destabilizes the position of the cytokinetic furrow. *Nature*, *476*, 462–466. http://dx.doi.org/10.1038/nature10286.
Sherrard, K., Robin, F., Lemaire, P., & Munro, E. (2010). Sequential activation of apical and basolateral contractility drives ascidian endoderm invagination. *Current Biology*, *20*, 1499–1510. http://dx.doi.org/10.1016/j.cub.2010.06.075.
Slattum, G., McGee, K. M., & Rosenblatt, J. (2009). P115 RhoGEF and microtubules decide the direction apoptotic cells extrude from an epithelium. *The Journal of Cell Biology*, *186*, 693–702. http://dx.doi.org/10.1083/jcb.200903079.
Stenn, K. S., & Paus, R. (2001). Controls of hair follicle cycling. *Physiological Reviews*, *81*, 449–494.
Suzanne, M., Petzoldt, A. G., Spéder, P., Coutelis, J. B., Steller, H., & Noselli, S. (2010). Coupling of apoptosis and L/R patterning controls stepwise organ looping. *Current Biology*, *20*, 1773–1778. http://dx.doi.org/10.1016/j.cub.2010.08.056.
Suzanne, M., & Steller, H. (2013). Shaping organisms with apoptosis. *Cell Death and Differentiation*, *20*, 669–675. http://dx.doi.org/10.1038/cdd.2013.11.
Tamada, M., Perez, T. D., Nelson, W. J., & Sheetz, M. P. (2007). Two distinct modes of myosin assembly and dynamics during epithelial wound closure. *The Journal of Cell Biology*, *176*, 27–33. http://dx.doi.org/10.1083/jcb.200609116.
Teng, X., & Toyama, Y. (2011). Apoptotic force: Active mechanical function of cell death during morphogenesis. *Development, Growth & Differentiation*, *53*, 269–276. http://dx.doi.org/10.1111/j.1440-169X.2011.01251.x.

Toyama, Y., Peralta, X. G., Wells, A. R., Kiehart, D. P., & Edwards, G. S. (2008). Apoptotic force and tissue dynamics during Drosophila embryogenesis. *Science*, *321*, 1683–1686. http://dx.doi.org/10.1126/science.1157052.

Wang, Y. C., Khan, Z., Kaschube, M., & Wieschaus, E. F. (2012). Differential positioning of adherens junctions is associated with initiation of epithelial folding. *Nature*, *484*, 390–393. http://dx.doi.org/10.1038/nature10938.

Yamaguchi, Y., & Miura, M. (2013). How to form and close the brain: Insight into the mechanism of cranial neural tube closure in mammals. *Cellular and Molecular Life Sciences*, *70*, 3171–3186. http://dx.doi.org/10.1007/s00018-012-1227-7.

Yamaguchi, Y., Shinotsuka, N., Nonomura, K., Takemoto, K., Kuida, K., Yosida, H., et al. (2011). Live imaging of apoptosis in a novel transgenic mouse highlights its role in neural tube closure. *The Journal of Cell Biology*, *195*, 1047–1060. http://dx.doi.org/10.1083/jcb.201104057.

INDEX

Note: Page numbers followed by "*f*" indicate figures and "*t*" indicate tables.

A

B

D

E

R

S

T

U

X

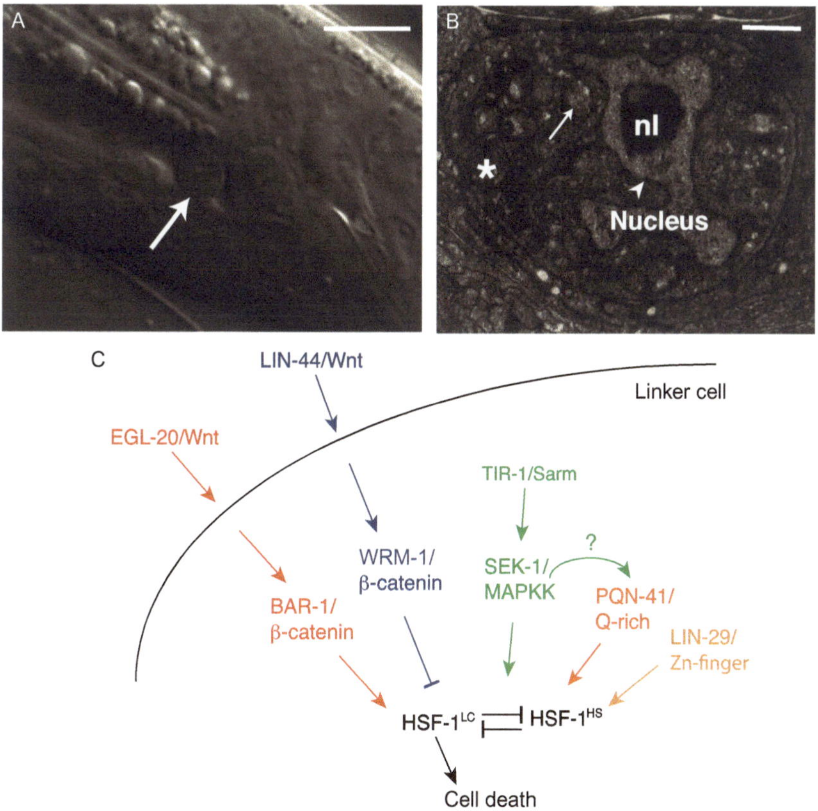

Jennifer Zuckerman Malin and Shai Shaham, Figure 3 Linker cell death in *C. elegans*. (A) DIC image of a dying linker cell (arrow). Scale bar, 5 μm. (B) Electron micrograph of a dying linker cell. Note crenelated nucleus (arrowhead), intact nucleolus (nl), and swollen organelles (asterisk and arrow). Scale bar, 700 nm. (C) A model for linker cell death. Proteins in the same regulatory pathway are color-matched. How regulators impinge on HSF-1 function is not known.

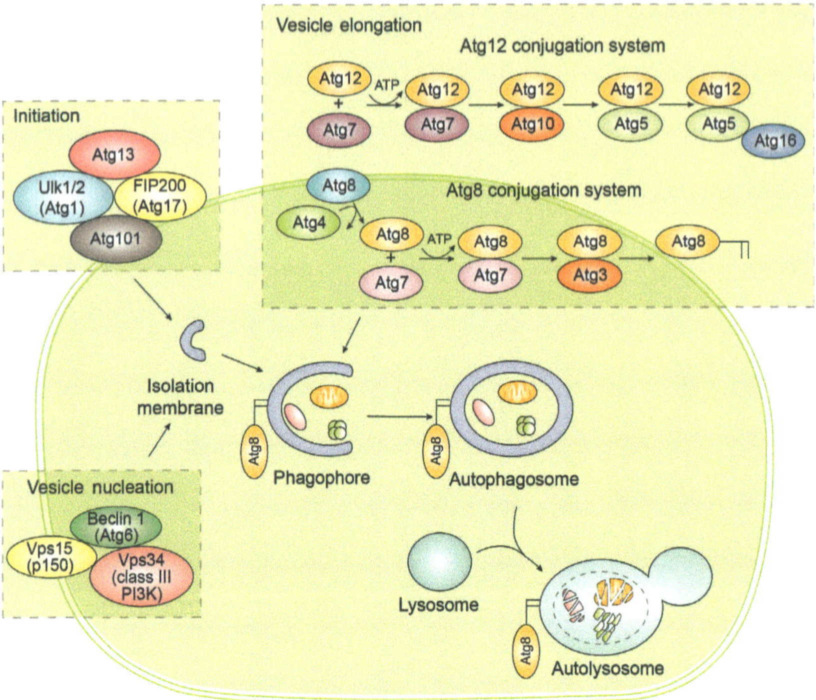

Allyson L. Anding and Eric H. Baehrecke, Figure 2 Regulation of autophagy. The initiation of autophagy requires Atg1 and its interacting partners. The Vps34 complex is responsible for vesicle nucleation, and this core complex can also associate with other proteins such as Atg14L and UVRAG. Vesicle elongation is regulated by the Atg12 and Atg8 ubiquitin-like conjugation systems that enable the tethering of Atg8 to the autophagosomal membrane and eventual autophagosome formation.

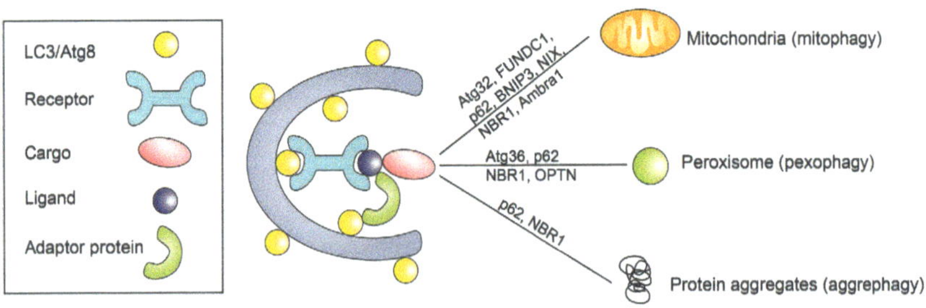

Allyson L. Anding and Eric H. Baehrecke, Figure 3 Selective autophagy. In selective autophagy, receptor proteins enable the interaction of the autophagosomal membrane and cargo to be degraded, sometimes with the help of ligands such as ubiquitin or adaptor proteins. Three types of selective autophagy, mitophagy, pexophagy, and aggrephagy, are represented here, along with some of the receptor proteins associated with these processes.

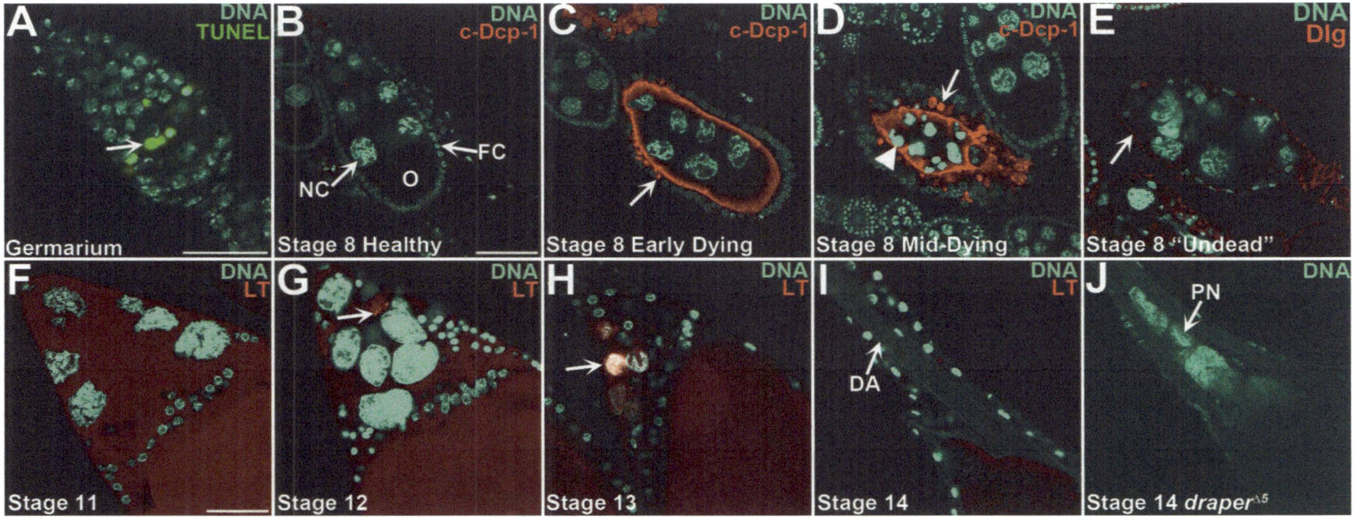

Jeanne S. Peterson *et al.*, Figure 3 Overview of programmed cell death in the *Drosophila* ovary. (A) Wild-type (w^{1118}) germarium stained with DAPI (cyan) to label DNA and TUNEL (green arrow) to label fragmented DNA. Scale bar = 25 μm. (B–D) Control (*GR1-GAL4 mCD8-GFP/TM3*) stage 8 egg chambers from starved flies were stained with DAPI (cyan) and cleaved (active) Dcp-1 antibody (c-Dcp1, red). (B) Healthy egg chamber has no apparent cleaved Dcp-1 staining. Each egg chamber is comprised of three cell types: nurse cells (NC), follicle cells (FC), and the oocyte (O). (C) Cleaved Dcp-1 staining (red) appears in an early phase dying egg chamber and is present in vesicle structures in the engulfing follicle cells (arrow). (D) A mid-phase dying egg chamber has condensed and fragmented nurse cell nuclei (arrowhead) and stains with cleaved Dcp-1 (arrow). Engulfing follicle cells have enlarged. (E) *Diap1* was overexpressed in the germline (*nanos-GAL4* > *UASp-Diap1*), and ovaries from starved flies were stained with DAPI (cyan) and anti-Discs Large (Dlg, red) to label membranes. The overexpression of *Diap1* in the germline results in an "undead" stage 8 egg chamber, where the nurse cell nuclei have failed to condense or fragment and the follicle cells have largely disappeared (arrow). Scale bar = 50 μm for (B)–(E). (F–I) Wild-type (w^{1118}) stage 11–14 egg chambers were stained with LysoTracker (LT, red) to label acidified structures. (F) Stage 11 egg chamber has no LysoTracker staining. (G) Stage 12 egg chamber has LysoTracker puncta (arrow). (H) Stage 13 egg chamber has LysoTracker-positive nurse cell nuclei (arrow). (I) Stage 14 egg chamber is characterized by the presence of dorsal appendages (DA, arrow) and has no LysoTracker staining. (J) Stage 14 egg chamber from *draper*$^{\Delta5}$ has persisting nurse cell nuclei (PN, arrow). Scale bar = 50 μm for (F)–(J).

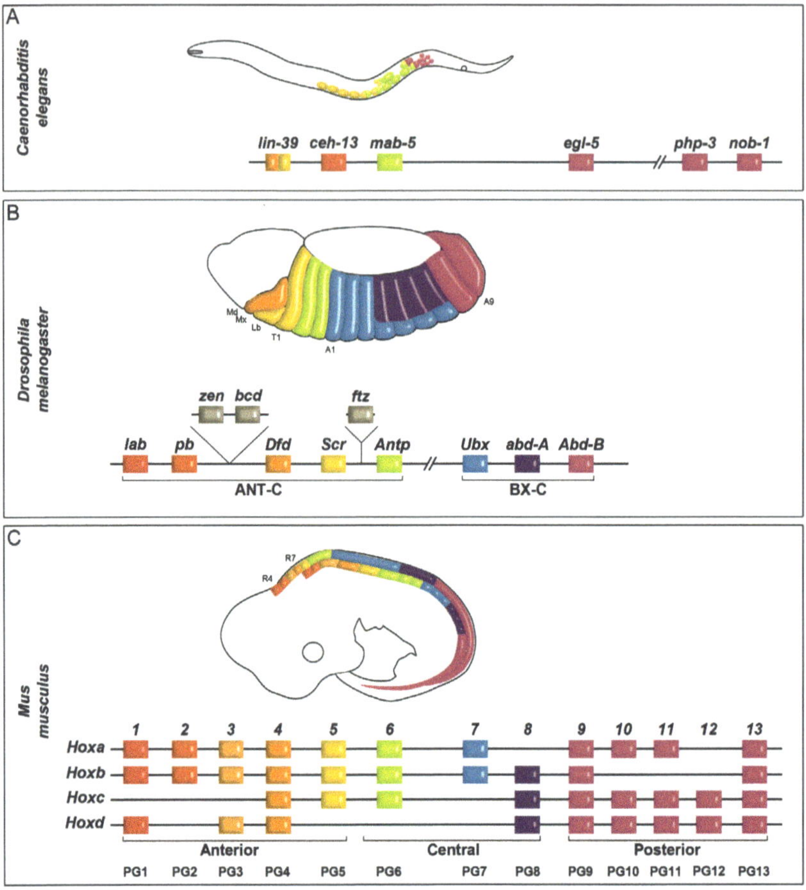

Katrin Domsch *et al.*, Figure 1 Schematic diagram of the *Hox* gene clusters and their organization in the genome (not in scale). The relative position of *Hox* genes within the clusters reflects their expression pattern along the anterior–posterior axis of the model organisms. *Hox* genes and their corresponding expression domains are always illustrated in the same color. (A) The *Hox* gene cluster in *Caenorhabditis elegans* (*C. elegans*) consists of the *Hox* genes *lin-39* (*abnormal cell lineage 39*), *ceh-13* (*homeobox 13*), *mab-5* (*male abnormal 5*), *egl-5* (*egg-laying defective 5*), *php-3* (*posterior Hox gene paralogue 3*), and *knob-like posterior* (*nob-1*). (B) The *Hox* gene cluster and its segmental expression pattern in the embryo of *Drosophila melanogaster* (stage 14). The *Antennapedia* complex (ANT-C) includes the more anterior *Hox* genes *lab* (*labial*), *pb* (*proboscipedia*), *Dfd* (*Deformed*), *Scr* (*Sex combs reduced*), and *Antp* (*Antennapedia*). The *Bithorax* complex (BX-C) includes the posterior most Hox genes *Ubx* (*Ultrabithorax*), *abd-A* (*abdominal-A*), and *Abd-B* (*Abdominal-B*). (C) *Hox* gene clusters and expression domains in the spinal cord of a mouse embryo (E12.5). Vertebrate *Hox* genes are organized in four different clusters (*Hoxa*, *Hoxb*, *Hoxc*, and *Hoxd*) and consist of up to 13 paralogous groups (PGs). The PGs can be clustered into three classes (anterior, central, and posterior). Orthologous genes and genes of the same PG are indicated in the same color. Md, mandibular; Mx, maxillary; Lb, labial; T, thoracic; A, abdominal. *Panel (B) Picture was adapted from Hartenstein (1993).*

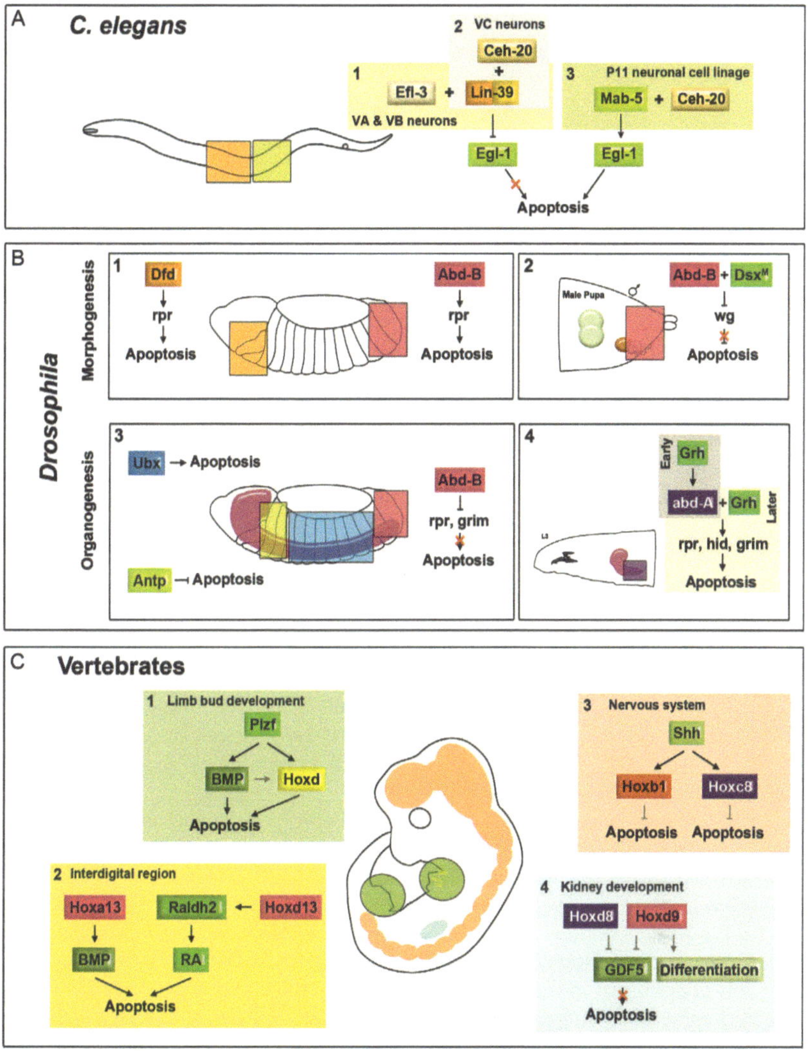

Katrin Domsch *et al.*, Figure 3 Key paradigms illustrating Hox and apoptosis interplay during the development and morphological diversification of *C. elegans* (A), *Drosophila* (B), and vertebrates (C). Arrows indicate activating and blunt ending arrows indicate inhibitory input: the black-colored arrows represent direct regulations, while the gray ones represent an indirect control or not yet identified mechanism. (A) Cell fate decisions in VA, VB, VC neurons, and P11 neuronal lineage in *C. elegans*. (A1) In the VA and VB neurons, Lin-39 and Efl-3 together inhibit Egl-1 and prevent apoptosis. (A2) In VC neurons, Lin-39 and Ceh-20 form a complex that represses *egl-1* and blocks apoptosis. (A3) In the P11 pioneer neuronal lineage, Mab-5 and Ceh-20 form a complex, which binds the *egl-1* regulatory region and activate apoptosis. (B) Hox and apoptosis interplay during *Drosophila* development. (B1) Shaping the *Drosophila* embryo is achieved through a direct binding of the Hox TFs Dfd and Abd-B to the *rpr* enhancer, thereby activating apoptosis in specific cells. (B2) During pupal stages, Abd-B interacts with the male-specific form of Dsx (Dsx^M) to repress *wg* expression and activate apoptosis giving rise to male-specific structures. (B3) During embryonic ventral nerve cord

(Continued)

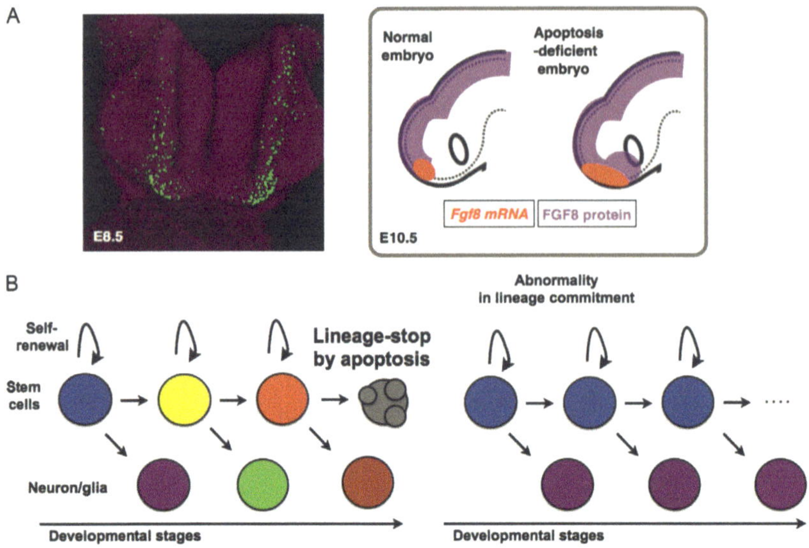

Yoshifumi Yamaguchi and Masayuki Miura, Figure 1 Examples of early neural cell death. (A) Elimination of morphogen-producing cells. (Left) A lot of apoptotic (TUNEL+) cells (green) appear in the boundary regions between the neural plate and the nonneural ectoderm. Purple: nucleus. (Right) Expansions of fgf8-expressing domain and distribution of FGF8 protein in apoptosis-deficient (apaf-1 KO or Caspase-9 KO) embryos (Nonomura et al., 2013). (B) Size regulation of specific cell lineage. Each colored cell expresses different combination of a series of transcriptional factors, which may constitute an intrinsic developmental timer.

Katrin Domsch *et al.*, Figure 3—Cont'd development in *Drosophila* (stages 11–15), repression of apoptosis in the thoracic segments by Antp and activation of apoptosis in the abdominal segments by Ubx control the cell fate of defined neurons. In the posterior segments of *Drosophila* embryos (stages 16–17), Abd-B represses *rpr* and *grim* and inhibits apoptosis in the MP1 and dMP2 pioneer neuronal lineages. (B4) In *Drosophila* larvae, Abd-A and its cofactor Grh reduce the number of neuroblasts in CNS by activating apoptosis via *rpr*, *hid*, and *grim* upregulation. (C) Hox-regulated apoptosis in vertebrates. (C1) During limb bud development, Plzf activates BMPs and *Hoxd* cluster gene expression and thereby triggers apoptosis, downstream of this cascade, in the interdigit regions. (C2) During digit formation, apoptosis within the interdigit region is triggered upstream of the cascade by Hoxa13, which activates BMP2 and BMP7, and by Hoxd13 which regulates Raldh2 and activates retinoic acid production. (C3) In vertebrate CNS, Hoxb1 in the hindbrain and Hoxc8 in the spinal cord block apoptosis and at the same time promote specification within the respective CNS regions. (C4) During kidney organogenesis, coordination of differentiation and apoptosis relies primarily on the function of Hoxd8 and Hoxd9, which block GDF expression and prevent the development of polycystic kidneys. Drosophila *pictures were adapted from Hartenstein (1993).*

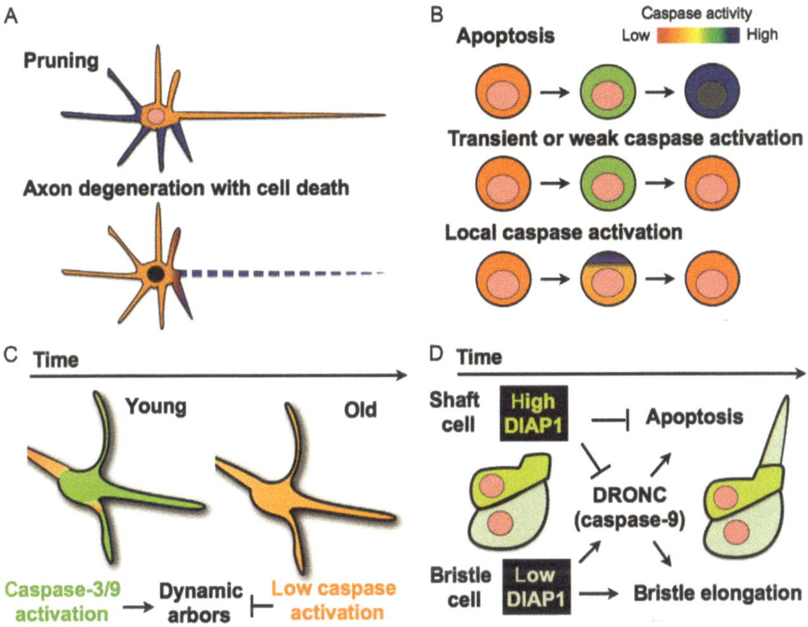

Yoshifumi Yamaguchi and Masayuki Miura, Figure 2 Nonapoptotic caspase activation in neural development. (A) Pruning of neurites without causing cell death and axon degeneration accompanying cell death. Blue indicates caspase activation. (B) Schematic illustration of distinct modes of caspase activation. (C) Local caspase activation observed in real time in young axonal arbors of developing zebrafish embryos (Campbell & Okamoto, 2013). (D) Nonapoptotic caspase activation through the regulation of DIAP1, a caspase inhibitor, in sensory bristle formation of fly in a cell-type-specific manner. DRONC activity is required for bristle elongation without causing cell death (Koto, Kuranaga, & Miura, 2009).

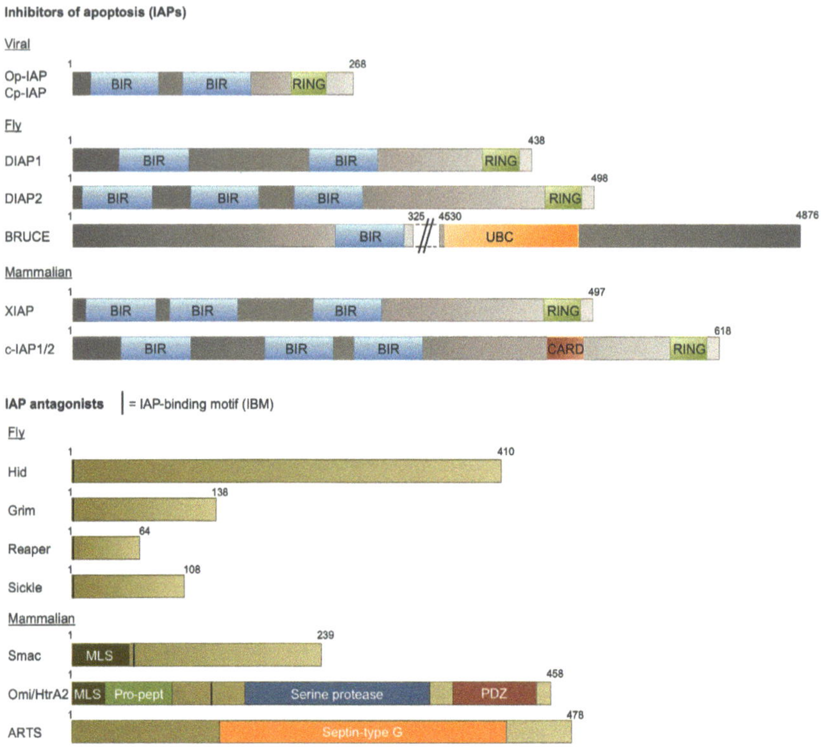

Deepika Vasudevan and Hyung Don Ryoo, Figure 1 Domain maps of IAPs and their antagonists from various model systems. All known IAPs contain at least one baculovirus IAP-repeat (BIR) domain. In addition, most have RING domains. BRUCE is the largest IAP (as indicated by the break in sequence in figure) and an exception in that it does not contain a RING domain but instead has an ubiquitin conjugation domain (UBC). Most IAP-antagonists contain a short 5–10 amino acid IAP-binding motif (IBM) at their N-terminii, usually immediately after the methionine, which is cleaved to expose the IBM. The mammalian IAP-antagonists, Smac, ARTS, and Omi/HtrA2, localize to the mitochondria for their function and hence contain a mitochondria localization sequence (MLS) among other domains. ARTS belongs is a noncanonical IAP-antagonist that does not have an N-terminal IBM and instead uses the C-terminal sequences to bind IAPs. *Domain maps to scale, source: www.uniprot.org.*

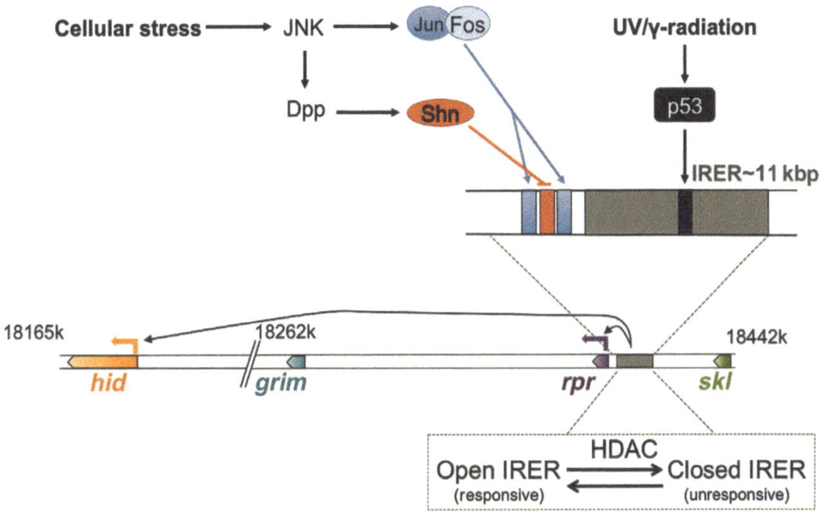

Deepika Vasudevan and Hyung Don Ryoo, Figure 3 A schematic showing the *Drosophila* H99 locus and its regulation. The 3L chromosome arm contains all four of the IAP-antagonists, *hid, grim, reaper* (*rpr*), and *sickle* (*skl*), clustered in the H99 locus. A ~12 kbp region upstream of *rpr* is the hub of most transcriptional regulation and controls transcription of *rpr* and *hid,* although the *hid* locus is more than 250 kbp away. It contains a ~11 kbp irradiation-responsive enhancer region (IRER) containing a 20 bp p53-binding site critical for radiation-induced apoptosis. The IRER region is heavily epigenetically regulated during development by histone deacetylases (HDAC) and methylases. The *rpr* upstream promoter region contains sites for binding several transcription factors, notably Jun/Fos and Schnurri (Shn), which mediate context-dependent JNK-driven apoptosis.

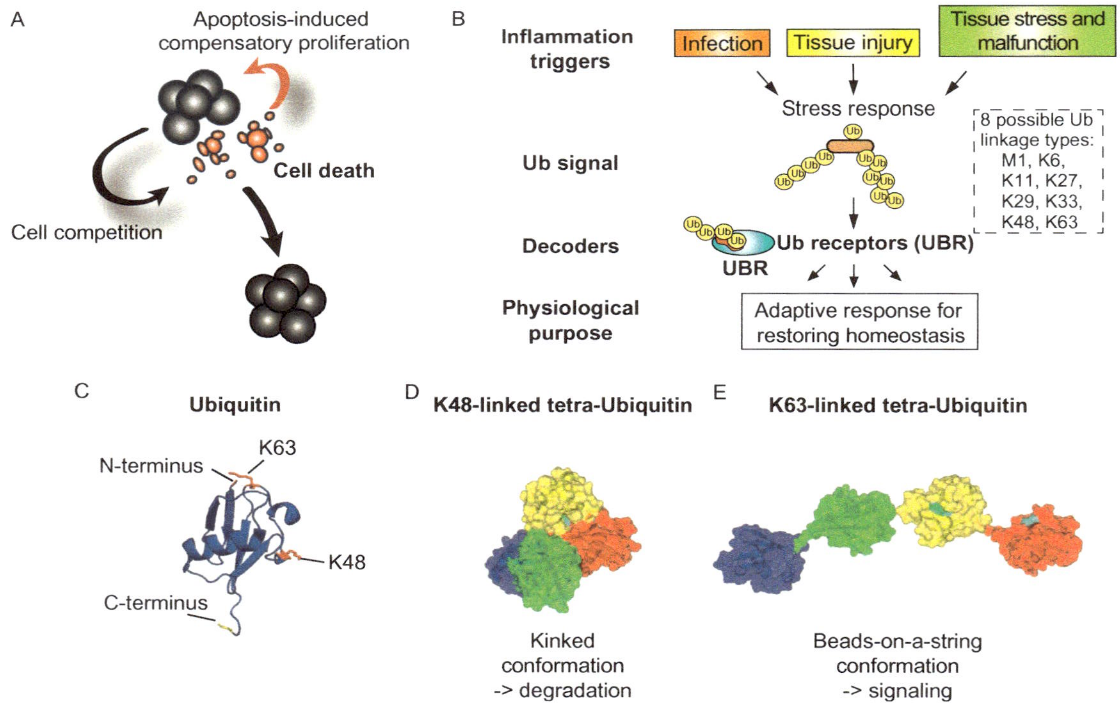

Pascal Meier *et al.*, Figure 1 Ubiquitylation as mediator and regulator of signal transduction in cell death, inflammation, and defense of homeostasis. (A) Tissue malfunction results in a secretory program and inflammatory response whose purpose is to restore homeostasis. Tissue homeostasis is regulated by a collective decision mechanism that influences cell death and proliferation across tissues. These include cell competition and apoptosis-induced compensatory proliferation. (B) Tissue stress response and inflammation underlie a common principle in which the conjugation of typical Ub chains produces robust networks that are decoded by Ub receptors whose actions serve to coordinate adaptation to tissue stress. (C) Ribbon structure of Ub. Lysines at position 48 and 63 of Ub are highlighted. Topology of K48- (D) and K63-linked (E) tetra-Ub chains.

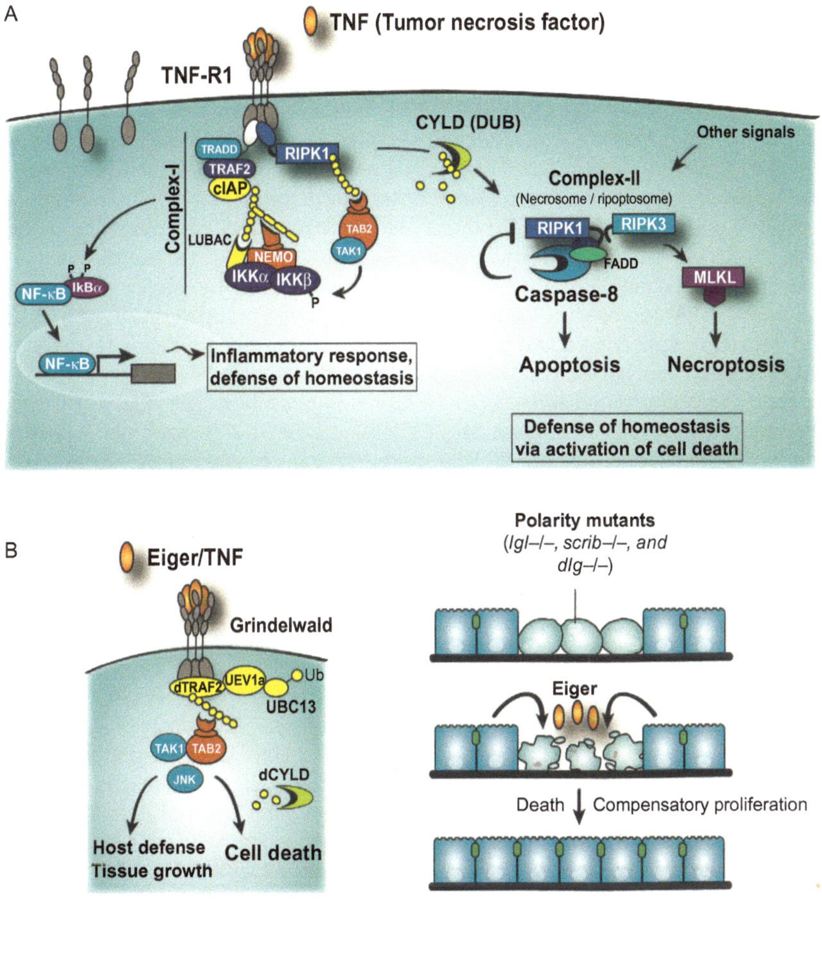

Pascal Meier *et al.*, Figure 2 See legend on next page.

Pascal Meier *et al.*, Figure 2 Evolutionary conservation of TNF-induced cell death and the defense of homeostasis. (A) TNF signaling and the transition of complex-I to complex-II. Upon TNF binding, cIAPs are recruited to the TNF-R1 signaling complex (complex-I) via TRADD/TRAF2. cIAPs ubiquitylate several molecules within the complex. RIPK1 ubiquitylation is the most readily observed. Ubiquitylation of components of complex-I, such as RIPK1, drives the recruitment of HOIL-1/HOIP/Sharpin that together form the Linear Ubiquitin Assembly Complex (LUBAC). LUBAC generates linear Ub chains on NEMO and RIPK1 that in turn recruits more NEMO molecules via its linear Ub-binding UBAN domain. NEMO is probably constitutively associated with IKKα/IKKβ and IKKβ is phosphorylated and activated by TAK1 that is independently recruited to ubiquitylated complex-I via its Ub receptors TAB2 and TAB3 that bind only to K63-linked Ub chains. Phosphorylated and activated IKKβ in turn phosphorylates IκBα, which leads to recruitment of a HECT E3 ligase. This E3 ligase promotes K48-linked ubiquitylation and proteasomal degradation of IκBα, allowing translocation of NF-κB subunits p50/p65 to drive production of cytokines. p50/p65 also promote expression of *IκBα*, to cause feedback inhibition, as well as genes such as *cFLIP* that are required to protect cells from complex-II-induced cell death. The numbered arrows provide a tentative indication of temporal sequence. Complex-II is most likely generated from complex-I, in an as yet undefined manner, and comprises RIPK1, FADD, and caspase-8. Deubiquitylation by CYLD is thereby a decisive step in the transition of complex-I to complex-II. Caspase-8 limits complex-II formation by cleaving and inactivating RIPK1. Consequently, loss of IAPs, LUBAC, or caspase-8 activity results in formation of complex-II that is able to drive necroptosis. Formation of complex-II, necrosome, or ripoptosome can also occur following stimulation of pattern recognition receptors or genotoxic stress. (B) Eiger-mediated signaling that regulates a variety of cellular and tissue processes, including the elimination of polarity mutant cells. Eiger mediates its effect through binding to its cognate receptor Grindelwald. This results in activation of JNK in a DTRAF2/Bendless/dUev1A-dependent manner. The *Drosophila* homologue of TAB2/3 (dTAB2) links TAK1 to the presumptive Ub chains conjugated by DTRAF2. dCYLD influences the decision as to whether JNK drives cell death or noncell death processes (see text for further details).

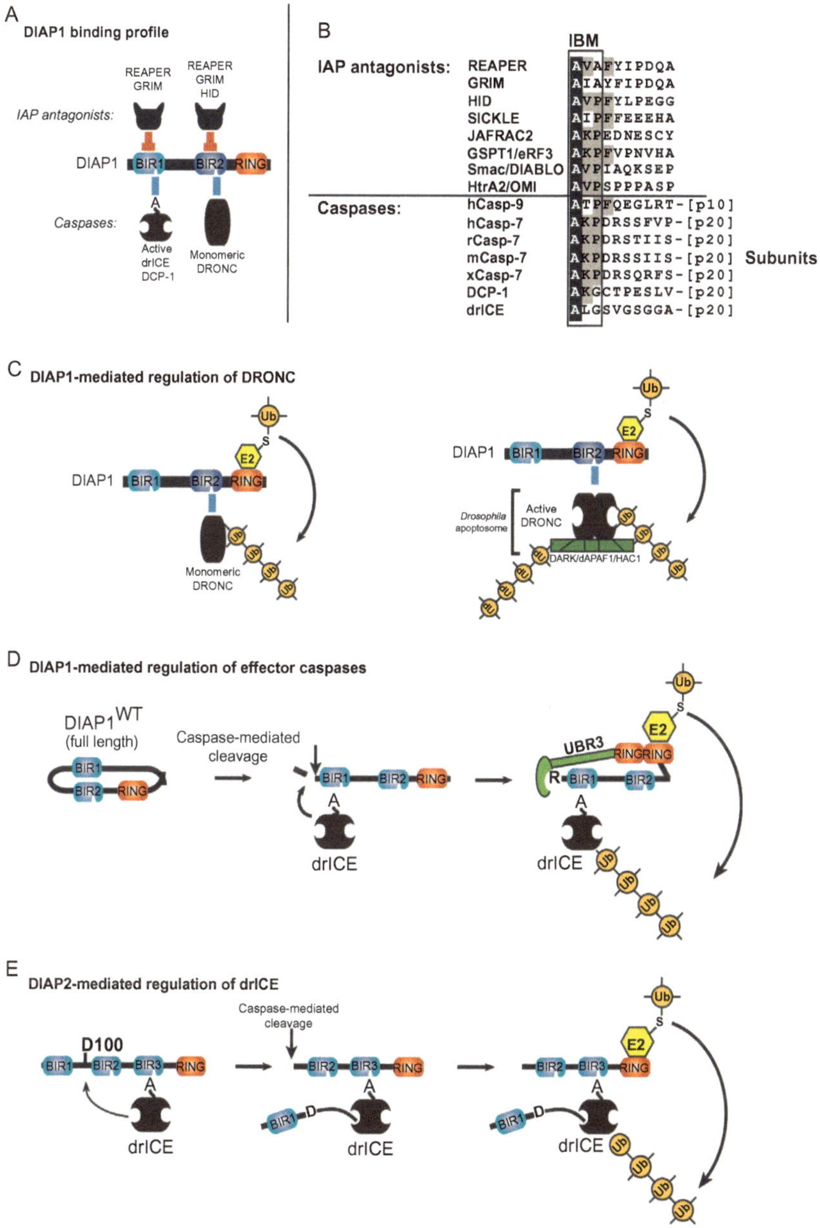

Pascal Meier *et al.*, **Figure 3** See legend on next page.

Pascal Meier *et al.*, Figure 3 IAP-mediated regulation of caspases in *Drosophila*. (A) Binding profile of DIAP1 with caspases and IAP antagonists. Direct physical interaction with the effector caspases drICE or DCP-1 and the initiator caspase DRONC is mediated through DIAP1's BIR1 and BIR2 domains, respectively. Following their activation, drICE and DCP-1 expose an NH_2-terminal IBM (depicted as A), which allows their binding to BIR1. (B) Sequence alignment of IBM-bearing proteins. Identical residues are highlighted in black. Residues conserved in four or more IBM proteins are indicated in gray. (C) DIAP1's BIR2-DRONC association is essential for DIAP1 to neutralize DRONC. Following binding, DIAP1's RING finger promotes Ub conjugation of DRONC, leading to its inactivation through nondegradative ubiquitylation of monomeric DRONC (left panel), and by targeting apoptosome-associated active DRONC for degradation (right panel). Mechanism of effector caspase (drICE) inactivation by DIAP1 (D) and DIAP2 (E). (D) Full-length wild-type DIAP1 is held in an inactive conformation and requires caspase-mediated proteolytic cleavage at residue 20 for its activation. After cleavage, BIR-mediated caspase binding occurs more efficiently. Cleavage also facilitates recruitment of N-end rule UBR E3 ligases, which together with DIAP1's RING domain promote ubiquitylation and inactivation of drICE and DCP-1. (E) drICE is also subject to regulation by DIAP2. drICE binds to the BIR3 of DIAP2 in an IBM-dependent manner, and following binding cleaves DIAP2 at D100. DIAP2 cleavage results in a covalent adduct between D100 and the catalytic machinery of drICE, trapping the caspase. Full inactivation of drICE is achieved through RING-mediated ubiquitylation.

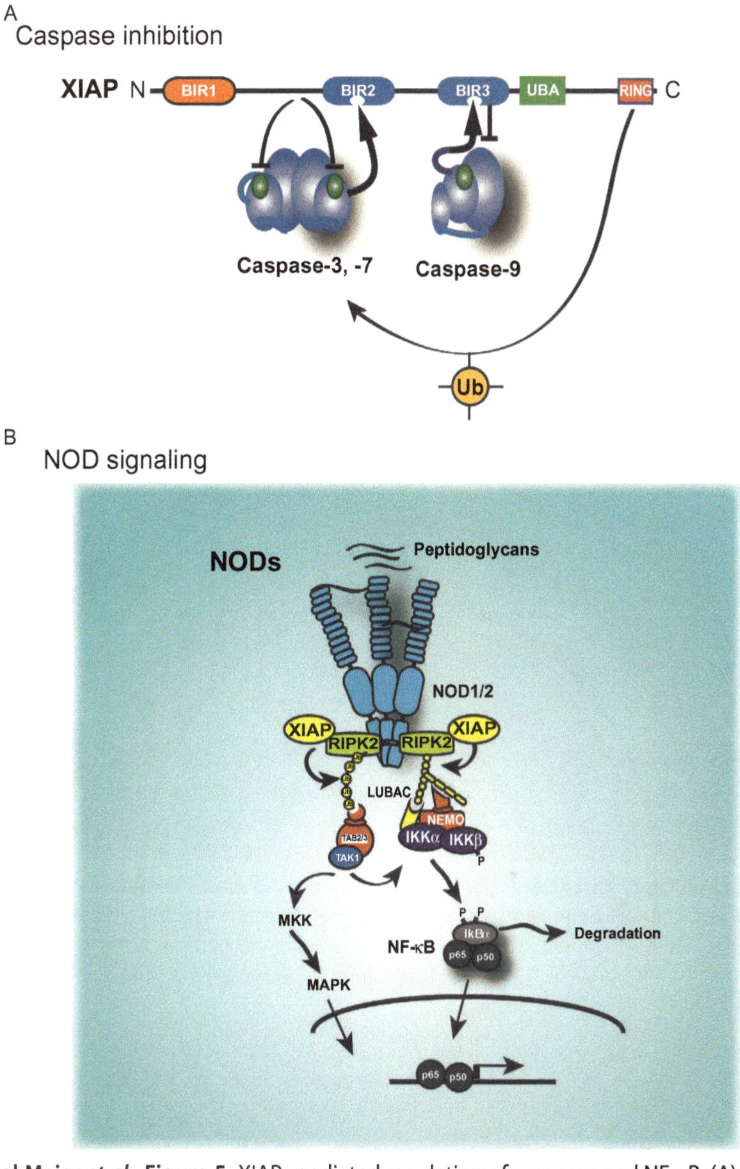

Pascal Meier *et al.*, Figure 5 XIAP-mediated regulation of caspases and NF-κB. (A) XIAP directly inhibits the effector caspase-3 and -7, and the initiator caspase-9. The sequence preceding the BIR2 domain of XIAP occupies the catalytic pocket of caspase-3 or -7, thereby blocking substrate entry. In addition, the BIR2 domain interacts with the IBM of caspase-3 or -7 that is exposed following their proteolytic activation (shown as an arrow). XIAP-mediated inhibition of caspase-9 requires proteolytic cleavage of caspase-9, which exposes an IBM that binds to the BIR3 of XIAP. Caspase-9 activity is blocked because XIAP prevents caspase-9 dimerization, a prerequisite for initiator caspase activity. The RING domain of XIAP also contributes to caspase inhibition. (B) NOD-mediated activation of NF-κB and MAPK signaling. Detection of bacterial peptidoglycans by NOD1 and NOD2 results in the formation of an oligomeric signaling complex that recruits RIPK2 and XIAP. XIAP-mediated ubiquitylation of RIPK2 allows the recruitment of TAB2/TAB3/TAK1 and IKKs, thereby triggering NF-κB and MAPK signaling.

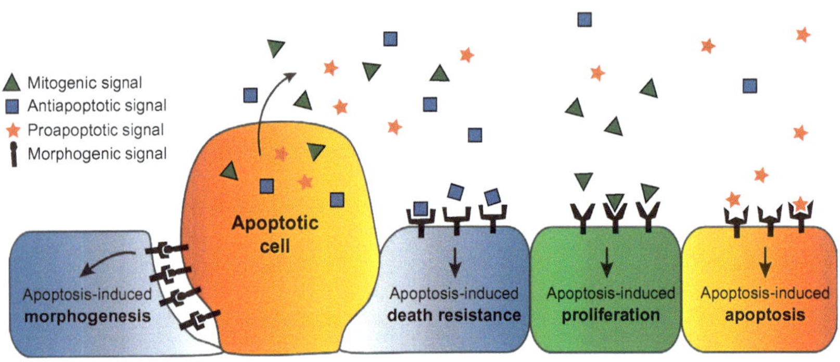

Caitlin E. Fogarty and Andreas Bergmann, Figure 2 Apoptotic cells direct growth control. Apoptotic cells can produce a variety of signals that can have an effect on the growth state of the surviving tissue. The outcome of this communication can depend on both the amounts of signal produced and the receptivity of the surviving cell to these different stimuli.

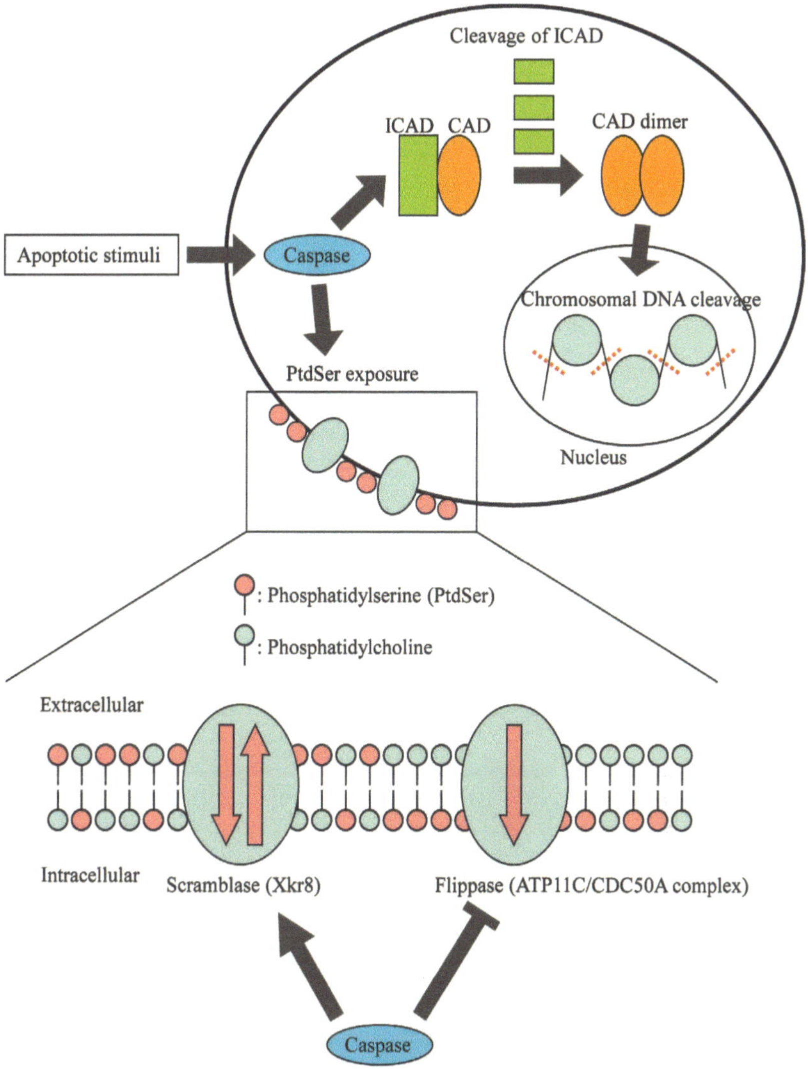

Satoshi Toda *et al.*, Figure 1 Apoptotic DNA degradation and PtdSer exposure. In healthy cells, CAD forms a heterodimer with ICAD. Apoptotic stimuli activate effector caspases that cleave ICAD at two positions, releasing CAD. The released CAD forms a homodimer that cleaves chromosomal DNA in the nucleus into nucleosomal units. In healthy cells, PtdSer is localized to the inner leaflet of the plasma membrane by the action of ATP-dependent flippase. When cells receive apoptotic stimuli, effector caspases cleave and activate Xkr8 scramblase, while cleaving and inactivating ATP11C flippase, quickly exposing PtdSer to the surface of the apoptotic cell.

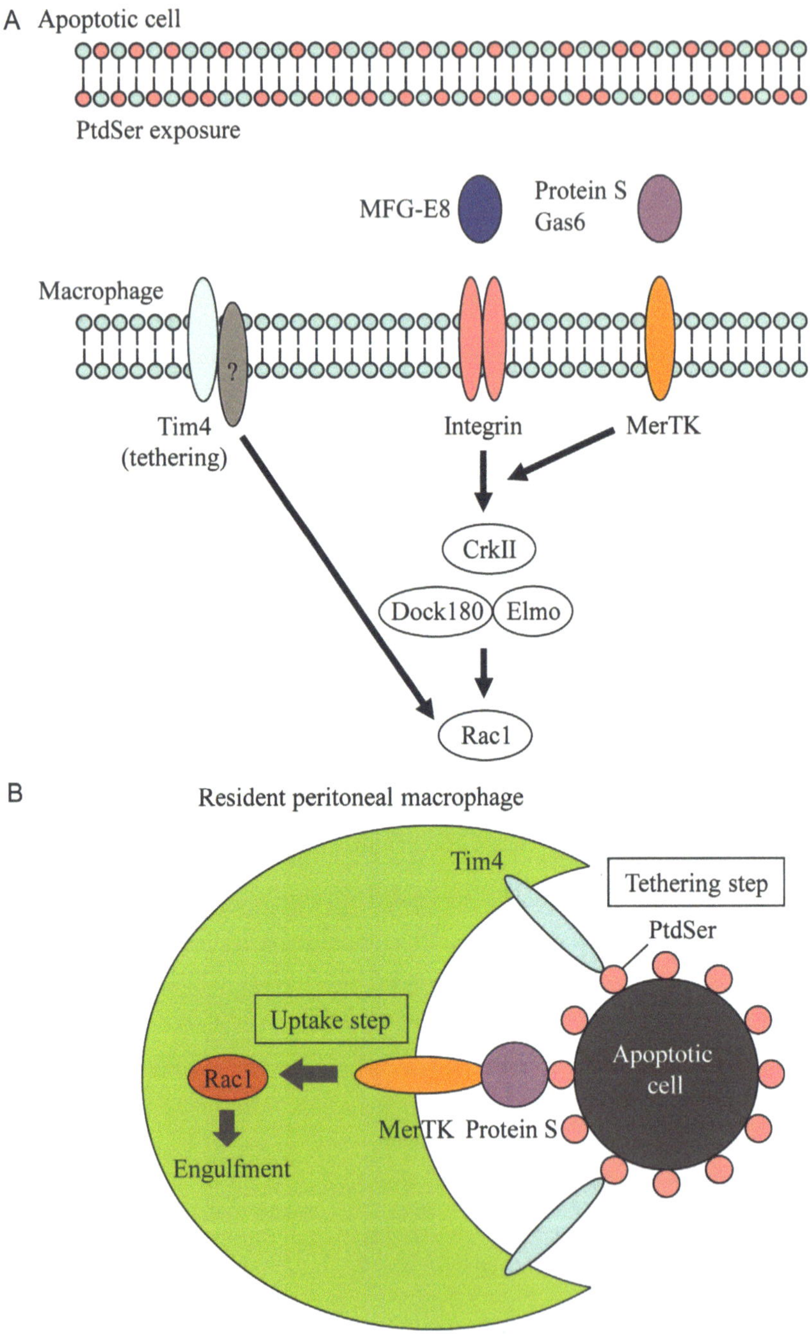

Satoshi Toda *et al.*, Figure 2 Engulfment of apoptotic cells. (A) The secreted proteins MFG-E8, Protein S, and Gas6 form bridges between apoptotic cells and phagocytes by recognizing and binding PtdSer on apoptotic cells, and binding integrins or TAM

(Continued)

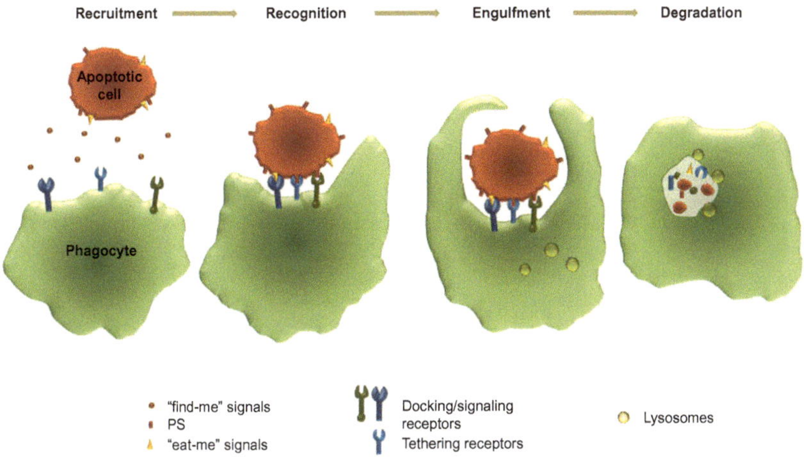

Jeny Shklover *et al.*, Figure 1 Schematic representation of different stages of apoptotic cell clearance. "Find-me" signals are secreted from apoptotic cells. "Eat-me" signals are exposed on the surface of apoptotic cells. Interactions between phagocytic receptors on phagocytic membrane and "eat-me" ligands on apoptotic cells lead to clustering of receptors, engulfment of apoptotic cell, phagosome formation, maturation, and finally degradation of the apoptotic cell.

Satoshi Toda *et al.*, Figure 2—Cont'd kinase receptors on phagocytes. Tim1 and Tim4 are type I membrane proteins that directly recognize PtdSer. During the engulfment of apoptotic cells, Rac1 is activated and promotes actin polymerization to form a phagocytic cup. Signaling molecules such as CrkII, Dock180, Elmo1, and GULP may act upstream of Rac1. (B) In a two-step process, Tim4 tethers apoptotic cells to the surface of mouse resident peritoneal macrophages, and Protein S binds to PtdSer on the tethered apoptotic cells to stimulate MerTK, which activates intracellular phagocytic signaling to engulf the apoptotic cell. Thus, two PtdSer-binding systems, the Tim4 and Protein S/MerTK systems, cooperate to engulf apoptotic cells.

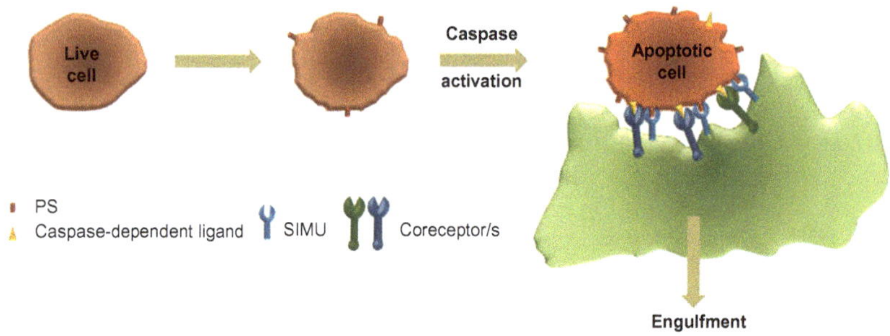

Jeny Shklover *et al.*, Figure 2 PS and caspase activation are required for engulfment. Schematic representation of PS and additional ligands on an apoptotic cell exposed following caspase activation that are recognized by different receptors on the phagocytic surface. Clustering of receptors leads to engulfment of the apoptotic cell by the phagocyte.

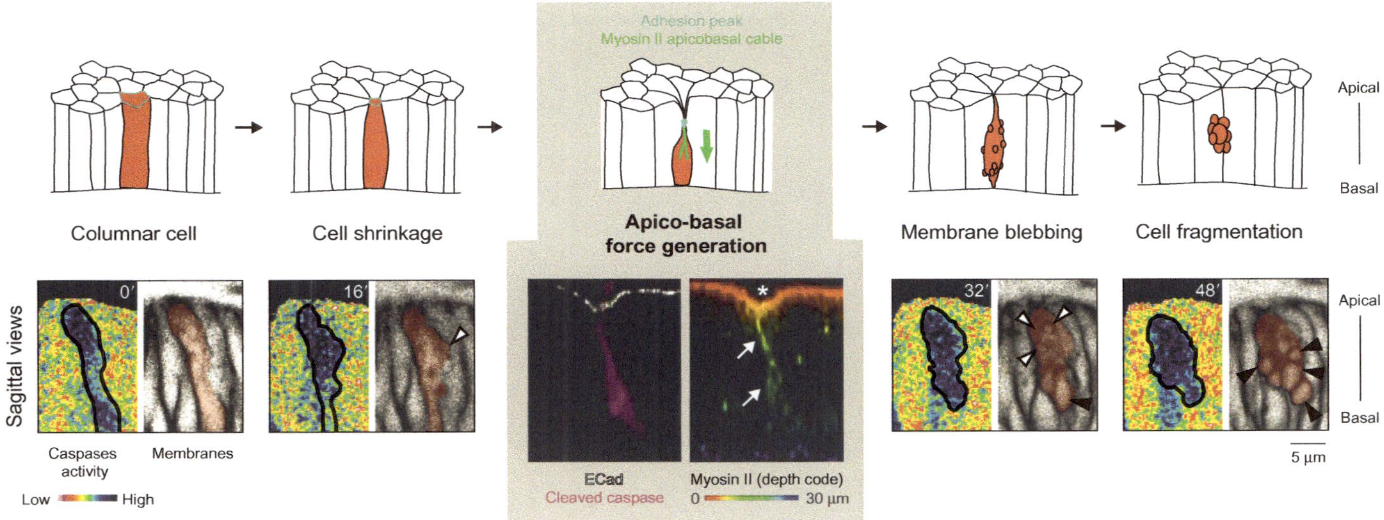

Bruno Monier and Magali Suzanne, Figure 2 *A new step during the early execution phase generates an apico-basal force within Drosophila epithelia*. Schematization (top) and corresponding confocal sagittal views (bottom) illustrating the successive phases of the execution phase in *Drosophila* epithelia. The new step generating an apico-basal force during apoptosis is boxed in gray. Bottom images either represent successive steps of a movie of an apoptotic cell in the pupal leg where caspase activity is monitored using a FRET probe and membranes using a lipid-binding dye or an apoptotic cell in the pupal wing costained for E-Cadherin (white), active caspases (magenta), and myosin II (red to blue color code indicating depth). White and black arrowheads point at blebs and apoptotic bodies, respectively. White arrows point at the myosin II cable (note the deformation of the apical surface of surrounding cells at that stage, asterisk). *Figure adapted from Monier et al. (2015).*

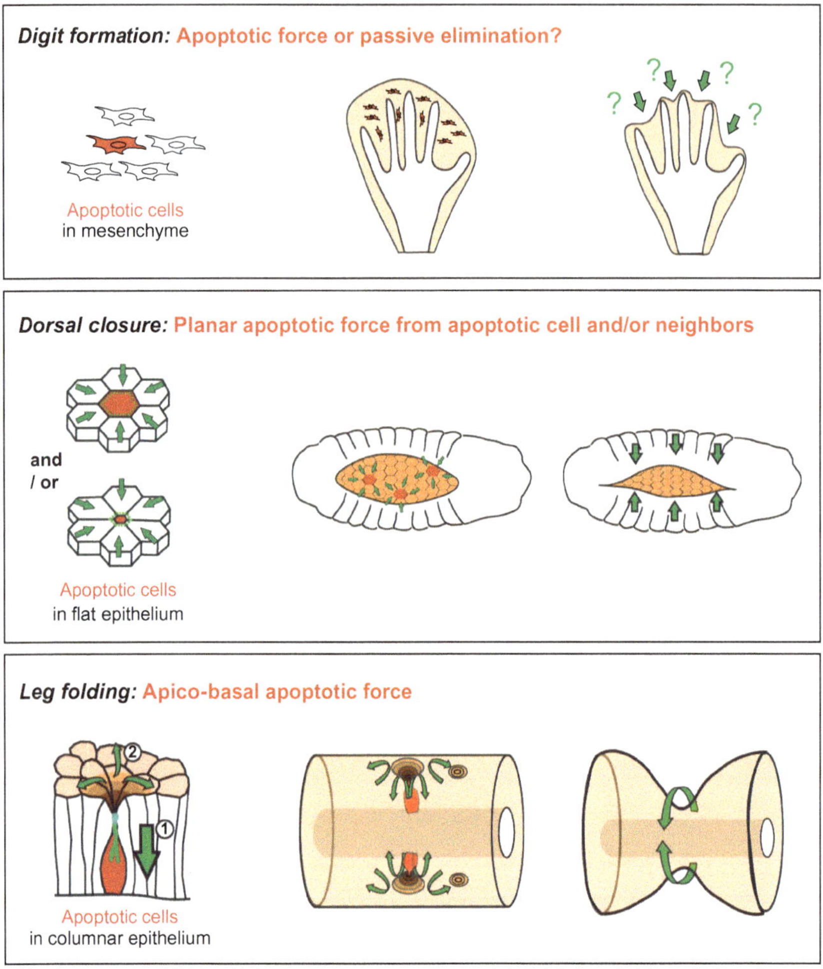

Bruno Monier and Magali Suzanne, Figure 3 *Consequence of the apoptotic force on tissue remodeling*. Schematization of the consequence of apoptosis on three different morphogenetic processes: digit formation (left), dorsal closure (center), and leg folding (right). The type of cells undergoing apoptosis together with the force exerted by apoptotic cells are schematized in the upper part, the whole tissue is schematized in the lower part. In digit formation, the participation of an apoptotic force to interdigital tissue regression (green arrows) has been proposed but not demonstrated. During dorsal closure, an apoptosis extrusion force accelerates the migration of the lateral epithelium (green arrows); however, the participation of apoptotic cells versus neighbors has not been tested yet. During leg folding, apoptotic cells generate an apico-basal force (1) transmitted to the rest of the epithelium through an increase in tissue tension (2) which leads to tissue constriction (the different forces are represented by green arrows).

FSC
www.fsc.org
MIX
Papier | Fördert
gute Waldnutzung
FSC® C083411

Kolibri 360 GmbH
Ferdinand-Jühlke-Straße 7
99095 Erfurt, Deutschland
produktsicherheit@kolibri360.de